2nd International Conference on Mechanical, Material and Aerospace Engineering (2MAE 2018)

MATEC Web of Conferences Volume 179 (2018)

Wuhan, China
10 - 13 May 2018

Editor:

Dan Zhang

ISBN: 978-1-5108-6802-1

Printed from e-media with permission by:

Curran Associates, Inc.
57 Morehouse Lane
Red Hook, NY 12571

Some format issues inherent in the e-media version may also appear in this print version.

Printed by Curran Associates, Inc. (2018)

For additional information, please contact EDP Sciences – Web of Conferences at the address below.

EDP Sciences – Web of Conferences
17, Avenue du Hoggar
Parc d'Activité de Courtabœuf
BP 112
F-91944 Les Ulis Cedex A
France

Phone: +33 (0) 1 69 18 75 75
Fax: +33 (0) 1 69 28 84 91

contact-edps@webofconferences.org

Additional copies of this publication are available from:

Curran Associates, Inc.
57 Morehouse Lane
Red Hook, NY 12571 USA
Phone: 845-758-0400
Fax: 845-758-2633
Email: curran@proceedings.com
Web: www.proceedings.com

TABLE OF CONTENTS

COMMITTEES

General Chair

Prof. Huafeng Ding, China University of Geosciences (Wuhan), China

Conference Committee Chair

Prof. Dan Zhang,York University, Canada

Program Committee Chair

Prof. Liu Qiang, Beihang University, China

Local Committees

Prof. Guojun Wen, China University of Geosciences (Wuhan), China

Prof. Li Bo, China University of Geosciences (Wuhan), China

International Technical Committees

Dr. Alex Ong Zhi Chao, University of Malaya, Malaysia

Dr. B. F. Yousif, University of Southern Queensland, Australia

Prof. M. Chandrasekaran, Vels University, India

Prof. Shinn-Liang Chang, National Formosa University, Taiwan

Prof. Tang Sai Hong, Universiti Putra Malaysia, Malaysia

Dr. Darwin Lau, The Chinese University, Hong Kong

Dr. Sayeed Hossain, University of Bristol, UK

Dr. Laurent Tranchant, Kyushu Institute of Technology, Japan

Dr. AloK Jha, Kyushu Institute of Technology, Japan

Dr. Dehong Huo, Newcastle University, UK

Dr. Jie Zhang, University of Bristol, UK

Prof. David Bassir, Shanghai Jiaotong University, China

edp sciences

MATEC Web of Conferences

Statement of Peer review

In submitting conference proceedings to MATEC Web of Conferences, I certify to the Publisher that I adhere to the **Policy on Publishing Integrity** of the journal in order to safeguard good scientific practice in publishing.

1. All articles have been subjected to peer review administered by the proceedings editors.
2. Reviews have been conducted by expert referees, who have been requested to provide unbiased and constructive comments aimed, whenever possible, at improving the work.
3. Proceedings editors have taken all reasonable steps to ensure the quality of the materials they publish and their decision to accept or reject a paper for publication has been based only on the merits of the work and the relevance to the journal.

Title, date and place of the conference

2018 2nd International Conference on Mechanical, Material and Aerospace Engineering (2MAE 2018)

May 10-13, 2018 Wuhan, China

Proceedings editor(s):

Prof. Dan Zhang, York University, Canada

Date and editor's signature

13-JUNE-2018

PREFACE

2MAE 2018 was held in Wuhan during May 10-13, 2018. 2MAE 2018 are organized by China University of Geosciences (Wuhan) and Hong Kong Society of Mechanical Engineers(HKSME). The conference provides a useful and wide platform both for display the latest research and for exchange of research results and thoughts in broad Mechanical, Material and Aerospace Engineering. The participants of the conference were from almost every part of the world, with background of either academia or industry, even well-known enterprise. The success and prosperity of the conference is reflected high level of the papers received.

The proceedings are a compilation of the accepted papers and represent an interesting outcome of the conference. This book covers 3 chapters: Mechanical; Materials; Aerospace.

We would like to acknowledge all of those who supported 2MAE 2018. Each individual and institutional help were very important for the success of this conference. Especially we would like to thank the organizing committee for their valuable advices in the organization and helpful peer review of the papers.

We sincerely hope that 2MAE 2018 will be a forum for excellent discussions that will put forward new ideas and promote collaborative researches. We are sure that the proceedings will serve as an important research source of references and the knowledge, which will lead to not only scientific and engineering progress but also other new products and processes.

Prof. Dan Zhang, York University, Canada

Global Existence of the Three Dimensional Heat-conductive Incompressible Viscous Fluids

Jie Wu[1,a]

[1]*School of Mathematical Sciences, University of Electronic Science and Technology of China, Chengdu 611731, China*

Abstract. In this paper, we consider the Cauchy problem of non-stationary motion of heat-conducting incompressible viscous fluids in $\mathbb{R}^3$. About the heat-conducting incompressible viscous fluids, there are many mathematical researchers study the variants systems when the viscosity and heat-conductivity coefficient are positive. For the heat-conductive system, it is difficulty to get the better regularity due to the gradient of velocity of fluid own the higher order term. It is hard to control it. In order to get its global solutions, we must obtain the a priori estimates at first, then using fixed point theorem, it need the mapping is contracted. We can get a local solution, then applying the criteria extension. We can extend the local solution to the global solutions. For the two dimensional case, the Gagliardo-Nirenberg interpolation inequality makes use of better than the three dimensional situation. Thus, our problem will become more difficulty to handle. In this paper, we assume the coefficient of viscosity is a constant and the coefficient of heat-conductivity satisfying some suitable conditions. We show that the Cauchy problem has a global-in-time strong solution $(\boldsymbol{u},\theta)$ on $\mathbb{R}^3\times(0,+\infty)$.

1 Introduction

We continue to consider the Cauchy problem of the system (1), the model of a heat-conducting incompressible viscous fluid is

$$\begin{cases} \partial_t \boldsymbol{u} + (\boldsymbol{u}\cdot\nabla)\boldsymbol{u} - div(\mu(\theta)\nabla\boldsymbol{u}) + \nabla P = 0, \\ \partial_t\theta + (\boldsymbol{u}\cdot\nabla)\theta - div(\kappa(\theta)\nabla\theta) - \mu(\theta)\left|\nabla\boldsymbol{u}\right|^2 = 0, \\ div\,\boldsymbol{u} = 0, \\ \boldsymbol{u}\big|_{t=0} = \boldsymbol{u}_0, \quad \theta\big|_{t=0} = \theta_0, \end{cases} \tag{1}$$

with $x\in\mathbb{R}^3$ and $t>0$. Here, the unknowns $\boldsymbol{u},\theta,P$ denote the velocity, temperature and pressure of the fluid, respectively. $\mu(\theta)$ is the coefficient of viscosity, we assume that $\mu(\theta)\equiv\mu$ is a constant, $\kappa(\theta)$ denotes the coefficient of heat conductivity and assumes that functions of temperature satisfying

$$\kappa(\cdot)\in C^1([0,+\infty)),\ 0<\underline{\kappa}\le\kappa(\theta)\le\bar{k},\ \kappa'(\theta)>0,\ \frac{\bar{\kappa}}{\underline{\kappa}}<\kappa_1, \tag{2}$$

[a] Corresponding author : dxtxwj@126.com

for some positive constant μ, $\underline{\kappa}, \bar{\kappa}, \kappa_1$.

There are many mathematical investigators for the variants system when the viscosity and heat-conductivity coefficient are positive constants in [1-3]. Benes [1] proved that the system has a local existence in time and global uniqueness of the strong solution in two dimension bounded domain. Kagei [2] considered adding the external force field $f(\theta)$ in the system the first equation of (1) , he give the attractors of the initial-boundary value problem in $\mathbb{R}^2$. Kakizawa [3] proved the exixtence of mild solution for general initial data in local time and global solution for the initial value under a smallness assumption in time. Shilkin [4] considered a simple uni-directional Poiseuille-type flow of a incompressible Newtonian fluid :

$$\begin{cases} \partial_t \boldsymbol{u} - div(\mu(\theta)\nabla \boldsymbol{u}) = f, \\ \partial_t \theta - \Delta\theta = \mu(\theta)|\nabla \boldsymbol{u}|^2. \end{cases} \tag{3}$$

He proved existence of weak solutions in spatial dimension three and a classical solution on an arbitrary interval of time in two dimension. Bulicek et al [5] investigated the maximal L^2 regularity theory of system (1) under κ is a positive constant and certain structural assumptions of $\mu(\theta)$ without the convective terms and [6] studied $\mu = \mu(\theta, \nabla \boldsymbol{u})$ the regularity properties of unsteady flows in a two dimension periodic establishing under certain structural assumption on the Cauchy stress. However, the system (1) is difficulty to get the better regularity of $\boldsymbol{u}$ and θ, moreover, in the most general case that become much more complication. Very recently, Ye [7] investigated global existence in time under some smallness condition of initial data or the $|\mu'(\theta)|$ is small enough, or the $\underline{\kappa}$ is suitably large in two dimensional case.

In this paper, we aim to study global existence under some smallness condition of initial data or the $\underline{\kappa}$ is suitably large in three dimensional situation.

2 A Priori Estimates

In order to obtain the global existence, the main purpose of this section is to prove the following key a priori estimate. For getting the priori estimates of Lemma 2.3-Lemma 2.5, we give the following two important inequalities.

Lemma 2.1 (Gronwall's inequality; see [8])

(i) Let $\eta(\cdot)$ be a nonnegative, absolutely continuous function on $[0,T]$, which satisfies for a.e. t the differential inequality

$$\eta'(t) \le \phi(t)\eta(t) + \psi(t), \tag{4}$$

where $\phi(t)$ and $\psi(t)$ are nonnegative, summable functions on $[0,T]$. Then

$$\eta(t) \le e^{\int_0^T \phi(s)ds}[\eta(0) + \int_0^T \psi(s)ds], \tag{5}$$

for all $0 \le t \le T$.

(ii) In particular, if

$$\eta'(t) \le \phi(t)\eta(t) \text{ on } [0,T] \text{ and } \eta(0) = 0, \tag{6}$$

then

$$\eta(t)=0, \text{ on } [0,T]. \tag{7}$$

Lemma 2.2 (Gagliardo-Nirenberg Interpolation Inequality ; see[9], Proposition 3)

Let j,k be any integers satisfying $0 \le j < k$, and let $1 \le S,Q \le \infty$ and $R \in \mathbb{R}$, $\frac{j}{k} \le \theta \le 1$, such that

$$\frac{1}{R} = \frac{j}{k} + \theta(\frac{1}{s} - \frac{k}{n}) + (1-\theta)\frac{1}{Q}. \tag{8}$$

Then

1. For any $h \in W^{k,S}(\mathbb{R}^n) \cap L^Q(\mathbb{R}^n)$, there is a positive constant C depending only on n, k, j, Q, S, θ such that the following inequality holds :

$$\left|D^j h\right|_R \le C\left|D^k h\right|_S^{\theta} \left|h\right|_Q^{1-\theta} \tag{9}$$

with the following exception: If $1 < S < \infty$ and $k - j - \frac{n}{S}$ is a nonnegative integer, then (9) holds only for θ satisfying $\frac{j}{k} \le \theta < 1$.

2. For any $h \in W^{k,S}(\Omega) \cap L^Q(\Omega)$ where $\Omega \subset \mathbb{R}^n$ is a bounded domain with smooth boundary,there are two positive constants C_1, C_2 such that the following inequality holds :

$$\left|D^j h\right|_R \le C\left|D^k h\right|_S^{\theta} \left|h\right|_Q^{1-\theta} + C_2\left|h\right|_Q \tag{10}$$

with the same exception as in part 1. In particular, for any $h \in W_0^{k,S}(\Omega) \cap L^Q(\Omega)$, the constant C_2 in (10) can be taken as zero.

Lemma 2.3 Assume that $(\boldsymbol{u}_0, \theta_0) \in H^1(\mathbb{R}^3)$. Let $(\boldsymbol{u}, \theta)$ be a solution of (1) on $\mathbb{R}^3 \times (0,T)$. Then,

$$\sup_{0\le t\le T} \|\boldsymbol{u}\|_{L^2(\mathbb{R}^3)}^2 + 2\mu\int_0^T \|\nabla \boldsymbol{u}\|_{L^2(\mathbb{R}^3)}^2 dt = \|\boldsymbol{u}_0\|_{L^2(\mathbb{R}^3)}^2. \tag{11}$$

Proof. Multiplying the first equation of (1) by $\boldsymbol{u}$ and integrating by parts over $\mathbb{R}^3$, we derive that

$$\frac{1}{2}\frac{d}{dt}\|\boldsymbol{u}\|_{L^2(\mathbb{R}^3)}^2 + \mu\|\nabla \boldsymbol{u}\|_{L^2(\mathbb{R}^3)}^2 = 0. \tag{12}$$

Integrating (12) with respect to time from 0 to T, we can get (11).

Lemma 2.4 Assume that $(\boldsymbol{u}_0, \theta_0) \in H^1(\mathbb{R}^3)$. Let $(\boldsymbol{u}, \theta)$ be a solution of (1) on $\mathbb{R}^3 \times (0,T)$ satisfying

$$\int_0^T \|\nabla \boldsymbol{u}\|_{L^2(\mathbb{R}^3)}^6 dt \le K_1 \text{ and } \int_0^T \|\nabla \boldsymbol{u}\|_{L^4(\mathbb{R}^3)}^4 dt \le K_2, \tag{13}$$

where $K = \max\{K_1, K_2\}$ is a constant independent of T.

Then,

$$\|\nabla\theta\|_{L^2(\mathbb{R}^3)}^2 \le \frac{C}{\underline{\kappa}}\exp\{\frac{1}{\underline{\kappa}^2}\}(\frac{1}{\underline{\kappa}} + \bar{\mu}^2\kappa_1^2) \tag{14}$$

and

$$\int_0^T \|\Delta\theta\|_{L^2(\mathbb{R}^3)}^2 dt \le \frac{C}{\underline{\kappa}^3}\exp\{\frac{1}{\underline{\kappa}^4}\}(1 + 4\mu^2\bar{\kappa}\kappa_1) \tag{15}$$

Proof. We use the similar technique as in [7] to introduce a new quantity $\hat{\theta} = \int_0^{\theta} \kappa(z)dz$, which satisfies the following Cauchy problem

$$\begin{cases} \partial_t\hat{\theta} + (\boldsymbol{u}\cdot\nabla)\hat{\theta} = \kappa(\theta)\Delta\hat{\theta} + \mu\kappa(\theta)\left|\nabla \boldsymbol{u}\right|^2, \\ \hat{\theta}(x,0) = \int_0^{\theta_0} \kappa(z)dz := \hat{\theta}_0(x). \end{cases} \tag{16}$$

Multiplying (16) by $\Delta\hat{\theta}$ and then integrating over $\mathbb{R}^3$, we derive from Hölder's inequality and Young inequality that

$$
\begin{aligned}
&\frac{1}{2}\frac{d}{dt}\left\|\nabla\hat{\theta}\right\|_{L^2(\mathbb{R}^3)}^2+\left\|(\kappa(\theta))^{1/2}\Delta\hat{\theta}\right\|_{L^2(\mathbb{R}^3)}^2\leq\int_{\mathbb{R}^3}\left|\boldsymbol{u}\nabla\hat{\theta}\Delta\hat{\theta}\right|dx+\mu\int_{\mathbb{R}^3}\left|\kappa(\theta)\left|\nabla\boldsymbol{u}\right|^2\Delta\hat{\theta}\right|dx\\
&\leq\left\|\boldsymbol{u}\right\|_{L^4(\mathbb{R}^3)}\left\|\nabla\hat{\theta}\right\|_{L^4(\mathbb{R}^3)}\left\|\Delta\hat{\theta}\right\|_{L^4(\mathbb{R}^3)}+\mu\bar{\kappa}\left\|\nabla\boldsymbol{u}\right\|_{L^4(\mathbb{R}^3)}^2\left\|\Delta\hat{\theta}\right\|_{L^2(\mathbb{R}^3)}\\
&\leq\frac{2}{\underline{\kappa}}\left\|\boldsymbol{u}\right\|_{L^4(\mathbb{R}^3)}^2\left\|\nabla\hat{\theta}\right\|_{L^4(\mathbb{R}^3)}^2+2\mu^2\bar{\kappa}\kappa_1\left\|\nabla\boldsymbol{u}\right\|_{L^4(\mathbb{R}^3)}^4+\frac{\underline{\kappa}}{4}\left\|\Delta\hat{\theta}\right\|_{L^2(\mathbb{R}^3)}^2\\
&\leq\frac{C_1}{\underline{\kappa}}\left\|\boldsymbol{u}\right\|_{L^2(\mathbb{R}^3)}^{\frac{1}{2}}\left\|\nabla\boldsymbol{u}\right\|_{L^2(\mathbb{R}^3)}^{\frac{3}{2}}\left\|\nabla\hat{\theta}\right\|_{L^2(\mathbb{R}^3)}^{\frac{1}{2}}\left\|\Delta\hat{\theta}\right\|_{L^2(\mathbb{R}^3)}^{\frac{3}{2}}+2\mu^2\bar{\kappa}\kappa_1\left\|\nabla\boldsymbol{u}\right\|_{L^4(\mathbb{R}^3)}^4+\frac{\underline{\kappa}}{4}\left\|\Delta\hat{\theta}\right\|_{L^2(\mathbb{R}^3)}^2\\
&\leq\frac{C_2}{\underline{\kappa}^4}\left\|\boldsymbol{u}\right\|_{L^2(\mathbb{R}^3)}^2\left\|\nabla\boldsymbol{u}\right\|_{L^2(\mathbb{R}^3)}^6\left\|\nabla\hat{\theta}\right\|_{L^2(\mathbb{R}^3)}^2+2\mu^2\bar{\kappa}\kappa_1\left\|\nabla\boldsymbol{u}\right\|_{L^4(\mathbb{R}^3)}^4+\frac{\underline{\kappa}}{2}\left\|\Delta\hat{\theta}\right\|_{L^2(\mathbb{R}^3)}^2
\end{aligned}\tag{17}
$$

This combined with (11) and Gronwall's inequality implies that

$$
\begin{aligned}
&\sup_{0\leq t\leq T}\left\|\nabla\hat{\theta}\right\|_{L^2(\mathbb{R}^3)}^2+\underline{\kappa}\int_0^T\left\|\Delta\hat{\theta}\right\|_{L^2(\mathbb{R}^3)}^2dt\\
&\leq\exp\left\{\frac{2C_2}{\underline{\kappa}^4}\left\|\boldsymbol{u}_0\right\|_{L^2(\mathbb{R}^3)}^2\int_0^T\left\|\nabla\boldsymbol{u}\right\|_{L^2(\mathbb{R}^3)}^6dt\right\}\left(\left\|\nabla\hat{\theta}_0\right\|_{L^2(\mathbb{R}^3)}^2+4\mu^2\bar{\kappa}\kappa_1\int_0^T\left\|\nabla\boldsymbol{u}\right\|_{L^4(\mathbb{R}^3)}^4dt\right)\\
&\leq C\exp\{\frac{1}{\underline{\kappa}^4}\}(1+4\mu^2\bar{\kappa}\kappa_1).
\end{aligned}\tag{18}
$$

By virtue of $\nabla\theta=\frac{1}{\kappa(\theta)}\nabla\hat{\theta},\Delta\theta=\frac{1}{\kappa(\theta)}\Delta\hat{\theta}-\frac{\kappa'(\theta)}{\kappa^3(\theta)}\left|\nabla\hat{\theta}\right|^2\leq\frac{1}{\underline{\kappa}}\Delta\hat{\theta}$ and (2), (18), we have

$$
\left\|\nabla\theta\right\|_{L^2(\mathbb{R}^3)}^2\leq\frac{C}{\underline{\kappa}^2}\exp\left\{\frac{1}{\underline{\kappa}^2}\right\}(1+\bar{\mu}^2\bar{\kappa}\kappa_1)\leq\frac{C}{\underline{\kappa}}\exp\left\{\frac{1}{\underline{\kappa}^2}\right\}(\frac{1}{\underline{\kappa}}+\bar{\mu}^2\kappa_1^2),\tag{19}
$$

and

$$
\int_0^T\left\|\Delta\theta\right\|_{L^2(\mathbb{R}^3)}^2dt\leq\frac{1}{\underline{\kappa}^2}\int_0^T\left\|\Delta\hat{\theta}\right\|_{L^2(\mathbb{R}^3)}^2dt\leq\frac{1}{\underline{\kappa}^3}\exp\left\{\frac{1}{\underline{\kappa}^4}\right\}(1+4\mu^2\bar{\kappa}\kappa_1).\tag{20}
$$

Lemma 2.5 Assume that $(\boldsymbol{u}_0,\theta_0)\in H^1(\mathbb{R}^3)$. Let $(\boldsymbol{u},\theta)$ be a solution of (1) on $\mathbb{R}^3\times(0,T)$ satisfying (13). Then, we have the following energy inequality

$$
\left\|\nabla\boldsymbol{u}\right\|_{L^2(\mathbb{R}^3)}^2+\mu\int_0^T\left\|\Delta\boldsymbol{u}\right\|_{L^2(\mathbb{R}^3)}^2dt\leq2C_5K+\left\|\nabla\boldsymbol{u}_0\right\|_{L^2(\mathbb{R}^3)}^2.\tag{21}
$$

Proof. Multiplying the first of (1) by $\Delta\boldsymbol{u}$ and integrating by parts over $\mathbb{R}^3$, we derive that

$$
-\int_{\mathbb{R}^3}\nabla\boldsymbol{u}\cdot\nabla\boldsymbol{u}_t dx+\int_{\mathbb{R}^3}\Delta\boldsymbol{u}\cdot\boldsymbol{u}\cdot\nabla\boldsymbol{u}dx-\mu\int_{\mathbb{R}^3}\left|\Delta\boldsymbol{u}\right|^2dx+\int_{\mathbb{R}^3}\Delta\boldsymbol{u}\cdot\nabla Pdx=0,\tag{22}
$$

And using Hölder's inequality and Young inequality, we have

$$
\begin{aligned}
&\frac{1}{2}\frac{d}{dt}\left\|\nabla\boldsymbol{u}\right\|_{L^2(\mathbb{R}^3)}^2+\mu\left\|\Delta\boldsymbol{u}\right\|_{L^2(\mathbb{R}^3)}^2\leq\int_{\mathbb{R}^3}\left|\Delta\boldsymbol{u}\right|\left|\boldsymbol{u}\right|\left|\nabla\boldsymbol{u}\right|dx\\
&\leq\left\|\boldsymbol{u}\right\|_{L^6(\mathbb{R}^3)}\left\|\nabla\boldsymbol{u}\right\|_{L^3(\mathbb{R}^3)}\left\|\Delta\boldsymbol{u}\right\|_{L^2(\mathbb{R}^3)}\leq C_3\left\|\nabla\boldsymbol{u}\right\|_{L^2(\mathbb{R}^3)}\left\|\nabla\boldsymbol{u}\right\|_{L^3(\mathbb{R}^3)}\left\|\Delta\boldsymbol{u}\right\|_{L^2(\mathbb{R}^3)}\\
&\leq C_4\left\|\nabla\boldsymbol{u}\right\|_{L^2(\mathbb{R}^3)}^{\frac{3}{2}}\left\|\Delta\boldsymbol{u}\right\|_{L^2(\mathbb{R}^3)}^{\frac{3}{2}}\leq\frac{\mu}{2}\left\|\Delta\boldsymbol{u}\right\|_{L^2(\mathbb{R}^3)}^2+C_5\left\|\nabla\boldsymbol{u}\right\|_{L^2(\mathbb{R}^3)}^6
\end{aligned}\tag{23}
$$

That is

$$\frac{d}{dt}\|\nabla \boldsymbol{u}\|_{L^2(\mathbb{R}^3)}^2 + \mu\|\Delta \boldsymbol{u}\|_{L^2(\mathbb{R}^3)}^2 \le 2C_5\|\nabla \boldsymbol{u}\|_{L^2(\mathbb{R}^3)}^6 \tag{24}$$

We apply Gronwalls inequality to (15) obtain (14).

3 Proof of Theorem 3.2

We purpose is to obtain the estimate about $\|\theta\|_{L^\infty(0,T;L^2(\mathbb{R}^3))}$. We give the following lemma.

Lemma 3.1 Assume that $(\boldsymbol{u}_0,\theta_0)\in H^1(\mathbb{R}^3)$ and (2), (13) hold. Let $(\boldsymbol{u},\theta)$ be a solution of (1) on $\mathbb{R}^3\times(0,T)$. Then

$$\sup_{0\le t\le T}\|\theta\|_{L^2(\mathbb{R}^3)}^2 + \int_0^T\|\nabla\theta\|_{L^2(\mathbb{R}^3)}^2\,dt \le C. \tag{25}$$

Proof. Multiplying the second equation of (1) with θ, we have

$$\int_{\mathbb{R}^3}\partial_t\theta\cdot\theta\,dx + \int_{\mathbb{R}^3}\boldsymbol{u}\cdot\nabla\theta\cdot\theta\,dx - \int_{\mathbb{R}^3}div(\kappa(\theta)\nabla\theta)\cdot\theta\,dx - \mu\int_{\mathbb{R}^3}|\nabla\boldsymbol{u}|^2\cdot\theta\,dx = 0. \tag{26}$$

We apply integrating by parts over $\mathbb{R}^3$, one has

$$\frac{1}{2}\frac{d}{dt}\|\theta\|_{L^2(R^3)}^2 + \int_{R^3}\kappa(\theta)|\nabla\theta|^2\,dx \le \mu\int_{\mathbb{R}^3}|\nabla u|^2\cdot\theta\,dx \tag{27}$$

That is

$$\begin{aligned}\frac{d}{dt}\|\theta\|_{L^2(\mathbb{R}^3)}^2 + 2\underline{\kappa}\int_{\mathbb{R}^3}|\nabla\theta|^2\,dx &\le 2\mu\int_{\mathbb{R}^3}|\nabla\boldsymbol{u}|^2\cdot\theta\,dx\\ &\le 2\mu\|\nabla\boldsymbol{u}\|_{L^4(\mathbb{R}^3)}^2\|\theta\|_{L^2(\mathbb{R}^3)}\\ &\le \mu^2\|\nabla\boldsymbol{u}\|_{L^4(\mathbb{R}^3)}^4 + \|\theta\|_{L^2(\mathbb{R}^3)}^2\end{aligned} \tag{28}$$

By the Gronwall's inequality and (18), we have

$$\|\theta\|_{L^2(\mathbb{R}^3)}^2 + 2\underline{\kappa}\int_0^T\|\nabla\theta\|_{L^2(\mathbb{R}^3)}^2\,dt \le e^T\left(\|\theta\|_{L^2(\mathbb{R}^3)}^2 + \mu^2K\right). \tag{29}$$

The proof of Lemma 3.1 is finished.

Theorem 3.1 (Local existence) Assume that $(\boldsymbol{u}_0,\theta_0)\in H^1(\mathbb{R}^3)$ and (2), (13) hold. Then there exists a small time classical solution $(\boldsymbol{u},\theta)$ to the Cauchy problem (1) such that

$$(\boldsymbol{u},\theta)\in L^2(0,T;H^2(\mathbb{R}^3)), \qquad (\nabla\boldsymbol{u},\nabla\theta)\in L^\infty(0,T;L^2(\mathbb{R}^3)). \tag{30}$$

Theorem 3.2 (Global existence) Assume that $(\boldsymbol{u}_0,\theta_0)\in H^1(\mathbb{R}^3)$ and (2), (13) hold. Then there is a global strong solution $(\boldsymbol{u},\theta)$ with the following regularity

$$(\boldsymbol{u},\theta)\in L^2(0,+\infty;H^2(\mathbb{R}^3)), \quad (\nabla\boldsymbol{u},\nabla\theta)\in L^\infty(0,\infty;L^2(\mathbb{R}^3)). \tag{31}$$

Proof. With all the priori estimates established at hand, due to Theorem 3.1, there exists a T_0 such that the system (1) has a local strong solution $(\boldsymbol{u},\theta)$ on $[0,T_0]$. We expect to extend the local solution to global one. We know that there exists a positive time $T_1\in(0,T_0]$ such that (13) hold for $T=T_1$. Set

$$\overline{T}:\sup\{T|\,(13)\text{ hold}\}. \tag{32}$$

It is clear that $\overline{T}\ge T_1$. Next, we claim that $\overline{T}=+\infty$.

Otherwise, $\bar{T} < \infty$. Then it follows from Lemma2.3-Lemma2.5 and (13) hold for all $0 < T < \bar{T}$. So, there exist some $\tilde{T} > \bar{T}$ such that (13) hold for all $0 < T < \tilde{T}$, which contradicts (32). Thus, $\bar{T} = \infty$. We can obtain $(\boldsymbol{u}, \theta)$ is a strong solution from Lemma 2.3-Lemma 2.5 on $\mathbb{R}^3 \times (0, \infty)$.

4 Conclusion

With all the priori estimates established at hand, we can obtain the global existence by extend the time. Meanwhile, we use the Gagliardo-Nirenberg interpolation inequality and some skills to obtain the higher regularity of $(\boldsymbol{u}, \theta)$.

References

1. Benes, M.: Strong solutions to non-stationary flows of heat-conducting viscous incompressible fluids with dissipative heating. Acta Appl. Math. 116, 237-254 (2011)
2. Kagei, Y.: Attractors for two-dimensional equations of thermal convection in the presence of the dissipation function, Hiroshima Math. J. 25, 251-311 (1995)
3. Kakizawa, R.: The initial value problem for motion of incompressible viscous and heat-conductive fluids in Banach space. Hiroshima Math. J. 40, 371-402 (2010)
4. Shilkin, T.: Classical solvability of the coupled system modelling a heat-convergent Poiseuille-type flow. J. Math. Fluid Mech. 7, 72-84 (2005)
5. Bulicek, M., Kaplicky, P., Malek, J.: An L^2-maximal regularity result for the evolutionary Stokes-Fourier system. Appl. Anal. 90, 31-45 (2011)
6. Bulicek, M., Malek, J., Shilkin, T.: On the regularity of two-dimensional unsteady flow of heat-conducting generalized Newtonian fluids. Nonlinear Anal. RWA 19, 89-104 (2014)
7. X. Ye, Global existence of heat-conductive incompressible viscous fluids. Acta. Appl. Math 148: 61-69 (2017)
8. Evans, L.C. Partial differential equations, Amer. Math. Soc. Graduates Studies in Math. 19. (1998)
9. Elio Espejo, Takashi Suzuki. Reaction terms avoiding aggregation in slow fluids. Nonlinear Anal. RWA 21, 110-126 (2015)

Universal CubeSat Platform Design Technique

Zhiyong Chen[1,a]

[1]*Interligent Manufacturing Key Laboratory of Ministry of Education, Shantou University, Shantou, China*

Abstract: This article summarizes cubesat technology, provides examples of their scientific impact, and describes the design and the manufacturing of a Cubesat platform. As for the design of the overall frame structure of the CubeSat, we have searched a lot of literature and consulted many predecessors' designs, and collected many satellite structure images. After analyzing the data, we aimed at all kinds of different structures' advantages and disadvantages, finally we got a best design.

It is a satellite of cubic shape (10 cm per side), weighing approximately 1kg, based on the creation of a central body made of different material using the 3D-Printing techniques. The 3D-Printing technique has several advantages including fast implementation, accuracy in manufacturing small parts and low cost. Moreover, concerning the construction of a small satellite, this technique is very useful thanks to the accuracy achievable in details, which are sometimes difficult and expensive to realize with the use of tools machine.

The structure must be able to withstand the launch loads. For this reason, several simulations using an FEM simulation and an intensive vibration test campaign will be performed in the system development and test phase.

1 Introduction

1.1 CubeSats History

CubeSats are a class of nanosatellites that use a standard size and form factor. The standard CubeSat size uses a "one unit" or "1U" measuring 10x10x10 cm and is extendable to larger sizes: 1.5, 2, 3, 6, and even 12U (Figure 1). Originally developed in 1999 by California Polytechnic State University at San Luis Obispo (Cal Poly) and Stanford University to provide a platform for education and space exploration. The development of CubeSats has advanced into its own industry with government, industry and academia collaborating for ever increasing capabilities. CubeSats now provide a cost effective platform for science investigations, new technology demonstrations and advanced mission concepts using constellations, swarms disaggregated systems.

[a] Corresponding author : 17zychen5@stu.edu.cn

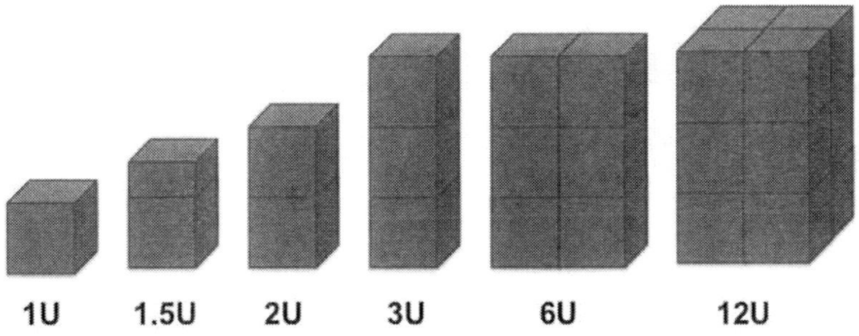

Figure 1. The standard CubeSat size uses a "one unit" or "1U" [1]

NASA Ames launched its first CubeSat, GeneSat, in December 2006. Since then Ames has launched 16 CubeSat spacecraft varying in size from 1U to 3U with an additional 12 CubeSats in development or awaiting launch. In 2014, two 6U Perseus-M CubeSats were launched for maritime surveillance, those two CubeSats represent the largest CubeSats flown as of 2015. The 2018 launch of the InSight lander to Mars, will include two 6U CubeSats called Mars Cube One (MarCO) [2-4].

NASA has selected 34 small satellites from 19 states and the District of Columbia to fly as auxiliary payloads aboard missions planned to launch in 2018, 2019 and 2020. Launch opportunities are leveraged from existing launch services for government payloads as well as via dedicated CubeSat launches from the new Venture Class Launch Services contracts. The proposed CubeSats come from educational institutions, universities, non-profit organizations and NASA centers [5].

This article summarizes CubeSat technology, provides examples of their scientific impact, and describes the design and the manufacturing of a CubeSat platform.

2 Main tasks

2.1 CubeSat Design

The CubeSat specification accomplishes several high-level goals. The main reason for miniaturizing satellites is to reduce the cost of deployment and are often suitable for launch in multiples, using the excess capacity of larger launch vehicles. The CubeSat design specifically minimizes the risk to the launch vehicle and payloads. Encapsulation of the launcher–payload interface takes away the amount of work that would previously be required for mating a piggyback satellite with its launcher [6].

Standard CubeSats are made up of 10×10×11.35 cm units designed to provide 10×10×10 cm or 1 liter of useful volume while weighing no more than 1.33 kg (2.9 lb) per unit. The smallest standard size is 1U [7]. Some Aerospace Corporation has constructed and launched two smaller form CubeSats of 0.5U for radiation measurement and technological demonstration [8]. In recent years larger CubeSat platforms have been proposed, most commonly 6U (10×20×30 cm or 12×24×36 cm) and 12U (20x20x30 cm or 24x24x36 cm, to extend the capabilities of CubeSats beyond academic and technology validation applications and into more complex science and national defense goals [9].

2.2 CubeSat Structure

As for the design of the overall frame structure of the CubeSat, we have searched a lot of literature and consulted many predecessors' designs, and collected many satellite structure images as shown below (Figure 2):

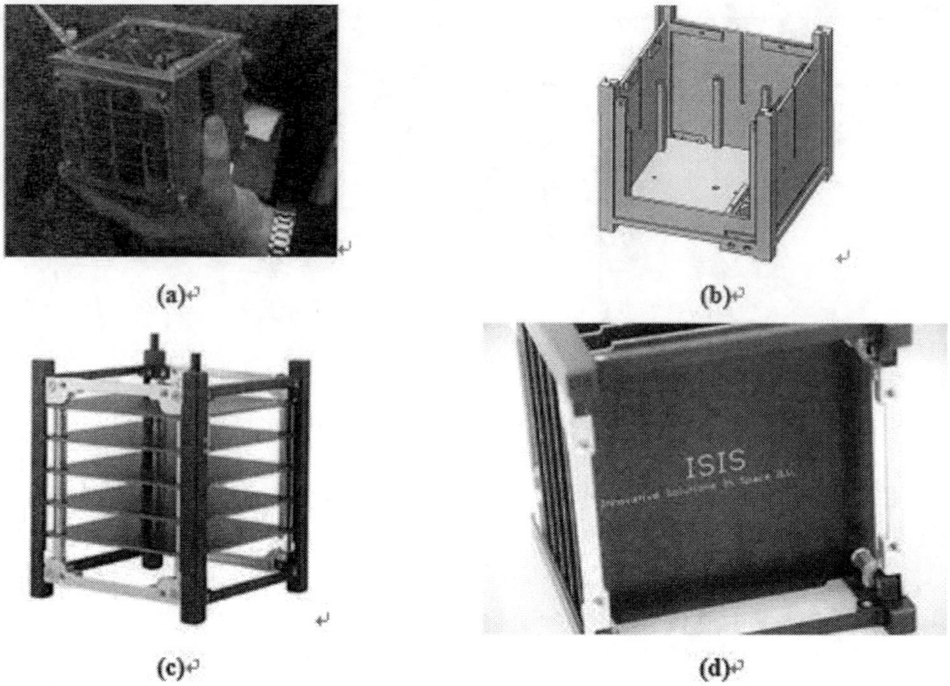

Figure 2. Some collected CubeSat structures [10-12].

In figure 2, the part (a) shows a relatively old cube model [10], the metal shell structure as a whole, contain the necessary collector and circuit board, it is the total weight of not more than 1 kg, but the metal shell weight is large, which means it cannot carry a large amount of load. Obviously, we want a bigger payload. And part (b) shows us a clear high-level design [11], the structure of the side panels can be separated from the base, which means that we can more easily placed the satellite payload, and because of the side panel and the base is not whole, we can use different materials to meet the demand of intensity. Part (c) and (d) are two pictures of the same cube [12]. This is a structure without the side panel, and using the lightweight composite materials, this means that the structure can guarantee the quality of strength requirement and reduce it, so we can carry more load effectively.

In order for this satellite structure to be universal, in other words, to be able to adapt to various carrier rocket connections, we designed specific joints for the structure. Meanwhile, we have specifically considered how to achieve larger payload volume while reducing the structure's mass.

Despite its small size, this single unit (1U) CubeSat should be equipped with several necessary subsystems, an active attitude control, a developed telecommunication system and high efficiency solar cells for power supply. The subsystems developed for this CubeSat have also been designed to be scaled up for larger satellites such as 2U or 3U CubeSats. The additional volume can be used for more payloads. So that the satellite can be used as a low cost platform for companies, institutions or universities to test components in space.

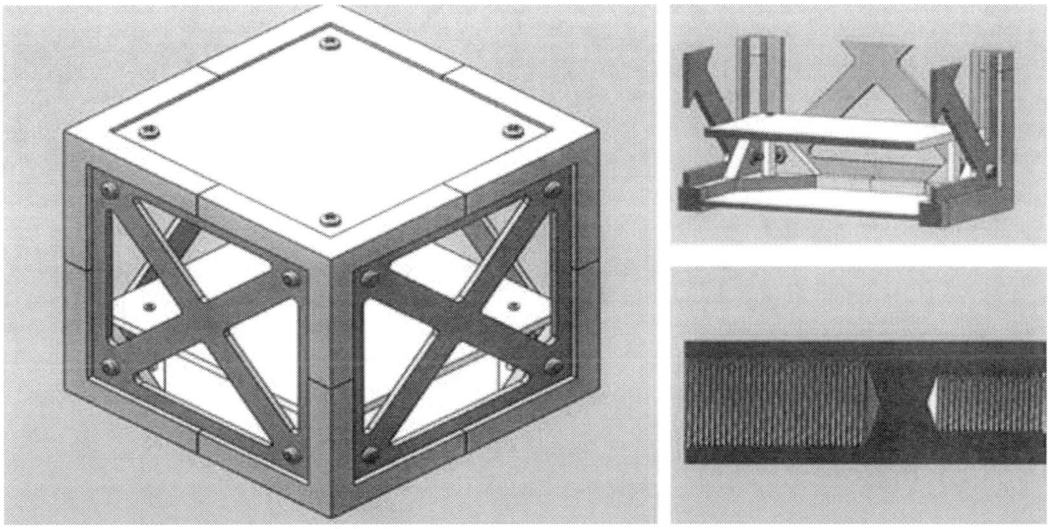

Figure 3. Completed structure design (left) & Structural design details & the plate support column of the electronic component (right).

After summarizing all the structures investigated, we have come up with the following structure design (Figure 3). As shown in the figure 3, the whole structure can be decomposed into two of the same top and bottom plates, four equal sides and eight exactly the same bases. It makes the manufacturing process of the structure simplified.

Such a design also facilitates the installation and disassembly of the whole structure. At the same time, we make the interior of the structure as large as possible to carry the payload. The design of the side plate is to reduce the mass under the premise of ensuring the strength of the structure. There are more details about the design structure (Figure 3).

3 Manufacturing technology

The CubeSat structure can be made of different material using the 3D-Printing techniques, for example, plastic materials, aluminum, gray cast iron, copper or other composite materials and so on it depends on our different design requirements (structural strength, mass requirements, etc.) for the structure. The 3D-Printing technique has several advantages including fast implementation, accuracy in manufacturing small parts and low cost. Moreover, concerning the construction of a small satellite, this technique is very useful thanks to the accuracy achievable in details, which are sometimes difficult and expensive to realize with the use of tools machine.

The advantage of the manufacturing technology is that it can produce product with very complex shape, and these products are impossible to implement in the form of low cost, or made in a short period with traditional manufacturing methods. In addition, this technique has changed the concept of general and secondary structures. Usually, the traditional manufacturing technology is aimed to remove material from the metal base, while 3D-printing, on the contrary, is covering the material layer by layer, so the parts are produced at the same time, the product is a whole. It reduces the number of parts, also reduces the screws and the needs of the adhesive, and simplifies the assembly process, at the same time, the reliability of the structure improved [11].

4 Simulations

The structure must be able to withstand the launch loads. For this reason, several simulations using an FEM simulation and an intensive vibration test campaign will be performed in the system development and test phase [13]. In addition, according to the requirements of the orbital environment, the most critical environmental tests and the critical components selection will be carried out.

Environmental testing, including the architecture of satellite vacuum test and thermal cycling test. Before and after the test, we hope that the structure mass won't be changed obviously, and during the pure cycle simulation of low earth orbit conditions (about an orbital period, temperature changes from - 20 °C to 80 °C), the quality of the satellite architecture will change no more than 1%. Moreover, a thermal shock test of structure should be taken, in order to response to sudden temperature changes (temperature range from - 20 °C to 60 °C).

The effects of radiation on satellite structures are thought to be negligible. According to the results of Ref. [14], in order to has serious damage to the structure, the total dose of radiation needs to be up to 10^9 rad (Si), however, any LEO orbit satellite 1 year experiment operating radiation (receiving less than 10^7 rad (Si), 2000 km height [15]) is far lower than this value. In addition, with regard to direct solar radiation, it is necessary to consider that the structure almost completely covers the solar cells, while solar cells block the main parts of the structure.

5 Conclusion

CubeSats are playing an increasingly larger role in exploration, technology demonstrations, scientific research and education at university. These miniature satellites provide a low-cost platform for national research missions, including planetary space exploration; Earth observation; fundamental Earth and space science; and technology demonstrations such as cutting-edge laser communications, energy storage, in-space propulsion and autonomous movement capabilities. They also provide educators an inexpensive means to engage students in all phases of satellite development, operation and exploitation through real world; hands-on research and development experience on NASA-funded ride-share launch opportunities.

A 1U Cubesat system has been developed aimed at education, low cost realization and research. The system includes innovative solutions, potentially improving the overall performance, flexibility and cost. Ground test results show that these are all feasible and promising. Under using structural and parametric conception to design universal multipurpose satellite platform CubeSat and improve modern technological process (3D printing and additive manufacturing) during it manufacture.

References

1. Elizabeth, *What are SmallSats and CubeSats?* (Mabrouk, Feb. 26, 2015).
2. MarCO, *Planetary CubeSats Become Real.* (www.planetary.org, Retrieved 2016-02-23).
3. Clark, Stephen. *Launch of NASA's next Mars mission delayed until at least 2018.* (Spaceflight Now, Retrieved 2016-02-23).
4. "CubeSat". Space.Skyrocket.de. Retrieved 2015-10-18.
5. NASA Announces Eighth Class of Candidates for Launch of CubeSat Space Missions, Editor: Shanessa Jackson, Feb. 21, 2017.
6. CubeSat - https://en.wikipedia.org/wiki/.
7. Mehrparvar, Arash, *"CubeSat Design Specification"* (PDF). (February 20, 2014).
8. AeroCube 6A, 6B (CubeRad A,B)., www.space.skyrocket.de , Retrieved 2015-10-18.
9. The CubeSat Program, CalPoly SLO. Retrieved March 25, 2017.
10. Kirk Woellert, Pascale Ehrenfreund, Antonio J. Ricco, Henry Hertzfeld. *Cubesats: Cost-effective science and technology platforms for emerging and developing nations.* Applied ScienceDirect. **47**(2011): 663
11. Jacopo Piattoni, Gian Paolo Candini, Giulio Pezzi, Fabio Santoni, Fabrizio Piergentili. *Plastic Cubesat: An innovative and low-cost way to perform applied space research and hands-on education.* Applied Elsevier. **81**(2012): 420
12. One-stop webshop for CubeSats and Nanosats - https://www.cubesatshop.com/.

13. Athirah, Nur; Afendi, Mohd; Hafizan, Ku; Amin, N.A.M.; Majid, M.S. Abdul. "Stress and Thermal Analysis of CubeSat Structure". Applied Mechanics and Materials. 554: **426–430**. Doi: 10.4028/www.scientific.net/amm.554.426.
14. NASA Space Vehicle Design Criteria (Structure), NASA SP-8053, June 1970.
15. W.J. Larson, J.R. Wertz, Space Mission Analysis and Design, third edition, Space Technology Library-Microcosm Press, California, 2004.

Numerical Simulation of Nozzle Flow Field with Secondary Injection Thrust Vector Control

H.R. Noaman[1,a], Tang Hai Bin [1] and Elsayed Khalil [2]

[1]*Beihang University, School of Astronautics, Beijing, China*
[2]*Military Technical College, Aerospace Department, Cairo, Egypt*

Abstract. Numerical simulations are performed to characterize the secondary injection thrust vector control. For this objective the following measurements were taken: considering the flow to be compressible and turbulent using Realizable k-ε turbulence model accompanied by enhanced wall treatment, the comparison between the CFD results and the experimental results shows a very good agreement. Then a parametric study on injection mass flow rate (changing secondary stagnation pressure) with the same injection location and injection angle is done. The results stated that increasing the injectant mass flow rate lead to shock impingement from opposite wall at secondary stagnation pressure 1.4 of the primary stagnation pressure.

1 Introduction

The propulsion system not only provides a thrust to a rocket or any flying vehicle but also can introduce a control mechanism to correct the vehicle's trajectory and attitude by systems of thrust vector control (TVC). With the control of thrust vector direction, pitching, yawing and rolling moments can be accomplished on the aerial vehicle; there are many techniques to generate the deflection of the thrust vector such as the usage of flexible nozzle joints, gimbaled nozzles, jetavators, jet vanes/tabs, secondary injection, and etc. From the different ways to deflect thrust vector of a flying vehicle, Secondary Injection Thrust Vector Control (SITVC) has been applied successfully since 1960's and is achieved by injecting secondary fluid to the supersonic flow inside the divergent part of the convergent-divergent nozzle. In contrast to mechanical thrust vector control systems, such as gimbaled nozzles, jet vanes/tabs, jetavators etc., that uses an actuation system to move the mechanical parts, SITVC does not use any movable parts and is governed by flow regulation, which minimize the losses of the axial thrust force during changing the thrust direction [1] The secondary injectant, (liquid or gas) can be supplied from the combustion chamber as a bleed or from a separate gas generator and it creates a complex flow field in the nozzle divergent part which contains not only a strong bow shock that creates asymmetry and a weak separation shock due to separation of the boundary layer upstream of the injector, but also downstream of the injector a Mach disk and reattachment region accompanied by recompression were created [2-4]. Figure1. [2] Schematically depicts the flow field structure inside the nozzle setup as a result of secondary injection. The causes of the deflection are as a result of the side force which produced by a combined effect of, a) interaction (induced) force, due to pressure rise along the wall, and b) jet reaction force, caused by the momentum

[a] Corresponding author : hossamnoaman84@outlook.com

of the injecting secondary fluid [5]. As the secondary injection in the supersonic flow creates a complex flow field, many past researches focused on both analytical models such as Blast-wave analogy model[4] to characterize the flow field due to secondary injection and experimental tests with cold flow [6,7] and real static firing [8,9].While analytical models deals only with very low injection flow rates and lack in general, the experimental tests of cold flow and static firing provides the main data of SITVC to be utilized for further analyses, but these tests gives only macroscopic performance predictions with high cost. On the other side the evolution of computational power and numerical methods, results in developing the computational Fluid Dynamics (CFD) to give a detailed microscopic description of flow properties. CFD became an attractive alternative to analytical models and a strong complimentary tool to experimental tests [10, 11].

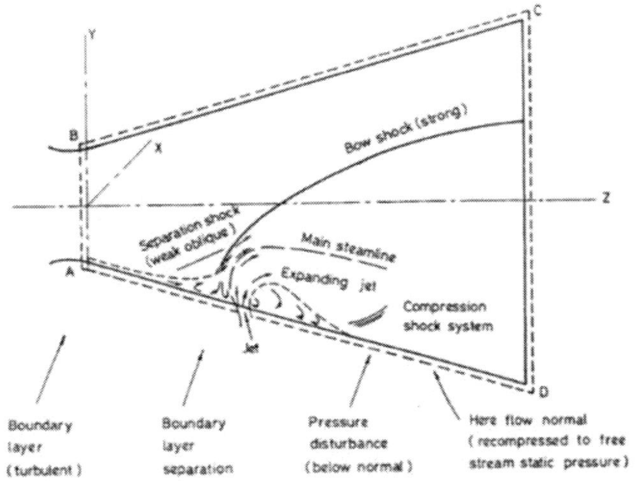

Figure 1. Flow field structure setup by secondary injection into primary nozzle flow [2]

In this paper the flow field investigation inside a nozzle in presence of secondary injection and analysis of SITVC performance parameters with variable injection mass flow rate (changing secondary stagnation pressure), for given injection location and injection angle were carried out.

2 Numerical Methodology

Numerical simulation of two dimensional flow field inside the nozzle used by Gushe [7] has been carried out with the aid of a steady Realizable k-ε turbulence model with enhancement wall treatment. The solved equations are the two-dimensional, compressible, Reynolds-averaged Navier–Stokes equations, which were discretized in finite volume form on each of the quadrilateral control volumes. A second-order upwind scheme was chosen for the spatial discretization; the implicit density-based algorithm was used to solve the equation, and Roe-averaged flux difference splitting (Roe FDS) was chosen for the flux type, Ideal gas (air) was considered for analysis.

The first cell height inside wall nozzle was taken to achieve wall Y plus about 1 in the boundary layer as shown in figure 2. The first cell height estimation is estimated using skin friction coefficient (C_f) for flat plate. Note that x is the distance from the nozzle inlet to injector location, where the separation of boundary layer occurs, and utilizing the definitions of wall Y plus and Uτ the first cell height is predicted. Then it is validated with numerical simulation.

$$C_f = 0.0576(\mathrm{Re_x}))^{-0.2} \qquad (1)$$

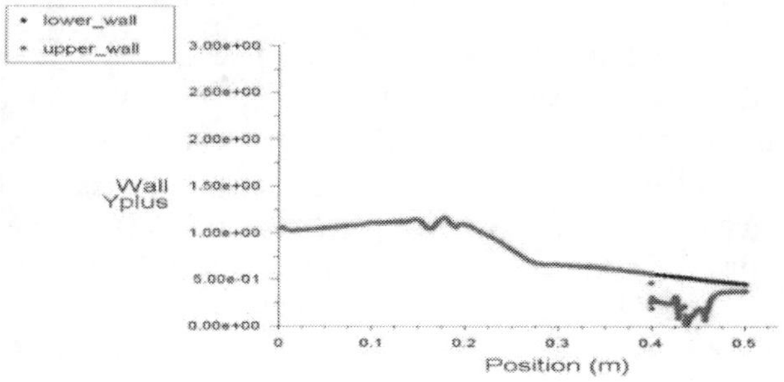

Figure 2. Wall Y plus-values along upper and lower wall

A typical structured grid system in the computational domain selected after grid independence study, where three grids with different aspects are considered as shown in Table1.

Table1. Computational grids for grid independence study

NO	Grid size	Injector grid	Height of first grid	y+
1	300×690	40×75	0.00002 in	0.45-1
2	200×640	40×75	0.00002 in	0.45-1
3	150×400	30×75	0.00004 in	0.84-1

Grid independent study shows that grid 1and 2 is a little better than grid 3 and for accuracy grid 2 was selected for the study.

The computational grid 2 are shown in figure 3.were the mesh around the injection slot and the walls were specially refined as shown in figure 4

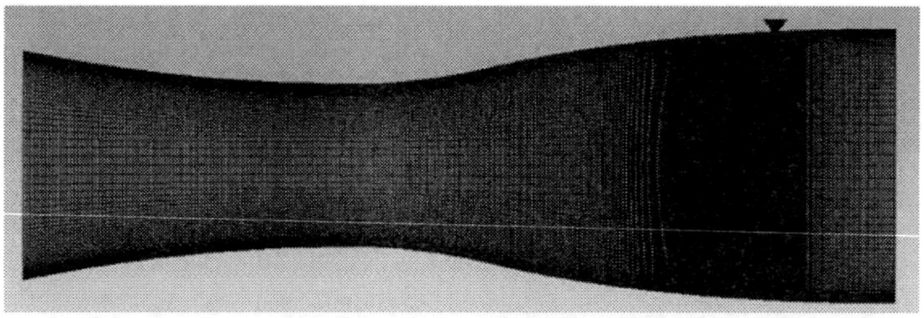

Figure 3. Computational structured Grid 2

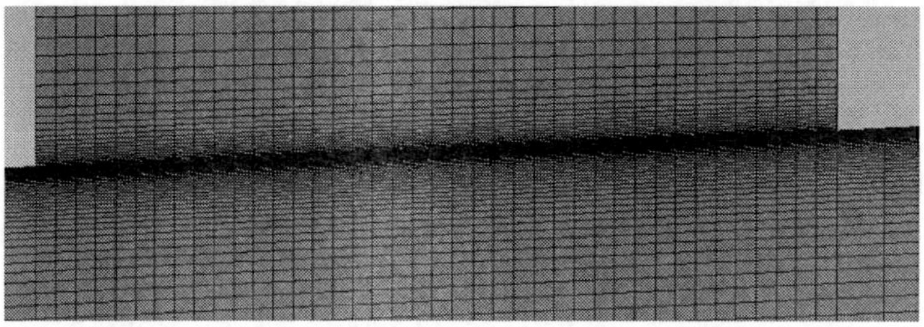

Figure 4. Close view at the injection grid 2

2.1 Boundary conditions

A pressure inlet boundary was used for nozzle inlet, secondary inlet and nozzle outlet where:

<u>Inlet:</u> Primary Stagnation Pressure, P_{op} = 100 Psig, Primary Stagnation Temperature, T_{op} = 258.5K

Outlet: Primary Nozzle Exit Pressure, P_{ep} = 101068Pa

Secondary inlet: - Secondary to primary stagnation pressure ratio P_{os}/P_{op}=0.6.

-Secondary injectant temperature T_{os}=272.2K.

- Secondary injection slot to primary throat area ratio A_s/A^*=0.05

-Injection location with respect to the primary flow Mach number Mp=1.904.

2.2 Validation of the numerical simulation

First numerical simulation of the flow field inside the nozzle used by Gushe [7] Without injection to calculate the primary axial thrust at no injection state and a comparison between the calculated Mach number [7] and the numerical one has been done as in figure 5.And shows a good accuracy with error less than 10%.

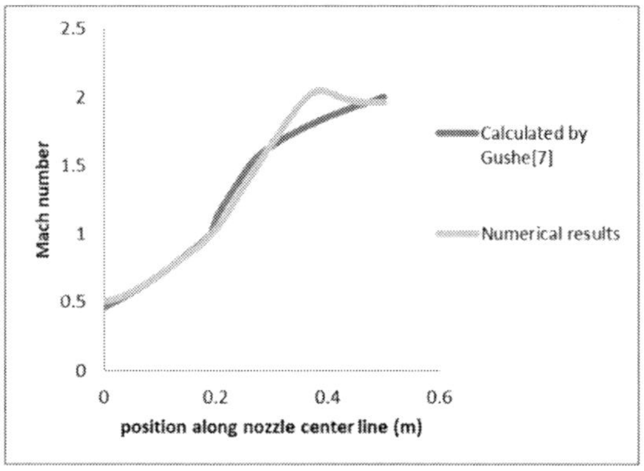

Figure 5. Comparison between calculated and numerical results for Mach number (no injection)

Then the simulation of the flow field of nozzle with the secondary injection slot carried out and a comparison between the experimental and numerical results has been done for the static pressure distribution along the upper wall where the injector have been installed using realizable *k-ε* model, *k-ω* model and the shear stress transport (SST) *k-ω* model it shows that all of the models behave similarly With an excellent accuracy in the unaffected zone and good accuracy in the high and low pressure zone less than 10% also a little difference between the experimental and numerical results at the boundary layer separation point figure 6. Also the comparison for the static pressure distribution along lower wall (no injector) has been done and shows an excellent accuracy for all models figure 7.

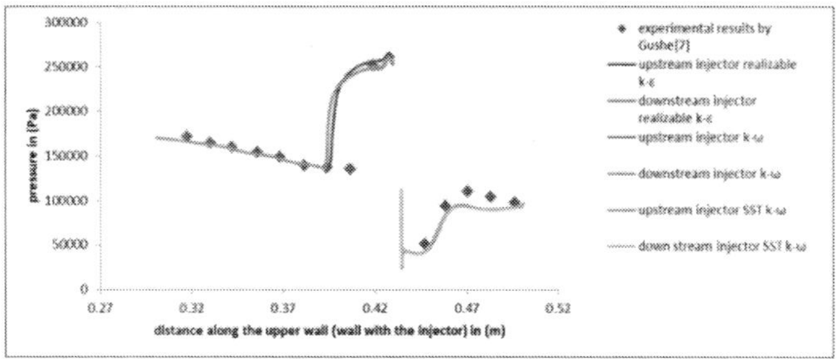

Figure 6. Comparison between Experimental and numerical results for pressure distribution upstream and downstream the injector on the upper wall using different turbulence models.

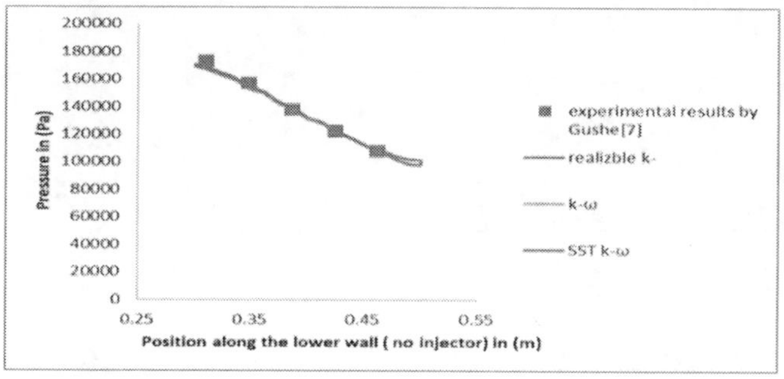

Figure7. Comparison between Experimental and numerical results for pressure distribution on the lower wall using different turbulence models

2.3 SITVC Performance Parameters

Secondary injection thrust vector control (SITVC) performance parameter investigated in present study includes [5-12]:

1-Axial Thrust Augmentation *ATA%*

$$ATA\% = (F_P - F_P^o)/F_P \tag{2}$$

2-Thrust ratio is the ratio of the side thrust force to axial thrust force *Fs/Fp* where:

*Side Thrust *Fs*:

$$F_s = F_i + F_j \tag{3}$$

F_i is the interaction Force (Side-Thrust-Pressure Component).

$$F_i = PA_{upperwall} - PA_{lowerwall} \tag{4}$$

F_j is the jet Reaction Force (Side-Thrust-Momentum Component).

$$F_j = \dot{m}_s V_{sy} + ((P_{es} - P_{as})A_s)cos\beta_{inj} \tag{5}$$

*Primary axial thrust *Fp*:

$$F_P = \dot{m}_e V_x + (Pe - Pa)Ae \tag{6}$$

3- System Specific Impulse Loss δIsp

$$\delta Isp = Ispsys - Isp^\circ \tag{7}$$

4- Specific Impulse Ratio (Amplification Factor) AK is the ratio of the side specific impulse to main axial specific impulse and it determines the amount of fluid to be injected to have a specified side force [6].

$$AK = Isps/Isp^\circ \tag{8}$$

where:

$$Isp^\circ = F_p/\dot{m} \tag{9}$$

$$Isps = F_s/\dot{m}_s \tag{10}$$

$$Ispsys = \frac{\sqrt{F_p^2+F_s^2}}{(\dot{m}_s+\dot{m}_p)g} \tag{11}$$

3 Results and Discussion

A parametric study on injection mass flow rate (changing secondary stagnation pressure) with the same injection location and injection angle is done as mass flow rate of the injectant varies directly with the injection total pressure.

$$\dot{m}_s = \frac{A_s P_{os}}{\sqrt{T_{os}}} \sqrt{\frac{\gamma}{R}\left(\frac{2}{\gamma+1}\right)^{\frac{\gamma+1}{\gamma-1}}} \tag{12}$$

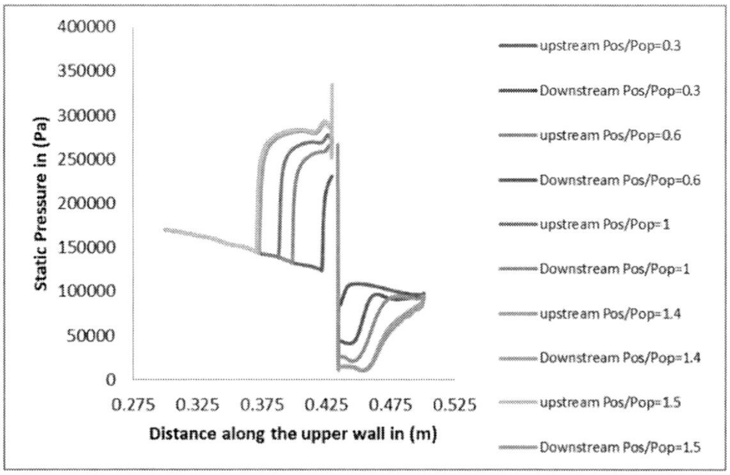

Figure 8. Effect of injection mass flow rate on injector wall static pressure distribution

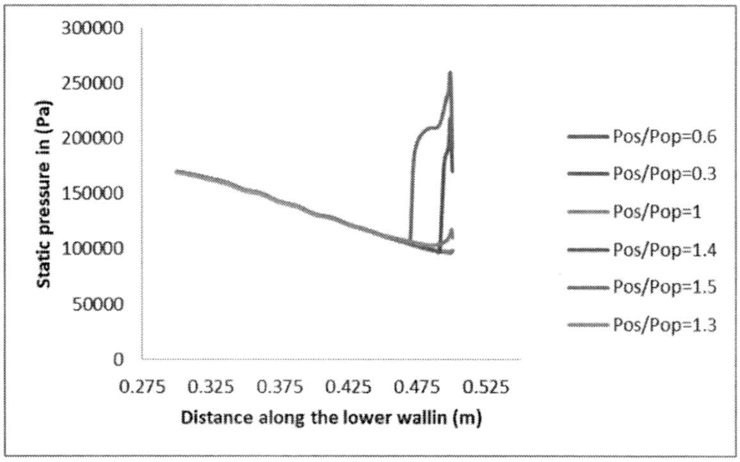

Figure 9. Effect of injection mass flow rate on lower wall static pressure distribution

The strength (shock angle) of the primary bow shock inside nozzle increases with the secondary mass flow rate (by increasing the secondary stagnation pressure) as shown in figure 8.

The origination point of the primary bow shock moves further upstream of the injector as the secondary mass flow rate increases (by increasing the secondary stagnation pressure). Leads to extended higher pressure region upstream of the injector as shown in figure 8.

Same as in primary bow shock, the strength of the separation increases as the secondary mass flow rate is increased (by increasing the secondary stagnation pressure). Stronger separation shock results into stronger injector upstream higher wall pressure region as shown in figure 8.

Higher secondary mass flow rates (resulting from higher secondary stagnation pressure) also cause extended lower pressure regions downstream of the injector. As the injection mass flow rate is lowered, the reattachment point moves upstream on the primary nozzle wall in the aft section of the injector as shown in figure 9.

In figure 9 the effect of increasing the injectant mass flow rate on the lower wall it shows that no changes in the static pressure along the lower wall till reaching Pos/Pop=1.3 a very small shock impingement arises on the exit of the nozzle and when Pos/Pop increased to 1.4 and then 1.5 a significant shock impingement is shown leads to rise in the static pressure on the lower wall which effects the side force. The complex flow field with shock impingement setup by secondary injection inside primary supersonic flow was shown in Figure 10.

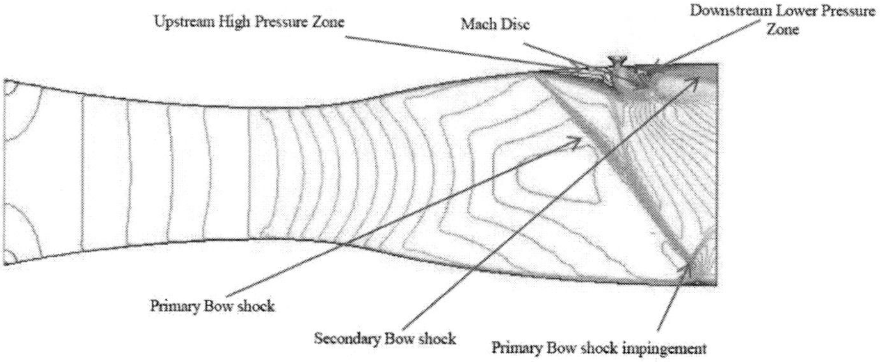

Figure 10. Flow field structure (Mach number contours) setup by secondary injection into primary nozzle flow.

The influence of injection stagnation pressure on the secondary injection performance parameters for a given injection location and angle of injection are shown in the figures from 11-14.

The observations are discussed as follows:

1 - Referring to figures 11-14, secondary mass flow rate is a function of secondary injection stagnation pressure. In other words, secondary mass flow rate varies directly with secondary injection stagnation pressure for a given injection area.

2 – Increasing the secondary mass flow rate (by increasing the secondary stagnation pressure), the axial thrust augmentation increases as can be observed in figure 14. Primary reason is addition & subsequent adiabatic expansion of the injectant gas. Thrust augmentation is also a weak function of strength of the primary & secondary bow shocks that deflect the primary flow direction in the primary nozzle. More augmentation to the axial thrust is resulted in case of stronger shocks.

3 - The thrust ratio increases linearly as the injectant mass flow rate increases (by increasing the secondary stagnation pressure) till reaching the critical secondary stagnation pressure where shock impingement occurs as observed in figure 11.

4 - The amplification factor in figure 12 increase linearly and rapidly for low mass flow rates for moderate and higher mass flow rates the rate of increase decreases till reaching the critical secondary stagnation pressure where shock impingement occurs as observed in figure 12. Amplification factor does not linearly depend on the secondary mass flow rate due to non-linearities of the governing factors. Amplification factor is a function of physical/flow properties of primary & secondary fluids, injection location, angle of injection and primary nozzle geometry.

5 - Side thrust & axial thrust augmentation is achieved at the cost of specific impulse. It can be observed in figures 13 that as the secondary mass flow rate increases (by increasing the secondary stagnation pressure), system specific impulse decreases almost linearly. The reason of this loss is the

inefficient expansion of the injected gas. It is concluded that shock impingement surely affects the secondary specific impulse & amplification factors.

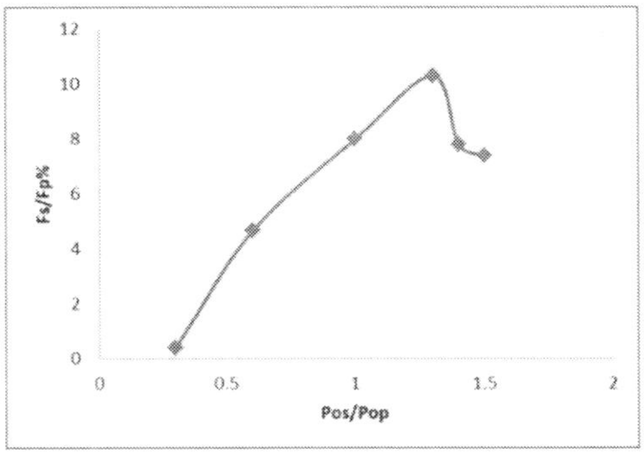

Figure11. Effect of secondary mass flow rate (by changing secondary stagnation pressure) on the thrust ratio

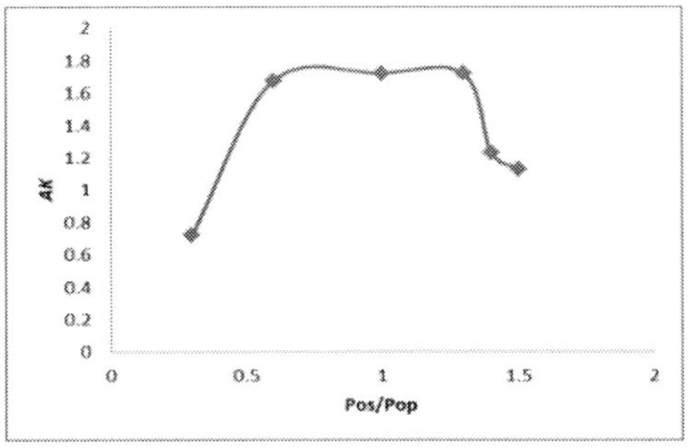

Figure12. Effect of secondary mass flow rate (by changing secondary stagnation pressure) on the amplification factor

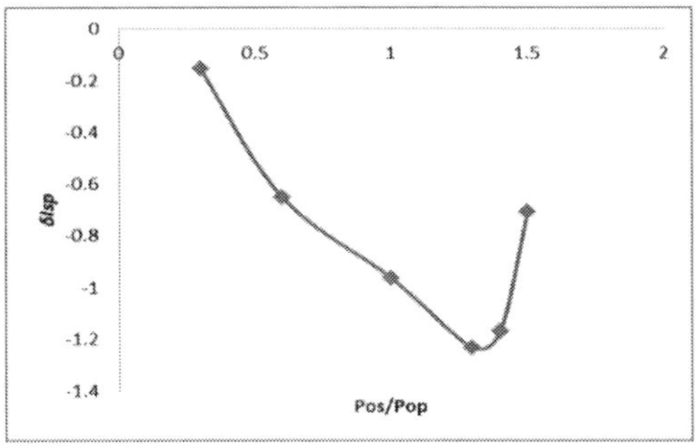

Figure13. Effect of secondary mass flow rate (by changing secondary stagnation pressure) on the specific impulse loss

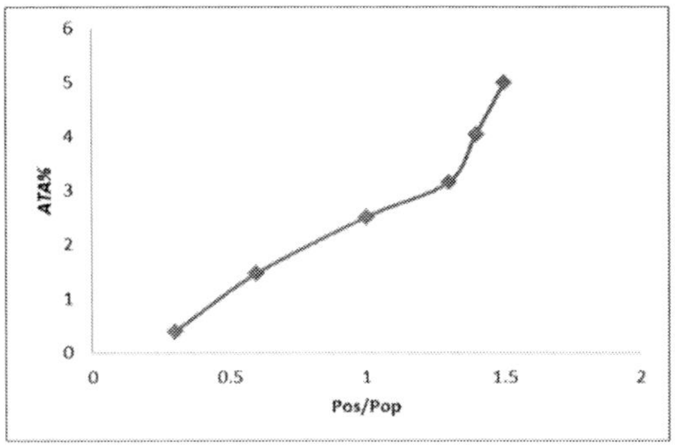

Figure14. Effect of secondary mass flow rate (by changing secondary stagnation pressure) on the axial thrust augmentation

4 Conclusion

The objective of the present paper was to investigate the performance and physics of SITVC. realizble k-ε model with enhanced wall treatment used and compared to the experimental results to validate it. Performance parameters of SITVC system have been investigated with respect to different secondary mass flow rate. It was found that Injectant mass flow rate is a main parameter that can provide a wide range of operational flexibility & controllability in SITVC systems and must be investigated to get the maximum injectant mass flow rate with safe injection to avoid shock impingement on the opposite side of the nozzle wall that leads to reduce net side force or in worst cases vectoring the system into undesired direction.

Nomenclature

A	Area (m^2)
$ATA\%$	Axial thrust augmentation percentage
AK	Amplification factor
A_s	secondary injection area (m^2)
C_f	Skin friction coefficient
F	Force (N)
F_s	Side force (N)
F_p	Primary axial thrust (N)
$F_p^{\,0}$	Primary axial thrust without injection (N)
M_p	Injection location (in terms of primary flow axial Mach number corresponding to injection point located on primary nozzle wall
$\dot{m}_e$	Total exit mass flow rate (Kg/s)
$\dot{m}_p$	Primary mass flow rate (kg/s)
$\dot{m}_s$	Secondary mass flow rate (kg/s)
P	Pressure (Pa)
P_{es}	Secondary exit pressure (Pa)
P_{as}	Ambient pressure at the secondary port before injection (Pa)
P_{os}	Secondary stagnation pressure (Pa)
P_{op}	Primary stagnation pressure (Pa)
R_{ex}	Reynolds number at the distance x
T_{os}	Secondary stagnation temperature (k)
$U\tau$	Friction velocity (m/s)
V_x	Velocity along x- axis (m/s)
V_{sy}	Secondary gas velocity along y- direction (m/s)
Isp_s	Secondary specific impulse (sec)
Ispsys	System specific impulse (sec)
$Isp_p^{\,o}$	Primary specific impulse without injection (sec)
δIsp	Specific impulse loss
β_{inj}	Wall angle at point of injection (deg)
γ	Specific heat ratio

References

1. Sutton, G.W., "*Rocket Propulsion Elements*", 7th ed., John Wiley & Sons, 2001.
2. Balu, R., "Analysis of Performance of a Hot Gas Injection Thrust Vector Control System", *Journal of Propulsion andPower*, Vol. 4, 1991, pp. 580-585.

3. Newton, J.F., Spaid, F.M., "Interaction of Secondary Injectants and Rocket Exhausts for Thrust Vector Control", *Journal of the American Rocket Society*, August 1962, pp, 1203-1211.
4. Broadwell, J.E., "Analysis of Fluid Mechanics of Secondary Injection for Thrust Vector Control", *AIAA Journal,* Vol. 1, No. 10, 1963, pp. 1067-1075.
5. Muhammad Usman Sadiq "Performance Analysis and Flowfield Characterization of Secondary Injection Thrust Vector Control (SITVC) for A 2DCD Nozzle" Msc. Thesis Faculty of the Viterbi School of Engineering University of Southern California, 2007.
6. Masuya, G., "Secondary Gas Injection into a Supersonic Conical Nozzle", AIAA Journal, Vol. 15, No.3, 1977, pp. 301-302.
7. Guhse, R. D., "An Experimental Investigation of Thrust Vector Control by Secondary Injection" NASA CR-297 (1965).
8. Zeierman, I., Timnat, Y. M., "Full Control of Solid Propellant Rockets by Secondary Injection", *Journal of Spacecraft and Rockets ,* Vol. 10, No. 3, 1973.
9. Inouye, T., "Experiments on Rocket Thrust Vector Control by Hot Gas Injection", *Journal of Spacecraft and Rockets*, Vol.3, No.4, pp. 737-739, 1966.
10. M.salehifar, M.Hojaji, A.Dartoomian, "CFD modeling for flow field characterization and performance analysis of HGITVC" applied thermal engineering 103,291-304,2016.
11. chuan Tian and Yijia Lu " turbulence models of separated flow in shock wave thrust vector nozzles" engineering applications of fluid mechanics vol7, no 2, 2013.
12. Erdem, E., Albayrak K., and Tinaztepe h. T., "Parametric Study of Secondary Gas Injection into a Conical Rocket Nozzle for Thrust Vectoring", Presented at 42nd Joint Propulsion Conference, July 09-12, 2006.

Battery group parameter selection and dynamic simulation of pure electric vehicle

Guo Minrui[a] and Cheng Lei

College of Mechanical and Energy Engineering, Huanghuai University, Henan Zhumadian, China

Abstract. The battery pack is one of the core components of pure electric vehicle, dynamic performance of the whole vehicle is closely related to the matching design of the battery, and is affected by the air resistance coefficient and the windward area of the whole vehicle. The dynamic indicators include maximum driving speed, 0-100km/h acceleration time and climbing grade, the battery parameters are designed and matched before the vehicle design, mainly analyze influence on the vehicle dynamic performance of the types of batteries, air resistance coefficient, windward area by the simulation software ADVISOR, and optimize the combination of these parameters. The results show that the dynamic performance of the vehicle reaches the initial design index and the dynamic performance of the vehicle is improved significantly.

1 The introduction

In 2015, Chinese new energy automobile production become the first, the direction of the development of auto technology will be safe, energy-saving, environmental protection, new energy vehicles will be the future development trend, this pointed out the direction of the investment and the development of new energy vehicles industry[1]. However, the research on pure electric vehicles in China is not comprehensive enough, especially in battery technology. The data shows that the investment of domestic battery industry is about 100 billion RMB in 2016[2]. The battery types used in electric vehicles are lead-acid batteries, nickel-series batteries, lithium ion battery, fuel cell, etc. China's electric vehicle technology development "12th five-year plan" has pointed out that the development of vehicle powered batteries with lithium ion batteries as the focus will be promoted. BYD K9A use lithium iron phosphate battery, which is nearly 2/3 of the cost of driving 300 km compared to fuel bus, and can run more than 270 kilometers on a single charge, with a maximum speed of 80Km/h[3]. The Nissan Leaf entered the Chinese auto market in 2011, using the latest lithium battery pack, with a nominal voltage of 345V and a maximum speed of 144km. Tesla is electric sports car brand established by Musk, Tesla Model 3 adopted more than 7000 section 18650 lithium-ion battery pack with a voltage of about 400V DC and accelerates in less than 6 seconds in a hundred kilometers in 2016[4]. The current status of the use of battery packs at home and abroad shows that there is still a certain gap between the domestic battery pack technology and foreign countries, especially in terms of the vehicle's driving force, such as the maximum travel speed and the

[a] Corresponding author : 814293109@qq.com

acceleration time of 100 km[5]. However, battery technology research and breakthroughs for vehicle power to improve is essential. As the core component of pure electric vehicle, how to design and select battery packs, shorten the development cycle and make it meet and exceed the design requirements is the key to the development of pure electric vehicles.

2 Battery pack parameter selection

Power battery is the only device to provide energy vehicle, the role is very important. Electric vehicle battery pack must have a higher power density, in order to achieve vehicle power performance requirements. At the same time should have a higher energy density in order to achieve adequate driving range. In addition, the power battery should have more charge and discharge times, high enough cost performance, ease of maintenance, high discharge efficiency and other characteristics. Battery parameters include the choice of type, voltage, capacity and battery number and a series of parameters.

2.1 Choice of battery type

Lithium iron phosphate battery use lithium iron phosphate as the cathode material. due to its better safety performance, there has been no explosion, high reliability, long cycle life, it can discharge with high current, starting current up to 2C, lead-acid batteries now do not have this performance, nickel-based battery larger environmental pollution, energy density of lithium-ion battery is about 100 ~ 150Wh / kg about 2 to 3 times higher than lead-acid batteries, and high temperature performance, large single cell capacity, no memory effect, light weight, environmental protection, so lithium iron phosphate battery is chose.

2.2 Selection of battery voltage

Battery voltage is mainly according to the rated voltage of the motor to determine the size, in addition to consider power consumption of lighting device, instrument accessories, so the total voltage of the battery should be slightly greater than the rated voltage of the driven motor. It is the series relationship between the individual lithium iron phosphate battery, which can improve the working voltage of the whole vehicle and prevent the shortage of battery voltage and affect the driving distance of the electric vehicle.

Table 1. Technical parameters of lithium iron phosphate battery

Number	Parameters	Values
1	Single battery capacity (Ah)	70
2	Single rated voltage (V)	3.2
3	Single battery internal resistance.	1
4	Single maximum voltage (V)	3.9
5	Single minimum voltage (V)	2
6	Battery voltage (V)	270
7	Number of battery	81

2.3 Selection of battery number

Combined with automobile design and automobile theory knowledge, the number of batteries is calculated by the continuous mileage calculation, the other parameters of the vehicle are well matched, and the parameters of the lithium iron phosphate battery pack are shown in Table 1.

3 Simulation analysis of vehicle power performance

The ADVISOR is a software for simulation analysis of traditional internal combustion engines, fuel cell vehicles, hybrid electric vehicles and pure electric vehicles, by ADVISOR software to simulate the dynamic performance of the vehicle, analyze the factors affecting the dynamic performance, thus reduce the development cycle of electric vehicle.

Based on inputting the vehicle technical parameters, firstly set up the initial capacity of the battery pack, minimum quantity, the quality of the car, shift delay time, the simulation parameters such as mileage, climbing on the constant speed. In this paper, the initial battery charge is 100%, the minimum charge is 30%, the shift delay time is 0.2s, and the simulated driving range is 0.402km. After these initial conditions are determined, simulation of the dynamic performance of the vehicle load of 130 kg (about two people) is carried out. This is shown in Table 2.

Table 2. Simulation of vehicle dynamic performance under different load quality

Number	Dynamic performance	The simulation results
1	Maximum driving speed (km/h)	133.3
2	0-100km/h acceleration time (s)	18.8
3	Climbing grade	16.7%

The dynamic performance is influenced by a variety of factors, which type of battery is selected, how the wind resistance coefficient and the windward area can be obtained through the simulation analysis below.

4 Simulation analysis and comparison of results

The dynamism of electric vehicles is the most important and basic performance of all kinds of vehicles. There are many factors affecting the dynamic performance of electric vehicles, such as the type of battery, air resistance coefficient, the windward area, the structure of the vehicle, the external force suffered, the quality of electric vehicles, motor performance, etc. The impact of the first three factors is analyzed in this paper.

(1) Maintain the same capacity of battery pack, respectively, using lead-acid batteries, nickel-metal hydride batteries, lithium-ion battery vehicle performance simulation of the vehicle, the simulation results in Table 3.

Table 3. The dynamic performance simulation results under different types of battery

Battery type	Maximum travel speed (km / h)	0-100km /h acceleration time(S)	Climbing grade
Lead-acid battery	128.5	31.9	10.6%
Ni-MH battery	131.3	22.6	13.5%

Lithium-ion battery	134.0	15.9	20.7%

According to the above simulation results, select the type of battery as abscissa, the vertical axis of Figure 1 is the maximum speed, the vertical axis of Figure 2 is 0-100km / h acceleration time, draw different simulation comparison chart.

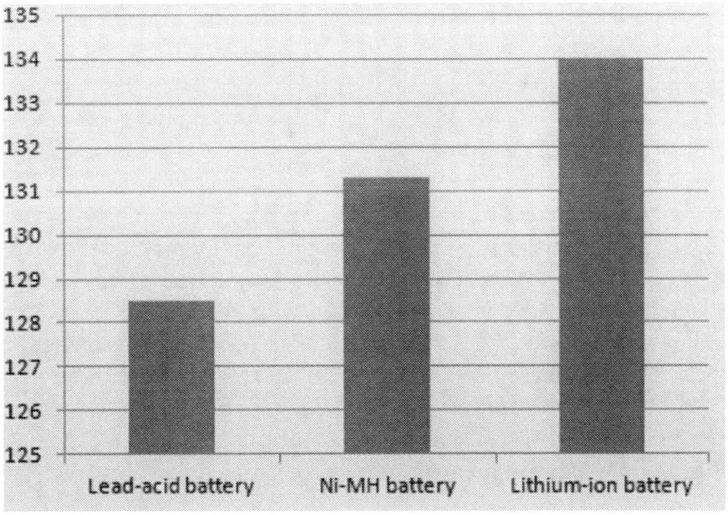

Figure 1. Comparison of maximum speed simulation (unit: km / h)

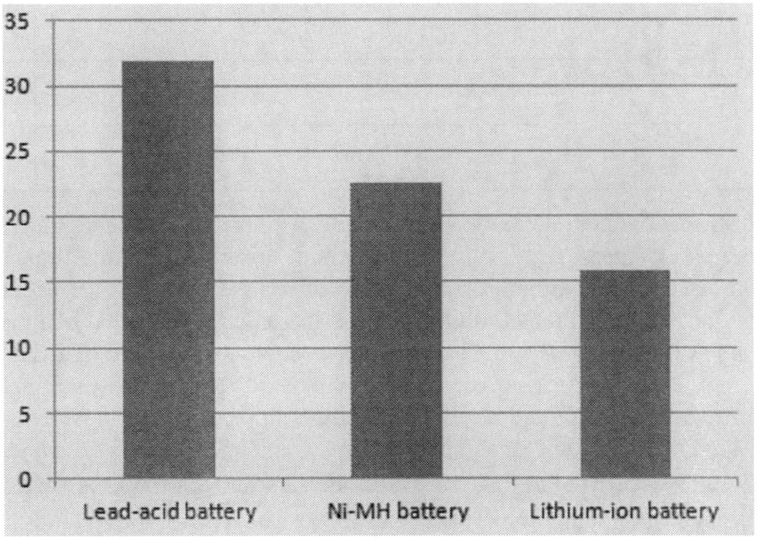

Figure 2. 0-100km / h acceleration time simulation comparison (unit: s)

From figure 1 and figure 2, it can be obtained that different types of battery are selected, and the dynamic performance of the vehicle is very different. The performance of li-ion battery in the maximum driving speed, 0-100km/h acceleration time and 40km /h is obviously better than that of lead-acid battery and ni-mh battery. In figure 1 ,the maximum speed of using lithium ion batteries increased by 4.3% than the lead-acid battery. In figure 2, the acceleration time of using lithium ion battery reduced by 16s than lead-acid batteries, by 6.7s than ni-mh battery. So the lithium-ion battery is increasingly widespread in the electric vehicle.

(2)Battery parameters remain unchanged, change the different air resistance coefficient, the selection of air resistance coefficient is 0.3, 0.35 and 0.4, and the simulation results of vehicle power performance are obtained, as shown in Table 4.

Table 4. The dynamic performance simulation results under different air resistance coefficients

Air resistance coefficients	Maximum travel speed (km/h)	0-100km /h acceleration time(s)	Climbing grade
0.3	137.2	15.7	20.2%
0.35	124.1	16.9	18.3%
0.4	118.3	17.3	16.8%

The larger the air drag coefficient, the smaller the maximum traveling speed, the longer the acceleration time, the smaller the climb grade. The maximum driving speed of air resistance coefficient is 0.3, which increases by 15.9% compared with the wind resistance coefficient as 0.4, the climb grade of 40km /h increases by 3.4%, while the dynamic performance of the air resistance coefficient as 0.35 is at a medium level. The car's surface and bottom can be machined as smooth as possible, and the air resistance coefficient can be optimized to reduce the coefficient of wind resistance.

(3) The impact of different windward area on the vehicle dynamic performance, as shown in Table 5:

Table 5. The dynamic performance simulation results under different air resistance coefficients

Windward area (m^2)	Maximum travel speed (km/h)	0-100km /h acceleration time(s)	Climbing grade
2.0	134.0	15.9	20.0%
2.3	129.8	16.3	18.9%
2.6	126.1	16.7	17.8%

From the analysis of Table 5, it can be seen that the dynamic performance of the vehicle decreases as the windward area increases. Especially the maximum driving speed, the impact of the windward area is more prominent. The maximum driving speed of windward area of $2.0m^2$ is 3.2% higher than the windward area of $2.3m^2$, and the climbing speed is increased by 1.1%. When the windward area is $2.0m^2$, the maximum driving speed is 6.3% higher than the windward area of $2.6m^2$, and the climbing speed is increased by 2.2%, taking into account the windward area by the space constraints, generally not easy to reduce. However, it should be kept to a minimum to ensure the best dynamic performance.

The initial design specifications of the pure electric vehicle dynamic performance are as follows: the maximum speed $v_{max} \geqslant$100km/h, 0-100km/h acceleration time $t \leqslant$20s, when the vehicle is at 40km/h uniform speed ,the climbing grade $\alpha \geqslant$ 15%. The above three kinds of parameters are optimized combination, the selected battery type of iron phosphate battery, the drag coefficient of 0.3, the windward area of $2.0m^2$, the simulation results in Table 2, indicate that the dynamic performance to meet the design requirements, the battery parameters of the pure electric car match reasonably, and the dynamic performance is remarkably improved.

5 Conclusion

This article focuses on analysis and comparison of different battery types, drag coefficient, the wind area for the dynamic performance simulation. After optimization, the simulation results of dynamic performance are obtained, which meet the requirements of dynamic performance index and significantly improve the dynamic performance.

References

1. Yuan Xiaodong, Power battery recycling, New economic guide (2016)
2. Zhao Guiliang, Status and performance analysis of urban pure electric buses(2016)
3. Hu Hua,Song Hui, *Electric car*(2012)
4. Li Zhe, Pure electric car's lithium iron phosphate battery performance research(2011)
5. Wang Guoqiang,Wang Zhixin. Research on optimization of new electric vehicle frame structure (2016)

Effects of the Wet Steam Non-Equilibrium Condensation on the Dynamics of the Excitation-Relying Bearing-Rotor system

Bing Guo [1,2,3], Weixiao Tang [1,2,3,a] and Tianhui Zhen [1,2,3]

[1] School of Mechanical Engineering, Shandong University, Ji'nan, 250061, P.R. China
[2] Key Laboratory of High Efficiency and Clean Mechanical Manufacture, Ministry of Education, Jinan, P.R. China
[3] National Demonstration Center for Experimental Mechanical Engineering Education, Jinan, P.R. China

Abstract. This paper investigated the effects of the wet steam non-equilibrium condensation on the dynamic characteristics of the bearing as well as the bearing-rotor system by constructing and analyzing a non-linear coupled model of the bearing-rotor system. An excitation-relying dynamic model of bearing is established based on the finite difference method, in which the excitation is converted from the pressure pulsation on the surface of rotor blades generated from the non-equilibrium condensation process. The Raccia transfer matrix method is implemented to analyse the dynamic behavior of this coupled system. Results show that the wet steam non-equilibrium condensation process would greatly reduce the bearing stiffness and damping and result in more intense vibration of the system, besides, its induced pulsed displacement would drive the excitation-relying bearing-rotor system more unstable.

1 Introduction

The instability of the bearing-rotor system would induce serious accidents of turbines. Especially for the wet steam turbine such as nuclear turbine and the last stage of low pressure gas turbine, the non-equilibrium condensation of wet steam (NECS) not only causes the pressure pulsation on the blade surface, but also affects the stiffness and damping of bearing by loading torque to the bearing. Therefore, it is has great engineering significance to analyze the dynamics of the turbines bearing-rotor system considering the effects of NECS.

Currently, in the design of bearing, only the gravity of the rotor is regarded as the bearing support reaction force to determine the stiffness and damping of bearing, a lot of studies on this subject used the computational fluid mechanics or the finite difference method have been carried out [1-5]. However, the calculated bearing stiffness and damping using above approaches is insufficient to reflect the reality, which is the bearing stiffness and damping will change during the NECS process. The effects of the NECS on the dynamic characteristics of the bearing as well as the bearing-rotor system have not been clarified clearly.

The aim of this paper is to investigate such effects by constructing and analyzing a non-linear coupled model of the bearing-rotor system considering the effects of NECS. An excitation-relying dynamic model of bearing is established based on the finite difference method (FDM), in which the excitation is equivalent from the pressure pulsation on the surface of rotor blades generated from the

[a] Corresponding author : tangwx@sdu.edu.cn

NECS process. The Raccia transfer matrix method (RTMM) is applied to investigate the dynamic behavior of this coupled system under the NECS excitation.

2 Calculation models

Taking a low pressure turbine that contains ten cascades as an example, the excitation-relying bearing-rotor system working in wet steam as shown in Figure 1. Inertial coordinate system (*O-XYZ*), bearing-rotor system coordinates (o_s-*xyz*) and blade coordinate system (o_i-$x_iy_iz_i$) are defined to describe the coupled system.

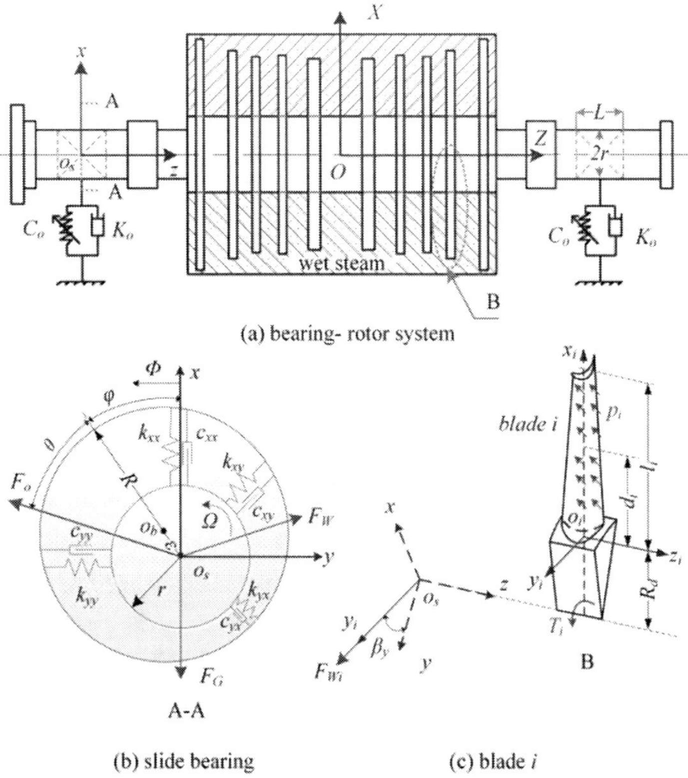

Figure 1. Dynamic model of the excitation-relying bearing-rotor system

The differential equations of motion of the excitation-relying bearing-rotor system can be expressed as

$$[M_r]\ddot{U}(t)+\{[C_r]+[C_o(F_W,t)]\}\dot{U}(t)+([K_r]+[K_o(F_W,t)])U(t)=F_W(t) \tag{1}$$

where M_r, C_r, K_r is the mass, damping and stiffness of the rotor (include the shaft and the blade). $F_W(t)$ is the NECS excitation. C_o and K_o are the excitation-relying damping and stiffness of the bearing, which can be deduced from the bearing support reaction force (i.e. the oil film force F_o) as given by

$$K_o=\begin{bmatrix} k_{xx} & k_{xy} \\ k_{yx} & k_{yy} \end{bmatrix}=\begin{bmatrix} \dfrac{\partial F_o\cos\Phi}{\partial x} & \dfrac{\partial F_o\cos\Phi}{\partial y} \\ \dfrac{\partial F_o\sin\Phi}{\partial x} & \dfrac{\partial F_o\sin\Phi}{\partial y} \end{bmatrix}; \quad C_o=\begin{bmatrix} c_{xx} & c_{xy} \\ c_{yx} & c_{yy} \end{bmatrix}=\begin{bmatrix} \dfrac{\partial F_o\cos\Phi}{\partial \dot{x}} & \dfrac{\partial F_o\cos\Phi}{\partial \dot{x}} \\ \dfrac{\partial F_o\sin\Phi}{\partial \dot{y}} & \dfrac{\partial F_o\sin\Phi}{\partial \dot{y}} \end{bmatrix} \tag{2}$$

where, $\Phi=\theta+\varphi$ is the opening angle of the oil film, θ is the angle between the bearing concentric line and the oil film force. x, y is the NECS excitation induced offset of the bearing center (O_b). $F_o\cos\Phi$ and $F_o\sin\Phi$ are the components of the oil film force in x and y deriection, can be expressed as functions of the oil film pressure (p_o) and the radius of the shaft journal (r) as follows

$$F_o \sin\Phi = \int_{-L/2}^{L/2}\int_0^{2\pi} -p_o \sin\Phi r d\Phi dz; \quad F_o \cos\Phi = \int_{-L/2}^{L/2}\int_0^{2\pi} -p_o \cos\Phi r d\Phi dz \tag{3}$$

The change of the oil film force (F_o) will result in the alter of the thickness h and the pressure p_o of oil film, the motion of oil film is governed by the Reynolds equation [6] as follows

$$\frac{1}{r^2}\frac{\partial}{\partial\Phi}\left(\frac{h^3}{\mu}\frac{\partial p_o}{\partial\Phi}\right)+\frac{\partial}{\partial z}\left(\frac{h^3}{\mu}\frac{\partial p_o}{\partial z}\right)=6\omega\frac{\partial h}{\partial\Phi}+12(\dot{y}\cos\Phi+\dot{x}\sin\Phi) \tag{4}$$

where, c and e are the radius gap and the eccentricity. μ and ω are the lubricant dynamic viscosity and the journal rotation angular velocity. The oil film force F_o can then be calculated by iterately solving Eqs.(3)-(4) use the FDM until the balance, which is the oil film force F_o is equal to the sum of the NECS excitation ($F_W(t)$) and the rotor gravity (F_G) as given by

$$F_o = F_W(t) + F_G \tag{5}$$

For a certain turbine, the rotor gravity can be easyly calculated, however, the NECS excitation is hard to be obtained. The approach applied here is to equivalent the pressure (p) generated from the NECS on the blade surface into the torque (T) acting on the axis [6] (as shown in Fig.1(c)), and then, convert the torque (T) into the NECS excitation ($F_W(t)$) by introducing the torque factors (α_x=3.5 and α_y=1.42) [7-8], the detailed deduction given as follows. The torque acting on the axis can be expressed as the resultant force moment of the y_i direction forces of each blade (Eq.(6)) because the x_i direction forces can be offset over the full cascade.

$$T(t) = \sum_i^n \iint p_i dS_i \cos\beta_{yi}(d_i + R_d) \tag{6}$$

where, for the blade i, d_i is the distance between the blade centroid and hub centroid (o_i). R_d is the radius of the shaft. S_i is the surface area of blade. β_{yi} is the angle between the y_i of the blade coordinates o_i-$x_iy_iz_i$ and y of the bearing coordinates o_s-xyz. n=10 is the number of cascade in the wet steam turbine. p_i is the pressure on the blade surface which can be obtained from the numerical simulation of the NECS performed by ANSYS CFX based on the non-isothermal classical nucleation theory [9-10] and the droplet growth model [11]. Then, the NECS excitation $F_W(t)$ can be calculated as gave by

$$F_W(t) = T\sqrt{\alpha_x^2 + \alpha_y^2} \tag{7}$$

Finally, the excitation-relying damping and stiffness of the bearing can be calculated by substituting Eq.(7) into Eq.(5) and combining with Eqs.(2)-(4). Once the NECS excitation $F_W(t)$ as well as the bearing stiffness and damping $K_o(F_W, t)$ and $C_o(F_W, t)$ are determined, the dynamic characteristics of the coupled bearing-rotor system can be obtained by calculating Eq.(1) based use the RTMM.

3 Calculation results and discussion

Take the blade at the last stage as an example, the time-varying pressure on the blade surface at different blade spans during the NECS process is obtained as shown in Figure 2. It can be found that the pressure shows great unsteady characteristic in the middle and lower area of the blade (3%~50% span), this pulsation will gradually disappear at the blade tip (90% span).

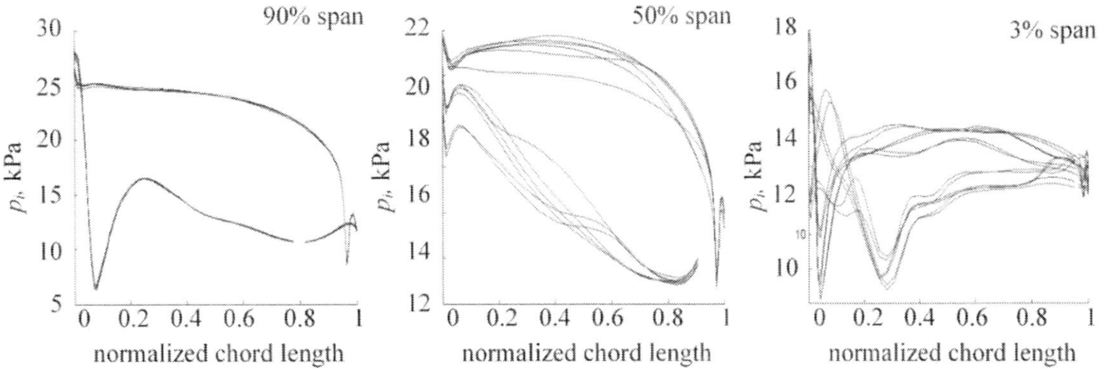

Figure 2. Pressure on the blade surfaces at different spans

The NECS excitation ($F_W(t)$) can be obtained by substituting the pressure on the blade surface into Eqs.(6)-(7) as shown in Figure 3. It can be found that the NECS excitation in x direction is fluctuating between about 6.3×10^5~7.3×10^5 N, and the NECS excitation in y direction is fluctuating between about 4.5×10^5~5.2×10^5 N.

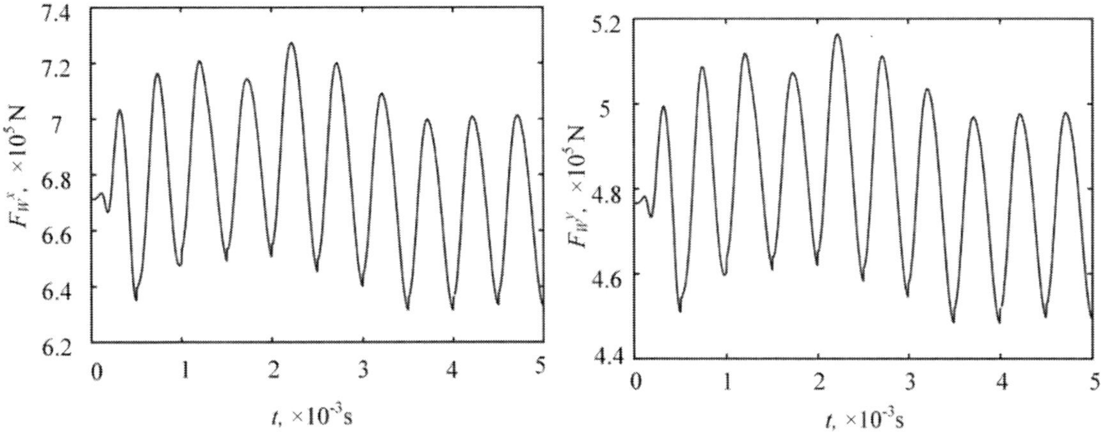

Figure 3. Components of the NECS excitation in x and y direction

Unsteady excitation-relying damping C_o and stiffness K_o of the bearing can be calculated use the FDM by substituting the NECS excitation $F_W(t)$ into Eq.(5) and combining with Eqs.(2)-(4). The eight components of the non-dimensional C_o and K_o are shown in Figure 4. Besides, in order to estimate the effects of NECS, the steady bearing stiffness and damping that calculated without the NECS excitation (i.e. $F_W(t)$=0) are shown in Table 1.

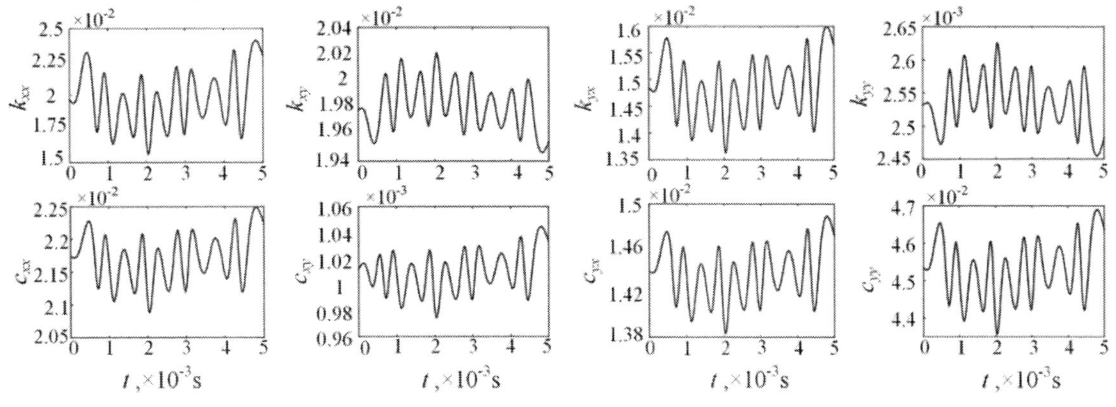

Figure 4. Components of the non-dimensional excitation-relying damping and stiffness of bearing

Table 1. Steady non-dimensional stiffness and damping of the bearing that calculated without the NECS excitation

k_{xx}	k_{xy}	k_{yx}	k_{yy}
0.0076	0.0246	0.049	0.019
c_{xx}	c_{xy}	c_{yx}	c_{yy}
0.0467	0.0021	0.0309	0.0974

The comparison between the bearing stiffness and damping calculated with and without the NECS excitation shows that most components of the excitation-relying damping and stiffness have great decrease (about 70%) except the k_{yy} which is increased by about 30%. Besides, the decreasing amplitude of the bearing stiffness is larger than the bearing damping, which means that the former is more sensitive to the NECS excitation.

Once the excitation-relying damping and stiffness of the bearing are obtained, the dynamic motion Eq.(1) can then be calculated by the RTMM. The axes contrails are drawn as shown in Figure 5 to show the stability of the bearing-rotor system.

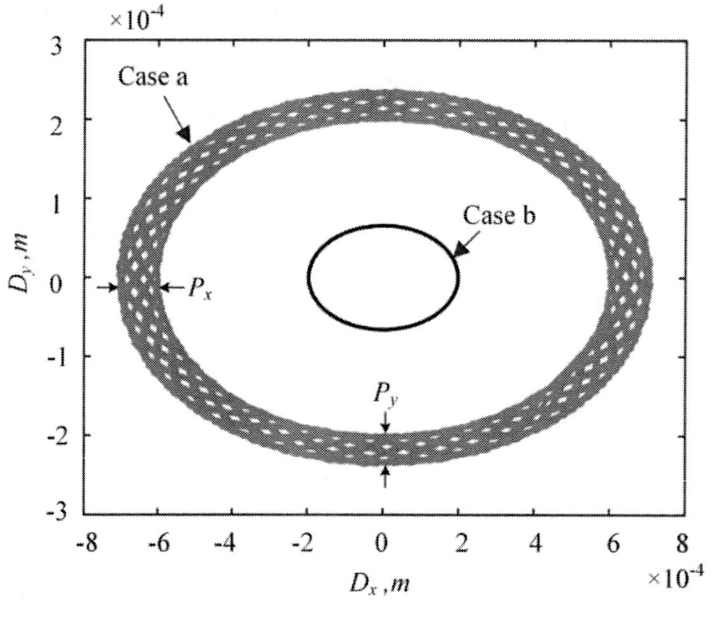

Figure.5 Axes contrails of the bearing-rotor system. Case a: with the NECS excitation, Case b: without the NECS excitation.

The comparison between Case a and Case b shows that the displacement of the excitation-relying bearing-rotor system is almost four times than the non-relying bearing-rotor system in which the effects of the NECS excitation on the bearing are ignored. Furthermore, axes contrails of Case a is pulsating with about 10-4m during the rotation of the system, in which the pulsation displacement in x direction (P_x) is slightly more than in y direction (Py). While axes contrails of Case b is steady during the rotation of the system. It can be concluded that when the effects of NECS excitation on the bearing are considered. The vibration response would greatly increase and the system has more tendency to instability.

In a sum, the NECS excitation would greatly reduce the bearing stiffness and damping and result in more intense vibration of the system. Besides, the pulsation displacement caused by NECS would drive the excitation-relying bearing-rotor system more unstable.

4 Conclusions

This paper constructed a non-linear coupled model of the excitation-relying bearing-rotor system considering the non-equilibrium condensation of wet steam to analyze the effects of wet steam excitation. An excitation-relying dynamic model of bearing was established based on the FDM, in which the excitation is converted from the pressure on the surface of rotor blades generated from the NECS process. The RTMM is applied to investigate the dynamic behavior of this coupled system. The following conclusions are draw:

(1) NECS could decrease the stiffiness and damping of the bearing. In the calculated case shown in this paper, mosts components of the bearing stiffness and damping are decreased by about 70%.

(2) When the effects of NECS excitation on the bearing are considered, the vibration response would greatly increase and the system has more tendency to instability. In the calculated case shown in this paper, the vibration response of the excitation-relying bearing-rotor system is almost four times than the non-relying bearing-rotor system, and the pulsation displacement (about 10^{-4}m) was generated in the former.

Acknowledgment:

This work was supported by the key research and development program of Shandong Province (Grant No. 2017GGX40116)

References

1. Yuansheng LI, Liangbo AO, Lei LI, YUE Zhufeng. Dynamic analysis method of dynamic character coefficient of hydrodynamic journal bearing. J. Mech. Eng. **46**, 48-53 (2010)
2. Qingshui Gao, Jiangang Yang. Research on the dynamic characteristics of hydrodynamical journal bearing based on CFD analysis. Lubrication Eng. **33**, 65-67 (2008)
3. Jihua Liu, Longxi Zheng, Gangqi Liu. Analysis of oil film parameter of journal bearing- rotor system based on finite difference method. Lubrication Eng. **38**, 52-56 (2013)
4. Zhiyong Yan, Liping Wang, Guang Qiao, Tiesheng Zheng. An analytical model for complete dynamical coefficients of a tilting-pad journal bearing. Tribology International. **43**, 7-15 (2010)
5. Cai Wan, Chang Jian, ChaoKuang Chen. Chaos and bifurcation of a flexible rub-impact rotor supported by oil film bearings with nonlinear suspension. Mechanism and Machine Theory. **42**, 312–333 (2007)
6. Bing, G., and Weixiao, T., 2018, "Wet Steam Non-equilibrium Condensation Flow Induced Vibrations of a Nuclear Turbine Blade," ASME J Gas Turb Pwr. doi:10.1115/1.4039834 (to be published)
7. B.O. Al-Bedoor, A.A. Al-Qaisia. Stability Analysis of Rotating Blade Bending Vibration due to Torsional Excitation. Journal of Sound and Vibration. **282**, 1065-1083 (2004)
8. Huer Sun, Zhaojian Yang, Qunlong Liang, Pang Xinyu. Transverse vibration response of rotor system under mutational torque excitation. J. of Vibration, Measurement & Diagnosis. **31**, 622-625 (2011)
9. Ford, I. Statistical mechanics of nucleation: a review. P I Mech Eng C-J Mec, **218**, 883-899 (2004)
10. Young, J. B. Two-dimensional, nonequilibrium, wet-steam calculations for nozzles and turbine cascades. J Turbomach, **114**, 569-579 (1992)
11. Young, J. B. The spontaneous condensation od steam in supersonic nozzles. Physicochem Hydrodyn, **3**, 57-82 (1982)

Numerical Analysis of the Strength of the Wet Steam Turbine Blade

Bing Guo[1,2,3], Weixiao Tang[1,2,3,a], Jin Liu[1,2,3] and Tianhui Zhen[1,2,3]

[1] School of Mechanical Engineering, Shandong University, Ji'nan, 250061, P.R. China
[2] Key Laboratory of High Efficiency and Clean Mechanical Manufacture, Ministry of Education, Jinan, P.R. China
[3] National Demonstration Center for Experimental Mechanical Engineering Education, Jinan, P.R. China

Abstract. A method to examine the influences of inlet wetness fraction and inlet angle on the strength performance of the rotating blade considering the effects of wet steam non-equilibrium condensation was proposed. A wildly verified numerical model was employed to simulate the wet steam flow to obtain the pressure distribution on the blade surface. Based on the finite element method, the strength analysis was carried out. Results show that the von-Mises stress of the rotor blade would increase with the inlet wetness fraction and the inlet angle. The sensitive range of studied inlet condition is located at $A_{in} \in [0,\ 15]$, $W_{in} \in [0.03,\ 0.04]$, in which the increases of the von-Mises stress are over 40%.

1 Introduction

Excessive stress may shorten the effective life of the rotating blades and threaten the safety of turbo-machinery. Especially for the turbine operating on wet steam, such as nuclear turbines and the last stage of low pressure gas turbines, in which the non-equilibrium condensation (NEC) process of the wet steam may greatly enhance the extent of flow non-uniformity to generate more intense fluid excitation and induce higher stress of the rotating blade. Hence, it is of great practical interest in the design of wet steam turbines to predict the strength performance of the rotating blades considering the effects of the NEC.

Currently, research on the fluid excitation generate from the NEC of wet steam is still insufficient. Generally, the condensation of wet steam is regarded as an equilibrium process, either or the condensation is ignored and the wet steam is simplified as an ideal gas. After above simplifications, the vortex shedding from the upstream blade is regarded as the major fluid excitation, i.e. the wake excitation. A lot of studies about the wake excitation and its induced vibration have been carried out, to name a few, Lau [1] used the Karman vortex street (KVS) to model the wake excitation and analyzed the vortex-induced vibrations of a 2D airfoil. García [2] numerically calculated the wake excitation acting on the rotor blade of the last stage of 110MW steam turbine. However, the strength performance of the rotating blade considering the effects of NEC hasn't been well clarified due to the difficulty in calculating NEC-induced excitation.

In this paper, the effects of wet steam NEC on the strength performance of the rotating blade are numerically studied. In which the NEC-induced excitation (pressure distribution on the blade surface

[a] Corresponding author : tangwx@sdu.edu.cn

during NEC process) can be obtained from the numerical simulation of the wet steam flow field based on the non-isothermal nucleation theory and the droplet growth model [3-5] that implemented in ANSYS CFX. Once the NEC-induced excitation is obtained, the strength performance of the rotating blade can then be analyzed use the finite element method (FEM) through the strength calculation under the pressure load with commercial software ANSYS Workbench. Besides, the influences of the wet steam inlet conditions (intake angle and wetness fraction) are analyzed.

2 Wet steam model

The effects of NEC on the strength performance of the rotating blade are determined through the pressure distribution on the blade surface during NEC process. To obtain the pressure distribution, a NEC model of wet steam is necessary. The wet steam is a mixture of two-phases, in which the primary phase is the gaseous-phase consisting of water-vapor (denoted by the subscript *v*), the secondary phase is the liquid-phase consisting of condensed-water droplets (denoted by the subscript *l*).

The numerical calculations can be carried out based on a compressible Navier-Stokes equations using Eulerian–Eulerian approach to govern the condensing flow. The equations are given by [6]

$$\frac{\partial}{\partial t}\begin{pmatrix}\rho \\ \rho V_i \\ \rho H\end{pmatrix}+\frac{\partial}{\partial x_i}\begin{pmatrix}\rho V_i \\ \rho V_i V_j + p - \mu_t \dfrac{\partial V_i}{\partial x_j} \\ \rho H + p - \lambda_t \dfrac{\partial T_v}{\partial x_j}\end{pmatrix}=0 \tag{1}$$

$$\frac{\partial \rho W}{\partial t}+\nabla\cdot\left(\rho \vec{V} W\right)=\Gamma \tag{2}$$

$$\frac{\partial \rho\varepsilon}{\partial t}+\nabla\cdot\left(\rho \vec{V}\varepsilon\right)=\rho I \tag{3}$$

in which, V is the flow velocity. Subscript i and j denote the tensor notation. ρ and H are the density and the entropy values of the mixture phase. W and ε are the wetness fraction and the number of droplets per unit volume. Γ and I are the liquid mass generation rate and the nucleation rate respectively. The NEC model is applied to describe the wet steam properties based on the nucleation theory and the droplet growth model. The nucleation rate I and the non-isothermal correction factor θ are used to calculate the droplet formation.

$$I=\frac{q_c}{(1+\theta)}\left(\frac{\rho_v^2}{\rho_l}\right)\sqrt{\frac{2\sigma}{M_m^3\pi}}\exp\left(-\frac{4\pi r_*^2\sigma}{3K_bT}\right) \tag{4}$$

$$\theta=2\frac{(\gamma-1)}{(\gamma+1)}\left(\frac{h_{lv}}{RT}\right)\left(\frac{h_{lv}}{RT}-0.5\right) \tag{5}$$

where q_c is the evaporation coefficient, k_b is the Boltzmann constant, M_m is the mass of one molecule, σ is the liquid surface tension, ρ_l is the liquid density, h_{lv} is the specific enthalpy of evaporation at pressure p and γ is the ratio of specific heat capacities. The droplet growth is modelled as given by

$$\Gamma=\frac{4}{3}\pi\rho_l I r_*^3+4\pi\rho_l\varepsilon\bar{r}^2\frac{\partial\bar{r}}{\partial t} \tag{6}$$

$$\frac{\partial\bar{r}}{\partial t}=\frac{p}{h_{lv}\rho_l\sqrt{2\pi RT}}\frac{\gamma+1}{2\gamma}C_P\left(T_l-T_v\right) \tag{7}$$

where r^* is the Kelvin-Helmholtz critical droplet radius.

Besides, the steam properties are based on the IAPWS standard. The NEC model is coupled with k–ω shear stress transport (SST) turbulence model, which is suitable for the turbo-machinery applications. The above models have been wildly validated against the experimental data from the nozzle flows [7] and the cascade flows [8] as well as the turbine flows [9].

3 Results and discussion

The last stage rotor blades of a wet steam turbine run at 1500rpm are taken for the investigated case. In which the calculation domain can be minimized as one blade passage using periodic boundary condition. In order to improve the calculation accuracy, the H-O-H structured grids are used as shown in Figure 1. ANSYS CFX is employed to solve the Reynolds-averaged Navier-Stokes equations and the k–ω SST turbulence model. The boundary conditions are given in table 1. The rotor blade is meshed use element soild185 in ANSYS Workbench. The material and geometric properties of the blade are given in table 2.

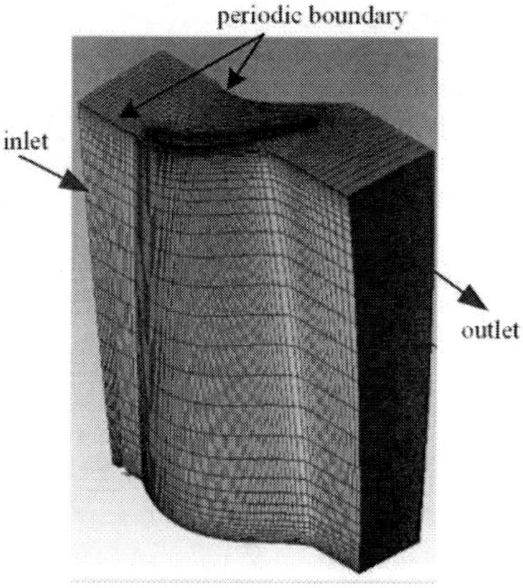

Figure 1. Calculation domain using periodic boundary condition

Table 1. Boundary conditions

Inlet total pressure	Inlet total temperature	Pressure ratio
39.597 kPa	300 K	1.71

Table 2. Material and geometric properties of the blade

Elasticity modulus	Density	Poisson's ratio	Length
198 GPa	7750 kg/m^3	0.3	1200 mm

The pressure distribution and its induced stress of the rotating blade at different intake angles (A_{in}=0°, 15°, 30°) and wetness fractions (W_{in}= 0.03, 0.04, 0.05) are simulated and calculated in ANSYS CFX and ANSYS Workbench respectively, as shown in Figures 2-5.

Figure 2 shows the pressure distribution on 50% blade span at different wetness fractions when the inlet angle is 0°. It can be found that the maximum of pressure is slightly increased with the wetness fractions. As a result, the maximum of von-Mises stress of the rotating blade is increased with the wetness fractions, as shown in Figure 3. Note that when the wetness fraction increased from 0.03 to

0.04, the von-Mises increased by 40.25%, while when the wetness fraction increases from 0.04 to 0.05, the von-Mises only increase by 0.54%.

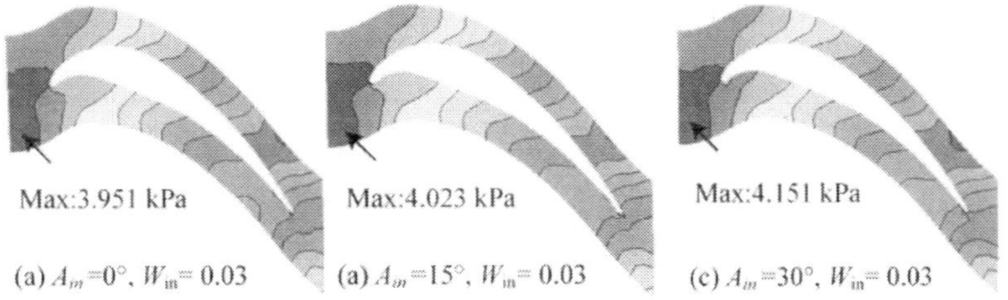

Figure 2. Comparison of the pressure distribution on the 50% blade span of three different wetness fractions.

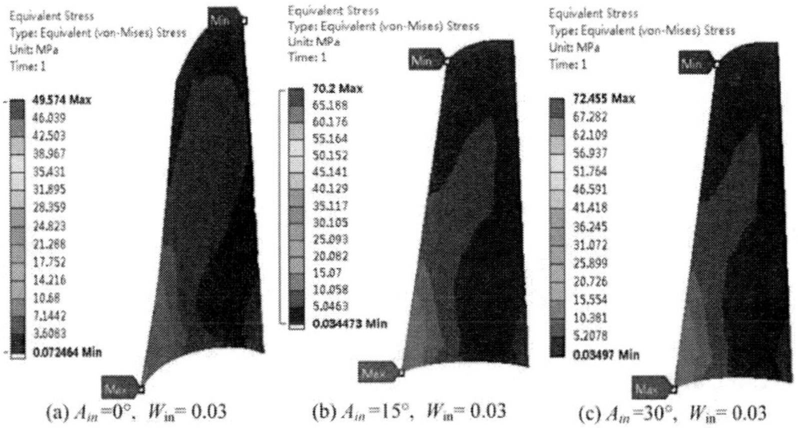

Figure 3. Comparison of the von-Mises stress on the surface of blade of three different wetness fractions.

Figure 4 shows the pressure distribution on 50% blade span at different inlet angles when the wetness fraction is 0.03. It can be found that the maximum of pressure is increased with the inlet angle. As a result, the maximum of von-Mises stress of the rotating blade is increased with the inlet angle, as shown in Figure 5. Note that when the inlet angle increased from 0° to 15°, the von-Mises increased by 41.61%, while when the wetness fraction increases from 0.04 to 0.05, the von-Mises only increase by 0.36%.

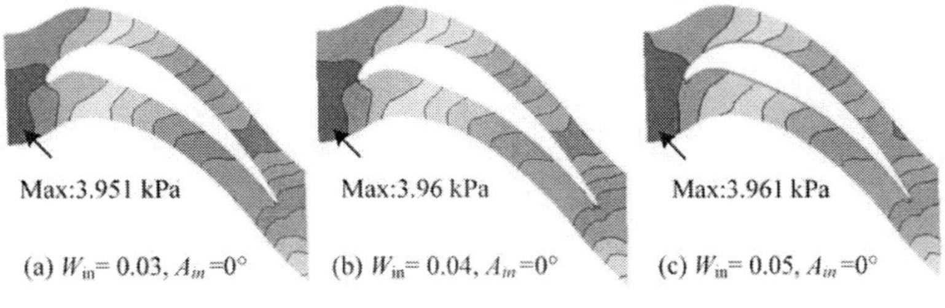

Figure 4. Comparison of the pressure distribution on the 50% blade span of three different inlet angles.

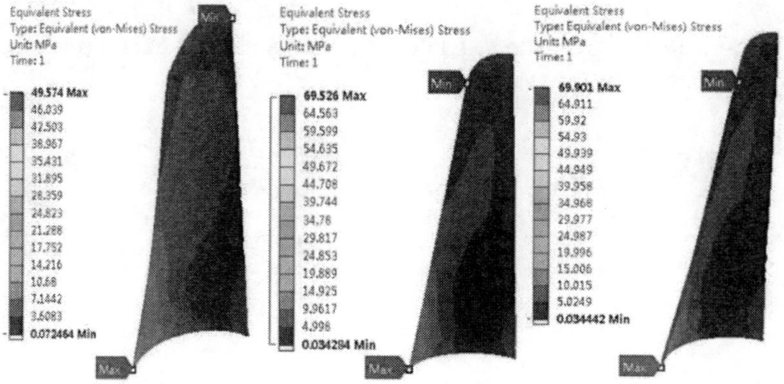

(a) W_{in}= 0.03, A_{in}=0°   (b) W_{in}= 0.04, A_{in}=0°   (c) W_{in}= 0.05, A_{in}=0°

Figure 5. Comparison of the von-Mises stress on the surface of blade of three different inlet angles.

The above results show that the higher inlet wetness fraction and/or bigger inlet angle would induce lager von-Mises stress of the blade. The maximum stress points are located in the trailing edge of the blade root, where should be to pay attention in the design.

4 Conclusions

This paper proposes a method to examine the influences of inlet wetness fraction and inlet angle on the strength performance of the rotating blade considering the effects of wet steam NEC. A wildly verified NEC model was employed to numerically simulate the wet steam flow with ANSYS CFX to obtain the pressure distribution on the blade surface. Applying the NEC-induced excitation, the strength analysis was carried out with ANSYS Workbench. Several conclusions were drawn as follows:

(1) Considering the effects of wet steam NEC, the von-Mises stress of the rotor blade would increase with inlet wetness fraction and let angle.

(2) The sensitive range of studied inlet condition is located at $A_{in} \in [0, 15]$, $W_{in} \in [0.03, 0.04]$, in which the increases of von-Mises stress are over 40%. It could be inferred that the stress of blade working in wet steam might be reduced by adjusting the inlet conditions.

(3) The presented method to calculate fluid excitation generated from wet steam NEC and its resulting stress of the blade could be extended into the design and analysis of wet steam turbines to meet the needs of engineering.

Acknowledgment

This work was supported by the key research and development program of Shandong Province (Grant No. 2017GGX40116)

Nomenclature

A	Inlet angle,	r^*	Kelvin-Helmholtz critical droplet radius, m
C_p	specific heat, J/(kg K)	S	super saturation ratio = $p/p_{sat}(T)$
H	total enthalpy of mixture phase, J/kg	T	temperature of mixture phase, K
h_{lv}	specific enthalpy of evaporation, J/kg	V	flow velocity, m/s
I	nucleation rate	W	wetness fraction
k_b	Boltzmann constant =1.3807×10^{-23}, J/K	γ	vapor specific heat ratio
L	blade height, m	Γ	liquid mass generation rate
p	pressure, Pa	ε	number of droplets per unit volume =

			$\beta\rho_g/(1-\beta)V_d\rho_l$
$p_{sat}(T)$	equilibrium saturation pressure, Pa	θ	non-isothermal correction factor
q_c	condensation coefficient	λ_t	effective thermal conductivity, W/(m K)
r	droplet radius, m	μ_t	effective dynamic viscosity, kg/(m s)
R	gas specific constant, J/(mol K)	ρ	density of mixture phase= $\rho_g/(1-\beta)$, kg/m^3
R_d	radius of shaft, m	σ	liquid surface tension, N/m

References

1. Lau, Y. L., Leung, R. C. K., So, R. M. C. Vortex-induced vibration effect on fatigue life estimate of turbine blades. Journal of Sound and Vibration **307**, 698-719 (2007)
2. García, J., Kubiak, J., Sierra, F., Urquiza, G., Rodríguez, J. Numerical analysis of the blade forces caused by wake/blade interaction in the last stage of a steam turbine. *Proceedings of power*, (St. Antonio, Texas, 2007)
3. Ford, I. Statistical mechanics of nucleation: a review. P I Mech Eng C-J Mec, **218**, 883-899 (2004)
4. Young, J. B. Two-dimensional nonequilibrium, wet-steam calculations for nozzles and turbine cascades. J Turbomach, **114**, 569-579 (1992)
5. Young, J. B. The spontaneous condensation od steam in supersonic nozzles. Physicochem Hydrodyn, **3**, 57-82 (1982)
6. K. Ishazaki, T. I., Daiguji, H. A High-Resolution Numerical Method for Transonic Non-equilibrium Condensation Flows Through a Steam Turbine Cascade. J. Turbomach, **1**, 479–484. (1995)
7. Moses, C. A., Stein, G. D. On the growth of steam droplets formed in a Laval nozzle using both static pressure and light scattering measurements. J Fluid Eng, **100**, 311-322 (1978)
8. White, A. J., Walters, P. T. Experimental validation of condensing flow theory for a stationary cascade of steam turbine blades. P. Roy. Soc. A-Math. Phy. **354**, 59-88 (1996)
9. Gardzilewicz, A., Kolovratnik, M. Numerical and experimental investigations of steam condensation in LP part of a large power turbine. J. of Fluids Eng., **131**, 041301(2009)

Load Planning of Transport Aircraft Based on Hybrid Genetic Algorithm

Yang Chenguang[1,a], Liu Hu[1] and Gao Yuan[2]

[1]*School of Aeronautical Science and Engineering, Beihang University, Beijing, 100191, China*
[2]*Power Electronics, Machines and Control Research Group, Faculty of Engineering, University of Nottingham, Nottingham, NG7 2RD, UK*

Abstract. Loading of transport aircraft attracts much attention as the airlift is developing rapidly. It refers to the process that various cargoes are loaded, in an appropriate manner, into kinds of transport aircrafts with constraints of volume, weight and gravity center. Based on two-dimensional bin packing with genetic algorithm (GA), a new hybrid algorithm is proposed to solve the multi-constraint loading problem of transport aircraft for seeking the minimum of fuel consumption. Heuristic algorithm is applied to optimize single-aircraft loading in GA decoding, and the procedure of hybrid GA is summarized for the multi-aircraft loading issues. In the case study, eight kinds of cargos are distributed in three different aircrafts. The optimal result indicates that this algorithm can rapidly generate the best plan for the loading problem regarding lower transport costs.

1 Introduction

Airlift is a transportation way with good properties like speed, safety and high-capacity. It has become one of the most important means of transportation especially for large and urgent cargoes. Similar to other planning researches, the motivations of studying transport aircraft loading here include the premier goal to reduce the cost of transportation. The former study of transport aircraft loading is mainly focused on the single aircraft loading [1] [2] [3]. In addition, the method for multi-aircraft cargo loading under different constraints is still primitive [4] [5]. Algorithms like "Best Fit Level Decreasing" (BFLD) [1], "Command Line Sequencing" (CLS) [2] etc. are usually utilized for transport aircraft loading. Vancroonenburg W [3] introduces a mixed integer linear programming model for aircraft load planning with multi-constraint. However, those methods are limited to the loading of single-aircraft freight. Roesener [4] developed a tabu search based heuristic to solve a multi-aircraft problem which accounts for aircraft specific constraints. However, the authors just focused on loading pallets for only one kind of cargo. Although Liu [5] proposed an algorithm that can automatically generate cargo loading plans for the transport aircraft fleet, the number of plans is far from generating an optimized airlift plan.

With heuristic algorithm (HA) utilized for single-aircraft loading, a new hybrid genetic algorithm (HGA) is proposed to solve multi-constraint loading problem of transport aircraft for minimizing the

[a] Corresponding author : transport_plane@163.com

fuel consumption. This algorithm can rapidly generate the optimized plans for transport aircraft loading.

2 Loading problem of transport aircraft

The problem of transport aircraft loading can be described as follows: With given cargoes and a fleet of transport aircrafts, how to reasonably arrange the cargo position on each aircraft under the constraints of volume, weight and center of gravity. The cost of transportation (referring to fuel consumption here) could be reduced by using some optimization methods. Transport aircraft loading is the design and optimization of two problems. One is for the loading position of cargos on single aircraft, and the other is to generate the loading order of transport aircrafts.

According to the characteristics and requirements of transport aircraft loading, design principles and simplifications can be given as follows:

• Since cargo is generally not stacked inside of the aircraft. Transport aircraft loading can be simplified as a two-dimensional packing problem [6].
• The shape of cargo hold of transport aircraft is simplified as a rectangular.
• The shape of cargo is simplified as a rectangular, and its center of gravity is its geometric center.

The optimization of aircraft loading is a Non-deterministic Polynomial (NP) complete problem [7]. Most of new GA methods have been challenged by the increasing complexity of formulation and computation; therefore, HGA is preferable because of its practicability for packing freight into a cargo hold.

On the one hand, with multiple constraints of space planning, it is not that easy to obtain optimal solution by exact algorithms. Therefore, HA is utilized to solve this problem. On the other hand, Genetic algorithm (GA) is a general and effective method to solve the loading order problem of transport aircrafts. HGA, which combines GA with a heuristic rule, is designed in this study for aircraft loading problem. Its basic thinking is that: decoding chromosome order using heuristic rule before fitness calculation; and then, an optimal solution (loading order of transport aircraft) is produced using GA.

3 Design of HA for single-aircraft loading

Single-aircraft loading can be simplified as a two-dimensional (top-view) packing problem. A single-aircraft loading algorithm has been proposed by Liu [5], this study improves it using heuristic algorithm. For each cargo to be loaded, there are two basic requirements that should be met: its height must be less than the height of the cargo hold; its total weight must be less than the remaining loading capacity of the transport aircraft.

It is noting that, regarding any part of the cabin hold for loading cargos, the loading priority order is sorted from maximum to minimum by their side lengths. The loading procedure is concluded as follows:

1) Rank all cargos by their lengths (not widths) and choose an aircraft randomly.
2) A virtual cabin (VC) needs to be constructed for the first cargo (which can be loaded firstly following the length order), and it should be positioned at the left bottom of cargo hold.
3) Then, the following cargoes should be loaded in the right space of this VC. For this space, try all cargos one by one following the length order until one is available. After loading the second cargo, if there still be enough space left in the right of this VC, try all cargos in the same way until the smallest cargo cannot be loaded on the right.
4) Consider the space of top right corner as a sub VC, and compare its two side lengths: 1. if the vertical length is larger than the horizontal length, load the cargos following the length order in the sub VC from bottom to top (follow step 3 in horizontal direction); 2. if the vertical is smaller than the horizontal, all cargos left should be rotated 90 degrees, and they will be loaded following the same length order in the sub VC from right to left (if needed, add the second cargo level upward in this sub VC).

5) If some cargos still fail to be loaded after Step 4, a new VC needs to be built. They can be loaded following step 3-4.
6) If there is no enough space to contain another new VC which means all vertical side lengths of the left cargos are larger than the remaining vertical length in this aircraft, another transport aircraft should be utilized for the cargos.

As shown in Figure 1, cargos, including 1 A, 1 B, 3 C, 2 D, 2 E, are ready to be loaded in one aircraft. According to above rules, VC1 should be firstly constructed and Cargo A is loaded in the left bottom of this VC. Then load Cargo C on the right of A, followed by considering the top right corner of VC1 as a sub VC. Its vertical length is larger than the horizontal, the other Cargo C is loaded vertically and one Cargo E is loaded on the top of C based on the rules in Step 4. Finally, VC2 is built for loading remaining B, C, 2 D and E. Different from VC1, all left cargos need to be rotated 90 degree before loading in its sub VC, and then 1 D, 1 E are loaded from right to left there.

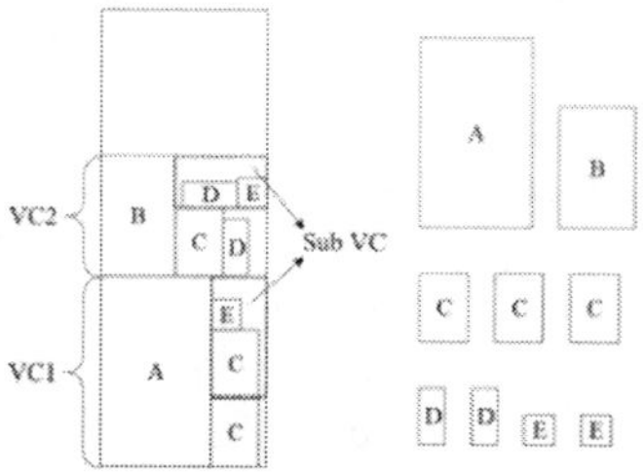

Figure 1. The HA of the single-aircraft loading.

The centers of gravity for all the cargos on the same aircraft should be arranged within a tolerance range for safety purpose. To satisfy the longitudinal balancing requirement, a method called "Pyramid Loading" [5] is utilized. During this balancing process, the cargos in one VC should be considered as a whole, their relative positions will not be changed.

For the vertical balance, firstly rank all VCs from heavy to light and they can be represented by VC1, VC2, VC3 etc. Then reload all VCs. Locate the heaviest one (VC1) in the middle of cargo hold. Put the second-heaviest (VC2) on the bottom of VC1 followed by putting VC3 on the top of VC1. Then VC4 is putted on the bottom of VC2, VC5 on the top of VC3. By that analogy, all VCs can be vertically positioned. If longitudinal balance is not met or any VC is out of the cargo hold, all VCs should be combined into one bigger VC and move it to a suitable position in the cargo hold.

For the horizontal balance, firstly calculate the rolling moment of all VCs, and rank the absolute values from large to small. Each VC need to be confirmed one by one, whether it should be symmetrically reversed from left to right, by the following three aspects: 1. rolling moment; 2. the sum of rolling moments of former VCs; 3. this VC contains only one kind of cargo or not. The procedure is shown in Figure 2.

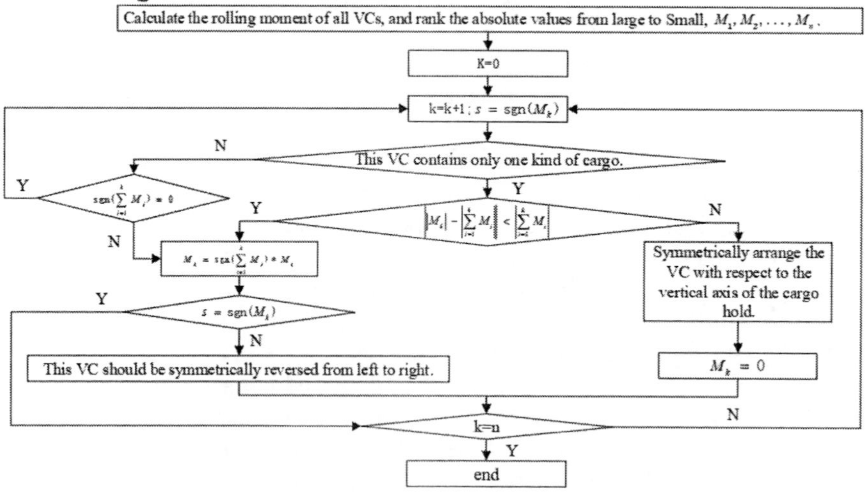

Figure 2. Horizontal balance procedure.

4 Design of HGA for multi-aircraft loading

HGA is used to solve the loading problem with more than one transport aircraft. The procedure of GA based on the heuristic algorithm is shown in Figure 3.

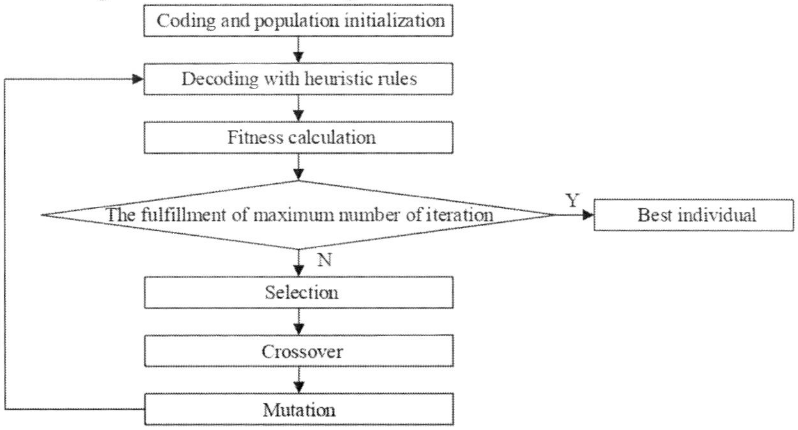

Figure 3. The basic procedures of HGA.

(1) Coding

Serial number coding is employed according to the characteristics of the loading. All the transport aircrafts are randomly arranged in a line, forming a chromosome. Their individual lengths equal to $L=\sum_{i=1}^{k} n_i$. For example, three types of aircraft (marked by A, B, C respectively, k = 3), and the numbers of each type of aircraft are 3, 4, 2. One individual serial number generated might be BCBAACBAB (including three "A", four "B" and two "C"), and the individual length is $L=9$.

(2) Population initialization

Assume that the scale of GA population in the execution is N. Repeat the individual coding for N times, individuals can be generated $x_i(i=1,2,\ldots,N)$, and the initial population $X=\{x_1,x_2,\ldots,x_N)$, whose scale is N*L, is finally confirmed.

(3) Decoding with heuristic rules

Based on the heuristic algorithm, all the cargos are loaded following a chromosome order of transport aircraft until the loading mission is complete. The total loading weight of each transport aircraft m_j should be recorded.

(4) Fitness calculation

The operational costs of the flights are calculated by empirical data under the flight mission profile. The fuel consumption of each loaded transport aircraft, $F_j(m_j)(j=1,2,\ldots,P)$, is a function of loading weight and it can be calculated by interpolation method based on the fuel consumption of some typical loading weights. According to loading plans mentioned above, the total fuel consumption of the cargo aircraft is $F=\sum_{j=1}^{P} F_j(M_j)$. The fitness value $f(x_i)$ of x_i individual can be recorded as the reciprocal of $(F-c_{\min})$. $c_{\min}$ is the lowest estimated magnitude of fuel consumption under current loading information.

(5) Genetic operation

Selection: roulette selection based on the elitist strategy.

Crossover: single-point crossover is applied, and the crossover probability is set as p_c; the crossover individuals x_i, x_j are randomly selected according to p_c; the crossover point in chromosome is randomly selected.

Mutation: randomly select the mutation position, and then change the generic code; the mutation probability is set as p_m . (The inversion operation can be added after mutation in order to improve the ability of local exploration.)

Termination condition: the maximum number of iteration.

5 Case study

The transport aircrafts in *A* airport are three Y-9, four IL-76 and two C-17 (Because the data of typical large transport aircraft Y-20 is still unknown, replace Y-20 with C-17). The parameters of the cargo to be transported are displayed in Table 1. The distance between A and B is 1300 kilometers. The optimal loading plan of the aircrafts transporting these cargos from A to B is to be solved for the minimum target of fuel consumption.

Table 1. Cargo parameters.

Cargo	Length (m)	Width (m)	Height (m)	Weight (t)	Number
Command vehicle	7.000	3.000	2.900	14.600	3
Communication vehicle	6.300	2.500	2.200	15.200	2
Ambulance	6.000	2.300	2.300	6.000	3
IC container	6.058	2.438	2.438	15.000	3
JY10 container	4.012	2.438	2.438	5.000	4
JY7 container	2.438	1.968	2.100	3.500	3
JY5 container	2.438	1.457	2.100	1.500	7
JY1 container	1.300	0.900	1.300	0.800	7

In this case study, HGA is developed in C language. For the initial parameters of GA, scale of population N=10, maximum number of iteration T=200, crossover probability $p_c = 0.6$ and mutation probability $p_m = 0.1$.

Optimal results using HGA is shown in Figure 4. The first point of convergence happens at the 73th generation which has a minimum cost value of 184.35. The final result of the optimized loading order is C-17, Y-9, C-17, IL-76, and the loading sequence of each cargo is monotonically decreasing regarding their lengths. The total fuel consumption is 184.35 t in this optimal plan. The optimal loading of each transport aircraft is shown in Figure 5. As the initial fuel consumption is 243.89 t, transport cost is reduced greatly by using HGA.

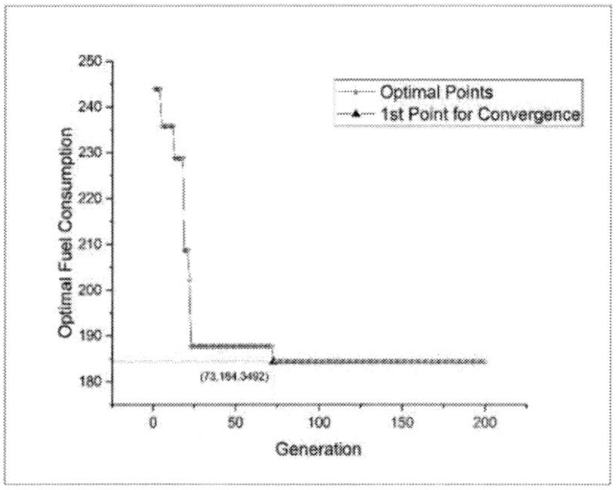

Figure 4. Optimal results using HGA.

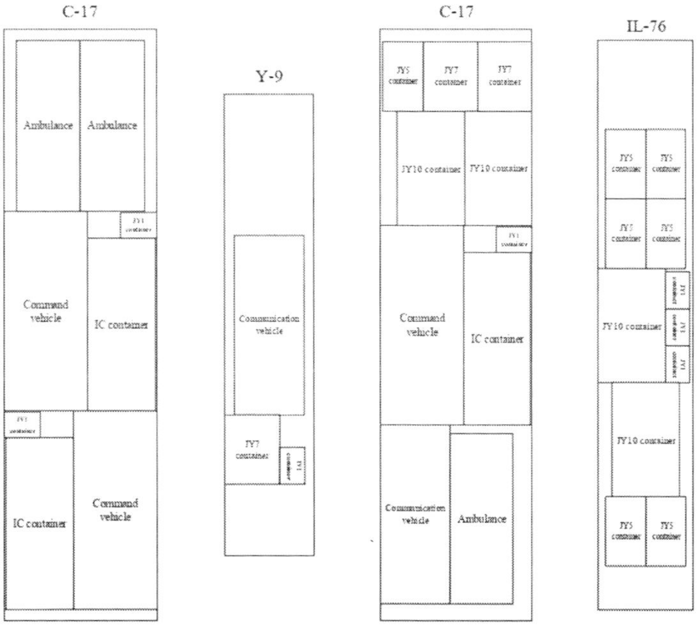

Figure 5. The optimal loading of each transport aircraft.

6 Conclusion

Considering various constraints of airlift loading, a new HGA is proposed for the aircraft loading problem. The interest of this algorithm is to combine HA of single-aircraft loading with GA based on two-dimensional bin planning, making easier the convergence of loops. The high efficiency and feasibility of HGA are proved in the case study. The optimal loading plan generated lowers the total fuel consumption and improves the efficiency of transportation, offering better auxiliary options for the load planning of transport aircraft.

References

1. Gehring H, Menschner K, Meyer M 1990 A computer-based heuristic for packing pooled shipment containers. European Journal of Operational Research, vol 44, no 2, pp 277-288.
2. Heidelberg, Kurt R 1995 A Bin Packing Algorithm for Cargo Conveyance Systems. IN Virginia Commonwealth University.
3. Vancroonenburg W, Verstichel J, Tavernier K, Berghe G V 2014 Automatic air cargo selection and weight balancing: a mixed integer programming approach. Transportation Research Part E Logistics & Transportation Review, vol 65, no 2, pp 70-83.
4. Roesener, August G 2006 An Advanced Tabu Search Approach to the Airlift Loading Problem. IN The University of Texas at Austin.
5. Liu Ningjun, Wang Lixin, Pan Wenjun 2013 An Optimal Method of Transport Aircraft Fleet Cargo Loading. Journal of Beijing University of Aeronautics and Astronautics, vol 39, no 6, pp 751-755.
6. Kurt R, Heidelberg, Gregory S, Parnell, James E, Ames IV 1998 Automated air load planning. Naval Research Logistics, vol 45, no 8, pp 751–768.
7. Garey M R, Johnson D S 1979 Computers and Intractability: A Guide to the Theory of NP-Completeness. W. H. Freeman Press, p 113.

Effect of Ellipsoidal Modulus and Internal Pressure on Bearing Capacity of Thrust-Bearing Aft Dome

Wang Peng[1], Fan Ruixiang[2,a], Huang Cheng[1], Yan Dongyang[1] and Wang Feifan[1]

[1]*Beijing Institute of Astronautical Systems Engineering, Beijing 100076, China*
[2]*China Academy of Launch Vehicle Technology, Beijing 100076, China*

Abstract. In order to analyse the effect of ellipsoidal modulus and internal pressure on bearing capacity of thrust-bearing aft dome, we obtain the stress and strain distribution and bearing capacity of 1.6, 1.4 and 1.0 modulus ellipsoidal aft dome under 0MPa~0.98MPa internal pressure and engine thrust by finite element method. We find as the modulus decreases, the bearing capacity of the ellipsoidal aft dome increases, and as internal pressure decreases, within the engineering range (0~0.98MPa), the bearing capacity increases. The conclusion can provide a guidance for the design of thrust-bearing aft dome.

1 Introduction

At present, the main structural forms of liquid rocket engine thrust transmission are of rods, beam, shell and load-bearing aft dome of tank, etc. [1]. Among them, the form of load-bearing aft dome means engine transmits its thrust directly or by a small frame to the aft dome. This form can obviously reduce the size of frame and even cancel the frame to reduce the mass of the structure, which is of great significance to the lightweight design of the rocket structure.

There are many applications of load-bearing aft dome. For example, Saturn V S-IVB stage transmits thrust to aft dome through a conical shell [2,3], as shown in figure 1. The Long March 4 third stage transmits the thrust to the aft dome of tank through a cylindrical shell frame called the engine cabin [4-6] ,as shown in figure 2.The upper stage Centaur transmits all thrust through a thrust barrel welded to the aft dome [2,7], as shown in figure 3.

There is little literature to discuss the Effect of the parameter designed on bearing capacity of Thrust-Bearing Aft Dome. And the design of Thrust-Bearing Aft Dome lacks guidance. So we use the finite element method to study the Effect of Ellipsoidal Modulus and Internal Pressure on bearing capacity of thrust-bearing aft dome .

[a] Corresponding author : fanruixiang_calt@163.com

Figure 1. Saturn S-IVB

Figure 2. Third stage of Long March 4

Figure 3. Centaur C upper stage

2 FEM MODELS

The model for analysis consist of aft dome, Y-ring, aft skirt, partial tank barrel and thrust ring (bearing thrust). Figure 4 shows an Ellipsoidal aft dome of 1.6 modulus. The specific dimensions are as follows: tank outer diameter: 3350mm, Ellipsoidal Modulus of aft dome: 1.6, 1.4 and 1.0(spherical aft dome), aft dome thickness: 5mm; tank barrel thickness: 8mm, height: 300mm; aft skirt thickness: 8mm, aft skirt and Y-ring total height: 300mm, Concave vertices of Y-ring away from the bottom of aft skirt: 185mm,as shown in figure 5; thrust ring Inner diameter: 1380mm, Outer diameter: 1430mm, the thrust ring outer height: 37mm,as shown in figure 6. Corner radius is 2mm in the model.

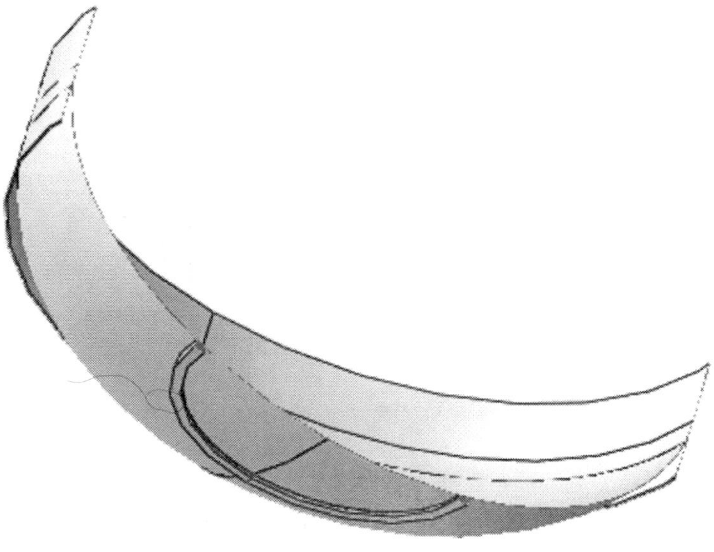

Figure 4. Symmetry plane section of 1.6 modulus aft dome

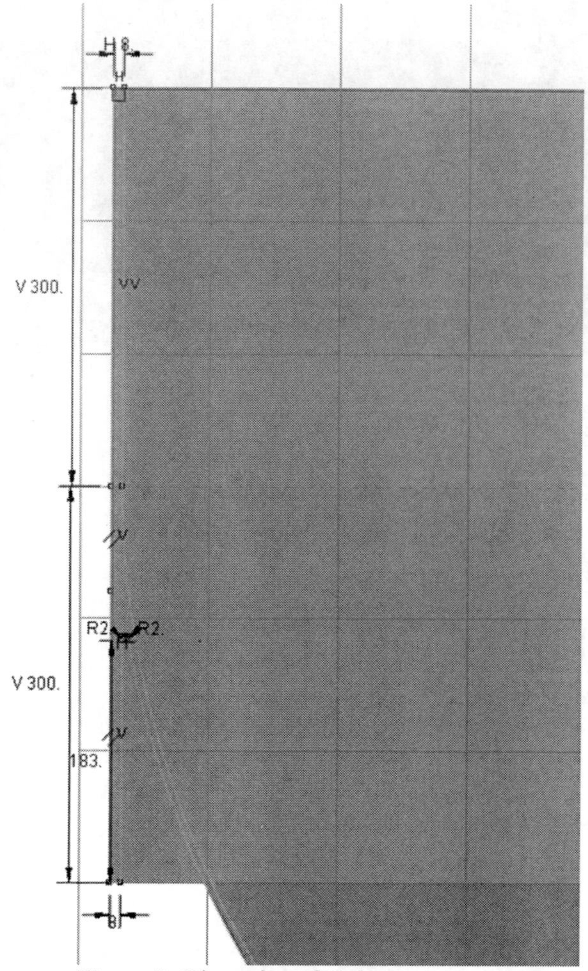

Figure 5. Dimension of Y-ring

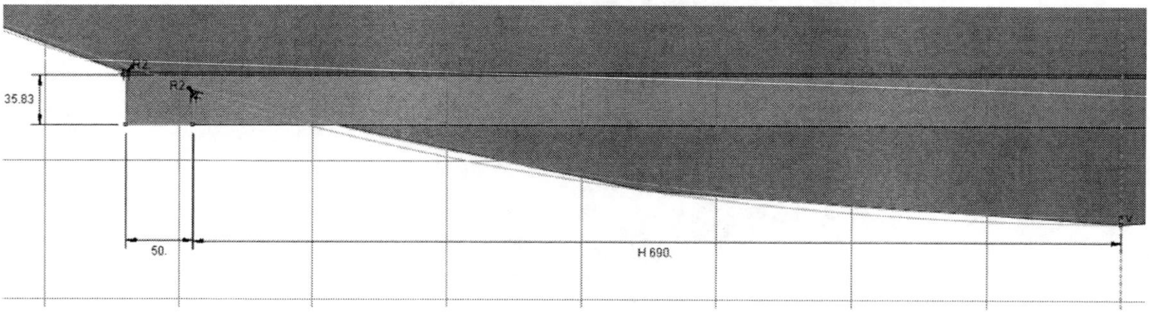

Figure 6. Position and dimension of thrust ring

Material parameters:

elastic modulus: 69000MPa, Poisson's ratio: 0.32, Density: 2.7e3 kg/m^3. Set nlgeom on in analysis step; the top of the barrel is fixed in the constraint. The grid uses C3D8I type.The internal pressure setting is shown in table 1. Thrust load is set to 2000kN.

Table 1. Internal pressure setting

Internal pressure（MPa）	0	0.08	0.18	0.28	0.38	0.48	0.58	0.68	0.78	0.88	0.98

3 RESULT

3.1 stress distribution

Within the range of load-bearing capacity, the stress distribution of the three moduli (1.6, 1.4, 1.0) aft dome under thrust and different internal pressure (0~0.98MPa) is the same. The overall trend of the stress is that the stress gradually increases from the centre of aft dome outward along the radial direction. A ring area with a relatively high stress appears at a radial distance of about 120 mm outside the thrust ring. When the load limit is reached, the maximum stress appears here. Figure 7 shows the Von Mises stress contour of 1.6 modulus aft dome under 1000kN thrust and 0.38MPa internal pressure, and Von Mises stress reaches 216.4MPa in the high stress area. The stress at the centre of aft dome is 104.9 MPa.

Figure 7. Von Mises stress contour of 1.6 modulus aft dome under 0.38MPa internal pressure and 1000kN thrust

At the same time, we obtain the stress distribution of aft dome when there is only internal pressure load. Figure 8 shows the Von Mises stress contour of 1.6 modulus aft dome under 0.38 MPa pressure. The maximum stress appears in the centre of the aft dome. And the stress reaches 104.9 MPa, which is the same in Figure 7. This shows that the stress at the centre of the aft dome is independent of the thrust and is only related to the internal Pressure.

Comparing the maximum stress area in Figure 7, the stress at the same area in fig 3-2 is 80 MPa when there is only internal pressure. It can be seen that the increase in thrust causes a significant increase in the stress in the area.

Figure 8. Von Mises stress contour of 1.6 modulus aft dome under 0.38MPa internal pressure

3.2 Deformation distribution

The Deformation distribution of the three moduli (1.6, 1.4, 1.0) aft dome under thrust and different internal pressure (0~0.98MPa) is the same. The aft dome deforms outward under the internal pressure, as shown in figure 9. It can be seen that the thrust ring has certain restraining effect on the expansion deformation of the aft dome. In the next step, the thrust load is applied and the thrust ring area deforms in the opposite direction against the internal pressure, forming the deformation in figure 10. Figure 10 is the displacement contour (U2-along rotary axis, 20x magnification) of 1.6 modulus aft dome under 0.38MPa internal pressure and 1000kN thrust.

Figure 9. U2 displacement contour of 1.6 modulus aft dome under 0.38MPa internal pressure (20 times magnification)

Figure 10. U2 displacement contour of 1.6 modulus aft dome under 0.38MPa internal pressure and 1000kN thrust (20 times magnification)

3.3 Structural failure mode analysis

Structural failure mode I: The aft dome is buckling due to thrust, resulting in stress in high stress area in Figure 3.7 to yield limit; In failure mode II, the stress in the central area of aft dome reaches the yield limit. In these analyses only does the failure mode I occurred when using the material properties of Aluminum alloy (yield limit 280MPa) .The occurrence of failure mode II is related to the internal pressure, but not to the thrust. It occurs only when the internal pressure is large, and does not occur in these analyses (the internal pressure is no larger than 0.98 MPa).

The failure mode I can be observed through the load-displacement curve at the thrust position, as shown in figure 11.When the thrust gradually increases to 1300kN, the axial displacement (U2) at the thrust area suddenly increases sharply, indicating that buckling occurs here and the bearing capacity Rapidly reduces .During the buckling, In figure 12, the high stress area is severely deformed, resulting in sharply increasing stress.

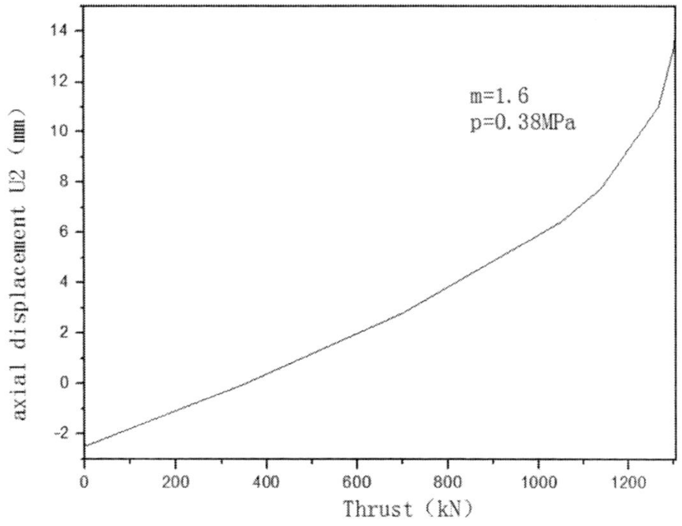

Figure 11. Load-displacement curve at the thrust position of 1.6 modulus aft dome under 0.38MPa internal pressure and thrust

Figure 12. Von Mises stress contour of 1.6 modulus aft dome under 0.38MPa internal pressure and 1000kN thrust (20 times magnification deformation)

3.4 Effect of Ellipsoidal Modulus and Internal Pressure on bearing capacity

We define bearing capacity as the maximum thrust, under which maximum stress in aft dome reaches 280MPa. In all analysis groups, the deformation of aft dome was within the allowable range when the yield limit was reached, and the maximum deformation was not more than 10 mm.

The results are shown in figure 13.

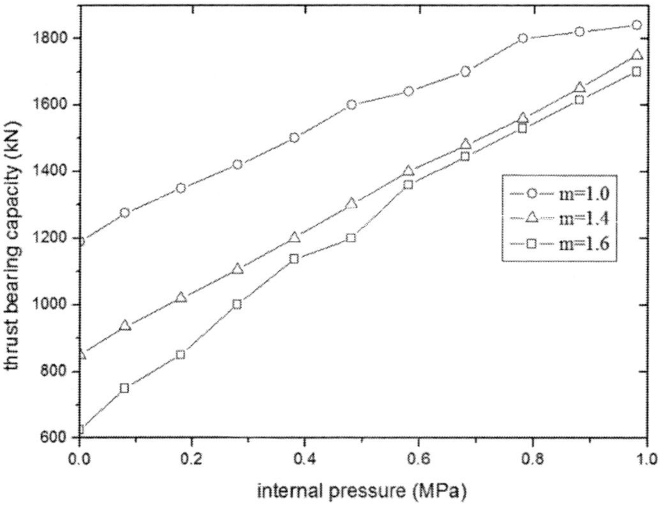

Figure 13. Thrust-bearing capacity of 1.6, 1.4 and 1.0 modulus aft dome under 0~0.98MPa internal pressure

we can see that for the material properties set in this analysis (elasticity modulus 69 GPa, Poisson's ratio 0.32, yield strength 280 MPa), reducing the ellipsoid modulus can improve the bearing capacity under different internal pressures, but At the same time reducing the modulus will increase the mass and the height of the aft dome, as shown in table 2; within the actual engineering pressure range (0~0.98 MPa), increasing the internal pressure of the tank can significantly increase the bearing capacity of the aft dome

Table 2. Mass and height of 1.6, 1.4 and 1.0 modulus aft dome

modulus	mass	Height of model
1.6	330kg	1047mm
1.4	344 kg	1196mm
1.0	381 kg	1675mm

4 conclusion

For the material properties (elasticity modulus 69GPa, Poisson's ratio 0.32, yield strength 280 MPa)

1.Reducing the ellipsoid modulus can improve the bearing capacity under different internal pressures, but at the same time it will increase the mass and height of the aft dome. So In the design, comprehensive consideration should be given to selecting an appropriate modulus.

2.Within the engineering pressure range (0~0.98 MPa), increasing the internal pressure of the tank can significantly increase the bearing capacity of the aft dome.

References

1. Zhu Ningchang, Liu guoqiu. Design of Liquid Propellant Rocket Engine, China Astronautic Publishing House, 3(2005)
2. Saturn V. Flight Manual[R]. MSFC-MAN-503, 1 November, (1968)
3. S-4B stage thrust structure influence coefficient.(1970)
4. Xu Xin-hua. Long March launch vehicle[J].Missiles and Spacecraft, 1(1991)
5. Zang Jia-liang. The surface tension tank design selected for the third stage's attitude control engine of Long March launch vehicle[J]. Aerospace China,8(1992)
6. Wu Rui-hua, Gu Song-an. Long March series launch vehicle introduction: Long March 4 series (3)[J]. Aerospace China, 2(1999)
7. San Diego, Titan 3E Centaur D-1T Systems Summary

Effects of Airflow Deflection Angle in Diffuser on Forced Response Caused by Impeller-diffuser Interaction in Centrifugal Compressors

Sun Peng[1,2,a], Wang Yanrong[1,2] and Zhang Xiaobo[1,2]

[1]School of Energy and Power Engineering, Beihang University, China
[2]Collaborative Innovation Center for Advanced Aero-Engine, China

Abstract. Aeroelasticity has always been a significant issue of compressors. Impeller-diffuser interaction (IDI) produces periodic aerodynamic excitation which forces the blades to vibrate in centrifugal compressors, and eventually causes the damage of blades. Therefore, it is of great importance to investigate the IDI in centrifugal compressors and the effects on the vibration of blades. In this paper, it is discussed in detail that the deflection angle of the airflow in diffuser affects the IDI of the centrifugal compressor and the vibration of blades. The results show that if deflection angle is positive, the interaction of the diffuser on impeller and the vibration of main blades are decreased. However, on condition that the deflection angle is excessive, the interaction of the diffuser on the flow field of impeller will be intensified, and the vibration of the main blades will be stronger. Therefore, it is necessary to reasonably control the range of the deflection angle so that centrifugal compressors are able to work in a safer environment.

1 Introduction

Compared with the parameters of axial compressors, the rotating speed of centrifugal compressors is much higher and the blades are slighter, which results in a much larger stress of blades. If blades vibrate due to aeroelastic problems during the operation of compressor, especially when resonance occurs, fatigue cracks may appear on blades, resulting in high-cycle fatigue (HCF) problems and the damage of blades. Therefore, it is necessary to investigate the factors that are likely to lead to aeroelastic problems, so as to avoid the failure of blades and compressors.

Forced response of a centrifugal compressor caused by IDI is one of the main problems of aeroelasticity. IDI in a centrifugal compressor can be defined as a group of mechanisms which cause the differences in flow phenomena and performance of the two components between isolated and coupled operation, including unsteady and other effects [1].

As for the research of IDI, Koth [2] compared the effects of vaned diffuser and vaneless diffuser on impeller, and found that steady simulations of flow field were approximately the same, but unsteady excitation was detected obviously under the case of the vaned diffuser. Lusardi [3] compared three different centrifugal compressors and concluded that unsteady periodic pressure occurs at the leading edge of diffuser, which perturbs the upstream flow field along the surface of blades. And the wave may have a phase difference between the pressure surfaces and suction surfaces of blades, which

[a] Corresponding author : sunpeng1994@buaa.edu.cn

can amplify the perturbation. Lusardi [3] and Borm [4] proved that some general softwares for CFD calculations (such as CFX, Fine Turbo, etc.) are able to well simulate the unsteady aerodynamic loads generated by IDI in centrifugal compressors and results are obtained accurately.

A number of researchers have worked on figuring out the parameters that affect the IDI of centrifugal compressors. Gould [5] proposed three main parameters that affect unsteady aerodynamic loads of the IDI: radial clearance, DeHaller number, and Mach number. The smaller radial clearance is set, the stronger the IDI is. Zemp [6] proved that radial clearance has a greater effect on the vibration of the blades under higher mass flow conditions compared to lower mass flow conditions. In addition, according to the results of experiments, Skoch [7] found that the diffusion ability of diffuser is greatly affected by the airflow deflection angle in the diffuser. However, for the vaned diffusers whose vanes are NACA-shaped, which are used more widely in centrifugal compressors, there is still a lack of relevant research. Besides, there are few studies focus on the relationship between the deflection angle and the vibration of impeller blades caused by IDI.

In order to figure out the effects of deflection angle of the airflow in diffuser on forced response of the impeller blades caused by IDI, a detailed study is conducted, including the simulation of flow field in centrifugal compressor, modal analysis and transient analysis of main blades of impeller. In the simulation, the deflection angle of the airflow in diffuser is determined by changing the camber line of the diffuser vanes, and then the purpose of changing the deflection angle of airflow in the diffuser is achieved.

2 Controlling equations

In terms of the simulations of the flow field of compressor, NUMECA software is used to solve the three-dimensional unsteady Reynolds Average Navier-Stokes (RANS) equations. Spalart-Allmaras model is used as turbulence model, and the Phase Lagged model is adopted on the interface of impeller and diffuser, which can make sure that the all computational domains are reduced to a single domain without changing the number of blades or vanes and considerably shorten the time for calculation [8]. Gerolymos [9] and Zante [10] have shown that accurate results can be obtained by the Phase Lagged model in the flow field calculation of a single-stage compressor, including a stage of stator and a stage of rotor. In terms of the structure of blades, the finite element method is used. ANSYS software is used to solve the three-dimensional structural dynamic equation:

$$\boldsymbol{M}\ddot{x}(t)+\boldsymbol{C}\dot{x}(t)+\boldsymbol{K}x(t)=\boldsymbol{F}(t) \tag{1}$$

where $\boldsymbol{M}$, $\boldsymbol{C}$, $\boldsymbol{K}$, and $\boldsymbol{F}$ represent mass matrix, damping matrix, stiffness matrix and force matrix respectively.

3 Introductions of model

A scaled model of the DDA 404-III centrifugal compressor impeller is used as the model in this paper, which contains 15 main blades and 15 splitter blades. The vaned diffuser contains 24 NACA-shaped vanes. It has been proved that the weight of centrifugal compressor is lighter with NACA-shaped vanes compared with wedge-shaped vanes, but the phenomenon of IDI is more severe, so it is more worth to investigate centrifugal compressor with NACA-shaped vanes. The model of centrifugal compressor is shown in Figure 1.

The deflection angle of airflow in diffuser is defined as

$$DA=\alpha-\beta \tag{2}$$

where α and β are the angles between the camber line of a vane and radial lines, as shown in Figure 2. r_1, r_2 and r_3 are the radii of the trailing edges of blades, the leading edges of vanes and the trailing edges of vanes. A series of diffuser airflow deflection angles are set in order to figure out the effects of deflection angle on the strength of IDI, the area affected by IDI and the vibration stress of blades. Four diffusers with different deflection angles are selected as cases for analysis, including deflection angles are 0°, 5°, 14° and 29°, as shown in Figure 3.

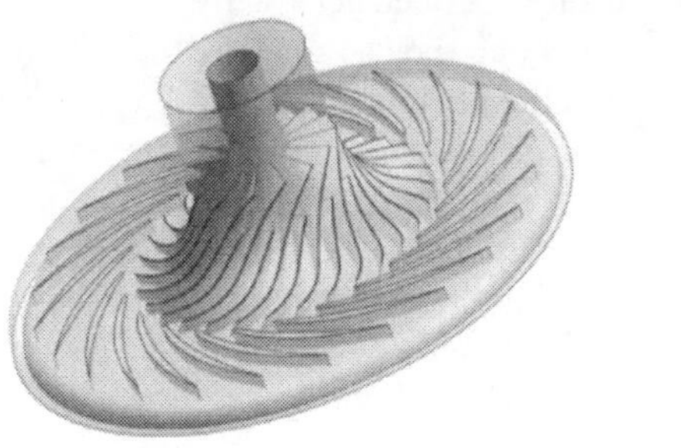

Figure 1. The model of centrifugal compressor

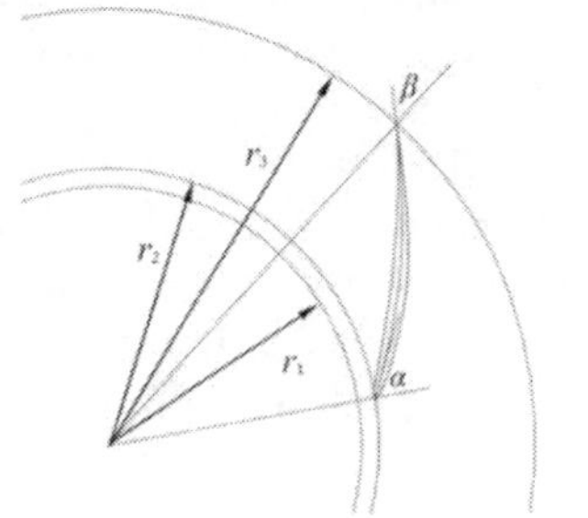

Figure 2. The definition of α and β

(a) DA=0° (b) DA=5° (c) DA=14° (d) DA=29°

Figure 3. Four cases with different deflection angles in diffuser

The boundary conditions of flow field are set, the inlet of which is 101.3kPa of total pressure and 288.15K of total temperature. Outlet total pressure has been set as 350kPa, making the efficiency of centrifugal compressor reaches the maximum. The grids of flow field are generated by Autogrid automatically, the total number of which is 1.8 million. The quality check is passed. In all cases, the total number of grids is kept close to avoid the difference of results caused by the density and quality of the grids. The tip clearance of the blades is set as a typical value, while the tip clearance of the vanes is set as 0.

4 The effects of deflection angle on flow field unsteadiness caused by IDI

Steady and unsteady simulations of flow field are carried out for all cases. The passing frequency can be derived with equation (3):

$$f_p = \frac{nN_s}{60} \tag{3}$$

where n is rotating speed, N_s is the number of vanes, which is 24 in this paper.

In order to figure out the law of flow field unsteadiness caused by IDI with different deflection angles, a series of monitoring points are selected, which located at 70% chord, 85% chord and trailing edge of main blades. The pressure changing with time at monitoring points are extracted, and the components of pressure at monitoring points are obtained after filtering and spectrum analysis. The results of spectrum analysis are shown in Figure 4.

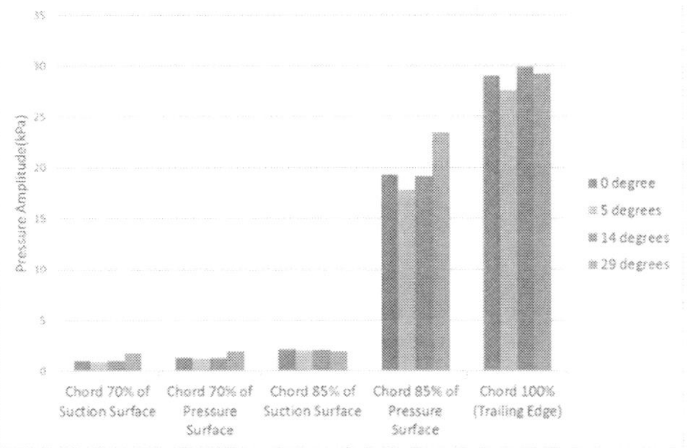

Figure 4. The results of spectrum analysis

According to the result, it can be concluded that:

1 IDI plays an important part in the flow field near the trailing edge of the blade. At the trailing edge, the pressure fluctuation amplitude is about 30 kPa. By contrast, the value of pressure in average at the trailing edge is 245 kPa, of which the amplitude reaches almost 1/8.

2 There is a big difference in the area of the flow field affected by IDI between suction surface and pressure surface of the main blade. The amplitude in the pressure surface at chord 85% of main blade is almost 20kPa, while it is reduced to less than 3kPa in the suction surface. Therefore, the effects of IDI on the pressure surface is wider than that on the suction surface. However, the area is still relatively limited. The flow field from the leading edge of the blade to chord 70% is hardly affected by IDI.

3 With the increase of the deflection angle, the pressure fluctuation amplitude at the 85% chord length of the pressure surface decreases at first and then increases, reaching the minimum when deflection angle equals to 5°. But when the airflow deflection angle is no more than 14°, the pressure fluctuation amplitude does not change much, but it is obviously smaller than the amplitude in the case where deflection angle is excessive. It can be confirmed that when the deflection angle is not quite large, there is almost no significant change in the strength of the IDI of a specific point and the area of flow field affected by the IDI. However, if the deflection angle is excessive, the pressure fluctuation amplitude and the area of flow field affected by IDI will both increase.

5 Model analysis of main blade

Previous studies have shown that the problems of vibration caused by aeroelasticity happened to mean blades more frequently. Therefore, the main blade is selected as the model for analysis. The finite element model of the main blade has been established, as shown in Figure 5. The titanium alloy TC4 has been selected as the material of the blade, the key parameters of which are shown in Table 1. The temperature field has not been taken into account. Model analysis of main blade has been carried out, according to the result of which Campbell diagram of main blade is plotted, as shown in Figure 6. The points where the stress amplitude is respectively large are selected as monitoring points for transient analysis carried out subsequently. There are 5 points are selected, as shown in Table 2. The distributions of the equivalent stress (SEQV) of the 6th mode and the 7th mode are shown in Figure 7, in which the position of the monitoring points are marked at the same time.

Table 1. Material Parameters of TC4

Material	Density(kg/m^3)	Elastic Modulus(GPa)	Poisson's Ratio
TC4	4440	109.1	0.34

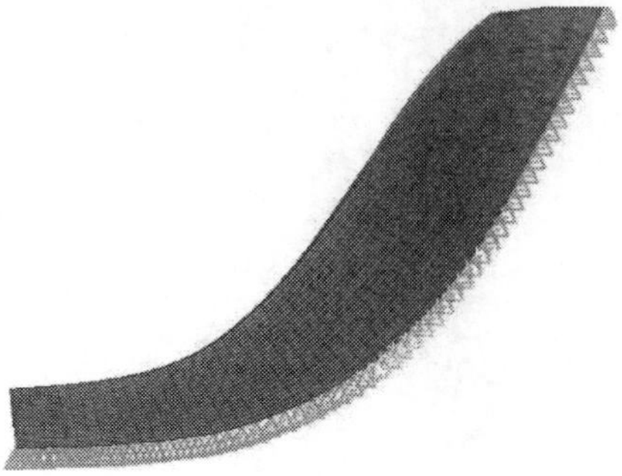

Figure 5. Finite element model of main blade

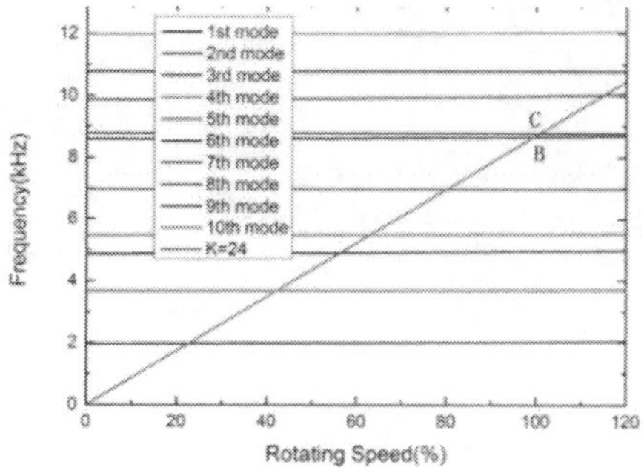

Figure 6. The Campbell diagram of main blade

Table 2. Selected points and working points

Mode	Point that intersects with line K=24	First principal stress	First principal stress	von Mises stress
6th Mode	B	No.1137	No.8	No.1137
7th Mode	C	No.1136	No.195	No.1126

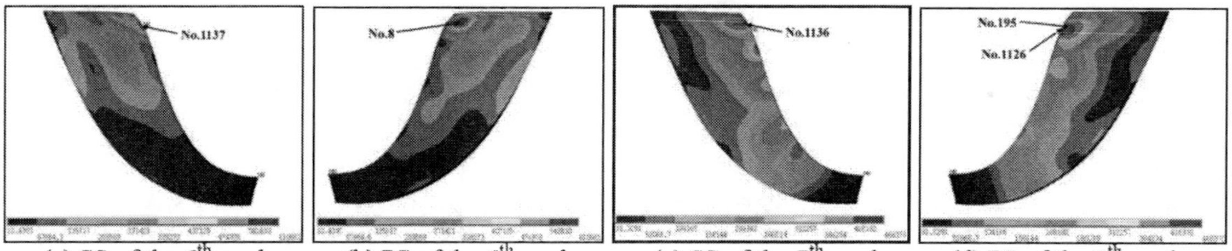

(a) SS of the 6th mode (b) PS of the 6th mode (c) SS of the 7th mode (d) PS of the 7th mode

Figure 7. The distribution of SEQV of main blade

6 The effects of deflection angle on the vibration amplitude of blades

The transient analysis results of monitoring points are extracted, and frequency components are derived by spectrum analysis with Fourier Transform method. The high-frequency components are omitted for the amplitudes are quite small. Only the amplitudes of top three low-frequency components are extracted, and the results are connected by smooth curves, as shown in Figure 8, where 1x, 2x, and 3x are used to represent the component of passing frequency, double of passing frequency, and triple of passing frequency respectively.

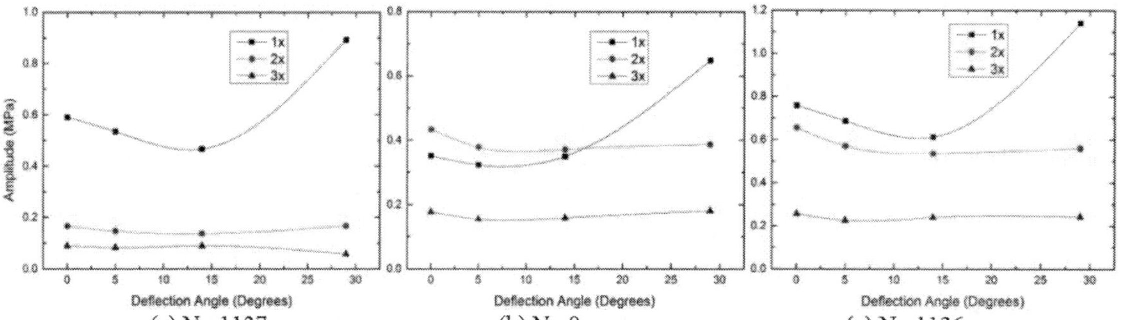

(a) No.1137 (b) No.8 (c) No.1136

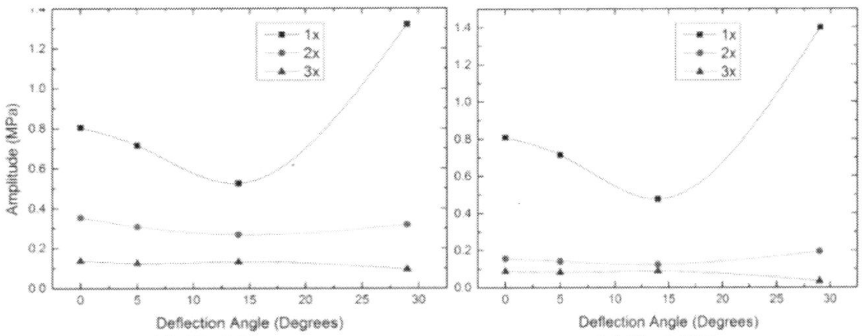

(d) No.195 (e) No.1126

Figure 8. Stress amplitudes of monitoring points

The results of all monitoring points show that the stress amplitude of 1x is greatly affected by deflection angle, but the stress amplitudes of 2x and 3x are hardly affected by deflection angle. With the increase of deflection angle, the stress amplitude of 1x has a tendency of decreasing at first and increasing subsequently. The value of the stress amplitude reaches the minimum when deflection angle is between 5° and 15°. In addition, the stress amplitude of 1x is much larger when the deflection angle is 29°, the value that actually exceeds the reference value. In other words, main blades are more likely to fail attributed to high vibration stress on condition that deflection angle is excessive in centrifugal compressor. Therefore, it should be fully taken into account that the deflection angle should not be designed to be excessive in the design of the diffuser, so that the probability of failure of main blades caused by massive stress can be reduced.

7 Conclusions

According to the simulation of flow field of a centrifugal compressor and transient analysis of blades, the effects of deflection angle of the airflow in diffuser on the vibration of main blades caused by IDI is clarified. The following conclusions are deduced from computed results:

1 IDI plays an important part in the flow field near the trailing edge of the blade, and the affect of IDI on pressure surface is greater than that on suction surface. On condition that deflection angle is excessive, pressure fluctuation amplitude increases considerably.

2 With the increase of the deflection angle, main blade stress amplitude of 1x has a tendency of decreasing slowly at first and increasing subsequently. The value of the stress amplitude reaches the minimum when deflection angle is between 5° and 15° in the centrifugal compressor introduced in this paper. While stress amplitudes of 2x and 3x are hardly affected by the deflection angle. Therefore, the deflection angle should not be designed to be excessive.

References

1. U.Z. Kai, H.E. Gallus, R. Niehuis, J. Tur. **125**, 545 (2003)
2. H. Koth, K. Oeder, J. E. Gas Tur. Pow. 688 (1987)
3. C. Lusardi, MIT thesis (2012)
4. O. Borm, H.P. Kau, ASME conf. (2012)
5. K. Gould, ASME conf. (2007)
6. A. Zemp, R.S. Abhari, J. Tur. **135**, 1507(2012)
7. G.J. Skoch, J. Tur. **127**, 107 (2004)
8. L. He, J. Tur. **112**, 714 (1990)
9. G.A. Gerolymos, G.J. Michon, J. Neubauer, J. Prop. Pow. **18**, 1139 (2012)
10. D.V. Zante, J. Chen, M. Hathaway, J. Tur. **130**, 1067 (2008)

One Micro-thrust Measurement Method Based on the Acoustoelastic Theory

Yang Ou[a], Jianjun Wu, Yu Zhang, Jian Li and Sheng Tan

College of Aerospace Science and Engineering, National University of Defense Technology, 410073, Changsha, P.R.China

Abstract: Concerning the traditional way of micro-thrust measurement cannot get the transient thrust value with high precision and the measuring process of piezoelectric sensors is sensitive to the electromagnetic interference, this paper builds one new measurement method basing on the acoustic elasticity theory. The acoustoelastic element is the core of the whole system as it can transform the thrust changes into the variation of acoustic time. Thus adopting a timer to record such variation so that the thrust changes in real-time could be obtained, which is helpful for studying the mechanism of micro thrusters. This method built in this paper is not only non-contact and non-destructive but also easy to operate with simple structure and high precision.

1. Introduction

As we known, large satellites often have long development cycle with high cost and complex technology, which greatly limits their application and development. Meanwhile, by providing a train of small, discrete impulse bits, micro-thrusters are ideal for such diverse applications as orbit raising, drag makeup, attitude control, constellation station keeping and deep-space missions. Micro-thrusters are especially well-suited for low-thrust, on-board propulsion because they use a non-toxic propellant and are highly reliable, solid-state, low power, and lightweight, as they require no propellant storage system [1].

Therefore, in recent years, many agencies, including the European Space Agency, Iranian Space Agency and Australian National University, have developed several forms of microsatellite systems with its advantages of micro size, light weight, short response time, high function density ratio and cost-effective. With the rapid expansion of micro satellite, micro satellite orbit maneuver, precise attitude control, atmospheric drag compensation and high precision formation flying ability all need such one thruster with small power, light weight and simple structure. The plasma thruster is not only the ideal propulsion scheme but also the key subsystem for attitude controlling, position keeping and orbit maneuver of micro satellite.

Through high precision and resolution measuring the micro thrust and micro impulse, it can get more discrete but precise enough pulse with the control of thrust. And then, such could guarantee the high precision attitude control, which contributes to the relative technologies. However the thrust produced by micro-thrusters is generally pretty small, which ranges from nN level to mN level. For example, the thrust produced by pulsed plasma thrusters is generally bovine micro scale. And ESA's

[a]Corresponding author : naihe620@163.com

LISA project uses a milli bovine scale thruster. In addition, the NASA ST7 mission uses a 2~20 N thruster with a precision of less than 0.1 N [2] [3].

Although it is relatively easy to measure the thrust in the range of mN magnitude, however, with thrust decreases, accurate measurements of the thrust of the nN and μN level become very difficult. On the one hand, the noise of the instrument could affect the accuracy; on the other hand, it will reduce the precision if transient force that passes to the platform remains pending. General speaking, it is urgent to further improve the accuracy of the thrust measurement, especial the thrust measurement in range of the μN level, as for the development of micro satellites [4].

The thrust measurement consists of average thrust and impulse measurement. The essence is to observe the mechanical behaviour of the thrust the thrust stand under the micro-thrust for measuring the thrust indirectly. Furthermore, according to different thrust range, measurement methods are different. For mN to N level, the weighing experiment rig and the tilting structure are adopted [5]. While for μN level, it is necessary to take the twist structure, double pendulum structure and the four arm counterweight structure into consideration.

At present, the micro thrust / micro impulse measurement methods with high precision are relatively scarce. In the past, impulse and instantaneous thrust measurement systems typically used an inductive proximity probe, and accelerometer, a linear voltage differential transformer, a differential plate capacitance system, or an optical angular readout. The most sensitive of these past devices appears to be the optical angular readout from Ref. 3 designed in 1968 at NASA Jet Propulsion Laboratory, which can measure thrust as low as 5 μN with loads up to 12kg. A comparable thrust stand was designed at Goddard Space Flight Centre in 1970 that measured thrust as low as 25μN, thrust stand deflection as low as 25 nm, and measured the thrust vector. Larger thrust and impulse bit measurement have been made as well. A 5 N thrust stand was developed by Hagg at NASA Lewis Research Centre in 1989 and impulse bits near 1 Ns were measured at Princeton University [2].

However, there is no high accuracy measurement system at present. It's urgent to raise the measurement accuracy of the measurement as exact as possible for promoting the development of micro-satellites. This paper would build a new thrust measurement system of micro-thrusters with high precision, which basing on the acoustic elasticity principle. This method can get the internal stress of the acoustic elastic element by measuring the variation of sound velocity. And then, the specific values of the thrust are obtained. The key components of this system include a lever table, an ultrasonic transducer, a data acquisition device, and a necessary signal processing system. This system, basing on the relationship between stress and acoustic speed, adopts one visualization method with the special elastic material which could achieve the aim of recording the variation of the micro-thrust.

2. Acoustoelastic theory

The acoustoelastic pulse-echo method that has gained wide application in industry was first described by Benson and Raelson in their fundamental 1959 study. The authors proposed a new method for measuring the stresses in an isotropic material. The phenomenon that the method is based on is in the difference of the velocities of transverse waves polarized parallel and perpendicular to the direction of stresses in the materials under load, i.e., the anisotropy of the stressed medium. This phenomenon has been termed the acoustoelastic effect and method for assessing the stress-strain state of structural elements is referred to as the acoustoelastic method [6].

And then, a number of studies have been done around the acoustoelastic method. Cantrell and Salama focused on inhomogeneous stress determination and on material characterisation for recent developments in non-linear acoustelasticity that implicates directly the thermoelastic properties and in appropriate cases the magneto elastic properties of multiphase materials. Dubuget proposed an original ultrasonic immersion device which permits the determination of the elastic constants by velocity measurements for tow aluminium alloys and the third-order elastic constants have been obtained from acoustoelastic responses [7]. However, seldom is such method applied to measuring the thrust, more application are in stress testing.

Acoustic elasticity is a branch of mechanics that studies the interaction of structures, fluids, and sound fields. It involves ships, underwater weapons, aircraft, automobiles, buildings, chemicals, marine and water conservancy projects, and many other fields. It will radiate sound waves to surrounding or inner media when elastic structures are forced to vibrate under the external thrust. Meanwhile, it could make the vibration or the thrust change as the acoustic wave loads in the elastic structure in the form of feedback. Therefore, it's the interaction of these that builds a coupled system between the excitation, structure and acoustics. And the main task of the acoustic elasticity theory is to detect and understand the relationship among the basic mechanical and acoustic, in addition to analyse the characteristics and regularities of the structure induced vibration and sound radiation. Such method is not only a non-destructive method for measuring stress, but also one new means to analysis the non-transparent material's stress as for its reliability, toughness, economical efficiency and easy-using. The wave velocity changes in accordance with the stress; therefore, we can get specific value of the stress by measuring the variation of wave velocity. What's more, if the same distance is set, we can get different sonic time in order to reflect the variation of the stress.

However, seldom is such method applied to measuring the thrust, more application are in stress testing. And the goal of this study is to obtaining the thrust data on the spatial and temporal dynamics of the variations when the micro-thrusters work.

As we can know from the former research, the transversal wave velocity, polarised along the direction of the stress, differs from that for which the polarisation is perpendicular to the stress. Therefore, by using measurement of the wave velocity in a material, in the absence or in the presence of stresses and with the knowledge of the different elastic constants, the applied stress on the solid can be determined. We note that the third-order elastic constants, the applied stress on the solid can be determined. We note that the third-order elastic constants are neglected in case of the linear elasticity. Moreover the exploitation of the acoustoelastic phenomenon is quite difficult since the influence of stresses on the ultrasonic velocity is very weak and also since the crystallite orientation or the texture and the microstructure have a strong influence on the ultrasound propagation.

The application of a stress on an isotropic and homogeneous body leads to a modification of its mechanical behaviour. In the presence of such a phenomenon, it has been shown that the propagation velocity of ultrasound waves propagating in the x-direction. And polarized along different directions are expressed by the following acoustoelastic equations [8] [9] [10].

Longitudinal wave propagating along the x-direction:

$$\rho_0 V^2_{xx(x)} = \lambda + 2\mu + \frac{\sigma}{3\lambda + 2\mu}\left[\frac{\lambda + \mu}{\mu}(4\lambda + 10\mu + 4m) + \lambda + 2l\right] \tag{1}$$

Transversal wave propagating along x and polarised along the y- or z-direction:

$$\rho_0 V^2_{xy(x)} = \rho_0 V^2_{xz(x)} = \mu + \frac{\sigma}{3\lambda + 2\mu}\left[4\lambda + 4\mu + m + \frac{n\lambda}{4\mu}\right] \tag{2}$$

In the above equations, ρ_0 is the density in the initial state. V_{xx}, V_{xy} and V_{xz} are, respectively, the propagation velocity of the ultrasound waves along the x-direction with a particle displacement along the x-, y- and z-directions. λ and μ are the elastic constants or Lame constants. l , m and n are the third-order elastic constants or Murnaghan constants. σ represents the axial stress [11] [12] [13].

From the equations, we know the propagation velocity of sound waves is inconsistent with the change of stress in different directions and the longitudinal wave shows the strongest sensitivity to the stress change along the direction of wave propagation. Therefore, the position of acoustic elastic element and acoustic emitter is set up to ensure that the longitudinal wave is parallel to the stress direction and satisfies the relationship (1):

$$\rho_0 V^2 = \lambda + 2\mu + \frac{\sigma}{3\lambda + 2\mu}\left[\frac{\lambda + \mu}{\mu}(4\lambda + 10\mu + 4m) + \lambda + 2l\right] \tag{3}$$

The velocity is generally obtained by measuring the propagation time of equidistant distances:

$$V = \frac{L}{t} \tag{4}$$

Where L is the distance and t is the propagation time.

From the mechanics of materials, we also get the relationship between the internal force and stress:

$$\sigma = \frac{F}{S} \tag{5}$$

Where F is the internal force and S is the cross-section.

Simultaneous these formulas, we can get the relationship between the force and the propagation time:

$$F = \frac{\mu S\left(\rho_0 L^2 - \lambda t^2 - 2\mu t^2\right)\left(3\lambda + 2\mu\right)}{t^2\left(4\lambda^2 + 15\lambda\mu + 4m\lambda + 4m\mu + 2l\mu + 10\mu^2\right)} \tag{6}$$

Control the loading so that the thrust is equal to the internal force so that we can get the thrust by measuring the propagation.

3. Basic structure

In this paper, the acoustic elastic element is regarded as the core of the system as it can reflect the variation of thrust through the change of time. And such measurement method is built according to the characteristics of micro thruster and acoustic elastic theory. On the basis of the system, we can obtain the real-time change of the thrust when using high-speed signal collector to record the value between the signals. This measurement is non-contact and non-destructive for a sensor or other measuring devices installed on the thrusters are not needful. Therefore, the system reduces the interference of mechanical vibration and power supply cables while solves problem like zero drift and unstable equilibrium position calibration, difficulty and low accuracy. General speaking, it has the advantages of simple structure, easy operation and high reliability while the most important thing is to provide a method to solve the problem of the traditional thrust measuring method which is difficult to measure the transient thrust[14] [15] [16].

The instantaneous thrust measurement system basing on the principle of acoustic elasticity consists of four parts, which are acoustic environment system, thruster system, control system, signal data collector and signal processing system. The overall composition is shown in Figure 1.

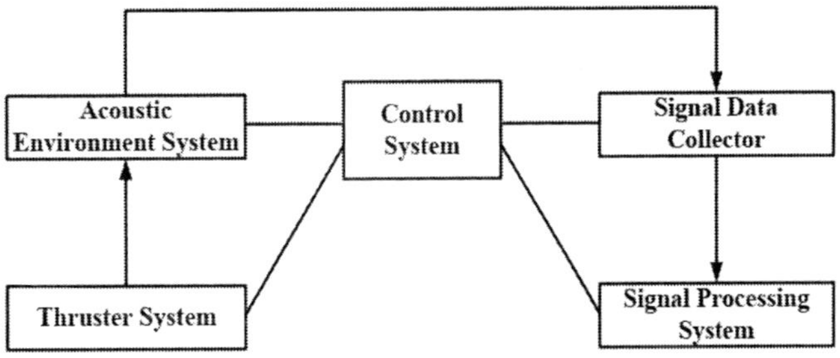

Figure 1. Overall composition.

3.1. Acoustic environment system

As shown in Figure 2, the acoustic environment system includes an organic glass wedge 1, a transmitting probe 2, a sound absorption material plate 3, an acoustic elastic element 4, a fixing device 5, and a receiving probe 7. In addition, the acoustic elastic element 4 can be replaced according to the different thrust so as to meet the measurement requirements. The transmitting probe 2 is used to produce sound waves while the receiving probe 7 mainly completes the filtering and wave acquisition. The Plexiglas wedge is a low velocity medium, and the acoustic elastic element is a high velocity medium. A sound absorbing material is used to absorb longitudinal waves reflected in a wedge by multiple reflections at the interface. The fixing device ensures the stability of each device [17] [18].

The transmitting probe transmits ultrasonic, and then it will produce LCR wave and refracted shear wave when the ultrasonic is incident on an interface between low wave velocity medium and high velocity wave medium at the first critical angle. The two waves have the same frequency and would be received by the receiving probe after multiple reflections. The LCR wave will precede the other ultrasonic clutter to the receiving probe due to its unique characteristics. The sound absorbing material board selects the sound waves which are not qualified so as to ensure that the sound waves needed for the measurement are left. The fixing device maintains the acoustic environment system stable and does not affect the normal operation of the acoustic component by movement.

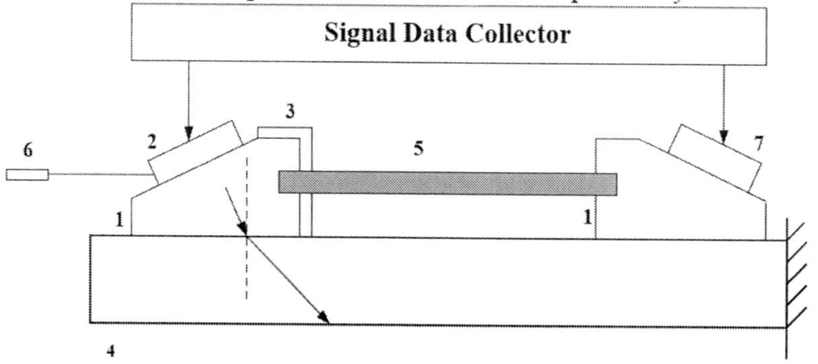

1-Organic Glass Wedge; 2-Transmitting Probe; 3-Sound Absorption Material Plate; 4-Acoustic Elastic Element; 5-Fixing Device; 6-Timing Trigger; 7-Receiving Probe

Figure 2. Acoustic environment system.

The main function of the system is to create an acoustic environment for the acoustic elastic element so that the stress change of the acoustic elastic element can be reflected by the wave velocity.

3.2. Thruster system

The thruster system is shown in Figure 3, which includes a micro-thruster 8 and a lever 9. The purpose of this paper is to realize the force measurement. Therefore, different types of thrusters can be replaced according to the requirements of the force measurement without considering the structure and function of the thruster. Pulsed plasma thruster is taken as an example. The lever 9 and the acoustic elastic element 4 are semi-rigid connected, and the lever transmits and amplifies the thrust.

The sound velocity variation of elastic element is directly related to the thrust's change. The lever can amplify the thrust in an appropriate ratio, and then provides a larger thrust for recording the velocity variation more precisely. Finally, real data is obtained by scaling the value according to the lever parameters.

The main function of this system is to generate thrust and to amplify and transfer the thrust to the acoustic elastic element so that it is in a stress state.

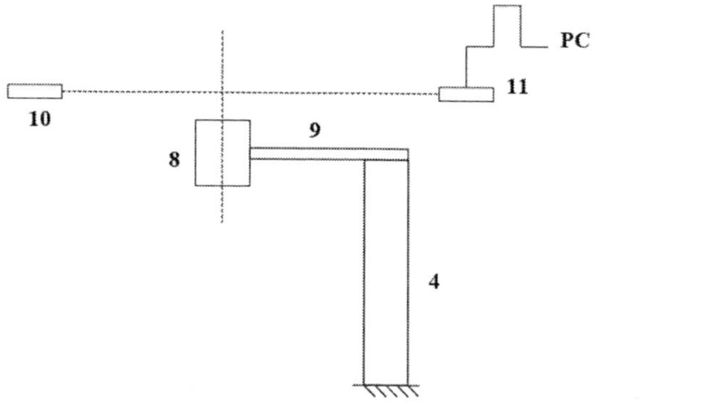

4-Acoustic Elastic Element; 8-Micro-thruster; 9-Level; 10-Electromagnetic Emitter; 11-ELectromagnetic Trigger

Figure 3. Micro-thruster system.

3.3. Control system

The ultrasonic wave emitted by the emission probe in Figure 2 does not exist continuously; otherwise the probe signal is chaotic. So it is necessary to make the transmitting probe timing trigger in a probe recognized frequency. The specific thrust calculated by this method is assumed to be approximately equivalent to the thrust loaded from the micro-thruster at the same moment as the deformation of elastic element is little. Therefore, a timing trigger is used here to realize the timing of transmitting ultrasonic waves.

Most of the thrusters are not continuous but pulsed. So it would be uneconomical if ultrasonic waves are still emitted when the micro-thruster doesn't work. This paper mainly aims at the plasma thruster which gets the thrust by spouting the plasma. As we known, plasma has shielding effect on electromagnetic signals. An electromagnetic emitter 10 is adopted. The moment when it shields the electromagnetic signal when the plasma is ejected, the electromagnetic trigger 11 triggers the suppression timing trigger 6 at the same time.

This system mainly aims to control the time interval of electromagnetic wave emission and the time of controlling electromagnetic wave emission.

3.4. Signal data collector

The whole system is arranged in the same distance so that the variation of the time can reflect the sound speed's change, and then it shows the changes of stress and thrust. The hardware measurement of sound propagation time is mainly based on the timer to obtain the time interval between the transmitting moment and the receiving moment, so as to calculate the propagation speed of the ultrasonic wave. High precision implementation of time interval measurement is the core part of the signal data collector. When the transmitting probe 2 emits an ultrasonic wave, the chip of the data collector carries on a record to it, and when the receiving probe 7 receives the ultrasonic wave, the chip collects the primary signal again[19] [20] [21].

3.5. Signal processing system

The measurement result getting from the signal data collector is not the thrust but the time, so the important function of the signal processing system is to convert the time, the sound velocity, the stress and the thrust. This step can be realized through MATLAB.

The propagation time of each moment corresponds to one thrust. Meanwhile, the control system and the signal data collector ensure that the conversion thrust is approximate equal to the average thrust during the set time interval [22].

4. Discussion

In this paper, the acoustoelectric method, which has the advantages of simple structure, easy operation and high reliability, is first applied to the field of instantaneous thrust measurement field. Moreover, this method provides a basic idea for measuring the thrust of micro-thrusters of, can realize high precision and accurate measurement of instantaneous thrust and impulse. Micro thrust measuring methods of the non-contact measurement, and non-destructive measurement methods, does not require a sensor or other measuring device is directly installed on the structure, can reduce the interference of mechanical vibration and power supply cables for thrust measurement, solve the present commonly used micro thruster and micro impulse measuring device of zero drift, balance position stable, low accuracy and calibration difficult problem.

References

1. Haag T W 1991 Thrust stand for high - power electric propulsion devices *Review of Scientific Instruments* vol 62 chapter 5 pp 1186-91
2. Cubbin E, Ziemer J, Choueiri E and Jahn R 1997 Pulsed thrust measurements using laser interferometry *Review of Scientific Instruments* vol 68 chapter 6 pp 2339-46
3. Hallock A, Polzin K and Emsellem GD 2012 Thrust Stand Measurements of a Conical Pulsed Inductive Plasma Thruster
4. Zakrzwski C, Benson S, Sanneman P and Hoskins A 2002 On-Orbit Testing of the EO-1 Pulsed Plasma Thruster
5. Santoyo and Fernando Mendoza 1999 Simultaneous three-dimensional dynamic deformation measurements with pulsed digital holography *Appl Opt* vol 38 chapter 34 pp 7056-7062
6. Vladimir K, Aleksander L, Viktor M, Amitriy T and LevV S 2017 An estimation of the strain-stress state under cyclic loading by the acoustoelasticity method *Physics and Mathematics* vol 3 pp 71-6
7. Astm S 2010 Standard Test Method for Verifying the Alignment of X-Ray Diffraction *Instrument for Residual Stress Measurement*
8. Soete W and R Vancrombrugge 1950 An industrial method for the determination of residual stresses *Proc. SESA* vol 8 chapter 1 pp 17-28
9. Eiras J, Vu Q, Lott M, Payá J, Garnier V and Payan C 2016 Dynamic acousto-elastic test using continuous probe wave and transient vibration to investigate material nonlinearity *Ultrasonics* vol 69 p 29
10. Si-Chaib M, Menad S, Djelouah H and Bocquet M 2001 An ultrasound method for the acoustoelastic evaluation of simple bending stresses *Ndt and E International* vol 34 chapter 8 pp 521-9
11. Longhui Yin 2012 Ultrasonic stress measurement based on the LCR wave(in Chinese)
12. Knapp C and Carter G 2003 The generalized correlation method for estimation of time delay *Transactions on Acoustics Speech & Signal Processing* vol 24 chapter 4 pp 320-7
13. Rose J and Nagy P 2004 Ultrasonic Waves in Solid Media *Journal of the Acoustical Society of America* vol 107 chapter 4 pp 1807-8
14. Cuihua Zhao 2013 Study on ultrasonic measurement of resudial stresses(in Chinese)
15. Haiquan Li, Lina Pan and Qingyan Liu 2000 Study and development of uitrasonic stress measurement system *Journal of Test and Measurement Technology* vol 22 chapter 7 pp 297-299(in Chinese)
16. Nikitina N and Ostrovsky L 1998 An ultrasonic method for measuring stresses in engineering materials *Ultrasonics* vol 35 chapter 8 pp 605-10
17. Pao Y 1987 Theory of Acoustoelasticity and Acoustoplasticity
18. Okada K 1981 Acoustoelastic determination of stress in slightly orthotropic materials *Experimental Mechanics* vol 21 chapter 12 pp 461-6
19. Chaki S and G Bourse 2009 Stress level measurement in prestressed steel strands using acoustoelastic effect *Experimental Mechanics* vol 49 chapter 5 p 673
20. Wei Zhao 2002 On Acoustoelasticity of Surface Acoustic Wave for Ultrasonic NDT
21. Dike J, Johnson G and Flower E 1988 Acoustoelasticity Using Longitudinal Waves for Residual Stress Evaluation *NDT and E International* vol 28 p 118
22. Yaojie Huang 2015 Research on measurement method of welding residual stress in steel based on acoustoelasticity principle (in Chinese)

Research on Interactive Routing of Cable in Virtual Environment

WAN Bi-le[1], HE Wen-xing[1,a] and CHEN Chang-yu[1]

[1]Beijing Institute of Spacecraft Environment Engineering, Beijing 100094, China

Abstract. On the basis of analyzing the defects of existing design methods of cable wiring, the feasibility of interactive cabling design in virtual environment is pointed out. Based on the discrete control point, the geometric shape modeling method and the topological structure matrix modeling method based on the wire unit are presented. Adopt the method of real-time interactive cable modeling and wiring, explains the wiring process and control point positioning technology, and through the stratified multistage collision interference detection technology improve the authenticity of the wiring process and results. After the wiring design is completed, the statistical summary of the model information is compiled to generate various inventory reports to guide the actual production.

1 Introduction

Cable has a large proportion in complex mechanical and electrical products. It is widely used in aerospace. The importance of cable is not only reflected in its proportion in products, but also in the impact of product design cost and assembly quality; unreasonable cable wiring will lead to a substantial increase in the cost of product design, and will directly affect the reliability of products. At present, in the process of complex product development, usually after the rigid structure design is completed, physical test equipment is used to determine the cable laying path length and the binding and fixing scheme in the physical prototype, there are three shortcomings in the design method of the cable wiring based on simulation test: ① It is a serial design method, after the establishment of the physical prototype, the wiring design can be carried out, affecting the product design cycle. ②It is a manual design method with low design efficiency and large labor intensity; ③The amount of extraction of physical simulation and preservation of the design information, bring about very low accuracy of results.

The virtual reality technology provides a feasible technical approach for the optimization of wire cable by its good human-computer interaction interface and highly realistic visual environment. Cable wiring design of the virtual environment, refers to in the virtual reality environment with virtual prototype as the space constraints, according to the electrical system connection diagram, in a virtual environment to determine the routing path and topology of cable. The result of the wiring design is the direction of the cable in the three-dimensional space and the exact length of each section of the cable. The form of the design is the three-dimensional design model and the corresponding list of reports in the virtual environment (such as the list of raw materials). In the virtual environment of cable wiring

[a] Corresponding author : hewenxing@163.com

design has three advantages. ① belongs to parallel design method of cable wiring design and product design of parallel structure, shorten the product development cycle; ②belongs to the design method of man-machine cooperation, people-oriented, people is the leading cable wiring design, supplemented by a computer, and provides decision support for the environment the digital designer; ③ the design results, the data is accurate, and easy to save and transfer.

At present, the research of cable modeling and wiring in virtual environment mainly focuses on two aspects,one is the cable modeling based on physical properties under the virtual environment, the goal of this research is to simulate the axial deformation process of circular section cable under the action of external force, such as based on the inverse kinematics model of the cable length, cable mass model based on spring model; The second is the wiring cable which under the virtual environment, including artificial intelligence and human-computer interaction and automatic reasoning method .[1-5]

The above research results have higher theoretical level, but it is difficult to get the actual application in the project, there have three reasons. ① The cable model based on physical attributes has higher requirements on computer hardware. The defects of modeling theory itself often lead to instability of model dynamic behavior. ② The artificial intelligence rationalistic method as a cable wiring the typical spatial planning problem, but because of unknown parameters, the feasible space is complex, it is difficult to determine the exact mathematical model of cable. ③ the current research results, the human-computer interaction method, only for the topology of a simple cable (such as single cable), rarely consider the complex parts (such as cable consists of a plurality of wires the wire harness) modeling method of topology, and in practice there are a lot of topological structure of extremely complex cable.

The research adopts man-machine interactive cable layout design method and focuses on two aspects: ① Cable modeling method of virtual environment, create a unified general cable model, virtual environment by setting different parameters can be in the cable model in virtual environment, and this model can from two aspects of geometric shape and topology truly reflect the complex cable in engineering practice; ② The fast interactive wiring design technology of the cable in virtual environment includes man-machine interaction technology, precise positioning technology and real-time collision interferometer technology in the process of cable fast wiring design.

2 Modeling method of cable in virtual environment

Under the environment of virtual cable wiring design is the first work of cable model is set up, in general, the cable model should include management property, electrical function, geometric shape, topology structure and so on four aspects of the information. The management attribute and electrical function belong to the description type information. This article is not discussed in detail, and the modeling method of the geometry and topology of cable is discussed.

2.1 Variable length modeling method of cable geometry based on discrete control points

The cable is a flexible body typical of the process, the core process of virtual environment is to adjust the design of the overall and local spatial position. So the geometric model of cable should support flexible cable shape characteristics, and has good maneuverability, combined with the characteristics of development tools, the variable length modeling cable geometry method.

Variable length cable geometry modeling method of the so-called, refers to the representation of the flexible cable is composed of a series of discrete section center point (usually circular section) is connected to the spatial continuous line segments. The center points as the control points of spatial position of the cable,and the control points of the operation to achieve the cable the model of spatial position transformation.

The control points can be divided into position control points and interpolation control, connection geometry adjacent control points of the control points are connected, defined as follows: ① Point position control: Control cable space position and attitude. The position control of a fixed and free two states, in the free state, according to actual needs, to dynamically change the position, and in the fixed state, it can't change its position. The interpolation control point: in order to control a bit more smoothly so as to improve the performance of the cable geometry model of cable the virtual degree introduced. The control points of connection: cable connecting adjacent control points (including position control points and interpolation control point) has a certain section characteristics of columnar swept volume.

In the wiring design process, when the control point of a cable model is manipulated, the position and pose of the adjacent control points do not change. It is only the control point that changes the position and length of the connector. Therefore, the length of the cable model is variable in the wiring design process. In engineering practice, the actual length of the cable is difficult to predict before completing the wiring design. Fig. 1 is a schematic diagram of a method for modeling cable length changes in a virtual environment. The left is the initial model of cable, including 5 control points and 4 control point connectors. The right is the result of the operation of the initial model in the wiring design process.

2.2 Matrix modeling method for complex cable topology based on traverse element

In the engineering practice, the cable parts are usual divided into four kinds: single cable, multi-core cable, flat cable and wire harness, and the wire is the basic unit of the cable. According to the different organization and arrangement of the wires in the four kinds of cables, the cable can be divided into two kinds according to the convenience of the data expression. One is a single wire, and the other is a wire harness. Harness includes trunk harness and branch harness. Flat cables and multicore cables are exceptions in the wiring harness, where flat cables are the main wire harness without a wire harness, and the wires in the main harness are flattened. Multicore cable is a wire harness for all branch wires that contain only one wire.

For the convenience of subsequent description, the concepts of wire elements, wires and harnesses are defined as follows: ① Wire element: A section of wire connected by a number of control points and control point connections. It is the basic unit of wire and the basic element of a wire harness; ② Wire: Wire consists of a plurality of unit from end to end; ③ Wiring harness: the main wire harness and the branch wiring harness are bundled together by a number of wire units belonging to different wires. The main wire harness includes a wire unit of all parts of the wire harness and wire harness from the trunk, branch wire harness or separated from other branches of a wire harness. Therefore, it contains only part of the wire harness element.

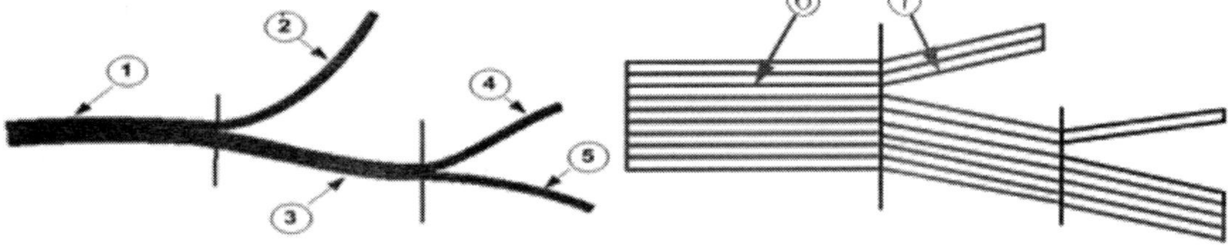

Figure1(a)Typical structure of wire harness (b) Relation between wire harnessand wire element

Figure 1(a) is a concise sketch of a typical structure of a cable component, ① of which is a trunk harness, and ② and ③ are branch wires drawn from ①, and ④ and ⑤ are branch harnesses derived from ③; The cable described in Figure 1 (b) is the same as in Figure 1 (a), and ① is bound by 10 wire elements belonging to 10 wires, ② are bundled by 3 wire units, and ③ are bound by 7 lead units; The ⑥ and ⑦ are 2 wire units belonging to the same wire, and the two wire elements are respectively bundled in the main wire harness ① and the branch wiring harness ②.

For any one wire harness parts are available from three different perspectives to split data on the wire harness, respectively from the perspective of perspective and wire unit perspective, obtained are the parts list of wire harness harness wire, wire list contains the element matrix.

$$
\begin{aligned}
\text{WIRE HARNESS Parts} &= \{\text{Wire harnessj}\}(j=1,2\ldots n) \\
&= \{\text{Wirei}\}(i=1,2\ldots m) \\
&= \{\text{Wire elementij}\}(i=1,2\ldots m,\ j=1,2\ldots n)
\end{aligned} \tag{1}
$$

Where n is the number of wire harness included in the cable parts, j is the serial number of the harness. M the number of wires contained in the cable parts, and I is the serial number of the wire; For the wire element, the I is the serial number of the wire to which the wire unit belongs, and j is the serial number of the wiring harness to which it belongs.

If the wiring J does not contain a wire element that belongs to the wire I, The element wire element ij in the wire element matrix is null, that is, the wire element ij does not exist. The relation between the harness and the element matrix of the wire is shown in equation (2); The relation between the wire and the element matrix of the wire is shown in equation (3)

$$\text{Wire harness } j = \{\text{Wire element } ij \mid \text{If the wire element } ij \text{ is present}\}(1<i<m) \tag{2}$$

$$\text{Wire } i = \{\text{ Wire element } ij \mid \text{If the wire element } ij \text{ is present}\}(1<j<n) \tag{3}$$

2 Fast interactive routing technology for cable in virtual environment

2.1 Real time creation of cable model in virtual environment

Generally speaking, the entity model in virtual environment mainly comes from two aspects. First, through standard neutral files (such as WRL or 3DS), the models built in other systems (such as 3DMAX) are imported into the virtual environment; Two is modeling through modeling engine in virtual environment. Cable is a flexible variable body, obviously second methods should be adopted, that is, modeling engine through modeling engine in virtual environment. In order to realize the wiring design in virtual environment, the cable modeling can be divided into two ways, one is the pre modeling, the original model firstly according to the input parameters of the designer create the cable, and then based on the original model of the operation to achieve the cable wiring design. The other is the real-time creation model, that is, the creation of the cable model and the wiring designed process synchronization, in the modeling at the same time, complete the cabling design of the cable. This paper focuses on the real-time modeling process of cable in virtual environment

Real time modeling of virtual environment and refers to the specific process under the control by the way of human-computer interaction in the constrained space model by fitting a rigid structure, real-time dynamic process of determining the topology and geometry of the cable. A specific process control refers to the order in which the process of modeling is to be followed. such as: Build backbone cable-> Create a path point on the backbone cable-> Add branch cable Creates a path point on the branch cable. The so-called real-time dynamic determination of the topology and geometry of the cable, means that each control point in the cable is defined in real time. As a result, the geometry of the cable, which is connected by the control point, varies dynamically as the control point changes (including changes in number and position). In .the process of modeling, the topological structure of the cable is also constantly changing (such as the increase and decrease of branch cables). Therefore, there are two key technologies for real-time modeling of cable in virtual environment. One is the control of modeling process, and two is the control point location.

In a virtual environment, polygon (usually triangular) patches are the basic unit of geometric shape of a model, and are often the smallest geometric elements that can be selected by human-computer interaction devices. The cables are usual laid on the surface of rigid structural parts, So the process of determining the initial position of the control points is the process of defining the position relationship between the control points and the triangular patches of the surface of the structural model. In general, the designer selects the patches on the structure model through human-computer interaction devices (such as mouse). The location of control points is defined by defining the position relation (fit and offset a certain distance) between control points and patches.

In a virtual environment, we can select a single slice, but it is difficult to select one point of the patch exactly, and we can only get the coordinates of the vertices of the patch. When the face film is small, the coordinate of any vertex can be taken as the approximate coordinate of the selected point for subsequent calculation. When the face is larger, the approximate method will lead to greater errors. In order to reduce the positioning error and ensure that the determined points are on the plane, approximate calculations of the selected point coordinates are carried out using a method as shown in Figure 2. ABC is a triangular patch on the surface of a structure in a virtual environment. D is the midpoint of the line segment , and E is the midpoint of the line segment . Formulas (4) and formulae (5) are formulas for calculating the coordinates of D points and E points respectively. Formulas (6), (7) and (8) are expressions of E, X, Y and Z coordinates respectively.

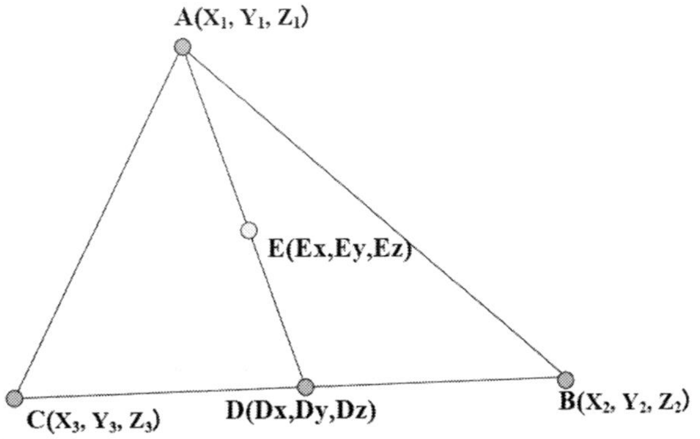

Figure 2. positioning method of cable control points

$$(Dx, Dy, Dz) = ((X_2 + X_3)/2, (Y_2 + Y_3)/2, (Z2 + Z3)/2) \tag{4}$$

$$(Ex, Ey, Ez) = ((Dx + X_1)/2, (Dy + Y_1)/2, (Dz + Z_1)/2) \tag{5}$$

$$Ex = X_1/2 + (X_2 + X_3)/4 \tag{6}$$

$$Ey = Y_1/2 + (Y_2 + Y_3)/4 \tag{7}$$

$$Ez = Z_1/2 + (Z_2 + Z_3)/4 \tag{8}$$

When defining the relation between the control point and the selected patch surface, the formula for calculating the position of the control point is formula (9). Where is the normal unit vector of the patch, and R is the radius of the control point, and represents the coordinates of points calculated according to formula (5); When the position between the control point and the selected patch is defined as offset distance D, the formula for calculating the position of the control point P is given as formula (10). Where is the normal unit vector of the patch, and R is the radius of the control point, and represents the coordinates of the points calculated according to the formula (5);

$$P(x, y, z) = \vec{n} * R + E(x, y, z) \tag{9}$$

$$P(x, y, z) = \vec{n} * (R + D) + E(x, y, z) \tag{10}$$

2.2 Multi slice real-time collision interference detection technology for cable in virtual environment

The cable is a flexible body, the wiring design process, model structure changing, and the system also need to deal with the events of the virtual environment to redraw the control point position calculation and refresh model, interpolation smoothing, and cable laying to tend to be formed by a rigid structure of the complex and narrow in the space. Therefore, in the virtual environment for cable wiring design, the collision interference detection algorithm speed and accuracy of higher requirements.

According to the above characteristics, a real-time collision detection algorithm for multi layer collision detection in virtual environment is proposed. Segmentation means that the cable model is divided into model segments consisting of several control points and control point connectors. When the interference is detected, the computation is reduced by segmentation of the model. In the case of single control point routing, we can first determine whether the model segment of the control point is in collision with the environment object. Returning if collision occurs, if no collision occurs, then the interference of other cable segments is checked The so-called "multi-level" means that in the process of conducting interference checks, the object and scope of interference checking should be refined and reduced from coarse to fine, so as to reduce the amount of calculation and improve the detection efficiency.

① Model pretreatment: Firstly, the cable model is divided into several model segments. The so-called models, refers to the number of control points and control points of a cable connector, and a plurality of geometry control points and control points of the model section contains the connecting body corresponding to a geometry.

② Model global bounding box detection, In the process of the model being manipulated, the interference between the model bounding box and the surrounding box of other objects in the environment is detected. If the interference does not occur, the model is adjusted according to the control intention of the designer. Enter step 3 if intervention occurs.

③ Model global bounding box detection: Turn on the cable model contains all models for interference checking bounding box level, if all the models are not interfere with objects in the environment, in accordance with the designer's intentions to model pose adjustment. If one of the model segments I and the other objects in the environment interfere with the interference of the box layer, we interrupt the step 3. In addition, the marking of the parts that interfere with the bounding box of the model segment is recorded and entered step 4 at the same time.

④ Rough detection of model segment surface level: Check all the patches contained in the model segment i, and the interference between the bounding box J of the part and the bounding box of the model segment, which is obtained in step 3. The model segment patches that do not interfere with the object J are filtered from the array of patches. Checking the interference between the patches in the object J and the bounding box of the i segment of the model segment, the patches in the object J are not filtered with the interference of the bounding box I of the model segment. If the patch array of the model segment I is empty with the part J patch array, the pose adjustment of the model is carried out according to the designer's manipulation intention. If not empty, go to step 5

⑤ Model segment patch level accurate detection: All the facets in the patch array of the model segment I obtained in step 4 are intersected with all the facets in the component j patch array. If all the patches do not intersect, then adjust the pose of the model according to the designer's control intention. If there is a pair of intersecting patches, the designer's control intent can not be executed and the model returns to the previous frame.

The example shows that the algorithm of segmented multi layer interference detection has the characteristics of fast speed and high precision. It can meet the requirements of cable layout design in the virtual environment, and ensure the authenticity of the wiring design process.

3 PROTOTYPE SYSTEM

A virtual assembly system is developed by using WorldToolKit9.0 toolkit and VC++ 6.0. Cable wiring as a module of the system, the main functions have been achieved at present. Such as: ① Cable modeling; ②Interactive routing design; ③ Interference check;④ information management.

4 Conclusion

Cable wiring is an important part in the development of aerospace products, and has a crucial impact on the reliability and development cycle of the entire product. Virtual reality technology provides a new technical route for cable routing in 3D visualization environment. The main results of this paper are as follows: ① A variable length cable modeling technique based on discrete control points is proposed, and the simplified expression of cable geometric appearance in virtual environment is realized. ② A matrix modeling method for complex cable topology based on wire element is proposed. The unified expression of topological structure of complex cable parts is realized and it lays the foundation for the automatic statistics of the cable model information (such as wire length and wire root number). ③ In this paper, we propose a multi segment interference detection technique for wired routing in virtual environments, which improves the routing process and the authenticity of routing results.

Acknowledgements

This research is supported by Natural Science Foundation of China (No. 51405025), and Beijing Spacecraft Smart Assembly and Equipment Engineering Research Center.

Referrence

1. E.Hergenröther,P. Dähne.Real-time virtual cables based on kinematic simulation[C]. In Proceedings of the WSCG 2000, 2000
2. Yili FU, Haibo FENG, Jianxun SUN, Yulin MA. Research on automatic laying of mechanical and electrical products pipeline based on Chaos Algorithm[J]. Computer integrated manufacturing system, 2007,13(3):497-501
3. G.Robinson,J.M.Ritchie,P.N.Day,etal.System design and user evaluation of Co-Star :An immersive stereoscopic system for cable harness design[J].Comoputer-Aided Design,2007,39(4):245-257
4. Fayuan WEI, Xinfa CHEN, Fengjun WANG. Virtual wiring of cable and its inverse kinematics simulation[J]. Journal of computer aided design and computer graphics, 2006,18(10):1623-1627
5. Wan Bile, Runxin NING, Jianhua LIU,Yi ZHENG. Research and implementation of cable modeling and routing in virtual environment [J].China Mechanical Engineering, 2006, 17(20), 2135-2138

Study on the Mechanism of a Carrier-based Engine Parts' Performance Decline and Its Impact on the Whole Engine Performance

Wu Heng[1,a], Li Benwei[1], Zhao Shufan[1] and Wang Yonghua[1]

[1]Aviation Foundation College, Naval Aviation University, Yantai Shandong 264001,China

Abstract. The technical approach "use – parts' change - engine performance change" has been adopted to study and analyse the gas unit parts' performance changes of an engine after a long time operation. The mechanism of performance decline of the turbine is analysed based on the numerical simulation, the impact of components' performance decline on overall engine performance is studied and the correlation analysis is carried out. The results show that the change of turbine tip clearance, roughness increase and surface change will lead to the enhancement of secondary flow and the increase of influence area, and the turbulence effect is strengthened, resulting in the decrease of turbine circulation capacity and efficiency. The booster ratio of high pressure compressor, the flow capacity of high pressure and low pressure turbines, the flow capacity and efficiency of fan are the major component parameters causing the overall engine performance's degradation. And it also provides theoretical basis for the prevention of engine performance's degradation and online washing of parts and the whole machine.

1 Introduction

The engine of a navy carrier aircraft should meet the requirements of the escape away for slipping off. Due to the particularity of marine shipborne environment, failure and performance degradation of the engine's components are more serious. The decrease of thrust and other indicators will directly affect the flight safety of the carrier aircraft. Therefore, it is urgent to study the performance degradation mechanism and reasons of the engine and using the corresponding maintenance and repair technology and making a reasonable arrangements for maintenance time to avoid or reduce the performance recession of engine.

A large number of studies on the performance degradation of aircraft engines at home and abroad can be divided into two ways: the first way is studying on the declining laws of engine performance based on the change of whole machine Performance and the second way is studying on the decline mechanism from parts to machine. The abroad research works in this field mainly include the influence of blade roughness and engine manufacturing tolerance on the performance of the engine's turbo and compressor[1-3], the impact of components' performance degradation on whole engine performance[4-5] as well as the issues of engine performance degradation and control strategy

[a] Corresponding author: wuheng89810@163.com

optimization[6-7]. In recent years, component performance change process has been analysed by means of experimental and numerical analysis at home[8-12].

In this paper, the technical approach "use – parts' change - engine performance change" has been adopted to study and analyse the gas unit parts' performance changes of an engine after running 100h. Based on the numerical simulation, the performance degradation mechanism of the fan and turbine are analysed, the influence of component performance degradation on the whole machine performance is studied and the correlation analysis is carried out, and the characteristic parameters of the gas unit parts are extracted.

2 Changes in engine gas unit components after prolonged use

Through the investigation and research, various data of a carrier-based aircraft engine after running 100h is obtained. After a long period of operation, the engine has occurred parts wear, cracks, flow channel gap increases, pollution, resulting in a certain degree of engine performance attenuation, this article analyses the possible performance degradation induced factors of the engine from the inspection of the engine gas unit components.

Through the observation, it is found that after a long period operation, there is no significant change in the appearance of the air intake casing, but the oxide layer of adjustable leaves' micro-arc is peeled off, and also there are local drum phenomenon, as is shown in Figure 1, the peeling off of the adjustable blade protective layer and local drum phenomenon directly affect the smoothness of the airflow channel, while the leaves' material is aluminium alloy so that salt spray corrosion is relatively poor, in ship-borne conditions the adjustable blade protection layer peeling off will cause further corrosion, change the fan flow conditions, affecting the overall performance of the fan.

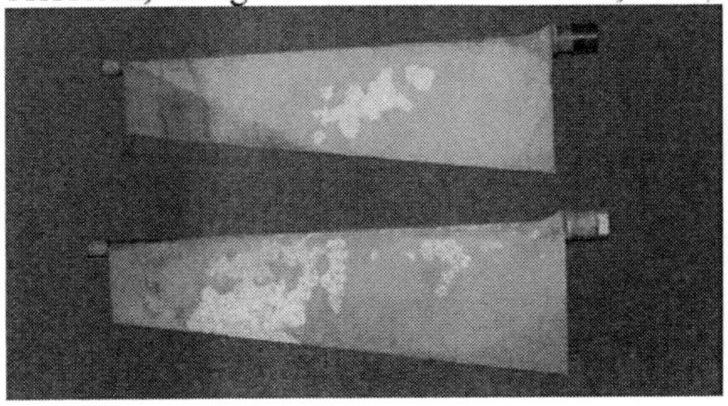

Figure 1. Adjustable leaf coating to peel off

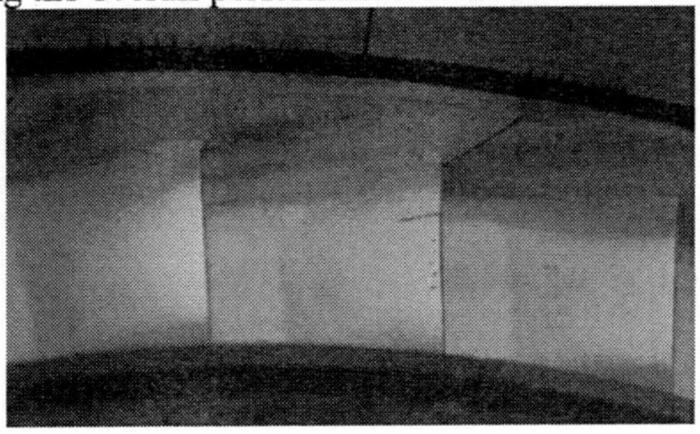

Figure 2. High pressure turbine guided blade fouling

Table 1. Main size parameters of high pressure turbine

Component name and drawing number	Test items	Drawing requirements	Actual inspection	Conclusion
On the component	Measure the tip diameter of the working blades	Φ727.2 – 726.9	Φ726.63	unqualified
	The first level	Φ556.1 – 555.6	Φ556.02	qualified
Outer diameter dimension of turbine front seal plate	The second level	Φ561.4 – 560.8	Φ561.30	qualified
	The third level	Φ566.6 – 565.9	Φ566.58	qualified

	The fourth level	Φ571.8 – 571.0	Φ571.86	unqualified

The surface properties of the high-pressure compressor blades and the performance of the main combustion chamber have no obvious change. The high-pressure turbine guide blades at the rear of the main combustion chamber have slight fouling and the roughness increases, as is shown in Figure 2. The trailing edge of the guide blade has a crack, the longest is up to 8mm (Specified ≯ 6mm). The high pressure turbine rear baffle is subjected to fluorescence permeation detection, and no crack defects are found, it is found that there is a lot of oil on the high pressure turbine disk by decomposition. The correlation measurement results in the process of high pressure turbine decomposition are shown in Table 1.

The oil on the high-pressure turbine wheel increases the possibility of further fouling, profile change, and roughness increase in the follow-up process. The tip diameter of the component is too small, it indicates that after a long operation the high pressure turbine blade tip gap is too large, which will increase the leakage, reduce the efficiency of the turbine. The low pressure turbine rotor uses a crowned blade, which can effectively reduce leakage of the working wheel, low-pressure turbine performance will not change significantly. Therefore, this paper mainly analyses the performance degradation mechanism of fan and high pressure turbine components.

3 Performance degradation mechanism of turbine components based on numerical simulation

The performance test of engine components is often costly, especially the blade angle adjustment is not in place, the roughness increases, the leaf type and other components performance analysis in non-normal conditions are very difficult, therefore in this paper how the change of parts' surface and surface conditions caused by the environment and the use of time will affect the component performance is analysed through CFD technology. Due to space limitations, a brief description of the numerical simulation method is given in this paper.

3.1 Numerical simulation method

3.1.1 Reconstruction and correction of geometric model

For the fan, compressor, turbine and other components' leaf type data is the engine's core secrets, through the investigation and research it can only get the basic geometric parameters rather than detailed aerodynamic design parameters of the first level of the fan and high pressure turbine. In this paper, GAMBIT is used to reconstruct the local linearity of each blade, and the geometrical information of GAMBIT is transformed into generalized geometric file IGES to the subsequent BLADEGEN processing software. Reverse design of the turbine blade is used to reconstruct the geometry model of the fan and turbine, the blade design pattern is transformed from "angle-thickness" to "suction surface-pressure surface" mode, and then the leaf type correction is realized by adjusting the pressure surface profile. The reconstructed high pressure turbine blade model is shown in Figure 3.

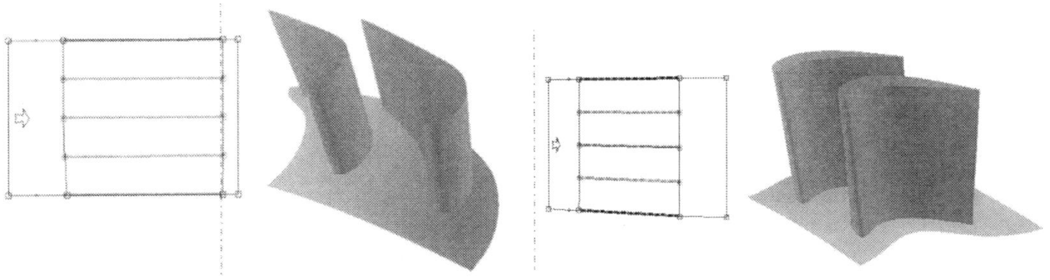

Figure 3. Reconstructed high pressure turbine blades

3.1.2 Calculate regions, grids, and theoretical models

Turbine calculation area is "0-level adjustable deflector blade +1-stage rotor + 1 level stator", in order to ensure the calculation accuracy, select the same fan scale for each level of calculation area, multi-channel calculation is used. Each calculation channel is "OH-shaped" mesh, and the non-matching grid interface is used for the moving leaves. Using the time-averaged turbulence model to calculate the specific heat flow field of variable specific heat:

$$\frac{\partial \rho}{\partial t}+\frac{\partial}{\partial x_i}(\rho u_i)=0 \tag{1}$$

$$\frac{\partial}{\partial t}(\rho u_i)+\frac{\partial}{\partial x_j}(\rho u_i u_j)=-\frac{\partial p}{\partial x_i}+\frac{\partial}{\partial x_j}\left[\mu_{eff}(\frac{\partial u_j}{\partial x_i}+\frac{\partial u_i}{\partial x_j})\right]-\frac{2}{3}\frac{\partial}{\partial x_j}\left[\mu_{eff}(\frac{\partial u_i}{\partial x_i})\right] \tag{2}$$

$$\frac{\partial}{\partial t}(\rho E)+\frac{\partial}{\partial x_i}(\rho u_i E)=-\frac{\partial p u_i}{\partial x_i}+\frac{\partial}{\partial x_j}\left[\mu_{eff}u_i(\frac{\partial u_j}{\partial x_i}+\frac{\partial u_i}{\partial x_j})\right]-\frac{2}{3}\frac{\partial}{\partial x_j}\left[\mu_{eff}u_i(\frac{\partial u_i}{\partial x_i})\right]+\frac{\partial}{\partial x_j}\left[k_{eff}(\frac{\partial T}{\partial x_j})\right] \tag{3}$$

$$p=\rho RT \tag{4}$$

The turbulence effect of the flow is calculated by using the $k-\omega$ turbulence model:

$$\frac{\partial}{\partial t}(\rho k)+\frac{\partial}{\partial x_i}(\rho u_i k)=\frac{\partial}{\partial x_j}(\Gamma_k\frac{\partial k}{\partial x_j})+G_k-Y_k+S_k \tag{5}$$

$$\frac{\partial}{\partial t}(\rho \omega)+\frac{\partial}{\partial x_i}(\rho u_i \omega)=\frac{\partial}{\partial x_j}\left[\Gamma_\varepsilon(\frac{\partial \omega}{\partial x_j})\right]+G_\omega-Y_\omega+S_\omega \tag{6}$$

Turbulent viscosity: $\mu_t=\alpha^*\frac{\rho k}{\omega}$ among them: $\mu_{eff}=\mu+\mu_t$.

The turbine is a typical pressure gradient flow field, so the improved wall function is used, and the wall grid is locally encrypted. The effect of fouling and corrosion on the blade and airflow channel is replaced by roughness equivalent, other import and export conditions are given in accordance with the design documents. The second-order finite volume method is used to solve the control equations, and the computational complexity and the calculation precision are taken into account. The coupling effect of the static and dynamic regions is calculated by the mixed plane method.

3.2 Simulation results analysis

The calculation results of the turbine flow field in design point are discussed. There are obvious secondary tip flow in the two conditions of the gap clearance and the excess, for the smaller gap of the leaves, only more than 90% of the leaf height has been influenced by the secondary flow, the area of the secondary flow increases when the gap is large. The secondary flow of the large tip gap creates a large low velocity zone at the tail of the turbine blade, affecting gas deflection and expansion, the secondary flow schematic diagram of a large turbine blade tip clearance is shown in Figure 4. In addition, the Mach number distribution indicates that the turbine is a guide supercritical high-compound turbine, both guide blade fouling and roughness increase will reduce its flow capacity, Mach number distribution of a large turbine blade tip clearance (90% leaf height) is shown in Figure 5.

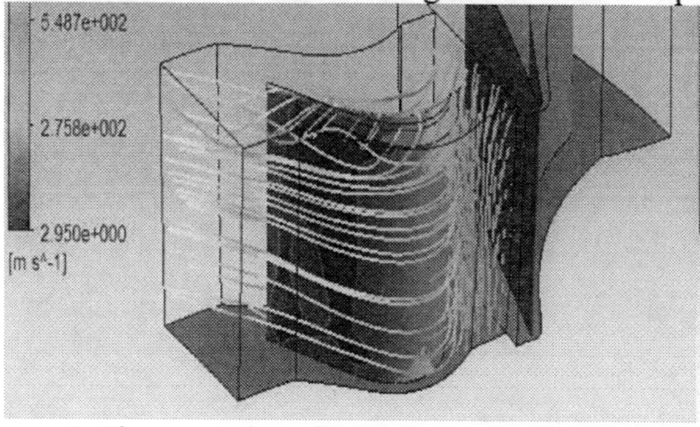

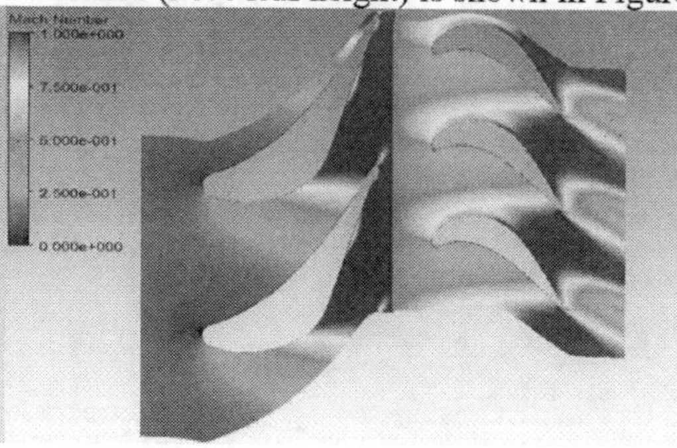

Figure 4. The secondary flow schematic diagram of a large turbine blade tip clearance

Figure 5. Mach number distribution of a large turbine blade tip clearance (90% leaf height)

Therefore, the change of turbine tip clearance, roughness increase and surface change will lead to the enhancement of secondary flow and the increase of influence area, and the turbulence effect is strengthened, resulting in the decrease of turbine circulation capacity and efficiency. When the fouling caused the change of leaf shape, turbine performance degradation is more serious.

4 Analysis on the effect of component performance decline on overall machine performance

4.1 Analysis on the law of change of whole machine performance

After a long period of operation, the engine performance changes mainly reflects in thrust drop and fuel consumption increases (where the fuel consumption is converted to the same thrust fuel consumption), the acceleration performance of the engine decreases slightly, but not obvious. Figure 6 ~ Figure 8 show the comparison of a variety of engine test parameters before and after a long time running.

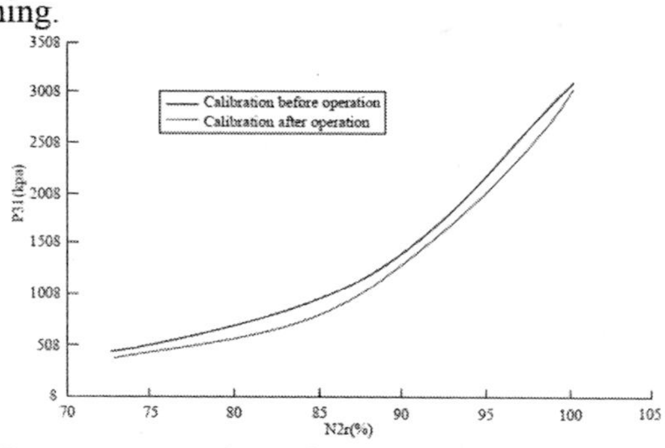

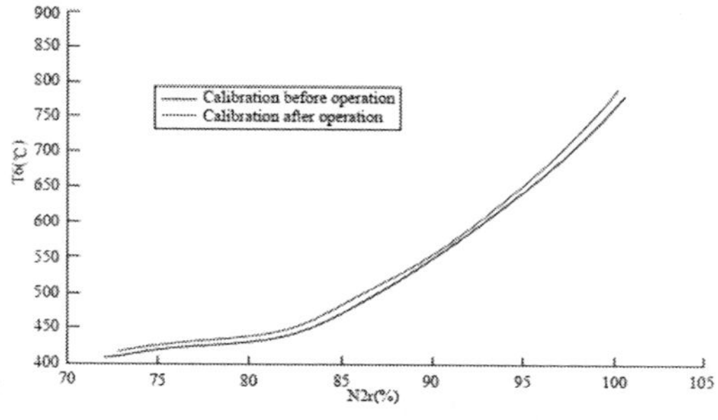

Figure 6. Comparison of engine total boost ratio **Figure 7.** Comparison of engine turbine rear temperature

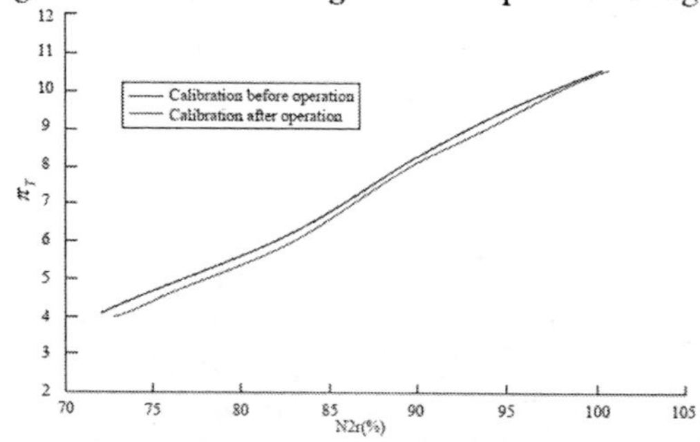

Figure 8. Comparison of turbo expansion ratio

As is shown in Figure 6 to Figure 8, after a long period of operation, the engine fan, high pressure compressor inlet guide vane, adjustable stator blade error, and fan blade surface roughness increases, may cause the degradation of efficiency and boost performance of the engine fan and compressor, also the degradation of the engine's total boost performance. The increase of turbine blade roughness and tip clearance may cause the decrease of turbine efficiency and the turbo expansion ratio, and increase of the exhaust temperature.

4.2 Correlation analysis of component characteristic parameters and engine performance

The calculation of engine performance parameters will produce some errors by using the aerodynamic thermodynamic performance model of a kind of carrier-based engine. However, the sensitivity and trend analysis are mainly carried out in this paper, so the model meets the requirements.

When the engine is running for a long time, the characteristic parameters of the rotating parts are susceptible to changes and affect the target performance parameters of the engine. Therefore, the high pressure compressor flow W_3, high pressure compressor efficiency η_{cH}, high pressure compressor boost ratio π_{cH}, fan flow W_{a22}, fan efficiency η_{cL}, fan pressure ratio π_{cL}, high pressure turbine flow W_{st45}, high pressure turbine efficiency η_{TH}, low pressure turbine flow W_{st5}, low pressure turbine efficiency η_{TL} are selected as the influencing factors for sensitivity analysis. Taking the actual situation and the difficulty of parameter measurement into account, the high pressure compressor outlet pressure P_3, low pressure turbine outlet temperature T_5, tail nozzle outlet temperature T_8, engine thrust F, and fuel consumption sfc are selected as the affected target performance parameters. Changes in the performance parameters of the engine components can cause changes in the performance of the components and thus affect the overall performance of the engine. However, different component characteristic parameters have different effects on different overall performance parameters. The correlation analysis method can be used to analyse the influence between parameters. Define the correlation coefficients as:

$$r_{ij} = |\Delta j / \Delta i| \tag{7}$$

Among them, Δj represents the amount of change of the jth target performance parameter and Δi represents the amount of change of the ith component characteristic parameter.

Firstly calculate the correlation coefficient between the engine component characteristic parameter and the target performance parameter. In this paper, a single parameter perturbation method is used. Reduce the component characteristic parameters by 2% successively, while keeping the remaining part characteristics unchanged, calculate the overall performance changes of the engine. For the correlation analysis, the correlation coefficient is normalized and the results are shown in Table 2.

$$\bar{r}_{ij} = r_{ij} / \sqrt{\sum_{k=1}^{10} r_{kj}} \quad (i = 1,2,\cdots,10, j = 1,2,\cdots,5) \tag{8}$$

Table 2. The correlation coefficient after normalization

%	P_3	T_5	T_8	F	sfc
W_{a22}	0.1281	0.0175	0.2017	0.1026	0.0821
η_{cL}	0.1052	0.1926	0.2456	0.0714	0.0372
π_{cL}	0.2205	0.0016	0.0016	0.0050	0.0805
W_3	0.0166	0.0394	0.0024	0.0585	0.0124

η_{cH}	0.0429	0.0009	0.0014	0.0010	0.0009
π_{cH}	0.1891	0.1295	0.0243	0.2160	0.3491
W_{st45}	0.0833	0.0723	0.1539	0.2301	0.2111
η_{TH}	0.1117	0.0181	0.0821	0.1962	0.0429
W_{st5}	0.0376	0.2527	0.1623	0.0931	0.1831
η_{TL}	0.0650	0.2753	0.1248	0.0260	0.0006

Define the sensitivity of the parameters:

$$k_i = \sum_{j=1}^{5} \overline{r_{ij}} ,(i = 1,2,\cdots,10) \tag{9}$$

The sensitivity coefficient of each component parameter is shown in Table 3.

Table 3. Sensitivity coefficient

	W_{a22}	η_{cL}	π_{cL}	W_3	η_{cH}	π_{cH}	W_{st45}	η_{TH}	W_{st5}	η_{TL}
k	0.5320	0.6520	0.3092	0.1293	0.0471	0.9080	0.7507	0.4510	0.7288	0.4917

The first five component performance parameters which affect the engine performance are: high pressure compressor boost ratio π_{cH}, low pressure turbine flow W_{st5}, high pressure turbine flow W_{st45}, fan efficiency η_{cL} and fan flow W_{a22}. Therefore, the prevention of engine performance degradation should start with the prevention of degradation on the engine's air compression system and turbine components performance.

5 Equations and mathematics Conclusion

Through the analysis of a carrier-based engine's performance change after prolonged use, the mechanism of performance degradation of fan and turbine are studied, and then the impact of component performance degradation on overall machine performance is analysed. The following conclusions can be drawn:

(1)the change of turbine tip clearance, roughness increase and surface change will lead to the enhancement of secondary flow and the increase of influence area, and the turbulence effect is strengthened, resulting in the decrease of turbine circulation capacity and efficiency. When the fouling caused the change of leaf shape, turbine performance degradation is more serious.

(2)the engine performance degradation mainly reflects in the decrease of thrust, increase of fuel consumption and exhaust temperature, the acceleration performance of the engine decreases slightly. High pressure compressor boost ratio, low pressure turbine flow, high pressure turbine flow, fan efficiency and fan flow are the main component performance parameters which lead to the engine performance degradation. Therefore, the prevention of engine performance degradation should start with the prevention of degradation on the engine's air compression system and turbine components performance. It also provides a theoretical basis for the engine parts and overall machine online washing.

Acknowledgments

Thanks for the financial support from special funds of Taishan Scholar Project.

References

1. P Luis, S Thorsten, The influence of blade roughness on the performance of a vertical axis tidal turbine,IJOME, 17: 136-146(2017).
2. S. S.Talebi, A. M.Tousi,The effects of compressor blade roughness on the steady state performance of micro-turbines,ATE, 115: 517-527(2017).
3. B Marco, B Fabio, S Daniele, Combined experimental and numerical investigations on the roughness effects on the aerodynamic performances of LPT blades,JOTS, 25(1): 32-42(2016).
4. A Castillaa, J Fernándezb, J.M. Montaneroc, Influence of the manufacturing process tolerance on the swirl number of a low-capacity engine,JOMS, 41: 157-164(2016).
5. T Mohammadreza, M Masdi, Z. A. A. Karim,Performance evaluation of a twin-shaft gas turbine engine in mechanical drive service,JOMSAT, 31(2): 937-948(2017).
6. K Vikas, Remaining life related issues being pursued at DRDO for indian air force,PE, 55: 428-432(2013).
7. A. T. James, S. L. Jonathan, A foreign object damage event detector data fusion system for turbofan engines,JOAIS, 2(7): 291-308(2005).
8. D Li, W Zhang, M. Z. Ji, Engine gas path performance model modification based ongenetic algorithm,GTXAR, 25(4):46-50(2012).
9. W Tan, X. C. Yu, D Li, Research of engine performance deterioration index forecasting based on CFD,GTT,30(1):48-52(2017).
10. D Li, W Zhang, M. Z. Ji, Numerical study of performance deterioration of engineguide vane and turbine blade,GTT, 28(1):35-41(2015).
11. S Wang, G. H. Wang, Q Han, Compressor performance deterioration caused by blade fouling,JOHEU,35(12):1524-1528(2014).
12. B. W. LI, D Li, W Shen, Research on turbine lamina roughness influence on its performance declination, ACT, 39(5): 26-29(2009).

Aeroelastic Stability of Labyrinth Seal with Different Structure Parameters

Wang Nengmao[a] and Wang Yanrong

Collaborative Innovation Center for Advanced Aero-Engine, School of Energy and power, Beihang University, 37 Xueyuan Road, Haidian District,Beijing 100191, China

Abstract. Numerical analysis of turbomachinery based on energy method is used to predict the aeroelastic stability of the straight-through labyrinth seal by solving aerodynamic work and damping. The aeroelastic stability of the labyrinth seal under different working conditions and vibration modes has been compared. It's found that the increase of pressure ratio leads to the greater possibility of aeroelastic instability. The periodic distribution of the aerodynamic work in the circumferential direction of the labyrinth seal corresponds to the number of vibrating nodal diameters. In order to investigate the influence of structure parameters, the effect of relative thickness of the tooth tip, the width of the seal cavity and the eccentricity of the rotor on the aeroelastic stability of the labyrinth seal has been studied. The result of numerical calculation shows that the change of the structural parameters can affect the aeroelastic stability of the labyrinth seal to a certain extent, and can be applied in the structural optimization.

1 Introduction

Labyrinth seals are widely used to limit the fluid leakage in aeroengines. With the improvement of performance requirements, the flutter problem of labyrinth seal has become increasingly serious (Ehrich, F., 1968[1]). Alford [2] studied the phenomenon of self-excited vibration of labyrinth seal, and believed that friction between the seal components could introduce energy into the vibrating system, increasing amplitude and causing fatigue failure. The changes of gas pressure in seal cavity can also affect the vibration of the system. It was found that the support position at upstream or downstream of labyrinth seal can be selected to prevent self-excited vibration to a certain extent.

The problem of aeroelastic instability was investigated during the development of F100 engines (Lewis, Platt and Smith, [3]). Through extensive engine tests the instability was finally attributed to the tight clearance of the first (upstream) knife edge and could be eliminated by adjusting it. Abbott [4] showed that when the mechanical frequency is close to the acoustic frequency, the labyrinth seal will have the greatest aerodynamic instability. The aerodynamic damping is negative if the structural frequency is less than the acoustic frequency when the labyrinth seal is supported on the low pressure side. The aerodynamic damping is positive if the structural frequency is less than the acoustic frequency when the labyrinth seal is supported at the high pressure side.

Srinivasan [5] determined the relationship between the aerodynamic forces and the modes, frequencies and geometrical parameters by analyzing the vibration modes of the labyrinth seal. The

[a] Corresponding author : wnm189981@163.com

relationship between the aerodynamic forces and the geometrical parameters was determined by solving the characteristic equation. And the stability of the seal components can be determined by this method. It is believed that increasing the stiffness of the rotating and stator components and decreasing the circumferential velocity of the air flow can increase the stability of the labyrinth seal.

In view of the complexity of flutter prediction, the early flutter studies use an empirical method [6] and experimental method [1]. Amirante, Hills and Barnes [7] used the torsion spring model to perform the finite element analysis of the interstage cavity and the aerodynamic numerical calculation. The finite element analysis software can simulate many characters of the labyrinth seal and save a great deal of computation time and test cost.

In recent years, the numerical method and the basic theory of aeroelasticity of labyrinth seal are mainly derived from the method of predicting flutter of turbomachinery. The energy method proposed by Bendiksen [8] and Carta [9] to calculate the unsteady aerodynamic forces on the oscillating cascade and the work done on the blades in one oscillation period. The aerodynamic work is taken as the criterion for the system to produce flutter.

The energy method developed from flutter prediction of turbomachinery is applied by Sayma [10] to the flutter analysis of the air-riding seal. It is believed that the large axial and radial deformation, vibration of the seal, thermal expansion of the gas, unsteady turbulence, high-speed flow of the seal clearance and other factors led to the interaction of the seal aeroelastic and thermoelastic phenomenon. Phibel [11] showed that the ratio of the mechanical vibration frequency to the acoustic frequency and the support position of the seal have a great influence on the flutter stability. The distribution of aerodynamic work was relevant to the modal shape and the distribution of unsteady pressure.

Phibel [12] also compared the effects of the CFD program and the three-control model and the two-control model on the numerical results of the aeroelastic stability of the labyrinth seal. Mare [13] used the energy method to calculate the aeroelastic stability of labyrinth seal, and studied the difference of the aeroelastic stability with differentnumber of teeth.

In this paper, based on the mature computational fluid dynamics (CFD) and computational structural dynamics, a numerical method (Zhang [14]) of fluid-structure coupled vibration developed to predict the aeroelastic stability of turbomachinery blades has been applied to the labyrinth seal. Influence and sensitivity of seal parameters on aeroelastic stability of labyrinth seal are investigated.

2 Numerical method

2.1 Fluid model

The most commonly used equations for the computational fluid dynamics control equations involved in aeroelastic stability analysis are Navier-Stokes equations, which are developed from the three conservation laws of mass, momentum and energy. In a real flow problem, the flow state is mostly turbulence, and the flow is very irregular in time and space.

Because the gas density is far less than the density of the solid structure, the additional stiffness caused by the gas is far less than the stiffness of the structure, so the additional stiffness of the gas is neglected (Srinivasan [15]) as done in the vibration characteristic for the labyrinth seal. The mechanical damping is related to structural materials, structural forms and many other factors. It is generally believed that mechanical damping of turbomachinery is much less than that of aerodynamic damping [16]. Therefore mechanical damping is neglected and only the aerodynamic damping is taken into account. The mode shape and natural frequency of labyrinth seal are largely not influenced by unsteady aerodynamic loads.

In general, the structural grids are much sparser than the fluid-domain grids, and the two sets of meshes cannot be completely overlapped at their interface. Therefore, the vibration displacements of the structure element node need to be interpolated to the fluid domain grid point on the fluid-solid coupling interface after modal analysis.

2.2 Overview of Energy method

In general, the mechanical damping of turbomachinery is far less than aerodynamic damping. Therefore, the influence of mechanical damping on aeroelastic stability is often neglected (Kielb and Imregun [16]) in the prediction of the aeroelastic stability of the labyrinth seal. The aerodynamic damping is used for judging whether flutter occurs in the labyrinth seal.

The excitation force (pressure) applied by the labyrinth seal to the flow field of the seal cavity is given by

$$p = p(x, y, z, t) = p(x, y, z, t+T) \tag{1}$$

The work of excitation force on the flow field in a vibration period

$$W = \int_0^T \int_S p\vec{v} \cdot \vec{n} dS dt \tag{2}$$

The natural frequency and mode shape of labyrinth seal in the flow field can be solved according to the modal analysis. Then unsteady flow field is analyzed by interpolating the vibrating displacement to the fluid grids as mesh deformation. Mesh deformation is an important component for solving problems with moving boundaries or moving subdomains. The motion of all remaining nodes is determined by the mesh motion model, which is currently limited to displacement diffusion. The displacements applied on domain boundaries or in subdomains are diffused to other mesh points. A transient periodic mesh motion is set that repeats itself at a given frequency.

The aerodynamic force and displacement of the coupled surface are used to calculate the aerodynamic work done by the airflow on the seal in a vibration period after the unsteady analysis, and then the aerodynamic damping is obtained.

The energy dissipated by the aerodynamic damping after the approximate treatment in a certain mode can be expressed as

$$W_{\text{aero}} = \oint c_{\text{aero}} \dot{q} dq = 2\zeta_{\text{aero}} \omega \int_0^T \dot{q}^2 dt = 2\pi \zeta_{\text{aero}} \omega^2 \left(q^{\text{cfd}}\right)^2 \tag{3}$$

By the definition of equivalent viscous damping, there is

$$-W = W_{\text{aero}} \tag{4}$$

$$-W = 2\pi \zeta_{\text{aero}} \omega^2 \left(q^{\text{cfd}}\right)^2 \tag{5}$$

Therefore, the modal aerodynamic damping ratio can be expressed as

$$\zeta_{\text{aero}} = \frac{-W}{2\pi\omega^2 \left(q^{\text{cfd}}\right)^2} \tag{6}$$

The modal aerodynamic damping ratio derived from the energy method is proportional to the unsteady aerodynamic work and is inversely proportional to the square of the amplitude and the natural frequency of labyrinth seal. That is, when the amplitude and the natural frequency are constant, the modal aerodynamic damping ratio would increase with the increase of aerodynamic work.

3 Case study

3.1 Labyrinth seal geometry

The aeroelastic stability of the labyrinth seal is predicted by the energy method. The structure of the model is shown in Figure1. The model is a four-tooth straight-through labyrinth seal consists of rotor and stator parts. The seal clearance is 0.5mm. It was pointed by Alford [2] that rotor seal had failed when supported at entrance side (high pressure side and upstream). So the boundary condition of supporting at entrance side was investigated.

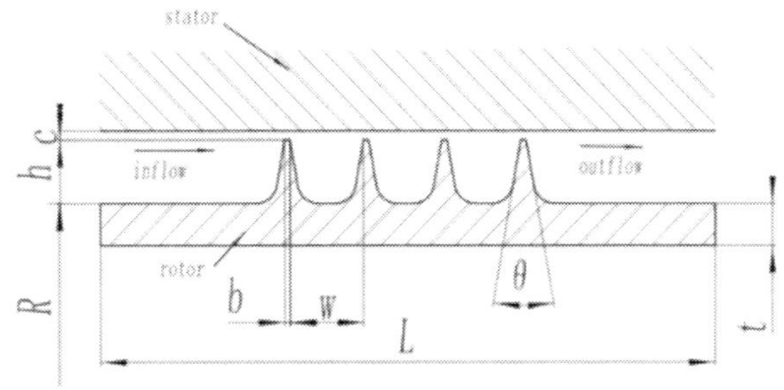

Figure 1. Labyrinth seal geometry.

3.2 Modal analysis

The following Figure 2(a) shows the finite element mesh model of the labyrinth seal for calculating the natural vibration characteristics. The number of nodes in the model is 70800, and the grid is extruded to 120 layers in the circumferential direction. The labyrinth seal is supported at the inlet high-pressure side and its natural vibration characteristics under the rotating condition are calculated. For a short thin-walled cylinder structure, such as labyrinth seal, the main vibration displacement is at the radial direction. Figure 2(b) and 2(c) shows the 2 nodal diameters mode and 3 nodal diameters mode of the labyrinth seal. The maximum amplitude appears at the free side of the seal.

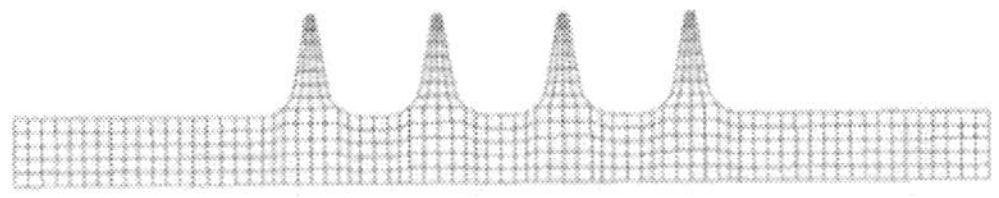

(a) Finite element model of the labyrinth seal

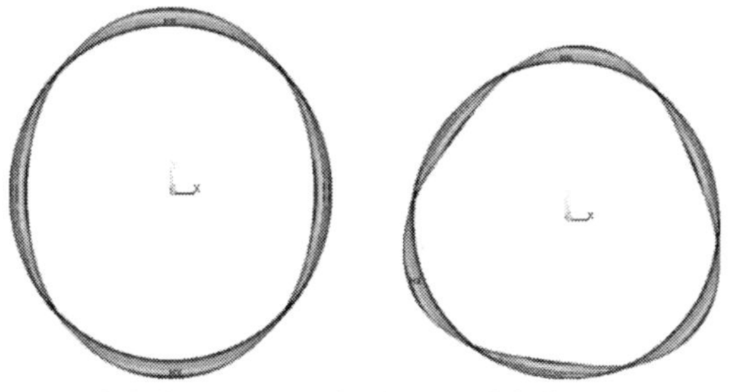

(b) 2 nodal diameters mode (c) 3 nodal diameters mode

Figure 2. Labyrinth seal FE model and 2~3 nodal diameters mode

The modal shapes of the labyrinth seal are obtained by finite element modal analysis, and the vibration displacement obtained by modal analysis can be interpolated to the fluid-structure coupling surface of the fluid domain by the developed data transfer algorithm. When the displacement interpolation is completed, the unsteady analysis of the cavity flow field is conducted by moving mesh technique under the displacement oscillation of the coupling surface. The steady-state analysis of the flow field is taken as the initial field of unsteady analysis to accelerate the convergence rate.

3.3 Steady flow simulation

The three-dimensional model is used to calculate the steady-state flow characteristics of the labyrinth seal cavity. In order to simulate the pressure ratio in this application, the fixed outlet pressure is set as the normal atmospheric pressure, and the total inlet pressure is adjusted for different pressure ratio from 1.2 to 3.6. The wall boundary is set to adiabatic and smooth. The ideal gas is selected as working fluid.

Figure3 lists Mach number and total pressure contour of the labyrinth seal. There are obvious stratification phenomena in the process of airflow passing through the teeth of each cavity. The upper part is the jet area and the lower part is the turbulence area. When the airflow is blocked by the tooth surface, a whirlpool is formed in the tooth cavity. The Mach number of the vortex flow close to the seal surface is higher and the Mach number of cavity center is the smallest. During the process of airflow passing through each seal, the pressure decreases gradually, and the pressure drop of the airflow through the first tooth is the largest.

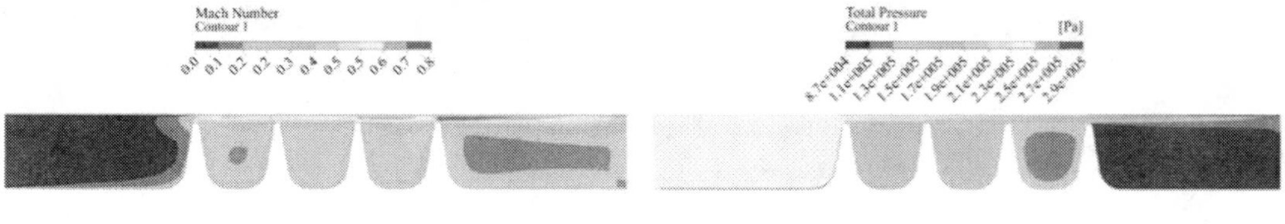

Figure 3. Predicted Mach number and total pressure contour at pressure ratio=2.4

3.4 Results and Discussions of Energy Method

The unsteady flow analysis is carried out by setting the labyrinth seal to vibrate with specified frequency and mode shape. The results of steady flow analysis are taken as initial conditions. The coordinates of the mesh nodes at the specified coupling surface of the flow field are set to oscillate with the law of sinusoidal motion.

The unsteady aerodynamic force and the displacement at each time step of the coupling surface in the flow field are derived for data post processing. And then the unsteady aerodynamic work and modal aerodynamic damping ratio in one vibration period are calculated. Figure 4 shows the aerodynamic work and modal aerodynamic damping ratios for 3~6 nodal diameters mode shape. By the definition of the energy method, negative aerodynamic work in a cycle indicates that the energy of labyrinth seal continuously release into the gas and the structure is stable.

Correspondingly, the positive modal aerodynamic damping ratio and negative aerodynamic work indicates that the aerodynamic damping of the labyrinth seal is consuming energy. It can be seen from Figure 4 that the aerodynamic work of the labyrinth seal changes almost lineally with the pressure ratio under different nodal diameter modes. When the pressure ratio is relatively small, the aerodynamic work is negative and the structure is stable. When the pressure ratio increases to a certain extent, the aerodynamic work will change from a negative value to a positive value and the structure becomes unstable. The aeroelastic stability of the labyrinth seal varies with the operating conditions. There are some differences in aeroelastic stability under different vibration modes. It can be seen that the 3 nodal diameters mode has better stability and the 5 nodal diameters mode is more prone to aeroelastic instability.

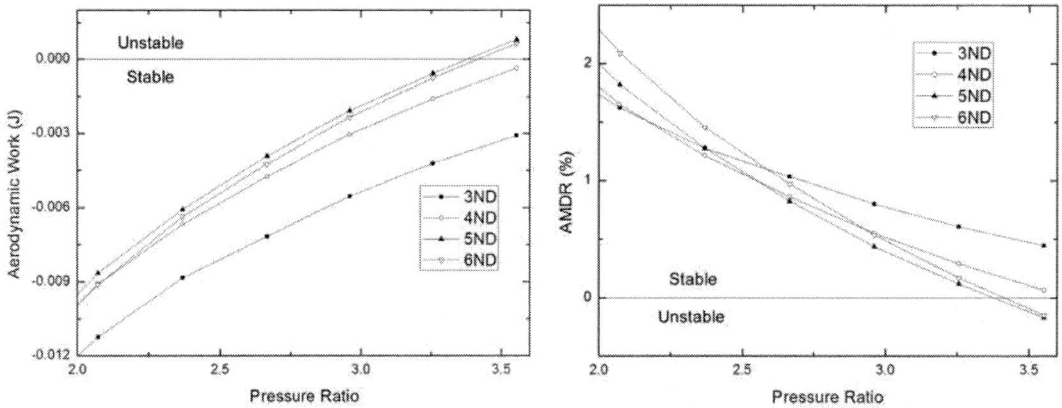

(a) Aerodynamic work of the labyrinth seal (b) Modal aerodynamic damping ratio

Figure 4. Aerodynamic work and AMDR of 3~6 nodal diameters at different pressure ratio.

4 Conclusion

In this paper, the aeroelastic stability of the labyrinth seal is predicted by the energy method. The influence on the aeroelastic stability of the labyrinth seal under different working conditions and vibration modes is investigated. The effects of the relative thickness of the tooth tip, the width of the seal cavity and the eccentricity of the rotor on the aeroelastic stability of the labyrinth seal are also studied. The conclusions are as follows:

The weak coupling calculation method based on energy method can analyze the solid domain and the fluid domain model separately, and transfer the displacement through linear interpolation, so that the calculation is simplified greatly. The vibration characteristic of labyrinth seal is similar to the short cylindrical shell that the radial displacement plays a major role. When supported at entrance side (high pressure side or upstream), the largest radial displacement is at the discharge side. However, the pressure at entrance side is higher. The aeroelastic stability of labyrinth seal can be solved at different natural frequency and working conditions. When the pressure ratio of the model increases, the area of the positive aerodynamic work of the coupling surface enlarges, resulting in the decrease of the aerodynamic damping ratio.

References

1. F. Ehrich, Aeroelastic Instability in Labyrinth seals, ASME Paper, No. 68-GT-32, (1968)
2. J. S. Alford, Protection of Labyrinth Seals From Flexural Vibration, ASME J. Eng. Power, **86**, (1964)
3. D. A. Lewis, C. E. Platt, B. E. Smith, Aeroelastic Instability in F100 Labyrinth Air Seal, AIAA J. Aircraft, 16, (1979)
4. D. R. Abbott, Advances in Labyrinth Seal Aeroelastic Instability Prediction and Prevention, ASME Paper, No. 80-GT-151, (1980)
5. A. V. Srinivasan, Aeroelastic Instabilities in Labyrinth Air Seal Systems, ASME Paper, No. 84-GT-169, (1984)
6. J. S. Alford, Nature, Causes, and Prevention of Labyrinth Air Seal Failures, AIAA J. Aircraft, **12**, (1975)
7. D. Amirante, N. J. Hills, Barnes, Thermo-Mechanical Finite Element Analysis/Computational Fluid Dynamics Coupling of an Interstage Seal Cavity Using Torsional Spring Analogy, ASME J. Turbomachinery, **134, (**2012)
8. O. O. Bendiksen , Aeroelastic Problems In Turbomachines, AIAA Paper, No. 90-1157, (1990)
9. F. O. Carta, Coupled Blade-Disk-Shroud Flutter Instabilities in Turbojet Engine Rotors, ASME J. Eng. Power, **89**, (1967)
10. A.I. Sayma, C. Breard, M. Vahdati, M. Imregun, Aeroelasticity Analysis of Air-Riding Seals for Aero-engine Applications, ASME J. Tribology, 124, (2002)
11. R. Phibel, L. D. Mare, J. S. Green, M. Imregun, Numerical Investigation of Labyrinth Seal Aeroelastic Stability, ASME Paper, No. GT2009-60017, (2009)
12. R. Phibel, L. D. Mare, Comparison Between a CFD Code and a Three-Control-Volume Model For Labyrinth Seal Flutter Predictions, ASME Turbo Expo, No. GT2011-46281, (2011)
13. L. D. Mare, M. Imregun, J. S. Green, A. I. Sayma, A Numerical Study of Labyrinth Seal Flutter, ASME J. Tribology, **132**, (2010)
14. Zhang, X. W., Wang, Y. R., and Xu K. N., Flutter Prediction in Turbomachinery with Energy Method, Proc. Inst. Mech. Engi., Part G: J Aero Eng, 225, (2011)
15. A. V. Srinivasan, Flutter and Resonant Vibration Characteristics of Engine Blades, ASME Paper, No. 97-GT-533, (1997)
16. R. E. Kielb, M. Imregun, Aeroelasticity in Axial Flow Turbomachines, NASA, No. TP-2006-212490, (2006)

Thermal Cycling Toughness and Strength Estimation of Cryogenic Filled Tank

Dmitry Uglanov[1,a], Dmitry Sarmin[1], Alina Akulova[1], Daria Aksenova[1] and Roman Panshin[1]

[1]*Samara University, Department of Heat Engineering and Heat Engines, 443086, Moskovskoye shosse, 34, Samara, Russia*

Abstract. At present, technologies based on liquid natural gas (LNG) using start to find a wide use. They also demand technologies of working fluid storage and transportation in cryogenic liquid condition. The working fluid will be used after its regasification from cryogenic liquid condition. When the gas cylinder is filled with gas, there is a danger of exceeding the temperature stresses in the cylinder walls. For the study, there is a method for calculating unsteady thermal conductivity for studying the temperature stresses in the wall of a gas cylinder during its interaction with an evaporating cryogen. This method is used for determining the thermoelastic stresses of the cryogenic filled tank. The CAE- calculation of the strength of the cryogenic filled tank is also done.

1 Introduction

Economic, exhibited and special requirements to gas and cryogenic liquids storage require creation of more perfect designs [1]. These requirements include abilities to endure high pressure, decreasing of the working fluid leakage, satisfaction of the accelerated fuelling requirement, meeting the safety requirement and having required durability.

Usually working fluids for technological and special purposes are stored in gaseous compressed form in high pressure tanks or in a liquid form with cryogenic temperatures.

For example, in case of natural gas using working fluid is filled in a tank to a 20 pressure of MPa and more. Fuelling is accomplished in the gas-filling stations by compressing. The time of the fuelling is equal to several minutes and is followed by tank heating which decreases the durability of the tank and may cause an under fuelling or increase the fuelling length. Fuelling by compressing require significant energy consumption.

At present, technologies based on liquid natural gas (LNG) using start to find a wide use [2-6]. There are also technologies of working fluid storage and transportation in cryogenic liquid condition. The working fluid will be used after its regasification from cryogenic liquid condition.

From aforementioned it is followed that it will be objectively useful and well-timed to develop and create universal equipment able to satisfy existing technologies of cryogenic liquid and compressed gases use.

For example, in gas-filling compressor stations it will be appropriate to combine the presently used tank filling method of compression with fuelling from cryogenic liquid condition.

[a] Corresponding author : dmitry.uglanov@mail.ru

2 The design of the cryogenic filled tank

The peculiarity of the cryogenic filled tank (Fig. 1), for example, for natural gas [7, 8] is that it can be refuelled to operating pressure (20-40 MPa), or liquefied natural gas (LNG). In the latter case, the cryogen is poured into an inner vessel, outside of which there is a multi-layer thermal insulation between it and the inner wall of the cylinder, which is necessary to prevent convection currents and to protect against heat stroke. At the same time, the heat exchange between the wall of the CNG balloon is caused only by the thermal conductivity of the gaseous methane located in the annular gas cavity.

Parameters of the working fluid in the tank was calculated by special methods [8]. It was shown that the heating of the liquid phase, its evaporation and redistribution of liquid and gaseous fractions is carried out in a relatively long period of time (up to two days). After this time, the parameters of the working fluid come to the equilibrium state of the entire balloon, with the equilibrium temperature reaching -80 ... -100°C, and the pressure does not exceed 2 ... 3 MPa.

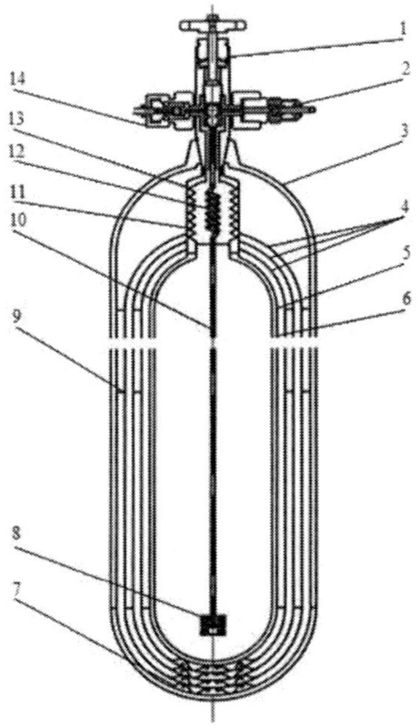

Figure 1. Cryogenic filled tank (CFT).
1 – cutoff valve; 2 – safety valve; 3- tank; 4 – insulation; 5 – insulation of the inner volume; 6 – inner volume; 7 – supporting space ring; 8 – filter; 9 – radial space rings; 10 – barrel; 11 – feeding tube; 12 – coil; 13 – bellows; 14 - flanged socket

Such parameters allow expecting small stresses in the tank wall, which increases the service life of the cylinder. Despite the fact that the problems of thermophysics in relation to the balloon can be considered to be solved in a certain amount, the issues of assessing the stress levels, and, accordingly, the thermal cycle strength requires a separate study. In the present article the solution of the problem of determination of stresses in the balloon wall in the conditions of its interaction with the evaporating cryogen is given.

3 The calculation of the strength of the cryogenic filled tank

The problem of determining the thermoelastic stresses is reduced to a joint solution of the strength problem and the problem of transient heat conduction. To solve this problem, the method of calculating transient heat conduction was used to study the temperature stresses [9]. The possibility of representing the problems of thermoelastic stresses simple and reliable functional operators makes it relatively easy to determine the fields of temperature stresses for any time-varying thermal effects.

After the tank is filled with a cryogen, the liquid begins to evaporate, and the evaporated fraction of the cryoproduct is redistributed along the layers of the gas cavity. The relative radius $p = \frac{r}{R}$ for the balloon wall with internal R_1 and outer R_2 radius is introduced. Then the differential equation of transient heat conduction inside the balloon wall under mixed boundary conditions of the third and second kind in dimensionless coordinates is given to the form of:

$$\frac{\partial T}{\partial Fo} = \frac{\partial^2 T}{\partial p^2} + \frac{1}{p} \cdot \frac{\partial T}{\partial p},$$

$$[T(p, Fo)]_{Fo-0} = T_0. \quad (1)$$

$T_o = T_o(c) = const$ – the initial wall temperature.

$$[\frac{\partial T}{\partial p} - Bi \cdot T(p, Bi)]_{p=1} = -Bi \cdot \varphi_2(Fo),$$

$$(\frac{\partial T}{\partial p})_{p=k} = \frac{R_1}{\lambda} \cdot \varphi_1(Fo). \quad (2)$$

Here $Fo = a \cdot \tau / R_1^2$; $Bi = \alpha \cdot R_1 / \lambda$; $k = R_2 / R_1 > 1$.

After the corresponding transformations [3], the relative excess wall temperature for the cooling case under consideration will be expressed as a ratio:

$$\theta(p, Fo, Bi, k) = \frac{T_{cp} - T(p, Fo)}{T_{cp} - T_0} = D(Bi, k) \cdot [\frac{(k-1) \cdot Bi + 2}{Bi} - \frac{(k-p)^2}{k-1}] \cdot \exp[-A(Bi, k) \cdot Fo]; \quad (3)$$

Tangential and radial stresses are found by the formulas with taking into account [3] in a cylindrical wall:

$$\sigma_\varphi = \frac{\beta \cdot E}{1-\upsilon} \cdot \frac{1}{p^2} \times [\frac{p^2 + 1^k}{k^2 - 1} \int_1^k T(p, Fo) p dp - T(p, Fo) p^2]; \quad (4)$$

$$\sigma_\varphi = \frac{\beta \cdot E}{1-\upsilon} \cdot \frac{1}{p^2} [\frac{p^2 - 1^k}{k^2 - 1} \int_1^k T(p, Fo) p dp - \int_1^k T(p, Fo) p dp]. \quad (5)$$

If we substitute the value of the current local temperature T (r, Fo) obtained from expression (3) into formulas (4), (5), then the dimensionless stresses:

$$\overline{\sigma_\varphi} = \frac{(1-\nu) \cdot \sigma_\varphi}{\beta \cdot E \cdot (T_{cp} - T_0)}; \quad (6)$$

$$\overline{\sigma_p} = \frac{(1-\nu) \cdot \sigma p}{\beta \cdot E \cdot (T_{cp} - T_0)}. \quad (7)$$

These equations are defined by the formulas:

$$\overline{\sigma_\varphi} = -D(Bi, k) \cdot N(p, k) \cdot \exp[-A(Bi, k) \cdot Fo]; \quad (8)$$

$$\overline{\sigma_p} = -D(Bi, k) \cdot M(p, k) \cdot \exp[-A(Bi, k) \cdot Fo]. \quad (9)$$

The values of $D(Bi, k)$ and $A(Bi, k)$ are calculated from the relations:

$$A(Bi, k) = \frac{10Bi[Bi \cdot (k-1)(k+3) + 12]}{(k-1)[Bi^2 \cdot (k-1)^2 \cdot (11k+5) + 10Bi \cdot (k-1) \cdot (5k+3) + 60(k+1)]}; \quad (10)$$

$$D(Bi, k) = 2{,}5 \cdot \omega \cdot Bi[Bi \cdot (k-1) \cdot (5k+3) + 12 \cdot (k+1)]; \quad (11)$$

$$\omega^{-1} = Bi^2 \cdot (k-1)^2 \cdot (11k+5) + 10Bi \cdot (k-1) \cdot (5k+3) + 60(k+1). \quad (12)$$

Formulas (10-12) allow to determine the stresses for any values of B_i and tank wall parameters k. By the above relations, the values of the parameters included in the calculation formulas were determined (Table 1).

Using these parameters, the values of the relative excess wall temperatures on the inner surface of the balloon were calculated, taking into account their time dependence$\theta(p,\ F_0,\ B_i,\ k)$, temperature distribution in the balloon wall $T(p,\ F_0)$ as a function of time, and also radial and the tangential stresses $\sigma_p(p)$, $\sigma_\varphi(p)$in the tank wall. The stresses created by the pressure of the evaporating gas (graph of the change in gas pressure with respect to time, is presented in the article "Numerical modeling of processes in a fuel balloon with cryogenic refueling with natural gas for non-drainage storage") are determined using the well-known expression 10. The results of the calculations are shown in Fig. 2. In accordance with Russian State Standard [5] the margin of safety is assumed to be n=2. Analysis of the obtained data allows to conclude that during the whole gasification process the stresses in the tank wall do not exceed the tensile strength, for example, for the tank BK-6601-400 TU, made of steel 45,σ_T=890 *MPa*. Thus, the stresses in the tank wall in this case are much lower, that is, by an order of magnitude than when refueled by compression in comparison with,σ_T.

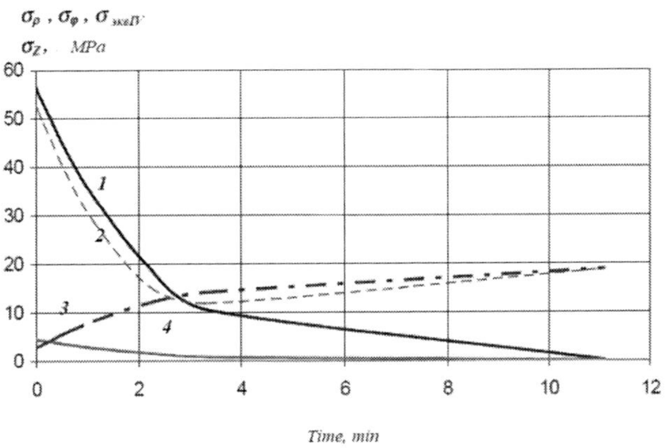

Figure 2. The graph of the change in the thermal radial σ_p(1), the equivalent stressesσ_{eqIV} (2), the stresses generated by the pressure of the evaporating gas (3) and the tangential stresses $\sigma_\varphi$in the wall of the cylinder (4).

Table 1. Calculation parameters, formulas and values.

Parameter	Value
R_1, mm	100
R_2, mm	106
K= R_1/ R_2	1,06
λ, W/(m·K)	16,3
α=λ/ρ·c, m^2/s	0,85 10-5
F_0= $\alpha\cdot\tau R_1^2$	0,00085·τ
α, W/(m^2·K)	22,9
B_i = α·R_1/λ	0,013
$B(B_i,k)$	0,066
$D(B_i,k)$	0,068
$A(B_i,k)$	2,27

$$\sigma_z = \frac{p_g \cdot D_2}{2(R_2 - R_1)}; \tag{13}$$

$$\sigma_{eqIV} = \sqrt{\sigma_z^{\ 2} + \sigma_p^{\ 2} + \sigma_\varphi^{\ 2} - (\sigma_z \sigma_p + \sigma_\varphi \sigma_p + \sigma_z \sigma_k}; \tag{14}$$

$$\sigma_{eqIV} \leq [\sigma]; \tag{15}$$

$$[\sigma] = \frac{\sigma_T}{n}. \tag{16}$$

If we take into account the practice of using the reservoir as a fuel tank for a car, even taking into account the occasional full gasification of LNG to the parameters of fuelling by compression, it can be argued that the cylinder resource is increased at least three times.

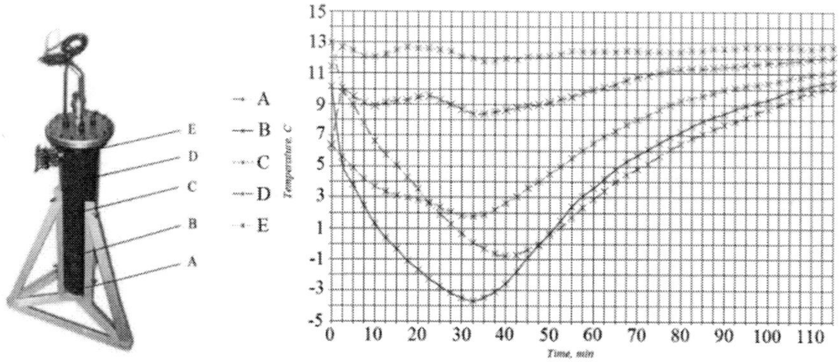

Figure 3. The distribution of the cylinder wall temperature in different sections, depending on the operating time on the flow rate.

4 Investigation of the temperature state of the cryogenic filled tank wall

Since one of the main functions of the internal thermal insulation of CFT is to prevent a thermal shock from the cryoproduct to the wall of the balloon, then the temperature of the CFT wall (in time) will be an important indication of this.

The temperature of the cylinder wall in different sections as a function of time was recorded by the Micron - M 7604 F thermal imager and is shown in Figure 4.

Measurements showed that the maximum change in wall temperature during the experiment was 120 min. has reached 19° C (from +14 °C to –4 °C), and it occurred for 30 minutes from the moment of hermetic sealing of the balloon, after which the temperature values in different sections equalized (Figure 3).

On the height of the balloon (vertical position) at the beginning of the process, the minimum temperature is observed in the middle of the balloon (+6 °C), at the end of the process - at the bottom of the balloon (+9°C).

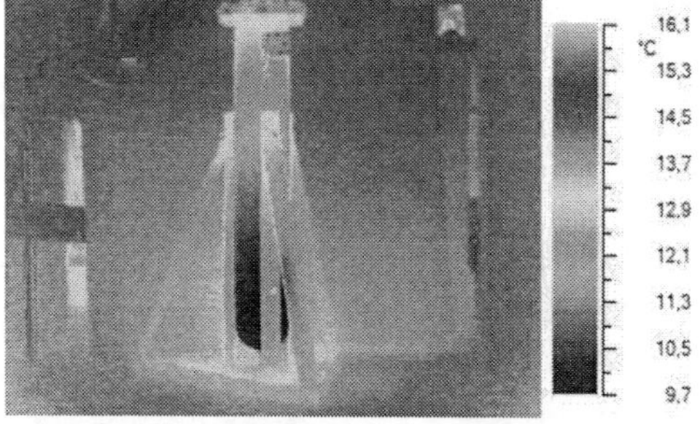

Figure 4. Initial and final state of the cylinder in the operating mode

The main conclusion is that the temperature level itself does not exceed 20°C,in height, and the rate of change of the gradient along the altitude ΔT_H is no more than 0.5 deg/min.

The temperature gradient over the wall thickness of $\delta \approx 5$ mm is not more than 0,05°C(calculations for the solution of the problem of non-stationary thermal conductivity of the system). The thermograms of the initial and final state of the cylinder wall temperature are shown in Fig. 4.

5 The CAE - calculation of the strength of the cryogenic filled tank

In the course of the research, a numerical assessment of the temperature influence on the elements of the prototype of the cryogenic refueling tank with the help of ANSYS SAE system was carried out.

It was created using ANSYS Meshing grid model design. By setting the grid type and cell sizes for each element of the model, a grid model acceptable for calculation was obtained. To calculate the unevenly heated cylinder, which is in the elastic-plastic state, its temperature-loaded depending on the distance to the bottom of the cylinder on the basis of earlier experimental studies (Figure 2). Upon completion of the calculations, stress fields for all elements of the model of the BKZ sample and its margin of safety are constructed. Control of the magnitude of the displacements allows to check the General plausibility of the calculations. Stresses, safety margins and displacements of all model elements are shown in figures 5.

Consider a cylindrical vessel (Figure 6) which is under the action of an internal pressure p. Radius of cylinder R, thickness h.

The axial component of the pressure forces will be equal to the product of the pressure p by the area of the projection of the surface, on a plane perpendicular to the given axis:

$$P = \pi R^2 p; \tag{17}$$

Thus, we obtain:

$$\sigma_m \, 2\pi R h = \pi R^2 p \,; \tag{18}$$

$$\sigma_m = \frac{pR}{2h} \,. \tag{19}$$

For a cylinder $\rho_m = \infty$, $\rho_t = R$. Therefore, from Laplace's formula we find:

$$\frac{\sigma_m}{\infty} + \frac{\sigma_t}{R} = \frac{P}{h} \,; \tag{20}$$

$$\sigma_t = \frac{pR}{h}. \tag{21}$$

those. The circumferential stress is twice the meridional.

For the sphere $\rho_m = R$, $\rho_t = R$, $\sigma_m = \sigma_t = \sigma$. Therefore, from Laplace's formula we find:

$$\frac{\sigma}{R} + \frac{\sigma}{R} = \frac{P}{h}; \tag{22}$$

$$\sigma = \frac{pR}{2h}. \tag{23}$$

We cut off the part of the cylinder by the cross section (Figure 7), the equilibrium equation for it (14):

$$\sigma_m \, 2\pi \, R \, h = P; \tag{24}$$

The shell, as a rule, is made of plastic materials, so we will use the third theory of limiting stresses.

$$\sigma_{eqIII}^{max} \le [\sigma]; \tag{25}$$

$$\sigma_{eqIII}^{max} = \sigma_1 - \sigma_3 \,. \tag{26}$$

The ABCD element, isolated from the cylindrical shell, is in a biaxial stress state (Figure 8):$\sigma_1 = \sigma_t, \sigma_2 = \sigma_m, \sigma_3 = 0$

Equivalent stress for the cylindrical part:

$$\sigma_{eqIII}^{max} = \sigma_1 - \sigma_3 = \sigma_t - 0 = \frac{pR}{h}. \tag{27}$$

For the spherical part of the shell, we find the equivalent voltage similarly:

$$\sigma_{eqIII}^{max} = \sigma_1 - \sigma_3 = \sigma - 0 = \frac{pR}{2h}. \tag{28}$$

Thus, for a cylindrical shell part:

$$\frac{pR}{h} \le [\sigma]. \tag{29}$$

For the spherical part of the shell:

$$\frac{pR}{2h} \le [\sigma]. \tag{30}$$

For the whole shell:

$$\frac{pR}{h} \le [\sigma]. \tag{31}$$

Since the outer shell of the cylinder is made of steel 30XMA, then for the ultimate strength we take the yield strength for this steel$\sigma_t = 392MPa$. From (27) we express h and take $\sigma_\tau = [\sigma]$. For p we take $p_{max} = 15MPa$.

$$h = \frac{pR_{б}}{\sigma_\tau} = \frac{15\times10^6\times0{,}13}{392\times10^6} = 0{,}0049\,\text{м}. \tag{32}$$

We take the wall thickness of 0.005 m.

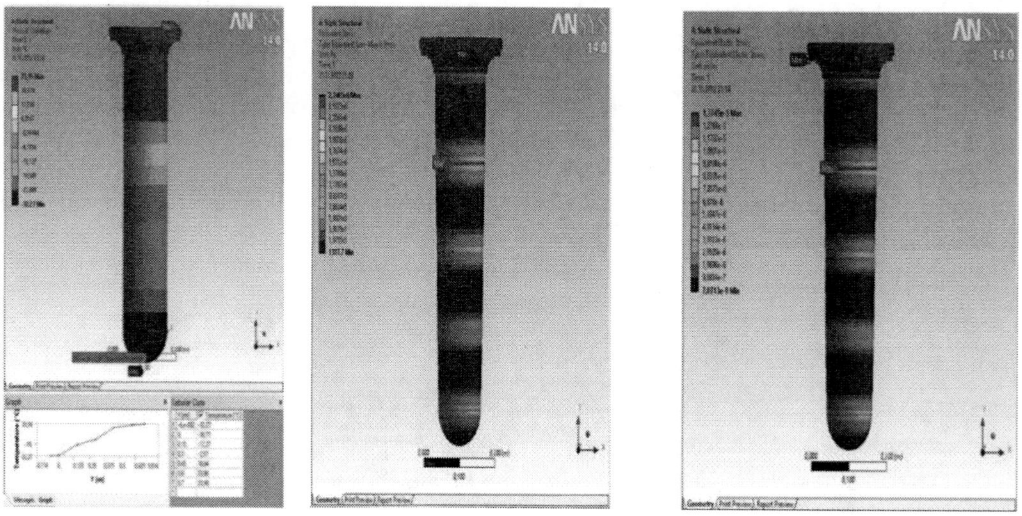

Figure 5. The graph of the change in the thermal radial (1), the equivalent stresses (2), the stresses generated by the pressure of the evaporating.

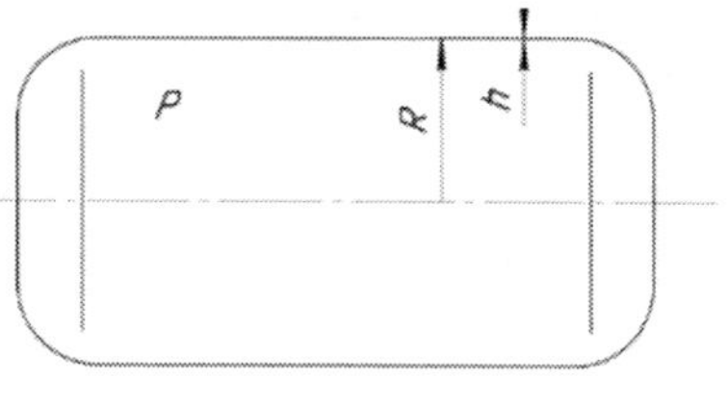

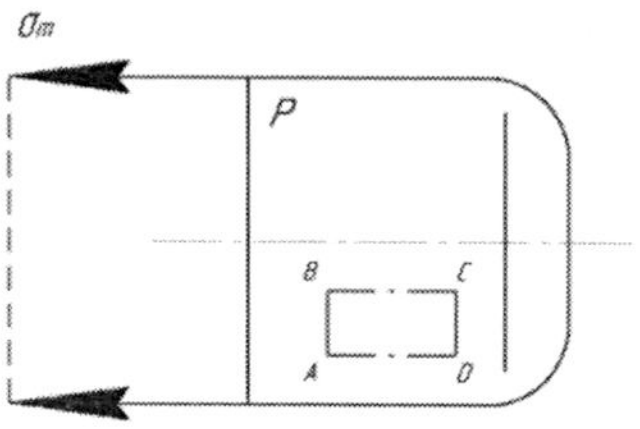

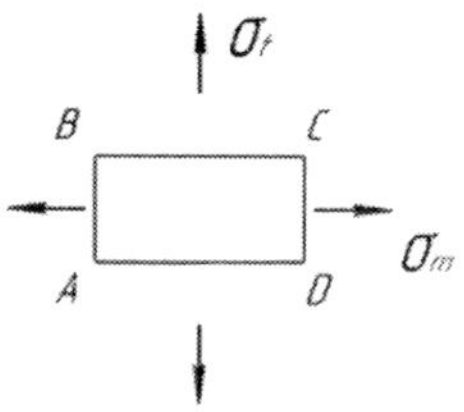

. **Figure 6.** Cylindrical vessel under the action of internal pressure p

Figure 7. Cross-sectioned part of the cylinder.

Figure 8. Isolated element from a cylindrical shell, in a biaxial stress state

We perform the verification calculation for strength:

$$\frac{pR_b}{h} = \frac{15\times10^6\times0.13}{0.005} = 390MPa; \tag{33}$$

$$390MPa < [392MPa]. \tag{34}$$

The condition of strength is observed.

6 Conclusions

From the result of the calculation shows that the maximum local stresses in the structure from thermal stresses occurring when filling the concert hall with cryogenic products does not exceed 2.8 MPa, which satisfies the conditions of strength and for a lot less compared to the stresses arising in cylinder running compression (pressure 15 MPa), it indicates significant potential to increase the resource of the container. The tank resource could be increased at least three times compared to standard tanks.

From the result of the calculation it can be seen that the maximum local voltage in the structure from the thermal stresses arising when filling the cryogenic filled tank with the cryoproduct does not exceed 2.8 MPa, which satisfies the strength conditions.

Compared with tanks filled with compression (pressure 15 MPa). This indicates a significant potential for increasing the life of the balloon.

References

1. S.-J. Jeon, B.-M. Jin, Y.-J. Kim, Consistent thermal analysis procedure of LNG storage Tank, Structural Engineering and Mechanics. **25**, 4 (2007)
2. B. Adouane, W. de Jong, J.P. van Buijtenen, G. Witteveen, Fuel–NOx emissions reduction during the combustion of LCV gas in an air staged Winnox-TUD combustor, Applied Thermal Engineering, **5**, 30(8) (2010)
3. A.K. Sen, G. Litak, B.-F. Yao, G.-X. Li, Analysis of pressure fluctuations in a natural gas engine under lean burn conditions, Applied Thermal Engineering, **4**, 30(8) (2010)
4. Q.Y. Li, Y.L. Ju, Design and analysis of liquefaction process for offshore associated gas resources, Applied Thermal Engineering, **8**, 30(8) (2010)
5. Kakumoto, Y., Koyamatsu, Y., Shiota, A., Qudaih, Y., Mitani, Y. Application of Geographic Information System to Power Distribution System Analysis, Energy Procedia, **6**, 100 (2016)
6. Burtsev, S.A., Karpenko, A.P., Leontiev, A.I. A method for distributed production of liquefied natural gas at gas-distribution stations, High Temperature, **4**, 54 (4), (2016)
7. Patent RU2163699.
8. E.V. Blagin, A.I. Dovgyallo, S.O. Nekrasova, D.V. Sarmin, D.A. Uglanov, Applied Thermal Engineering Estimation of the energy efficiency of cryogenic filled tank using in different system and devices, **8**, 101 (2016)
9. P. V. Tsoi, *Methods of calculation of heat and mass transfer problems* (Moscow: Energoatomizdat, 1984)

A Specific Sliding Mode Control for Autopilot Design of Bank-to-Turn Missiles

Zepeng Huai[1,a], Hongbo Wang[1], Min Gong[1], Pengfei Ren[1], Zijian Xu[1] and Lingxiao Cheng[1]

[1]*China Academy of Launch Vehicle Technology Tactical Weapons Division*

Abstract. This paper studies the BTT missile's sliding mode flight controller. Firstly, the control model is established according to dynamic inverse method and the general dynamic equation of the missile flying in the atmosphere. Then, the sliding mode controller is designed and it adopts the time-varying sliding mode surface, the self-defined reaching law and the adaptive control parameters. Finally, three sets of simulation are carried out to prove the flight controller designed in this paper for BTT missile can follow the guidance instructions and meet operational needs.

1 Introduction

The Back-to-turn steering is realized by rolling and pitching the missile while maintaining (ideally) a zero sideslip angle[1]. The most important advantage of BTT missile is high maneuverability[2], which is desperately needed under certain flying conditions such as air-to-air combat[3]. The bank-to-turn steering brings BTT missile many advantages, but also leads to the strong coupling between pitch channel and roll channel[4]. In actual flight, the sideslip angle is a small value around zero, and it cannot strictly guarantee the sideslip angle is zero. So the dynamics model of BTT missile is strongly nonlinear equations, this poses a severe challenge to the control system.

This paper firstly builds the control model by the general dynamics equation of the cruise missile, which include the dynamic inverse process. Then a specific sliding mode controller is designed. Finally, the effectiveness of the controller is verified through simulation.

2 Control model of BTT missile

The dynamics model of missiles flying in the atmosphere is established as follows:

[a] Corresponding author : hzphtyy@163.com

$$\dot{V} = \frac{P\cos\alpha\cos\beta - X - mg\sin\theta}{m}$$
$$\dot{\theta} = \frac{P(\sin\alpha\cos\gamma_v + \cos\alpha\sin\beta\sin\gamma_v) + Y\cos\gamma_v - Z\sin\gamma_v - mg\cos\theta}{mV}$$
$$\dot{\psi}_v = \frac{P(\sin\alpha\sin\gamma_v - \cos\varepsilon\alpha\sin\beta\cos\gamma_v) + Y\sin\gamma_v + Z\cos\gamma_v}{-mV\cos\theta}$$
$$\dot{w}_x = \frac{M_x + (J_y - J_z)w_y w_z}{J_x}$$
$$\dot{w}_y = \frac{M_y + (J_z - J_x)w_x w_z}{J_y}$$
$$\dot{w}_z = \frac{M_z + (J_x - J_y)w_x w_y}{J_z}$$
$$\dot{\vartheta} = w_y\sin\gamma + w_z\cos\gamma$$
$$\dot{\psi} = \frac{w_y\cos\gamma - w_z\sin\gamma}{\cos\vartheta}$$
$$\dot{\gamma} = w_x - \tan\vartheta(w_y\cos\gamma - w_z\sin\gamma)$$
$$\sin\beta = \cos\theta\left[\cos\gamma\sin(\psi - \psi_v) + \sin\vartheta\sin\gamma\cos(\psi - \psi_v)\right] - \sin\theta\cos\vartheta\sin\gamma$$
$$\sin\alpha = \frac{\cos\theta\left[\sin\vartheta\cos\gamma\cos(\psi - \psi_v) - \sin\gamma\sin(\psi - \psi_v)\right] - \sin\theta\cos\vartheta\cos\gamma}{\cos\beta} \quad (1)$$
$$\sin\gamma_v = \frac{\cos\alpha\sin\beta\sin\vartheta - \sin\alpha\sin\beta\cos\gamma\cos\theta + \cos\beta\sin\gamma\cos\vartheta}{\cos\theta}$$

Generally for BTT missile, the controlled variable of system, which means the guidance command, is $\begin{bmatrix} n_y & \beta & \gamma_V \end{bmatrix}^T$. n_y is pitch overload under the ballistic coordinate system:

$$n_y = \frac{P\left(\sin\alpha\cos\gamma_V + \cos\alpha\sin\beta\sin\gamma_V\right) + Y\cos\gamma_V - Z\sin(\gamma_V)}{mg} \quad (2)$$

System(1) is a strongly nonlinear and time-varying system in which the pitch channel, yaw channel and roll channel have a strong coupling. It is not conducive to directly analyse the law of system response under closed-loop controlled conditions. So we need to deal with equation(1) and derive the control model which can be directly used for solving control problems.

The essence of the dynamic inverse method to deal with the nonlinear model is the feedback linearization, that is, by introducing a new control quantity and designing the relationship between the new control quantity and former control quantity, the new control quantity and the controlled variable show a linear relationship. However, this linearity is pseudo-linear. In other words, the dynamic inverse method does not fundamentally change the nonlinear nature of the system. Instead, the nonlinear part is included in the control quantity design. That is, the nonlinearity of the feedback is used to solve the nonlinearity of the system.

Equation(1) is dealt with by dynamic inverse method as follows:

It is found that six equations in system(1) forms partial closure, which are

$$\begin{aligned}
\dot{w}_x &= \frac{M_x + (J_y - J_z)w_y w_z}{J_x} \\
\dot{w}_y &= \frac{M_y + (J_z - J_x)w_x w_z}{J_y} \\
\dot{w}_z &= \frac{M_z + (J_x - J_y)w_x w_y}{J_z} \\
\dot{\vartheta} &= w_y \sin\gamma + w_z \cos\gamma \\
\dot{\psi} &= \frac{w_y \cos\gamma - w_z \sin\gamma}{\cos\vartheta} \\
\dot{\gamma} &= w_x - \tan\vartheta(w_y \cos\gamma - w_z \sin\gamma)
\end{aligned} \tag{3}$$

Write it as a matrix form:

$$\begin{aligned}
\dot{\boldsymbol{\theta}} &= F_\theta \boldsymbol{\omega} \\
\dot{\boldsymbol{\omega}} &= J^{-1}\boldsymbol{F}_w + J^{-1}\boldsymbol{U}_M
\end{aligned} \tag{4}$$

where, the parameters are

$$\boldsymbol{\theta} = \begin{bmatrix} \vartheta \\ \psi \\ \gamma \end{bmatrix} \quad \boldsymbol{\omega} = \begin{bmatrix} \omega_x \\ \omega_y \\ \omega_z \end{bmatrix} \quad \boldsymbol{U}_M = \begin{bmatrix} M_x \\ M_y \\ M_z \end{bmatrix} \quad \boldsymbol{F}_\omega = \begin{bmatrix} (J_y - J_z)w_y w_z \\ (J_z - J_x)w_x w_z \\ (J_x - J_y)w_x w_y \end{bmatrix}$$

$$F_\theta = \begin{bmatrix} 0 & \sin\gamma & \cos\gamma \\ 0 & \dfrac{\cos\gamma}{\cos\vartheta} & -\dfrac{\sin\gamma}{\cos\vartheta} \\ 1 & -\tan\vartheta\cos\gamma & \tan\vartheta\sin\gamma \end{bmatrix} \quad J = \begin{bmatrix} J_x & 0 & 0 \\ 0 & J_y & 0 \\ 0 & 0 & J_z \end{bmatrix} \tag{5}$$

Take the derivative of both sides of $\dot{\boldsymbol{\theta}} = F_\theta \boldsymbol{\omega}$:

$$\ddot{\boldsymbol{\theta}} = F_\theta J^{-1}\left(\boldsymbol{F}_w + \boldsymbol{U}_M\right) + \boldsymbol{B}_w \tag{6}$$

where,

$$\boldsymbol{B}_w = \begin{bmatrix} \dot{\gamma}\left(\omega_y \cos\gamma - \omega_z \sin\gamma\right) \\ \dfrac{\dot{\vartheta}\left(\omega_y \cos\gamma - \omega_z \sin\gamma\right)\tan\vartheta - \dot{\gamma}\left(\omega_y \sin\gamma + \omega_z \cos\gamma\right)}{\cos\vartheta} \\ \dot{\gamma}\left(\omega_y \sin\gamma + \omega_z \cos\gamma\right)\tan\vartheta - \dot{\vartheta}\dfrac{\omega_y \cos\gamma - \omega_z \sin\gamma}{\cos^2\vartheta} \end{bmatrix} \tag{7}$$

According to equation(6), we take the new control quantity as follows:

$$\boldsymbol{v} = \ddot{\boldsymbol{\theta}} \tag{8}$$

The new control quantity is the output of controller and also is the input of the inverse system. The relationship between the former control quantity and the new control quantity $\boldsymbol{v}$ is as follows:

$$\boldsymbol{U}_M = JF_\theta^{-1}\left(\boldsymbol{v} - \boldsymbol{B}_w\right) - \boldsymbol{F}_w \tag{9}$$

Therefore, what we should do below is to design the new control quantity $\boldsymbol{v}$.

For BTT missile, the controlled variable is usually $\begin{bmatrix} n_y & \beta & \gamma_V \end{bmatrix}^T$. So we need derive the relationship between the controlled variable $\begin{bmatrix} n_y & \beta & \gamma_V \end{bmatrix}^T$ of system(1) and the controlled variable $\boldsymbol{\theta}$ of system(4), which includes two instruction translations:

Instruction translation1: the guidance command $\begin{bmatrix} n_{y_d} & \beta_d & \gamma_{V_d} \end{bmatrix}^T$ is translated to the expected input $\boldsymbol{\theta}_d$ of system(4).

Instruction translation2: the actual output $\boldsymbol{\theta}$ of system(4) is translated to actual output $\begin{bmatrix} n_y & \beta & \gamma_V \end{bmatrix}^T$ of system(1).

System(1) is shown in Figure 1.

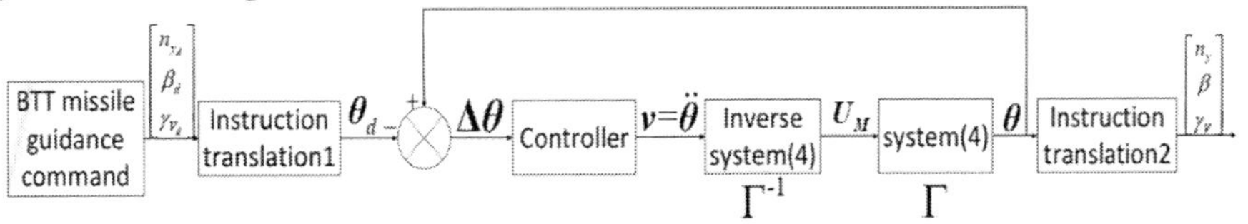

Figure 1 Structure of system(1)

The formulas of two instruction translations are shown below:

1) Instruction translation1

The missiles in equilibrium state satisfy the equation:

$$\begin{cases} n_y = n_{y_d} \\ M_z = 0 \end{cases} \tag{10}$$

According to equation(10) and aerodynamic data, the expected pitch overload n_{y_d} can be translated to expected attack angle α_d.

Then translate $\begin{bmatrix} \alpha_d & \beta_d & \gamma_{V_d} \end{bmatrix}^T$ to $\boldsymbol{\theta}_d = \begin{bmatrix} \vartheta_d & \psi_d & \gamma_d \end{bmatrix}^T$ by equation(11).

$$\begin{cases} \sin\vartheta_d = \cos\beta_d \cos\alpha_d \sin\theta + \sin\alpha_d \cos\theta \cos\gamma_{V_d} + \sin\beta_d \cos\alpha_d \cos\theta \sin\gamma_{V_d} \\ \cos\vartheta_d \cos\psi_d = \cos\beta_d \cos\alpha_d \cos\theta \cos\psi_V + \sin\alpha_d \left(\sin\psi_V \sin\gamma_{V_d} - \sin\theta \cos\psi_V \cos\gamma_{V_d}\right) \\ \qquad - \sin\beta_d \cos\alpha_d \left(\sin\psi_V \cos\gamma_{V_d} + \sin\theta \cos\psi_V \sin\gamma_{V_d}\right) \\ \cos\vartheta_d \sin\psi_d = \cos\beta_d \cos\alpha_d \cos\theta \sin\psi_V - \sin\alpha_d \left(\cos\psi_V \sin\gamma_{V_d} + \sin\theta \sin\psi_V \cos\gamma_{V_d}\right) \\ \qquad + \sin\beta_d \cos\alpha_d \left(\cos\psi_V \cos\gamma_{V_d} - \sin\theta \sin\psi_V \sin\gamma_{V_d}\right) \\ \cos\vartheta_d \cos\gamma_d = \cos\alpha_d \cos\theta \cos\gamma_{V_d} - \cos\beta_d \cos\alpha_d \sin\theta - \sin\beta_d \sin\alpha_d \cos\theta \sin\gamma_{V_d} \\ \cos\vartheta_d \sin\gamma_d = \cos\beta_d \cos\theta \sin\gamma_{V_d} - \sin\beta_d \sin\theta \end{cases} \tag{11}$$

The following equation need introduce trajectory inclination angle θ and trajectory deflection angle ψ_V. Because of the uncertainty of the quadrant of ψ and γ, each angle needs two equations to determine its sign and magnitude.

2) Instruction translation2

The formula calculating pitch overload n_y, sideslip angle β and aerodynamic roll angle γ_v is as follows:

$$
\begin{aligned}
n_y &= \frac{P(\sin\alpha\cos\gamma_V + \cos\alpha\sin\beta\sin\gamma_V) + Y\cos\gamma_V - Z\sin(\gamma_V)}{mg} \\
\sin\beta &= \cos\theta\left[\cos\gamma\sin(\psi-\psi_v) + \sin\vartheta\sin\gamma\cos(\psi-\psi_v)\right] - \sin\theta\cos\vartheta\sin\gamma \quad (12) \\
\sin\gamma_v &= \frac{\cos\alpha\sin\beta\sin\vartheta - \sin\alpha\sin\beta\cos\gamma\cos\theta + \cos\beta\sin\gamma\cos\vartheta}{\cos\theta}
\end{aligned}
$$

So the complete closed-loop control system(1) is shown in Figure 2.

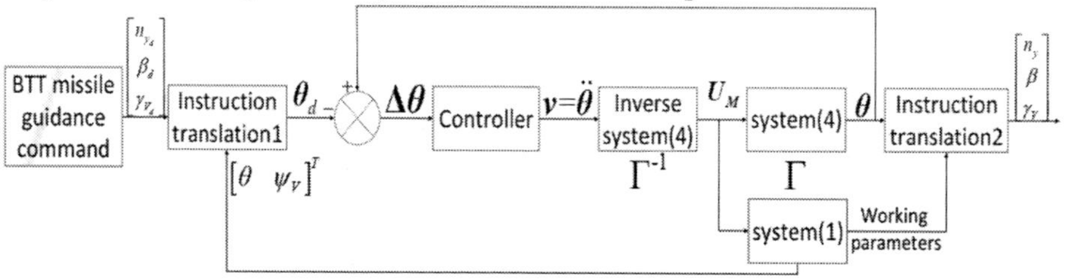

Figure 2 Control model of system(1)

The process of building control model has been completed in this section, and the controller is to be designed in the next section.

3 Sliding Mode Controller Design

The essence of design of controller is the calculation of control quantity, which is $\boldsymbol{v}$ in this paper. And the sliding mode control(SMC) can calculate a appropriate control quantity which can make system has a good dynamic performance.

SMC not only includes the idea of error feedback, but also includes the idea of designing error reduction process. When there is an error between the expected state and the actual state of control system, the control quantity given by classical PID control just makes the actual state approaching the expected state and the process of convergence cannot be strictly designed and controlled. On the contrary, the control quantity given by SMC not only makes the actual state approaching the desired state, but also makes the process of reaching the expected state is strictly designed, that is, the point representing the actual state of the system approaches the sliding mode surface at the rate of the sliding mode reaching law and finally slides on the sliding mode surface, which is the expected state of the system. In this process, we can flexibly design sliding mode surface and reaching law, of course, the premise is to ensure that the system will eventually stabilize in sliding mode surface. The following is shown in Figure 3.

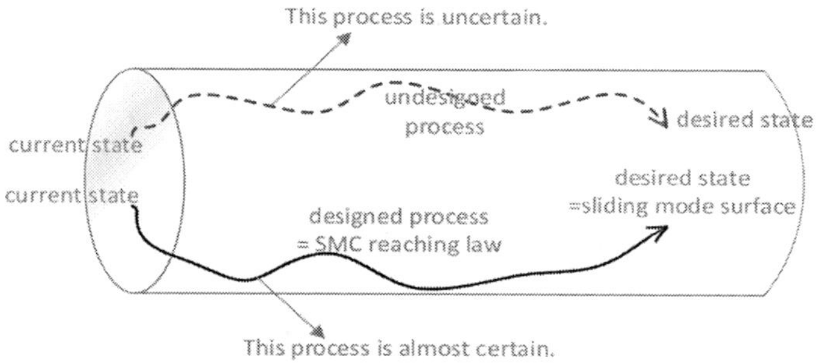

Figure 3 The difference between classical control and sliding mode control

The definition of error feedback is:

$$\begin{cases} e_\vartheta = \vartheta - \vartheta_d \\ e_\psi = \psi - \psi_d \\ e_\gamma = \gamma - \gamma_d \end{cases} \tag{13}$$

The sliding mode surface is designed as time-varying sliding mode surface:

$$\boldsymbol{S} = \begin{bmatrix} \left(\frac{\mathrm{d}}{\mathrm{d}t} + \lambda_\vartheta\right) e_\vartheta + A_\vartheta e^{-a_\vartheta t} \\ \left(\frac{\mathrm{d}}{\mathrm{d}t} + \lambda_\psi\right) e_\psi + A_\psi e^{-a_\psi t} \\ \left(\frac{\mathrm{d}}{\mathrm{d}t} + \lambda_\gamma\right) e_\gamma + A_\gamma e^{-a_\gamma t} \end{bmatrix} \tag{14}$$

As for sliding mode reaching law, we adopt the self-defined reaching law:

$$\dot{\boldsymbol{S}} = \begin{bmatrix} -\varepsilon_\vartheta \mathrm{sgn}(S_\vartheta) - k_\vartheta \left(e^{S_\vartheta} - 1\right) \\ -\varepsilon_\psi \mathrm{sgn}(S_\psi) - k_\psi \left(e^{S_\psi} - 1\right) \\ -\varepsilon_\gamma \mathrm{sgn}(S_\gamma) - k_\gamma \left(e^{S_\gamma} - 1\right) \end{bmatrix} \tag{15}$$

The parameters above are all positive numbers except A_ϑ, A_ψ and A_γ.

Take the derivative of equation(15), and bring the results into equation(16):

$$\begin{bmatrix} \ddot{\vartheta} \\ \ddot{\psi} \\ \ddot{\gamma} \end{bmatrix} = \begin{bmatrix} \ddot{\vartheta}_d \\ \ddot{\psi}_d \\ \ddot{\gamma}_d \end{bmatrix} - \begin{bmatrix} \lambda_\vartheta \dot{e}_\vartheta \\ \lambda_\psi \dot{e}_\psi \\ \lambda_\gamma \dot{e}_\gamma \end{bmatrix} + \begin{bmatrix} A_\vartheta a_\vartheta e^{-a_\vartheta t} \\ A_\psi a_\psi e^{-a_\psi t} \\ A_\gamma a_\gamma e^{-a_\gamma t} \end{bmatrix} + \begin{bmatrix} -\varepsilon_\vartheta \, \mathrm{sgn}(S_\vartheta) - k_\vartheta \left(e^{S_\vartheta} - 1\right) \\ -\varepsilon_\psi \, \mathrm{sgn}(S_\psi) - k_\psi \left(e^{S_\psi} - 1\right) \\ -\varepsilon_\gamma \, \mathrm{sgn}(S_\gamma) - k_\gamma \left(e^{S_\gamma} - 1\right) \end{bmatrix} \tag{16}$$

Equation(16) is the sliding mode controller of BTT missile.

Then prove two conclusions and analyse the reason why the sliding mode surface and reaching law are designed as shown above.

Conclusion1: If system(1) adopts equation(16) as the controller, the closed-loop system is globally exponentially stable.

The following is the proof:

Define Lyapunov function as follows:

$$V = \frac{1}{2} \boldsymbol{S}^T \boldsymbol{S} \tag{17}$$

Obviously V is positive definite function. Take the derivative of equation(17):

$$\dot{V} = \boldsymbol{S}^T \dot{\boldsymbol{S}} = S_\vartheta \dot{S}_\vartheta + S_\psi \dot{S}_\psi + S_\gamma \dot{S}_\gamma \tag{18}$$

Because the parameters in $\dot{\boldsymbol{S}}$ are all positive numbers, $S_\vartheta \dot{S}_\vartheta \le 0$, if and only if $S_\vartheta = 0$, $S_\vartheta \dot{S}_\vartheta = 0$. It has similar conclusions for other two directions of $\boldsymbol{S}$. So $\dot{V}$ is negative definite function.

According to Lyapunov second stability criterion, system(1) exists a positive definite function V and $\dot{V}$ is negative definite function, so system(1) is asymptotic stability. And also when $\|\boldsymbol{S}\|_2 \to \infty$, $V \to \infty$. So system(1) is global asymptotic stability.

Conclusion2: If parameter A_i satisfies $A_i = -\lambda_i * e_{i_0}$, $i = \vartheta, \psi, \gamma$, e_{i_0} is the error at time zero, the system will always be on the sliding mode surface from the initial moment, that is, for any time $t \in [0, \infty)$, it always exists $\boldsymbol{S} = \boldsymbol{0}$.

The following is the proof:
If

$$A_i = -\lambda_i * e_{i_0} \tag{19}$$

When $t = 0$,

$$S_i = 0 + \lambda_i e_{i_0} + A_i e^{-a_i * 0} = \lambda_i e_{i_0} - \lambda_i e_{i_0} = 0 \tag{20}$$

This lead to

$$V_0 = V|_{t=0} = 0 \tag{21}$$

In conclusion1, it has been proved that $\dot{V} \le 0$, and also $V \ge 0$, so $V \equiv 0$. Also because $V = \frac{1}{2}\boldsymbol{S}^T \boldsymbol{S}$, so

$$\boldsymbol{S} \equiv \boldsymbol{0} \tag{22}$$

It is worth mentioning that $\boldsymbol{S} \equiv \boldsymbol{0}$ is the result under the ideal condition. If some practical limitations such as the delay of rudder and the error of numerical calculation are considered, $\boldsymbol{S}$ is not always $\boldsymbol{0}$ but changes in a small range around $\boldsymbol{0}$, which means $\boldsymbol{S}$ is very close to zero vector.

The following analyses the advantages of time-varying sliding mode surface and self-defined reaching law:

According to conclusion2, when parameter A_i satisfies $A_i = -\lambda_i * e_{i_0}$, the sliding mode control above can guarantee the system state is always on the sliding mode surface, which can eliminate the reaching phase of sliding mode control. This can enhances the global robustness when the system has parametric uncertainty or external disturbance[5].

In this paper, the approach of controlling system(1) is achieved by controlling its subsystem(4), which needs calculating input instruction of subsystem(4) by real-time flight parameters. Being different from the system whose input instruction is imported directly, system(1) would produce a greater degree of oscillation, which presents a high demand on the sliding mode reaching law. Typical reaching law cannot reduce the chattering of SMC. Therefore, the self-defined reaching law is designed as follows:

$$\dot{S}_i = -\varepsilon_i \operatorname{sgn}(S_i) - k_i \left(e^{S_i} - 1\right) \tag{23}$$

where, the switching function $\operatorname{sgn}()$ is replaced by the saturation function $\operatorname{sat}()$ in order to further reduce the chattering:

$$\dot{S}_i = -\varepsilon_i \operatorname{sat}(S_i / \eta_i) - k_i \left(e^{S_i} - 1\right) \tag{24}$$

where, the saturation function $\operatorname{sat}()$ is:

$$\operatorname{sat}(S_i / \eta_i) = \begin{cases} \operatorname{sgn}(S_i) & |S_i| > \eta_i \\ S_i / \eta_i & |S_i| < \eta_i \end{cases} \tag{25}$$

The superiority in reducing the sliding mode chattering by self-defined reaching law comparing to typical reaching law is reflected through simulation in section5.

4 Adaptive control parameters

Different parameters in section3 have different control effects. It should be divided into two situations:

Situation1: When the error between the actual state and expected state is biggish, the settling time and overshoot of the step response is determined by parameters λ , a , ε and k together. In this situation, the actual state should reach expected state fast, which requires that parameters of the sliding surface and approach law are biggish.

Situation2: When the actual state is close to expected state, the settling time and overshoot of the step response is mainly determined by λ and a . In this situation ε and k have little influence on the settling time and overshot, and they start having great influence on the degree of chattering. If the value of ε and k are still biggish, the chattering phenomenon will be severe. So it requires both ε and k are smaller to reduce the chattering.

Synthesize these two situations and the control parameters can be designed in the following way. The parameters included in sliding mode surface can be designed as fixed value while the parameters included in sliding mode reaching law can be designed as variable value, which means ε and k are the adaptive parameters and their value are depended on the error between the actual state and expected state. Only in this way, it can be guaranteed that not only the controller meets the performance requirements but also the chattering degree is acceptable. The superiority of adaptive parameters comparing to fixed parameters is reflected through simulation in section5.

And the missile parameters used in this paper's simulation are given in Table 1.

Table 1 Missile parameters

Missile Parameters	Value
P	50000
m	500
J_x	7.7
J_y	1500
J_z	1400

After design, the control parameters are given in the Table 2.

Table 2 Control parameters

Control Parameters	Value	Control Parameters	Value	Control Parameters	Value
λ_ϑ	0.69	a_ϑ	1.5	k_ϑ	0.02* coe
λ_ψ	0.7	a_ψ	4	k_ψ	0.05* coe
λ_γ	0.7	a_γ	2.5	k_γ	0.05* coe
A_ϑ	-0.69* e_{ϑ_0}	ε_ϑ	1e-4* coe	η_ϑ	0.005
A_ψ	-0.7* e_{ψ_0}	ε_ψ	1e-4* coe	η_ψ	0.005
A_γ	-0.7* e_{γ_0}	ε_γ	1e-4* coe	η_γ	0.005

coe is adaptive coefficient. After design, its optimal value is as follows:

$$\text{coe}=1+\frac{999}{n_{y_c}-3.212}\left|n_y-n_{y_c}\right| \tag{26}$$

The number 3.212 is because that according to the aerodynamic data used in this paper, the pitch overload is 3.212 when the attack angle and sideslip angle is zero, the speed is 500 m/s and the air density is 1.17 kg/m^3 .

5 Simulation

Take guidance commands as $n_{y_d} = 5$, $\beta_d = 0$, $\gamma_{V_d} = 60^\circ$. And three sets of simulations are performed below.

Simulation1: the response under the condition of adaptive parameters, time-varying sliding mode surface and self-defined reaching law.

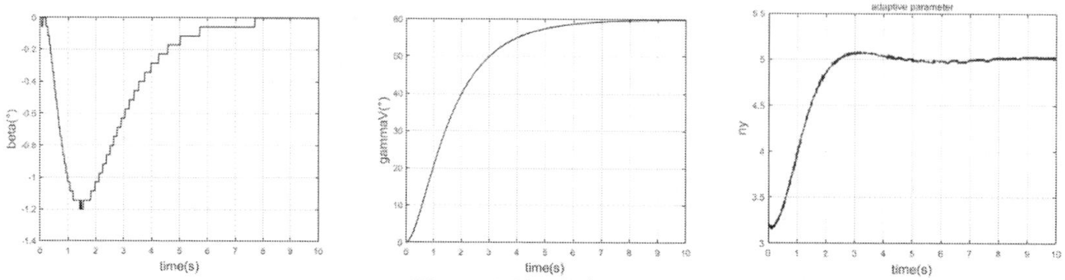

Figure 4 Simulation1

Simulation2: the response under the condition of fixed parameters($coe \equiv 1000$), time-varying sliding mode surface and self-defined reaching law.

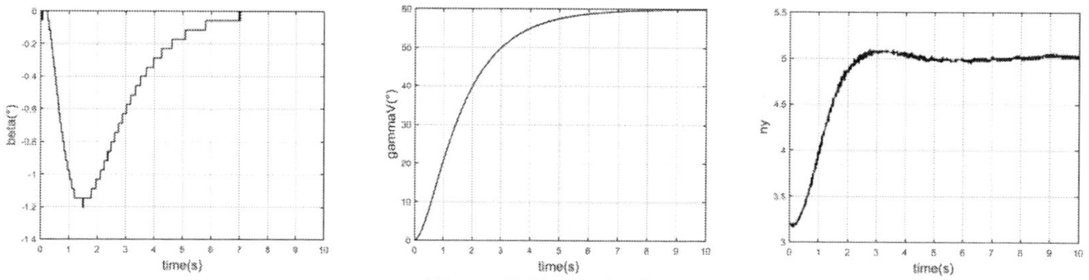

Figure 5 Simulation2

Simulation3: the response under the condition of adaptive parameters, time-varying sliding mode surface and typical reaching law.

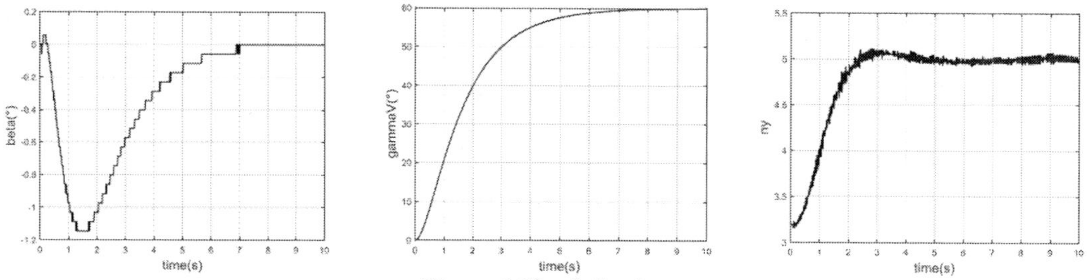

Figure 6 Simulation3

It can be found that:

1) The chattering degree of fixed parameters is more severe than the chattering degree of adaptive parameters according to Figure 4 and Figure 5;

2) The chattering degree of typical reaching law is more severe than the chattering degree of self-defined parameters according to Figure 4 and Figure 6;

3) According to Figure 4, when time of simulation is more than 5 seconds, the actual pitch overload will stabilize at 5, the sideslip angle will be less than 0.1° and the aerodynamic roll angle will be more than 58° and stabilize at 60° .

4) During the whole turning process, the maximum of sideslip angle is 1.2° . And this maximum is produced by a great maneuvering condition, which means if the missile is flying under a smaller maneuvering condition, the sideslip angle will be smaller than 1.2°. And in most time of flying, the

guidance command of missile is smoother, so the sideslip angle just has a very little error comparing with zero. So the sideslip angle corresponds to the feature of BTT missile.

Therefore, we can make a conclusion that:

The response speed is fast, the overshoot is small and the sideslip angle maintain near zero under the condition of time-varying sliding mode surface, self-defined reaching law and adaptive parameters.

6 Conclusion

This paper studies the sliding mode flight controller of BTT missile. In the second section, the control model is established according to the general dynamic equation of missile and the dynamic inverse method. Since the control model studied in this paper is derived from the general dynamic equations of missile, the control model is applicable for any missile that flies in the atmosphere, which means what this paper studies has a wider application range in engineering. In the third and fourth sections, the sliding mode controller is designed. The main contents of the design include time-varying sliding mode surface, self-defined reaching law and adaptive control parameters. The time-varying sliding mode surface is designed to satisfy the control performance requirements. Self-defined reaching law and adaptive parameters are used to solve the chattering problem. In the fifth section, three sets of simulations are carried out. The result is the sideslip angle is very close to zero, both the pitch overload and aerodynamic roll angle are able to follow the guidance instructions faster and with small overshoot, and the chattering is also very weak, which are all proves the good performance in the time domain response and the effectiveness of reducing sliding mode chattering. The control method studied in this paper is universally applicable to any BTT missile. That is, for BTT missiles with different aerodynamic layouts, the structure of the sliding mode controller need not be changed and only the control parameters need to be changed.

References

1. Hui Li, Chengqi Guan, Danjie Zhou, Fengyi Li, Xu Zha, CMCE 2010, **1**(2010)
2. Jun Yang, Shihua Li , 24th CCDC, **A**(2012)
3. Liqun Ma, Chaoyang Duan, Gongping Zhang, 29th CCDC, **A**(2017)
4. Zhaoqing Xu, SICE 2015, **A**(2015)
5. Jili Zhu, Xiangdong Liu, Liang Wang, Binglong Cong, Journal of Astronautics, **32**(2011)

Increasing the Energy Efficiency of the Power Supply System on Liquefied Natural Gas through the Use of a Combined Cycle Gas Turbine Plant

S V Zaika[a], D A Uglanov, E A Shatokhin, S S Zaika and A A Shimanov

Samara University, Department of Heat Engineering and Heat Engines, 443086 Moskovskoye shosse, 34, Samara, Russia

Abstract. The use of LNG for the production of additional energy is topical today. Very often LNG is used for gasification of settlements and industrial enterprises. LNG makes it possible to gasify objects remote from main pipelines for long distances by creating an LNG reserve directly from the consumer, avoiding the construction of expensive piping systems. However, in this case, there are large losses of cryoproduct and its low-potential heat. In this article, we propose a technical solution for compensating for the loss of cryoproducts during storage using an installation consisting of two circuits, one of which is a loss compensation loop and the other is a circuit for generating additional energy. The conducted researches showed that the use of this system allowed to return the previously used for liquefying electric power in the amount of 24%.

1 Introduction

At present, natural gas is becoming one of the most important sources of energy, as its reserves are huge, and it is environmentally friendly fuel in comparison with petroleum products. Natural gas in a compact for transportation and storage form can be contained in compressed (gaseous) and liquefied states. The most relevant at the moment is the use of liquefied natural gas (LNG) as an energy carrier for the same purposes as conventional natural gas, since this allows to reduce the mass-dimensional characteristics of tanks for storage and transportation [1].

In a liquefied state, natural gas at a pressure of 1 atmosphere is 640 times denser than under normal conditions, which significantly reduces the volume and weight of cylinders for storage and transportation, and also makes it technically possible for the accumulation, storage and delivery of large gas masses to the consumers at the required time [2]. For example, only in a liquefied state, methane makes it economically feasible to solve the problem of delivery to remote areas.

LNG enables gasification of facilities remote from long-distance pipelines over long distances by creating an LNG reserve directly from the consumer, avoiding the construction of costly pipeline systems [3].

Typically, gasification of LNG is mainly due to the heat of the environment. It is considered that the use of heat from the environment is not an energy-consuming process. However, it is worth mentioning that for the liquefaction of LNG, considerable energy was previously expended (about 1

[a] Corresponding author: zaika@ssau.ru

kWh of electricity per 1 kg of LNG), which for just this gasification experiment is simply reset to three times. Thus, liquefied natural gas (like any other cryoproduct) contains an energy potential that could be used to return it to its original gaseous state, and therefore the very process of LNG reagation has a significant potential for energy saving [4].

This energy is called cold energy and now there are a lot of science work in field of its using.

2 Application of LNG for gasification of settlements, industrial enterprises and production of additional electric energy

In case of need for gasification of remote settlements, industrial enterprises and autogas filling stations, supply of which is impossible through gas pipelines, the only way is to deliver natural gas in a liquefied state with its subsequent transfer due to evaporation into the gaseous state [5].

When supplying LNG to the gas distribution point and then through the local highways networks to residential houses, industrial enterprises and for refueling vehicles at the gas station, a regasification process in the evaporator is usually carried out, usually atmospheric type (Figure 1). This process is usually due to the supply of heat to the environment. However, with the regasification process, it is possible to return some of the energy expended for the liquefaction of natural gas.

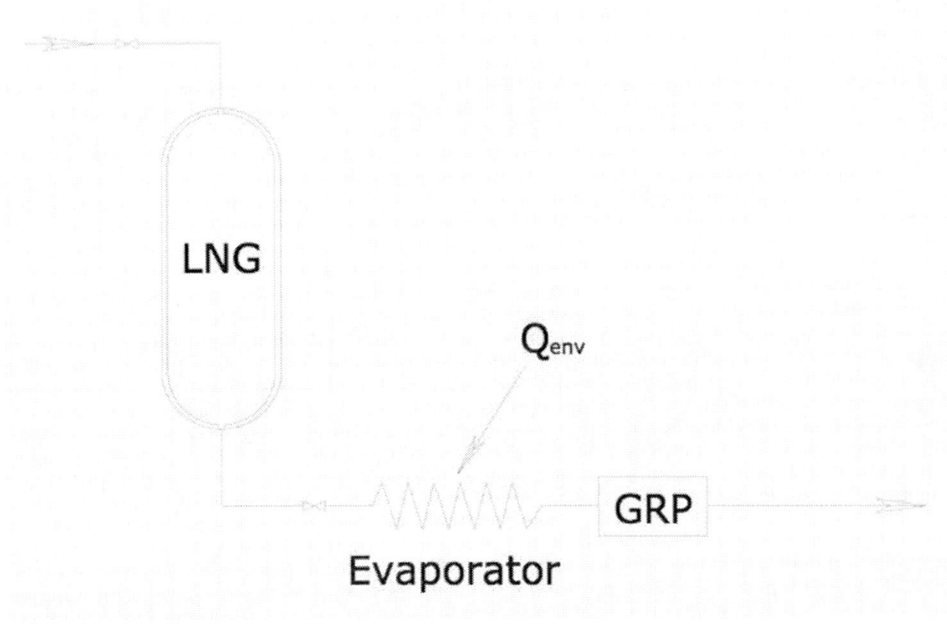

Figure 1. Diagram of the plant for regasification of LNG.

Based on the fuel consumption of the heat power plant and the combined-cycle plant G = 0.06 kg/s, it is necessary to have an LNG stock of 163 135.2 kg, taking into account the losses of the cryoproduct for evaporation. In Figure 2 (a, b, c) three different options for organizing an LNG storage facility are presented. The first combination consists of 2 tanks with a capacity of 496 m^3 (248 m^3 volume of one tank), the second one consists of 4 tanks with the capacity of 448 m^3 (112 m^3 volume of one tank), the third one - from 8 tanks with the capacity of 396 m^3 (49.50 m^3 volume of one tank). The organization and selection of more profitable LNG storage options is based on the calculation of the required fuel stock for the plant operation, the climatic conditions for the least evaporation losses during the idle time of the cylinders.

The number of tanks of certain volumes was selected on the basis of the mass of fuel required to operate the system for 32 days, which is 163 135.2 kg or 400.76 m^3 in the liquefied state (scheme b).

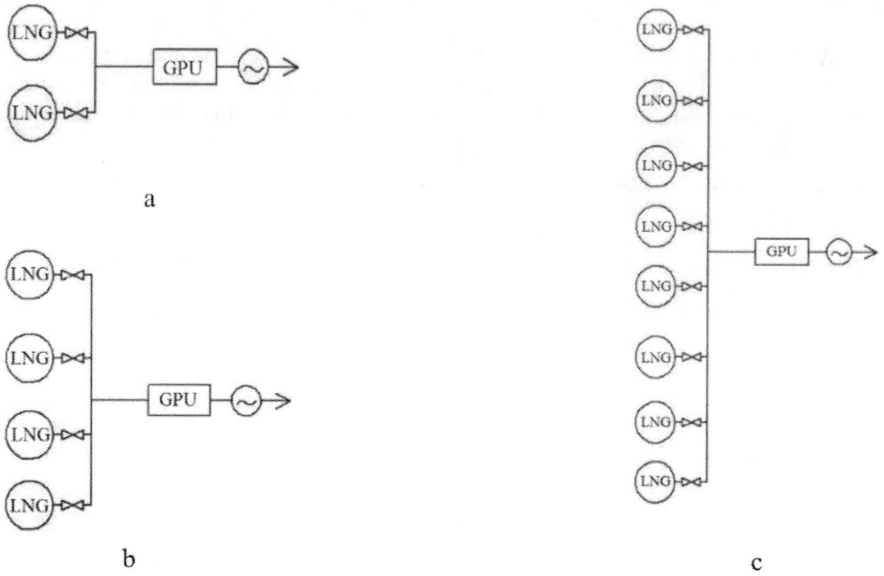

a - with two tanks, b - with four tanks, c - with eight tanks.
Figure 2 – Variant of organization of the LNG storage facility.

In scheme b (Table 1), each tank is designed for 8 days of continuous operation. While the first tank is used, the second tank is idle for 8 days, the third – for 16 days, and the fourth – for 24 days. Based on daily losses of selected reservoirs, the value of losses for the entire operation time of the plant is determined. Similarly, the waiting time for the remaining LNG storage options, which is presented in Table 2, has been calculated.

Table 1. Tank activation.

	Waiting time (inactivity), day								Total time
Number of tanks	1	2	3	4	5	6	7	8	
Scheme a	0	16	-	-	-	-	-	-	16
Scheme b	0	8	16	24	-	-	-	-	48
Scheme c	0	4	8	12	16	20	24	28	112

During operation, tanks are used alternately. Losses in the used tank are not taken into account, while in the remaining tanks losses occur while they are in the standby mode. The calculation of loss values is performed depending on the selected storage scheme and the characteristics of the tanks involved (Table 2).

Table 2. Losses of LNG for all time in all tanks in the schemes of organization of the storage park a, b, c.

The scheme	a	b	c
Number of tanks	2	4	8
V, m^3	248	112	49,50
M, kg	84 240	42 212	21 060
% per day	0,09	0,12	0,12
Losses in mass in one tank per day, kg	75,816	50,65	25,272
Losses for all time in all tanks, kg	1213,06	2431,2	2830,46

To compensate for the loss of the cryoproduct and its low-potential heat, a low-temperature power plant based on the steam-powered cycle was calculated.

Thus, to provide consumers with electric and heat energy, a combined-cycle plant (CCGT) with a capacity of 1 MW with a circuit for generating additional energy on the basis of a heat power plant was designed as an energy installation.

In this paper, we propose a technical solution for reducing the loss of cryoproducts during storage by using a plant consisting of two circuits, one of which is a loss compensation loop and the other is a contour for generating additional energy. The circuit for generating additional energy can be based on a steam power plant (Figure 3).

The principle of operation of the loss compensation circuit during evaporation is as follows. Liquefied natural gas is charged into the tank with a capacity in accordance with the selected scheme of the LNG storage facility (Table 2) at a temperature of 138 K and a pressure of 0.6 Mpa. The evaporated natural gas flows through the drain valve to the receiver. Accumulated losses per day are compressed by a centrifugal compressor up to 1 Mpa. The operating time of the compressor is 20 minutes. The drive of a centrifugal compressor is an electric motor powered by a thermoacoustic engine, a steam power plant and a Stirling engine, depending on the chosen scheme. After the compressor, the heated gas is cooled by liquid methane in the heat exchanger-evaporator and then throttled to 0.65 Mpa. Liquefied gas returns to the tank.

The contour of generating additional energy in the energy system is supposed to use a thermal power plant as the main power converter. The heat in the plant is supplied from the exhaust gases of the hcp in the heat exchanger-evaporator, and the condensation of the working fluid takes place in the heat exchanger-condenser due to the low-potential heat of LNG.

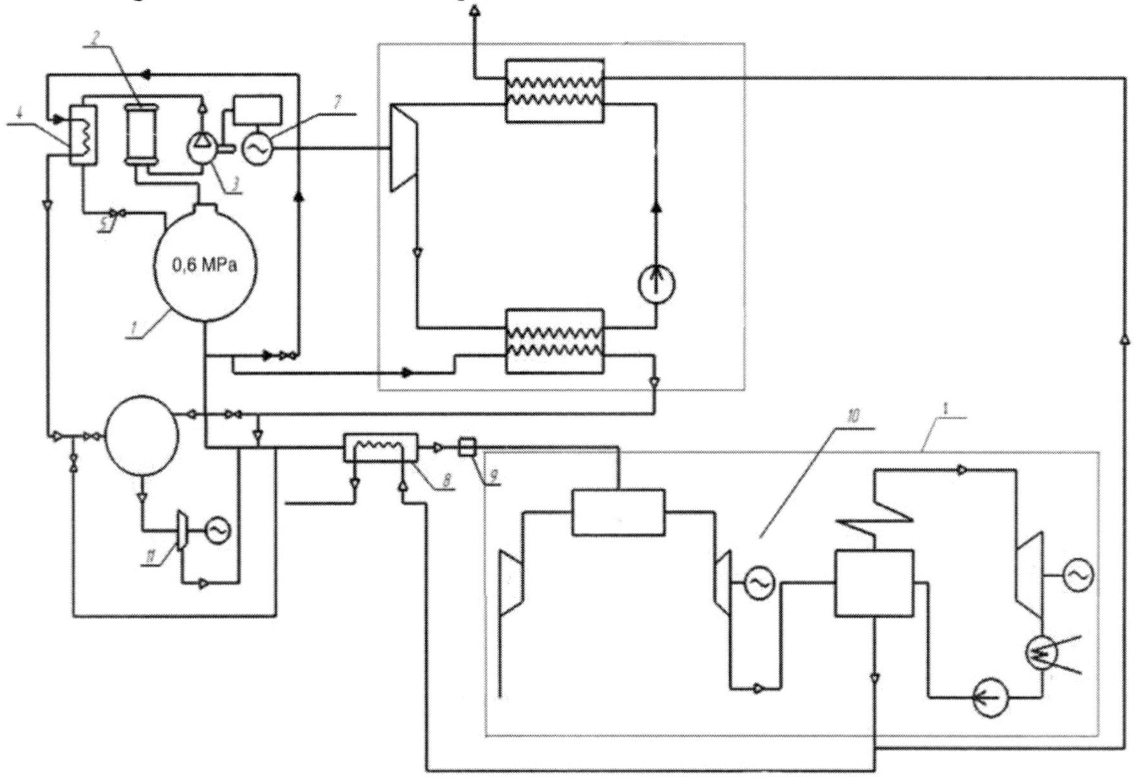

1 – tank with liquefied natural gas (LNG); 2 – gas receiver; 3 – the compressor; 4 – heat exchanger-evaporator; 5 – throttle valve; 6 – steam power plant; 7 – electric generator; 8 – heat exchanger-utilizer; 9 – reducer; 10 – Combined-cycle plant (CCGT); 11 – expansion turbine.

Figure 3. Schematic diagram of the installation based on the thermal power plant and CCGT unit.

3 Calculation of the heat-power and combined-cycle plant

It is the main circuit for the production of electrical energy and has the designation I (Figure 3).

The calculation of the cycle of a gas turbine unit (GTU) is made by characteristic points and is presented in Table 3. The GTU cycle is presented (Figure 4). Calculation of the cycle of a combined-cycle turbine by characteristic points is presenter in Table 4.

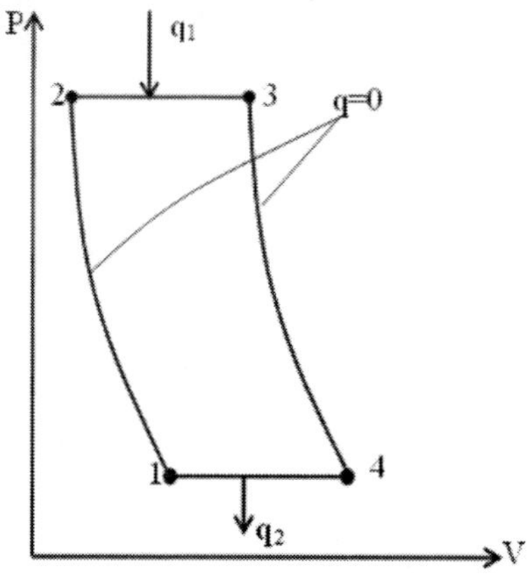

Figure 4. Gas turbine system cycle.

Table 3. Calculation of cycle GTU by characteristic points.

Initial data:	Results of calculation:
$P_1 = 0,1$ MPa	$P_2 = 1,57$ MPa
$T_1 = 288$ K	$P_3 = 1,57$ MPa
$T_3 = 1500$ K	$P_4 = 0,1$ v
$T_4 = 683$ K	$T_2 = 547$ K
Estimated capacity GTU was $W_{GTU}=897,1\ kW$	

Table 4. Calculation of the cycle of a combined-cycle turbine by characteristic points.

Thermodynamic parameters	1	2	3	4	5	6
t, K	683	373	373	390	468	468
P, bar	14	1,0	1,0	2,1	14	14
I, kJ/kg	3278	2678	434,5	530	NA	2797
V, m^3/kg	0,222	1,685	NA	NA	NA	0,141
s, kJ/kg*K	7336	7378	1362	1442	2345	6451
The calculated power W_{ST}=111,2 kW.						

$$W_{STU(1)} = W_{GTU} + W_{ST} = 1008,3\ kW. \quad (1)$$

Similarly to the steam turbine unit, the turbine of the additional electric energy installed in the power generation circuit, which has the designation II, in which different working bodies, ethane or ethylene (and) were used, was calculated from characteristic points.

Turbine power of the circuit for generating additional electric energy:

- on ethane 51,4 kW;
- on ethylene 41,1 kW.

As calculations have shown, the greatest amount of additional electric energy can be obtained by using ethane as a working fluid.

The total capacity of the entire system will be 813.8 MWh for the entire period of the system operation.

4 Conclusion

Based on the conditions of the task, it is necessary to use 163135.2 kg of LNG for 32 days (768 hours), it is also necessary to note that 1 kilowatt-hour is needed to liquefy 1 kilogram of natural gas, accordingly, 163 135, 2 kWh of electrical energy, which is included in its cost. The calculated system, namely the circuits for generating additional electrical energy and compensating for losses will make it possible to obtain 51.4 kW or 39,475.2 kWh of electrical energy during the entire operation of the system.

Thus, the calculations show that the use of this system will allow to return the previously used energy for the liquefaction of 24%, thereby compensating for the partial costs of the fuel purchased.

References

1. Burtsev, S.A., Karpenko, A.P., Leontiev, A.I. A method for distributed production of liquefied natural gas at gas-distribution stations. (2016) High Temperature, 54 (4), pp. 573-576.
2. A.M. Arkharov, I.D. Kunis. *Cryogenic refueling systems of launch rocket and space complexes*, publishing house MSTU. NE Bauman, 2006.
3. G. Zhang, W. Xu, Y. Yang, D. Zhang, Energy Procedia, *Utilization of LNG cryogenic energy in a proposed method for inlet air cooling to improve the performance of a combined cycle*, **Volume 61**, p. 899-903, (2014).
4. Y. Xiong, P. Luo, B. Hua, Energy Procedia, *A novel CO2-capturing natural gas combined cycle with LNG cold energy utilization*.**Volume 61**, pages 899-903, (2014).
5. E.V. Blagin, A.I. Dovgyallo, S.O. Nekrasova, D.V. Sarmin, D.A. Uglanov, Applied Thermal Engineering *Estimation of the energy efficiency of cryogenic filled tank using in different system and devices*. 2016. **Volume 101**, 25 May 2016, Pages 537-544

Prediction Method of Equipment Degradation State Based on Improved RVM

Lu Cheng[1,a], Wang RuiQi[1], Xu TingXue[1] and Chen YuQi[1]

[1] *Coastal Defense College, NAAU Yantai, China*

Abstract. In order to improve the prediction accuracy of the relevance vector machine model, an improved method for equipment condition prediction is proposed. First of all, an improved kernel function of variance Gauss kernel (VGKF) is constructed to improve the global performance and generalization ability of the kernel function. Then, by using the method of selecting the number of adjacent points in the chaotic sequence local prediction method, the H-Q criterion was used to optimize the embedding dimension of the training space to avoid the blindness of subjective selection. Through the prediction example of terminal guidance radar equipment test parameters, the effectiveness and superiority of the improved RVM were verified.

1 Introduction

State prediction technology is the key technology of condition-based maintenance strategy. It integrates the current and historical status monitoring data of equipment to determine its status and development trend in the future, so as to provide decision-making information for maintenance support[1].

In recent years, the research of equipment condition prediction mainly focuses on grey theory, artificial neural network, cloud theory and correlation vector machine[2-5], but the equipment history test data usually has the characteristics of small sample, nonlinear and dynamic uncertainty, and the traditional theory and method have some limitations.

Relevance vector machine[6-7] is a new machine learning algorithm based on sparse Bayes learning theory proposed by TIPPING in 2001. Compared with SVM, its kernel function is not limited by Mercer conditions, so it is more flexible to select; the parameter setting is simple, and the complex additional parameters are avoided; the sparsity of the solution is much higher than that of SVM, and it has better generalization ability.

Based on the above analysis, in order to further improve the prediction accuracy of the correlation vector machine model, an improved RVM equipment condition prediction method is proposed. Finally, the prediction performance of the improved RVM is verified by an example of terminal guidance radar equipment.

2 RVM Prediction Model

[a]Corresponding author: kvcelu@163.com

RVM is a nonlinear sparse Bayesian learning theory. Its good generalization performance and less correlation vector make it a better application in prediction field. The principle of the regression model is as follows:

Let be a training dataset with the sample number of N, $\boldsymbol{x}_i \in \mathrm{R}^D$ is D dimension input vector, $t_i \in R$ is output scalar. Assuming that they are all independently distributed, the relationship between the two is represented as follows:

$$t_i = y(\boldsymbol{x}_i, \omega) + \varepsilon_i \tag{1}$$

Among them, $\varepsilon_i \sim N(0, \sigma^2)$ is sample Gaussian noise. An expression similar to SVM, which can be represented by a series of kernel functions:

$$y(\boldsymbol{x}, \omega) = \sum_{i=1}^{N} \omega_i K(\boldsymbol{x}, \boldsymbol{x}_i) + \omega_0 \tag{2}$$

Among them, $\boldsymbol{\omega} = [\omega_0, \omega_1, \cdots, \omega_N]^T$ is the weight parameter vector, and $K(\bullet,\bullet)$ is the kernel function. So it can be deduced that $p(t_i|x_i) = N(t_i|y(\boldsymbol{x}_i, w), \sigma^2)$, t_i satisfies $N(y(\boldsymbol{x}_i, w), \sigma^2)$. In order to facilitate the expression and introduce a super parameter, the likelihood function of the entire training sample data set is represented as:

$$. p(\boldsymbol{t} \mid \boldsymbol{\omega}, \beta) = (\frac{\beta}{2})^{N/2} \exp\left\{-\frac{\beta}{2}\|\boldsymbol{t} - \boldsymbol{\Phi}\boldsymbol{\omega}\|^2\right\}. \tag{3}$$

Among them, $\boldsymbol{t} = [t_1, t_2, \cdots, t_N]^T$, $\boldsymbol{\Phi} \in \mathrm{R}^{N\times(N+1)}$ is the design matrix, defined as $\boldsymbol{\Phi} = [\mathrm{f}(\boldsymbol{x}_1), \mathrm{f}(\boldsymbol{x}_2), \cdots, \mathrm{f}(\boldsymbol{x}_N)]^T$, the base function vector is $\mathrm{f}(\boldsymbol{x}_i) = [1, K(\boldsymbol{x}_i, \boldsymbol{x}_1), \cdots, K(\boldsymbol{x}_i, \boldsymbol{x}_N)]^T$, $i = 1, 2, \cdots, N$.

The training of sample data using RVM is to obtain the posterior distribution of the weight parameter vector $\boldsymbol{w}$. It is assumed that the prior distribution of ω_j satisfies the Gaussian distribution with a mean value of 0 and a variance of .., expressed as:

$$p(\boldsymbol{\omega}|\boldsymbol{\alpha}) = \prod_{j=0}^{N} N(\omega_j|0, \alpha_j^{-1}) \tag{4}$$

Among them, each α_j in the hyper parametric $\boldsymbol{\alpha} = [\alpha_1, \alpha_2, \cdots, \alpha_N]$ is independent of each other and is only related to the corresponding weight ω_j. By means of formula (3) and formula (4), the mean matrix $\boldsymbol{\mu}$ and the covariance matrix å, respectively:

$$å = (\beta\boldsymbol{\Phi}^T\boldsymbol{\Phi} + A)^{-1} \tag{5}$$

$$\boldsymbol{\mu} = \beta å \boldsymbol{\Phi}^T \boldsymbol{t} \tag{6}$$

among them, $A = diag(\alpha_0, \alpha_1, \cdots, \alpha_N)$. To obtain the posterior distribution of $\boldsymbol{\omega}$, two impact parameters β and α_j must be optimized, the specific method is to maximize the edge likelihood function $p(\boldsymbol{t}|\boldsymbol{\alpha}, \beta)$. After obtaining the objective function on both sides of the $p(\boldsymbol{t}|\boldsymbol{\alpha}, \beta)$ equal sign, the partial derivative of α_j and β is obtained according to the objective function, and the derivative is 0, and the iterative formula of two parameters is obtained:

$$\alpha_j = \frac{1}{\mu_j^2 + \sum_{jj}} = \frac{\gamma_j}{\mu_j^2}, j = 0, 1, \cdots, N \tag{7}$$

$$\beta = \frac{N - \sum_{j=0}^{N} \gamma_j}{\|\boldsymbol{t} - \boldsymbol{\Phi}\boldsymbol{\mu}\|^2} \tag{8}$$

$$\gamma_j = 1 - \alpha_j \sum_{jj} \tag{9}$$

Among them, μ_j is the j element of $\boldsymbol{\mu}$, and Σ_{jj} is the j diagonal elements of covariance matrix å.

The training of the RVM model is to update the $\boldsymbol{\mu}$ and å by iterative computation until the parameter converges until the parameter converges or reaches the maximum number of iterations. In the convergence of the parameter estimation, $\boldsymbol{\alpha}_{MP}$ and σ_{MP} are obtained by the maximum likelihood method. Given the new input value, $\boldsymbol{x}_*$, the predicted mean y_* and variance σ_*^2 are:

$$y_* = \boldsymbol{\mu}^T \phi(\boldsymbol{x}_*) \tag{10}$$

$$\sigma_*^2 = \sigma_{MP}^2 + \phi(\boldsymbol{x}_*)^T \text{å} \phi(\boldsymbol{x}_*) \tag{11}$$

3 Parameter Optimization of RVM Regression Model

3.1 Improvement of Gauss kernel function

Under the prior structure, the RVM algorithm removes the irrelevant sample points by using the auto correlation judgment theory, thus realizing the sparseness of the model [8]. Although compared with SVM, the kernel function of RVM is independent of the restriction of Mercer condition, but the performance of the predefined kernel function will directly affect the accuracy of RVM regression prediction.

The commonly used kernel functions include Gauss kernel function, Laplace kernel function, Cauchy kernel function and so on. Gauss kernel function is the most widely used kernel function, and its standard form is defined as $K_{std}(\boldsymbol{x}, \boldsymbol{x}_i) = \exp\left\{-\frac{\|\boldsymbol{x}-\boldsymbol{x}_i\|^2}{2b^2}\right\}, i = 1, 2, \cdots N$, among them, the parameter b is the nuclear width, and N is the training sample number.

As a kernel function based on the distance equation, the test points of Gauss kernel function tend to become sparse when they are mapped from the source space to the high-dimensional feature space. The above can be changed only when the nuclear function is not only able to maintain a rapid attenuation at the test point, but also to maintain a certain attenuation at infinity. However, the Gauss kernel function only satisfies the former, so it is a typical local kernel function. In recent years, many scholars composed of mixed kernel by constructing Gauss kernel function and global kernel function to improve the performance of kernel function [9-11], but in this process, multiple parameter variables are introduced, which greatly increases the computational complexity of RVM. Based on the Gauss kernel function, this paper improves the performance of RVM by constructing an improved variance Gauss kernel function.

It can be observed that in the standard form of Gauss kernel function, the characteristics of all elements of the scale factor uses the same measure. This feature also makes the characteristics of large numerical scale range more dominant, and this limitation seriously affects its performance in practical applications. To overcome this problem, we normalize all the feature elements by zero mean, and mark them as unit standard deviations:

$$\bar{x}_i = \frac{x_i - \mu_i}{\sigma_i} \tag{12}$$

Among them, μ_i and σ_i are the mean and standard deviations of the eigenvector, and the formula (12) is substituted into the standard form, and the result is rewritten into the form of the kernel function, so a new kernel function is generated:

$$K(\boldsymbol{x}, \boldsymbol{x}_i) = \exp\left\{-\frac{\|\boldsymbol{x}-\boldsymbol{x}_i\|^2}{2\sigma_i^2 \bullet b^2}\right\}, i = 1, 2, \cdots N \tag{13}$$

Since σ_i^2 is the variance of the vector i, we call formula (13) the variance Gauss kernel function, so that the eigenvectors are characterized by different scaling factors. Further analysis makes it possible for the function to maintain a slow decay rate while away from the sample points, consider putting the Euclid distance equation on the denominator of the exponential function so that the improved VGKF expression follows:

$$K(x,x_i)=\exp\left\{-\frac{2\sigma_i^2\bullet b^2}{\|x-x_i\|^2}\right\},i=1,2,\cdots N \tag{14}$$

The kernel function constructed in this way not only meets the two conditions mentioned above, but also avoids the introduction of unknown parameters. Moreover, the normalization process of data is not needed before the input of model data.

3.2 Optimization of training sample space dimension

The key to realize the time series prediction of $\{x_1,x_2,\cdots,x_n\}$ is to find the accurate mapping of the sliding time window between $X_t=\{x_{t-1},x_{t-2},\cdots,x_{t-m}\}$ and output $Y_t=\{X_t\}$, the mapping is $f:R^m\to R$, $t=m+1,m+2,\cdots,n$, m is used to predict embedding dimension. Therefore, the determination of parameter m is crucial for predicting the accuracy of modeling. In order to overcome the blindness of subjective m selection, the local chaotic time series prediction of the number of neighboring points selection method based on the idea of training sample space[12], the optimal embedding dimension of RVM were optimized, the method is suitable for small amount of data, has simple calculation and high precision.

According to the idea of small sample training in local method[13-14], the predictive value of target points can be inferred by the motion trend of adjacent phase points, that is to approximate the evolution value of current vector $X(n)$ and future value $\hat{x}_{n+T}$ by mapping, the approximate process of this mapping is achieved by relying on the d adjacent point $X(n)$ of $X(n_i)(i=1,2,\cdots,d)$. In this paper, we consider the use of neighborhood points to train RVM, and propose a method for optimizing the spatial dimension of RVM training samples. $X(n_i)(i=1,2,\cdots,d)$ is used as input sample, and $x_{n_i+1}(i=1,2,\cdots,d)$ is used as output value to train RVM regression prediction, then the weight coefficients and correlation vectors are obtained. In this way, the local linear model is replaced by the RVM regression model, and the good approximation performance of RVM is fully utilized. After training, an optimal function is obtained to fit $f(\bullet)$, and the prediction value of the subsequent data can be obtained by calculating $\hat{x}_{n+T}$.

In the process of determining the number of adjacent points, this paper is based on the Hannan-Quinn criterion[15], and the process is as follows:

First, a broad range of d is given to the $d\in[d_{\min},d_{\max}]$, and the HQ criterion values for each d value are calculated sequentially:

$$C(d)=\ln\sigma(d)^2+(d+1)\frac{H\ln(\ln N)}{N} \tag{15}$$

$$\sigma^2(d)=\frac{\frac{1}{S}\sum_{j=1}^{S}(x_j-x_j')^2}{\overline{x}^2} \tag{16}$$

Among them, H is the constant representing the weight, $H>2$ is generally used, N is the number of samples to be predicted, σ^2 is the fitting variance, x_j is the sample data point, $\overline{x}$ is the sample mean, and x_j' is the sample prediction value, When the minimum value of formula (15) is reached, the corresponding d_{opt} value is the optimum number of adjacent points. At the same time, the model achieves the best balance of accuracy and complexity in theory.

In this process, the spatial dimension of training samples is optimized by explicit $m = d_{opt}$, which overcomes the shortcomings of previous subjective training based on subjective experience and FPE criterion, and has certain reference and practical value.

4 Forecasting Examples and Analysis

Take the missile in the storage state of the troops as an example, from the beginning of 2009 to the end of 2016, the terminal guidance radar unit test was carried out every quarter, and 32 sets of test data of 2 key parameters were obtained, as shown in Table 1. Based on this, the validity of the improved RVM model was verified.

Table 1. Test Data of Key Parameter

Number	Prediction parameter	
	c_1	c_2
1	4.47	28.35
2	4.46	28.69
...	...	...
31	4.13	34.37
32	4.02	34.95

The previous 25 sets of data were used as training data for RVM, and the latter 7 sets of data were used to test the predictive effect of the model. After the embedding dimension is determined by the H-Q criterion, the RVM is trained by 21 training samples after phase space reconstruction. Using the improved RVM and the standard RVM model (the Gauss kernel width is set to 4), 2 performance parameters are predicted, as shown in Figure 1 and figure 2.

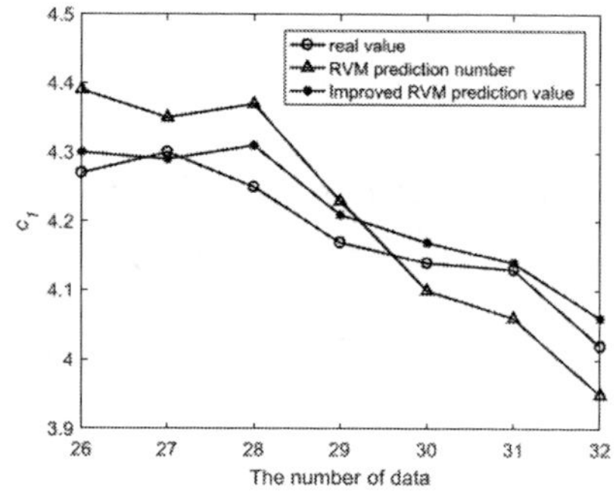

Figure 1. Prediction curves of parameter c_1

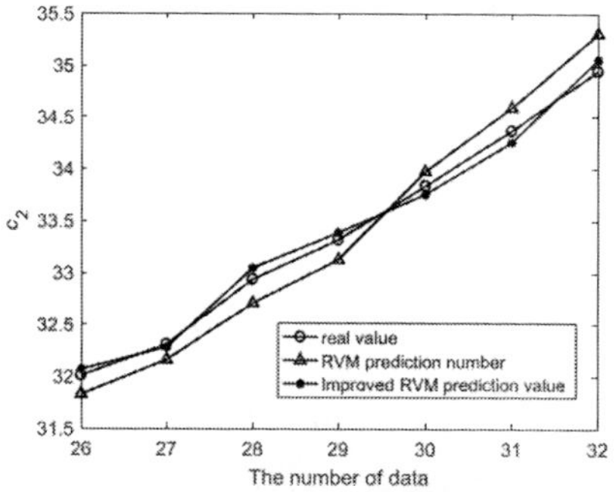

Figure 2. Prediction curves of parameter c_2

Furthermore, the average prediction error of the two is compared, and the average absolute percentage error and the correlation vector are used as the predictive performance evaluation indicators, as shown in table 2.

Table 2. Comparison of Prediction Results

Prediction method	c_1		c_2	
	MAPE/%	**nRV**	**MAPE/%**	**nRV**
RVM	1.81	6	0.63	5
Improved RVM	0.75	4	0.24	4

As can be seen from table 2, the improved RVM's prediction MAPE is lower than the standard RVM model, and the number of correlation vectors is also reduced. It shows that it has better prediction performance, which verifies the effectiveness and superiority of the proposed method.

5 Conclusion

In order to improve the prediction accuracy of equipment status, the RVM model is improved in this paper. By constructing the variance Gauss kernel function, the global performance and generalization ability of the kernel function are improved. Then, the H-Q algorithm is used to optimize the embedding dimension of the training space to avoid the blindness of its subjective selection. The performance prediction example of the terminal guidance radar proves that the improved RVM has higher prediction accuracy, and it is easy to calculate and has a good prospect of engineering application. At present, the prediction accuracy of RVM model is limited by single kernel function, In the future, the RVM based on the combination kernel function will be the focus of the next research.

References

1. Ma Sa Sa, Jia Xisheng, Xia Lianghua, "Review of military equipment maintenance engineering CBM," Journal of Institute of equipment, vol.19, pp. 111-116, February 2008.
2. Yingfeng L, "The Application of Improved Grey Model in Power System Load Forecast," Power System and Clean Energy, vol.3, pp. 10-11, 2009.
3. DongQin L, "Application of Support Vector Machine Regression Method in Prediction of Ship's Price Index," Ship & Ocean Engineering, vol.38, pp.104-106, February 2009.
4. Zhu Jianping, "Research on Application of neural network in power load forecasting," Science and technology information, vol.13, pp. 32, August 2015.
5. Zhao Wenbang, Wang Shaohua, Pu Ying, etc, "Prediction of equipment technical status based on cloud theory," Fire Control & Command Control, vol.38, pp.62-65, July 2013.
6. Tipping M E, "Sparse Bayesian learning and the relevance vector machine," Journal of machine learning research, vol.1, pp.211-244, June 2001.
7. Samui P, Dixon B, "Application of support vector machine and relevance vector machine to determine evaporative losses in reservoirs," Hydrological Processes, vol.26, pp.1361-1369, September 2012.
8. Zhou Jianbao, Research on prediction method of remaining life of lithium ion battery based on RVM, Harbin: Harbin Institute of Technology, 2013.
9. Yu J, "A nonlinear kernel Gaussian mixture model based inferential monitoring approach for fault detection and diagnosis of chemical processes," Chemical Engineering Science, vol.68, pp.506-519, January 2012.
10. Wei L, Wei B, Wang B, "Text Classification Using Support Vector Machine with Mixture of Kernel," Journal of Software Engineering and Applications, vol.5, pp.55, December 2013.

11. Duan Qing, Zhao Jianguo, Ma Yan, "Optimized combination kernel function correlation vector machine model for power load forecasting," Journal of electrical engineering and control, vol.14, pp.33-38, June 2010.
12. Du Jie, Cao family, Liu Zhijian, etc, "Multi step prediction model of local high-order Volterra filter for chaotic time series," Acta phys Sinica, vol.58, pp.5997-6005, September 2009.
13. Qing-Fang M, Yue-Hui C, Yu-Hua P, "Small-time scale network traffic prediction based on a local support vector machine regression model," Chinese Physics B, vol.18, pp2194, June 2009.
14. Bing Q, Gong B, Yang Z, et al. "Short-Term Traffic Flow Local Prediction Based on Combined Kernel Function Relevance Vector Machine Model," Math. Probl. Eng, vol.9, pp.1-9, Octorber, 2015.
15. Hannan E J, Quinn B G, "The determination of the order of an autoregression," Journal of the Royal Statistical Society. Series B (Methodological), pp.190-195, 1979.

Hertz Model Based Contact Modeling for Joints with Clearance

Jielong Wang[1,a] and Bing Shi[2]

[1]No.423 Bld.101, Future Science Park, Xiaotangshan, Changping Dist., Beijing, 102211, China
[2]No.518 Bld.101, Future Science Park, Xiaotangshan, Changping Dist., Beijing, 102211, China

Abstract. This paper presents the techniques of contact modeling for revolute joints in flexible multibody systems, in which the dry clearance revolute joints have been coupled with the flexibility of connected bodies. The contact model for revolute joints takes into account the relative planar motion caused by the clearance between the outer and inner races. This model applies a penalty method to simulate the phenomenon of inner-penetration between contact/impact bodies. The relationship between the normal contact force and the inner-penetration is described by the nonlinear Hertz model with energy dissipation. Meanwhile, the friction force can be predicted from continuous Coulomb's law. Finally, an example of flexible multibody systems has been simulated by using the developed contact models.

1 Introduction

Engineering practices show that the clearances in the revolute joints are inevitable to allow for the assembly of the pair elements. Furthermore, manufacturing tolerances[1], wear[2] and thermal effects[3] also cause clearances. The problems of modeling simple mechanical systems with clearance joints were extensively investigated[4-8]. A typical problem is the slider-crank mechanism, in which only the radial clearance located in the planar revolute joint is considered[9]. Since rigid modeling of the links will affect the prediction of the system real dynamic behaviour, the modelling of mechanism flexibility is also required. A comprehensive review can be found in Ref.[10], where the commercial software MSC ADAMS was applied to simulate the crank-slider mechanism with rigid/flexible links and planar joint clearances. More recently, the combined influence of the axial and radial clearances in the revolute joint[11, 12] has been studied. Ref.[13] predicts the dynamic behaviour of slider-crank mechanism with single and two revolute clearance joints. It concludes that the dynamic response in a mechanism with two clearance joints is not a simple superposition of that in mechanism with one clearance joint, and then proposes that all the joints in a multibody system should be modeled as clearance joints. The coupling problem of joint clearances and link flexibilities of the complicated mechanism is still an open topic. The innovation of this paper is the implementation of a contact model for clearance joints with the purpose of providing a general solution to the coupling problem. The new contact model is also defined in the absolute nodal coordinate systems. It can be used to connect any rigid/flexible bodies with arbitrary topologies.

[a] Corresponding author : wangjielong@comac.cc

When modeling joints with clearance, an appropriate contact/impact force model is important. Gilardi and Sharf[14] distinguish the associated force calculation approaches into two different categories, impulse-momentum and continuous. The first approach is only practical if there is only a single contact. A serious disadvantage is that it will stop the simulation when the contact/impact occurs. In second approach, the simulation is not stopped, and the contact force in each contact pair is predicted at any time during the simulation[15]. Based on a relationship between contact force and penetration, for example, by Flores etc.[16], this approach performs the contact calculation in three steps: contact detection, calculation of the contact normal force, and calculation of the friction force[17]. In this paper, an algorithm based on the continuous approach is developed to evaluate the contact/impact force based on nonlinear Hertz model of Hunt and Crossley[18], which is suitable for multiple impact pairs. When the bodies are in contact/impact, the friction force can be predicted from the continuous Coulomb's law. It is known the Coulomb's friction law[19] describes the transitions from sticking to sliding and vice-versa, but these discrete transitions can cause difficulties for their numerical simulations. Many continuous friction laws[20, 21] have been advised to approximate the discrete transitions. This paper regularizes the friction coefficient of Coulomb's law to obtain a continuous friction law, which can describe both sliding and rolling/sticking behaviour of revolute joints.

In summary, this paper focuses on studying the coupling effect of joint clearances and link flexibilities of the complicated mechanism, and designs a detailed contact model for revolute joint with clearance. The nonlinear Hertz model with energy dissipation describes the relationship between the normal contact force and the inner-penetration. The continuous Coulomb's law are implemented for the simulation of friction phenomena.

2 Contact models for revolute joints with clearance

With the unilateral contact condition[19], a contact model for revolute joints is developed in this section to capture the phenomena of contact/impact due to clearance.

2.1 Contact kinematics of revolute joints with clearance

For the modeling of clearance, the scheme of contact detection should be designed at first. A straightforward strategy is predicting the relative distance between the contact point-pair candidates with the unilateral contact condition and then make a judgment of weather the contact is happened or not.

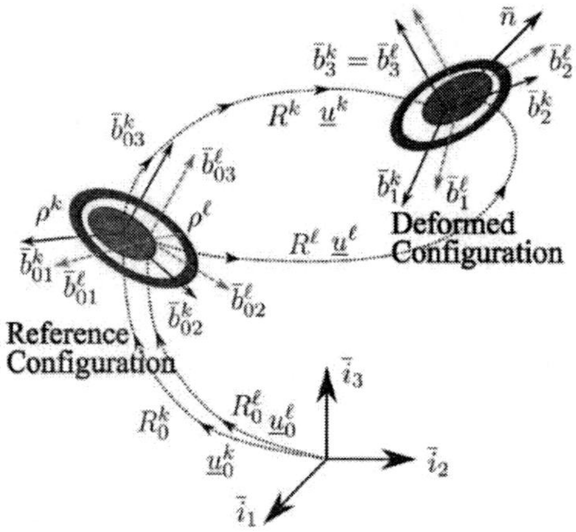

Figure 1. Configuration of revolute joint with clearance.

Observation of the geometric configuration of revolute joint with clearance, as depicted in Fig.1, shows that the contact point-pair candidates could be identified along the normal direction $\boldsymbol{n}$ defined as

$$\boldsymbol{n} = \boldsymbol{u} \,/\, \|\boldsymbol{u}\| \tag{1}$$

where $\| \bullet \|$ presents the 2-norm of vector. The positions of contact point-pair candidates on inner and outer races k and l, respectively, are written as

$$z^k = \boldsymbol{u}_0^{\,k} + \boldsymbol{u}^k + \rho^k \boldsymbol{n} \text{ and } z^l = \boldsymbol{u}_0^{\,l} + \boldsymbol{u}^l + \rho^l \boldsymbol{n} \tag{2}$$

where ρ^k and ρ^l are the radii of inner and outer races. The relative distance between the contact candidates predicts from

$$q = \boldsymbol{n}^T (z^k - z^l) \tag{3}$$

in details,

$$q = \boldsymbol{n}^T \boldsymbol{u} + \rho^k - \rho^l \tag{4}$$

During contact/impact, the inner and outer bodies will exert tension on one another and the inner-penetration will be happened. Obviously, if $q<0$, the contact/impact occurs and the inter-penetration between the contacting bodies is denoted by a negative value of q as $a = -q$. This work develops a penalty approach by adding spring and damper in the contact interface between inner and outer bodies to the modeling of unilateral contact conditions. Denoting the magnitude of normal contact forces as f_n, the normal contact forces applied to bodies k and l can be written as

$$\boldsymbol{F}_n^{\,k} = f_n \boldsymbol{n},\ \ \boldsymbol{F}_n^{\,l} = -f_n \boldsymbol{n} \tag{5}$$

In the case of friction in the contact interface there are possibilities of sticking and sliding. When sliding occurs, the relative velocity between material points which coincide with the contact candidates will compute from time derivatives of positions, z^k and z^l, as

$$\boldsymbol{v}^r = \boldsymbol{v}^l + \boldsymbol{\omega}^l \times \rho^l \boldsymbol{n} - \boldsymbol{v}^k - \boldsymbol{\omega}^k \times \rho^k \boldsymbol{n} \tag{6}$$

The component of relative velocity in the null space of normal direction $\boldsymbol{n}$ is readily obtained from

$$\boldsymbol{v}_t = (I - \boldsymbol{n}\boldsymbol{n}^T)\, \boldsymbol{v}^r \tag{6}$$

and the normalization of $\boldsymbol{v}_t$ produces the tangent vector $\boldsymbol{t}$ indicating the direction of sliding. The friction force and moment applied to bodies k and l are obtained from the associated virtual work done by the friction force, f_t, like

$$\boldsymbol{F}_t^k = f_t \boldsymbol{t},\ \boldsymbol{M}_t^k = f_t \rho^k \boldsymbol{n} \times \boldsymbol{t},\ \boldsymbol{F}_t^l = -f_t \boldsymbol{t},\ \boldsymbol{M}_t^l = -f_t \rho^l \boldsymbol{n} \times \boldsymbol{t} \tag{7}$$

2.2 Normal contact force based on Hertz model

For penalty method, the relationship between normal contact force and inner-penetration will be given by a constitutive law during contact. Conventional approach applies Hertz model[22], a spring model as the constitutive law, to derive the magnitude of normal contact force f_n from a potential, $V(a)$, like

$$f_n = \partial V / \partial a \tag{8}$$

In a generic sense, this work separates the normal force of contact into its elastic and dissipative components and a suitable expression is given as

$$f_n = f_e + f_c = \partial V/\partial a(1 + \mu\dot{a}) \tag{9}$$

where $\mu\dot{a}$ accounts for energy dissipation during contact. In the case of a Hertzian contact, the nonlinear modeling of elastic contact force is applied

$$f_e = k_h\, a^{3/2} \tag{10}$$

then the damping contact force becomes

$$f_c = \mu\, k_h a^{3/2}\dot{a} \tag{11}$$

The definitions of stiffness parameter k_h and damping coefficient μ can be found in Refs[18, 22].

2.3 Modeling of friction force

This work models the friction force based on the Coulomb's friction law, which states the friction force f_t is proportional to the magnitude of normal contact force f_n like

$$f_t = -\mu_k f_n \tag{12}$$

where μ_k is the coefficient of dynamic friction. If the relative velocity vanishes, sticking or rolling takes place. In this case, the friction force becomes

$$|f_t| \leq \mu_s f_n \tag{13}$$

where μ_s is the coefficient of static friction. This highly nonlinear phenomenon of friction coefficient switching between sliding and sticking conditions makes it very hard to deal with classic Coulomb's law numerically[23]. In current implementation, the friction coefficient is regularized to obtain a continuous friction law[24, 25], successfully treat sticking, sliding and transition regions, like

$$f_t = -\mu_k f_n \tanh(v_t/v_c) \tag{14}$$

where v_c is a characteristic velocity.

3 Bio-inspired flight flapping wing

This example deals with the kinetic problem of a stick model of flying robot, which is designed to simulate the flapping motion of birds. The internal structure of wing has been simplified to the four-bar linkages. Due to symmetry, only a half of the basic geometric configuration is depicted in Fig.2. Note that 16 points are numbered in the figure to determine the geometric topology of the four-bar linage, where the left linkage will connect to the right one through points 13 and 16. The whole structure is clamped at point 16. The coordinates of all the points are given in table 1 with the unit of length cm.

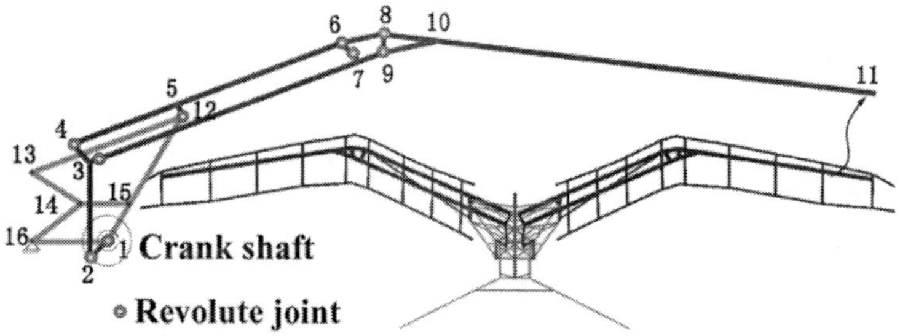

Figure 2. Configuration of four-bar linkage.

In current implementation, the numerical model of four-bar linkage is consisted of beam elements[26] and revolute joints. A user-defined function, $\phi = \omega t$, with constant angular speed ω = 12.56 rad/s, prescribes rotation of crank shaft, and the kinematic constraint[27] is enforced to the kinematic model for revolute joint located at point 1.

Table 1. Coordinates of the four-bar linkages control points.

No.	Left wing	No.	Right wing
L 1	(-4.91071, 0.00000, 0)	R 1	(4.91071, 0.00000, 0)
L 2	(-2.25663, 2.38975, 0)	R 2	(2.25663, 2.38975, 0)
L 3	(-2.22156, 9.08609, 0)	R 3	(2.22156, 9.08609, 0)
L 4	(-0.76893, 12.3488, 0)	R 4	(0.76893, 12.3488, 0)
L 5	(-11.4767, 12.7227, 0)	R 5	(11.4767, 12.7227, 0)
L 6	(-44.0461, 13.8600, 0)	R 6	(44.0461, 13.8600, 0)
L 7	(-45.4988, 10.5974, 0)	R 7	(45.4988, 10.5974, 0)
L 8	(-47.5828, 13.3630, 0)	R 8	(47.5828, 13.3630, 0)
L 9	(-48.0479, 10.7251, 0)	R 9	(48.0479, 10.7251, 0)
L10	(-52.9394, 13.4378, 0)	R10	(52.9394, 13.4378, 0)
L11	(-100.256, 14.0985, 0)	R11	(100.256, 14.0985, 0)
L12	(-11.5390, 10.9380, 0)	R12	(11.5390, 10.9380, 0)
L13	(0.00000, 8.75726, 0)	R13	(0.00000, 8.75726, 0)
L14	(-4.91071, 4.39570, 0)	R14	(4.91071, 4.39570, 0)
L15	(-7.36607, 4.39570, 0)	R15	(7.36607, 4.39570, 0)
L16	(0.00000, 0.00000, 0)	R16	(0.00000, 0.00000, 0)

The beam elements have shared the same material properties as following: mass per unit length, ρ = 0.1032 kg/m, bending stiffness, $I_{22} = I_{33}$ = 29.856 N • m^2, torsional stiffness, J = 23.303 N • m^2. Two different test cases have been performed. For case I, all the revolute joints have been modeled by using kinematic constraints. The governing equations of motion for discrete model will be DAE of index-3 with degrees of freedom, N = 1368, containing 42 Lagrange multipliers. The 2-stage Radau IIA algorithm[28] is applied to integrate the governing equations and the predicted trajectory of four-bar linkage is shown in Fig.3.

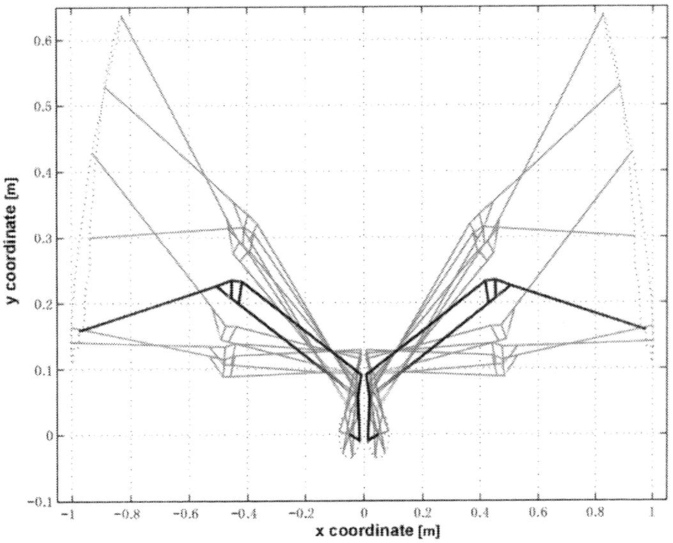

Figure 3. Trajectory of four-bar linkage.

The cyclic flapping of four-bar linkage apparently mimics the flapping motion of birds. For case *II*, the kinematic constraints at point 4 for both left and right linkages are replaced by two contact models, leaving the rest unchanged. The contact models with the same properties have been applied, and their physical properties are defined as: the radii of inner and outer races are 50 and 50.1 mm, respectively. The stiffness parameter for Hertz model is $k_h = 1 \times 10^9$ N/m$^{3/2}$. The Coulomb's law has been applied to predict the friction force, where $\mu_k = 0.6$, and $v_c = 2$ m/s. The numerical simulation results show that the clearance joints seriously affect the dynamic response of the flexible linkage and degrade the dynamic performance of connected bodies. Fig.4, presents the time histories of dynamic response of point 11 located at the tip of linkage. The apparent deviation of the simulation results of case *I* and *II* can be observed from the figure.

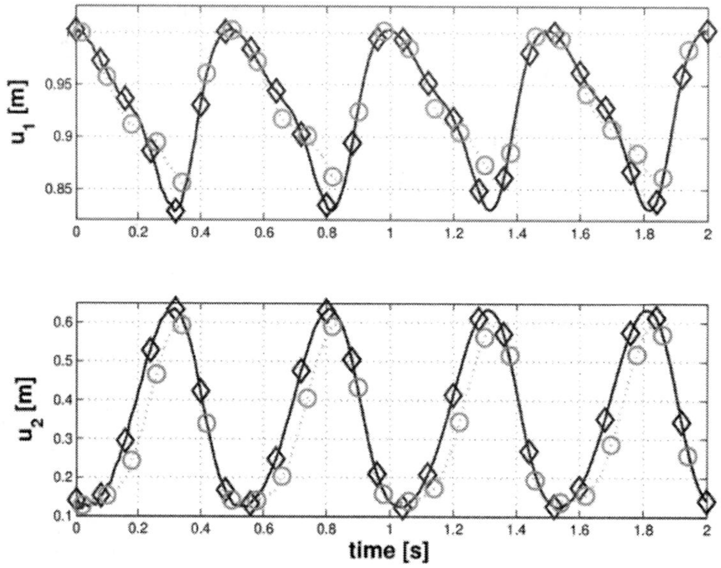

Figure 4. Time histories of position for point 11 at the tip of linkage, contact model: ○, kinematic constraints: ◇.

Acknowledgements

This paper is supported by Beijing Natural Science Foundation of China No. 3172014.

References

1. P. Flores, J.C.P. Claro, A systematic and general approach to kinematic position errors due to manufacturing and assemble tolerances. Proceedings of ASME 2007 International Design Engineering Technical Conferences and Computers and Information in Engineering Conference, Las Vegas, NV, 4-7 Sep. (2007).
2. E. Askari, P. Flores, D. Dabirrahmani, R. Appleyard, Dynamic modeling and analysis of wear in spatial hard-on-hard couple hip replacements using multibody systems methodologies. Nonlinear Dynamics, **82** (1-2), 1039--1058 (2015).
3. S. Bing, J. Ye, Dynamic analysis of the reheat-stop-valve mechanism with revolute clearance joint in consideration of thermal effect. Mechanism and Machine Theory, **43**(12), 1625--1638 (2008).
4. H. Funabashi, K. Ogawa, M. Horie, A dynamic analysis of mechanisms with clearances. Bulletin of the JSME, **21**(161), 1652--1659 (1978).
5. P. Flores, J. Ambrósio, Revolute joints with clearance in multibody systems. Computers & structures, **82**(17), 1359--1369 (2004).

6. P. Flores, P., Ambrósio, J., Claro, J.C.P., Lankarani, H.M.: Spatial revolute joints with clearances for dynamic analysis of multi-body systems. Proceedings of the Institution of Mechanical Engineers, Part K: Journal of Multi-body Dynamics, **220**(4), 257--271 (2006).
7. Q. Tian, Y. Sun, C. Liu, H. Hu, P. Flores, ElastoHydroDynamic lubricated cylindrical joints for rigid-flexible multibody dynamics. Computers and Structures, **114**, 106--201 (2013).
8. P. Flores, R. Leine, C. Glocker, Modeling and analysis of rigid multibody systems with translational clearance joints based on the nonsmooth dynamics approach. Multibody System Dynamics, **23**(2), 165--190 (2010).
9. S. Mukras, N.H. Kim, N.A. Mauntler, T.L. Schmitz, W.G. Sawyer, Analysis of planar multibody systems with revolute joint wear. J. Wear, **268**(5-6), 643--652 (2010).
10. B.A.M. Abdallah, I. Khemili, N. Aifaoui, Numerical investigation of a flexible slider-crank mechanism with multijoints with clearance. Multibody System Dynamics, **38**(2), 173--199 (2016).
11. F. Marques, F. Isaac, N. Dourado, P. Flores, An enhanced formulation to model spatial revolute joints with radial and axial clearances. Mechanism and Machine Theory, **116**, 123--144 (2017).
12. N. Akhadkar, V. Acary, B. Brogliato, Multibody systems with 3D revolute joints with clearances: an industrial case study with an experimental validation. Multibody System Dynamics, **81**,1--34 (2017).
13. H. Tan, Y. Hu, L. Li, A continuous analysis method of planar rigid-body mechanical systems with two revolute clearance joints. Multibody System Dynamics, **40**(4), 347--373 (2017).
14. G. Gilardi, I. Sharf, Literature survey of contact dynamics modelling. Mechanism and Machine Theory, **37**(10), 1213--1239 (2002).
15. C. Pereira, J. Ambrósio, A. Ramalho, Dynamics of chain drives using a generalized revolute clearance joint formulation. Mechanism and machine Theory, **92**, 64--85 (2015).
16. P. Flores, J. Ambrósio, On the contact detection for contact-impact analysis in multibody systems. Multibody System Dynamics, **24** (1), 103--122 (2010).
17. F. Marques, P. Flores, H.M. Lankarani, On the frictional contacts in multibody system dynamics. Multibody Dynamics, Springer International Publishing (2016).
18. K.H. Hunt, F.R.E. Crossley, Coefficient of restitution interpreted as damping in vibroimpact. Journal of Applied Mechanics, **42**(2), 440--450 (1975).
19. F. Pfeiffer, C. Glocker, Multibody Dynamics with Unilateral Contacts. John Wiley & Sons, New York (1996).
20. J.C. Oden, J.A.C. Martins, Models and computational methods for dynamic friction phenomena. Computer Methods in Applied Mechanics and Engineering, **52**(1-3), 527--634 (1985).
21. J.E. Shigley, C.R. Mischke, Mechanical Engineering Design. McGraw-Hill Book Company, New York (1989).
22. J. Alves, N. Peixinho, M.T. Silva, P. Flores, H. Lankarani, A comparative study on the viscoelastic constitutive laws for frictionless contact interfaces in multibody dynamics. Mechanism and Machine Theory, 85 (Supplement C), 172--188 (2015).
23. Z. Qi, X. Luo, Z. Huang, Frictional contact analysis of spatial prismatic joints in multibody systems. Multibody System Dynamics, **26**(4), 441--468 (2011).
24. O.A. Bauchau, C. Ju, Modeling Friction Phenomena in Flexible Multibody Dynamics. Computer Methods in Applied Mechanics and Engineering, **195**(50-51), 6909--6924 (2011).
25. E. Pennestrí, V. Rossi, P. Salvini, P.P. Valentini, Review and comparison of dry friction force models. Nonlinear Dynamics, **83**(4), 1785--1801 (2016).
26. J. Wang, Implementation of geometrically exact beam element for nonlinear dynamics modeling. Multibody System Dynamics, **35**(4), 377--392 (2015).
27. J. Wang, Modified models for revolute joints coupling flexibility of links in multibody systems. Multibody System Dynamics, 1-19 (2018).
28. J. Wang, Application of Radau IIA algorithms to flexible multibody system with holonomic constraints. Nonlinear Dynamics, 88(4), 2391-2401, (2017).

Dynamic Analysis of Shaft Structure Based on ANSYS/LS-DYNA

WANG Yu-ning [1,a], Zhang Zhao-gang [1], Lin Yi-zhen [1] and Yin Ming-ang [2]

[1]*Industrial Engineering Office, Engineering Technology Center, Shenyang Aircraft Industry (Group) Co., Ltd, China*
[2]*Modern Drive, Mechanical Engineering and Automation, Northeastern University, China*

Abstract. In this study, a part of the shafting structure of the accessory transmission system consisting of shaft-gear-bearing is studied. ANSYS/LS-DYNA explicit dynamics simulation software is used to simulate the dynamics of the shaft structure. It compares and analyzes the overall shaft structure, gears and bearings. The simulation results show that: The place where the equivalent stress of the shaft structure is relatively large is mainly concentrated at the tooth root and the tooth surface, where the stress is mainly in the vicinity of two or three teeth. With the change of time, the maximum equivalent stress value and position of the rolling elements are also correspondingly changed, it provides a reference for the design and optimization of gear and bearing in the actual assembly process for transmission system.

1 Introduction

The research on the dynamic characteristics and dynamics of the transmission system is gradually developed from static analysis to dynamic analysis. In recent years, the research on the mechanical characteristics and working performance of gears has mainly focused on two aspects: one is the study of tooth surface contact stress and strain when the gears are engaged, and the second is the study of gear pair dynamics during the transmission process [1-3]. The dynamic characteristics of gears are affected by a variety of random factors, and the impact of assembly errors on the dynamic contact problems of gears is of utmost importance. The bearing is an important component that transmits motion and bears load. It not only acts as a load supporter, but also serves as a moving joint. It has the advantages of easy replacement, low friction, etc., and can slide friction between a fixed body and a moving body that rotates with each other. The force becomes rolling friction, and the friction loss is small, mainly used in the automotive industry, aerospace and other fields [4-5]. In mechanical transmission, the mechanical properties of the bearing have a great influence on the dynamic characteristics of the entire system. In the moving state, the mechanical relationship between the internal parts is complex, mainly including the contact between the rolling elements and the inner and outer rings, and the rolling elements are in contact with the cage. Under no-load conditions, the rolling elements come in contact with the points. As the load increases, the point contact becomes surface contact. The position, size, shape, and frictional force distribution of the contact area are unknown before the analysis. It changes with the load and belongs to the nonlinear contact problem. The traditional research on the bearing characteristics is only in one aspect. It cannot verify the actual

[a] Corresponding author: wyn20090101@163.com

situation and the interaction with other mechanical systems [6-10]. Therefore, it is very important to establish a traditional system for shafting and to be able to effectively simulate the interaction between parts of the mechanical system.

2 Experiment procedure

This study uses a part of the shafting structure of the accessory transmission system consisting of shaft-gear-bearing as the research object. The main geometric parameters of the gear are shown in Table 1. The main geometric parameters of the bearing are shown in Table 2.

Table 1. basic parameters of gear.

Basic parameters	Driving gear	Driven gear
modulus	4	4
number of teeth	25	23
pressure angle (°)	25	25
tooth width (mm)	30	28
addendum coefficient	1	1
headspace coefficient	0.25	0.25
modification coefficient	0	0

Table 2. The size of the deep groove ball bearings (mm).

Basic parameters	Driving gear	Driven gear
inner diameter	30	25
outer diameter	72	62
bearing width	19	17
ball diameter	10.5	9.25
inside the outer ring	56.25	48.13
inner ring outer diameter	45.76	38.88
outer channel radius of curvature	5.51	4.86

The material has a modulus of elasticity of 211Gpa, a Poisson ratio of 0.3, and a gear torque of 210Nm. The solid assembly model of the shafting structure established in Pro/E is shown in Fig.1. The physical model after importing ANSYS/LS-DYNA is shown in Fig. 2.

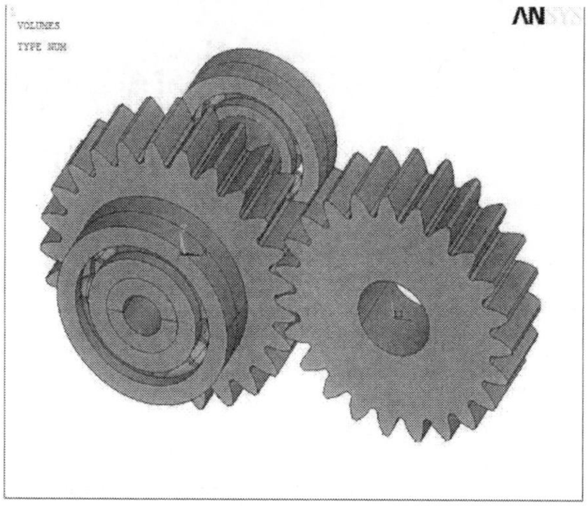

Fig.1 The Pro/E assembly drawing of shafting structure **Fig.2** The solid model of shafting structure

In the process of meshing the shafting structure, according to the characteristics of gear meshing and bearing contact, the contact position should be refined. For the overall use of Solid 164 units, sweeping and intelligent methods are used to divide. For the inner ring of the gear, the Shell163 unit is divided by the mapping method; Because the cage model is more complex, it is divided in the form of a free grid, and the dynamic finite element model of the shaft system is finally shown in Fig. 3. In ANSYS/LS-DYNA software, there is no contact unit, only the surfaces that may be contacted, the contact type and contact related parameters are defined, and the frictional force is taken into consideration. This ensures that no contact occurs between the contact surfaces during the calculation process. penetrate. This study selects the automatic contact of the surface and defines the contact pairs of the shafting system by means of the node component. The definition of nine contact pairs is shown in Fig.4.

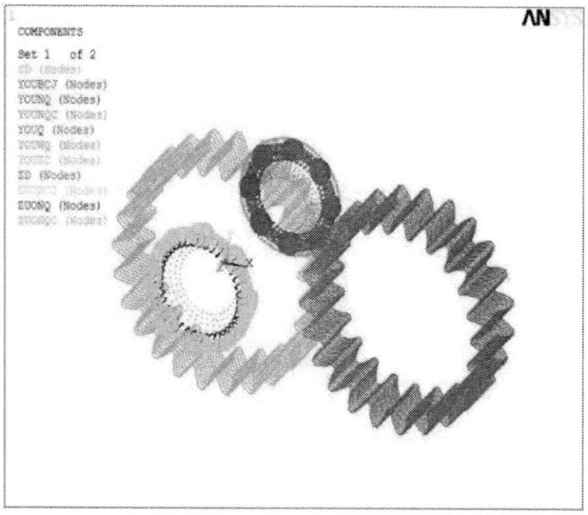

Fig. 3 The dynamic analysis finite element model of shafting structure **Fig.4** The dynamic contact

For the driving wheel with bearings, due to taking into account the analysis of the force of the driven gear during the software analysis. The drive gear shaft is set as an elastic body, and both ends of the shaft connected to the inner ring of the bearing are set as a rigid body. At the same time, in order to ensure that the shaft and the bearing have synchronous rotation speed, the inner ring of the bearing is affixed to this part of the shaft and is set as a rigid body, and a resistance moment is applied to the rigid body. Therefore, a resistive torque is applied to both the inner ring of the bearing and the shaft, and the rotational speed is applied to the driven wheel. The solving process requires solving control, including calculating the termination time, file output interval, and output file control.

3 Results and analysis

From the LS-DYNA's postprocessor LS-PREPOST, the time-dependent curve of the equivalent stress of the shafting structure is shown in Fig.5.

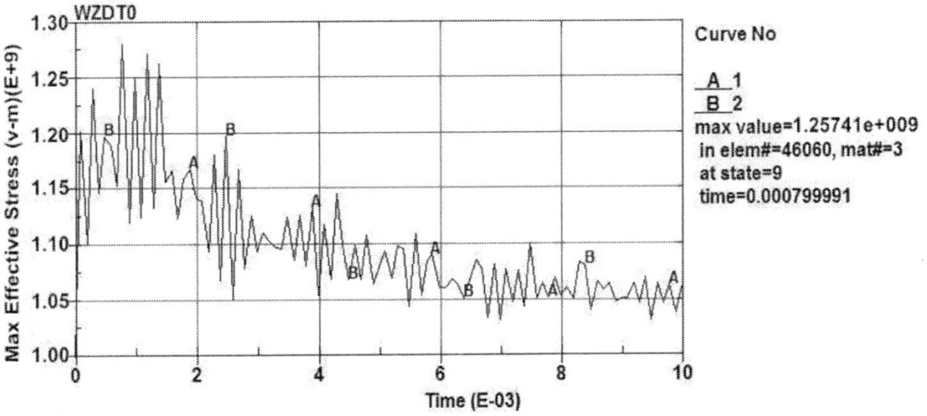

Fig.5 Equivalent stress with the change in time

Select the equivalent stress cloud at the time of 0.0091s, as shown in Fig.6.

Fig.6 Equivalent stress nephogram of shafting structure at 0.0091s moment

From Figure 6, we can see from the dynamic simulation results more in line with the actual situation, where the stress is relatively large mainly in the tooth root and tooth surface contact, where the stress is mainly in the vicinity of two or three teeth, For places where the contact is far away, the stress is zero, which is consistent with the St.Venern's law in elastic mechanics (indicating that when the area of a small part of the elastic body or the internal load of the volume is changed by the static equivalent, the stress distribution in the vicinity will be significant. Changes can be ignored in the distance.) In order to analyze the change in the equivalent stress of a pair of gears at different positions in the meshing process, three units are extracted in the middle position of a certain tooth and the tooth roots on the driving wheels of a pair of meshing gear pairs, as shown in Fig.7. The time-dependent change of the equivalent stress of the three elements during the meshing process is shown in Fig.8.

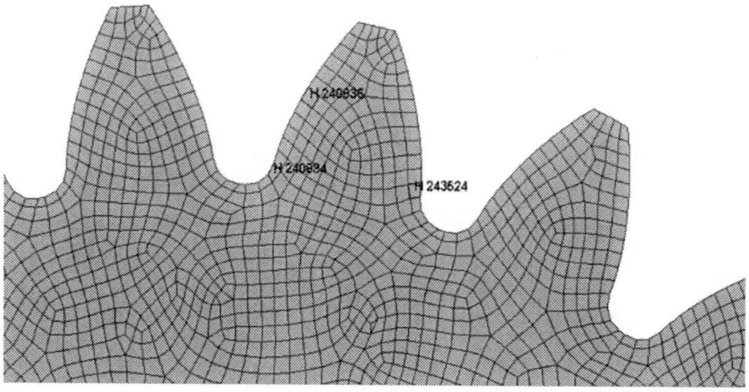

Fig.7 Schematic diagram about the driving gear of tooth surface

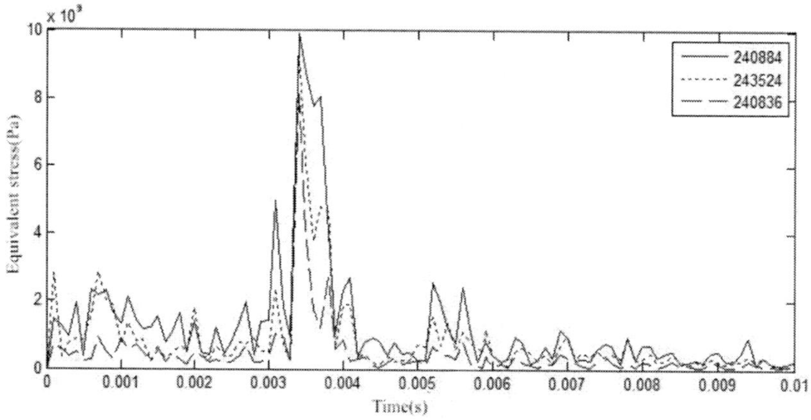

Fig.8 Equivalent stress of element curve

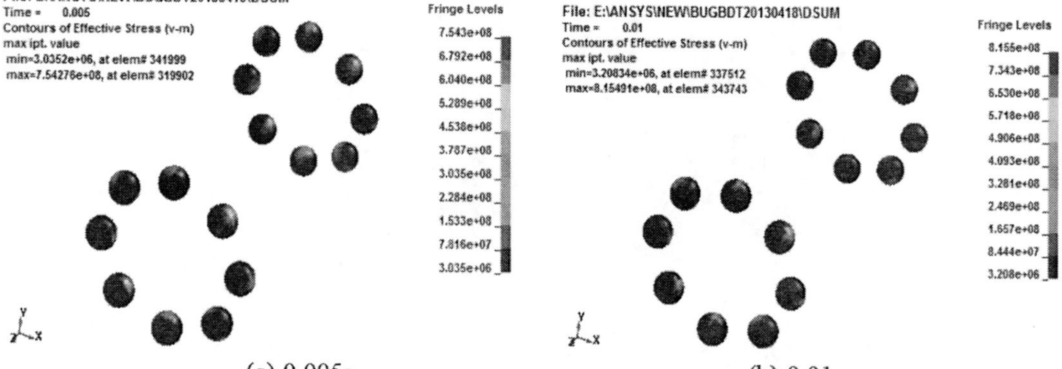

(a) 0.005s (b) 0.01s

Fig.9 Equivalent stress nephogram of rolling with the change in time

From Fig.8, it can be seen that when it enters the meshing position, its equivalent stress abruptly changes and gradually becomes larger, that is, the equivalent stress near the contact line is greatest when the gear is in contact, and the equivalent stress is when the contact is withdrawn. It will be substantially reduced to zero. When the driving wheel and the driven wheel come into engagement, the equivalent force of the driving wheel gradually decreases, because the engaged unit gradually approaches the tooth root.

Take two moments to analyze the results. Fig.9 is the equivalent stress diagram of the rolling bearing.

As can be seen from Fig.9, during the initial rotation of the bearing, the load is mainly concentrated on the rolling elements in the lower half of the bearing, and the force on the lowest rolling element is relatively large, and then decreases sequentially. At 0.005s, the maximum equivalent stress of the rolling bearing is 754 MPa. The maximum equivalent stress at 0.01s is 815 MPa. By comparison, it can be found that the maximum equivalent stress of the rolling bearing in the ideal model occurs at the contact area between the rolling element and the inner and outer rings of the bearing. In the course of operation, the maximum equivalent stress value and position of rolling element operation also change with time.

4 Conclusions

An experimental analysis method for explicit dynamics of shafting structures is proposed. The variation of the maximum equivalent stress with the time during the operation of the shafting structure was obtained. In the process of contact, the equivalent stress in the vicinity of the contact line of the gear is the largest, and when it comes out of contact, the equivalent stress of the gear will be substantially reduced to zero. When the driving wheel and the driven wheel come into engagement, the equivalent force of the driving wheel gradually decreases. The maximum equivalent stress of the rolling bearing in the ideal model occurs at the contact area between the rolling element and the inner and outer rings of the bearing. In the course of operation, the maximum equivalent stress value and position of rolling element operation also change with time.

References

1. Zhang Jing. Dynamic Simulation and Experimental Study of Gear Transmission under Different Factors, Chongqing: Chongqing University(2012)
2. Huang Linlin. Nonlinear analysis of roller bearing stiffness and its application in engineering, Dalian: Dalian University of Technology(2009)
3. Zhao Dan. Parametric design and finite element analysis of rolling bearings, Qingdao: Shandong University of Science and Technology(2008)
4. Tong Baohong, Liu Ying, Su Yu. Several problems in numerical simulation of dynamic contact characteristics of rolling bearings, Chinese Journal of Mechanical Engineering(2012)
5. Wang Zepeng, Hu Renxi, Kang Shiting. An example of non-linear finite element analysis of ANSYS 13.0/LS-DYNA, Beijing: Mechanical Industry Press(2011)
6. Y. Modarres-Sadeghi, H. Mukundan, J. M. Dahl, etal. The effect of higher harmonic forces on fatigue life of marine risers, Journal of Sound and Vibration(2010)
7. Jangyeol Yoon, Wanki Cho, Juyong Kang, etal. Design and evaluation of a unified chassis control system for rollover prevention and vehicle stability improvement on a virtual test track, Control Engineering Practice(2010)
8. Li Yuchun. Basic Theory and Engineering Practice of ANSYS11.0/LS-DYNA, Beijing: China Water Conservancy and Hydropower Press(2008)
9. Wan Changsen. Analysis Method of Rolling Bearing, Beijing: Mechanical Industry Press(1987)
10. Mechanical Design Manual Editorial Committee. Mechanical Design Manual: Rolling Bearing, Beijing: Mechanical Industry Press(2007)

MATEC Web of Conferences 179, 01020 (2018) https://doi.org/10.1051/matecconf/201817901020
2MAE 2018

Research of the Frictional Characteristics of Piston Ring - Cylinder Liner Based on GT-Suite

SHI Chao-fan[a], WANG Pu-kai, LIU Jian-min, KANG Qi and DONG Yi

Department of Vehicle Engineering ,Academy of Army Armored Forces ,100072, Beijing ,China

Abstract. Establish the piston ring-cylinder liner frictional model using the software of GT-Suite combine both of the frictional condition and lubrication, analyze and compute the thickness of oil film between piston ring and cylinder liner, the distribution of oil stress, and the force and power of the friction in rated condition. Then, mainly analyze the thickness of oil film and power of friction at the first piston ring under different temperatures of lubrication oil and different speeds. The results indicate that: The lubrication effect at first piston ring is poor, and the consumption of frictional power is high. The frictional power is significantly reduced with the oil temperature increases. However, the thickness of the oil film will decrease with the oil temperature increases, which is bad to the lubricational condition. Considering both of the lubrication condition and the frictional power, the frictional characteristics are better when the oil temperature is maintained at 80-90℃. With the increase of rotational speed, the oil film thickness and frictional power increase, but comparing with the influence on the frictional power, the change of rotational speed has less influence on the oil film.

1 Introduction

Piston ring is one of the most demanding parts of working condition in diesel engine, which not only withstanding high temperature and high pressure gas, but also sliding with a high speed along the cylinder liner, and being constraint for a heavy diesel engine becoming a high power density engine.

Therefore, studying the friction characteristics of piston ring-cylinder liner has the instructive meaning for improving the reliability and life span of diesel engine.

Scholars [1-5] conducted an in-depth study on the friction, wear and lubrication between the piston ring and cylinder liner. The analysis model covered kinetics, friction, structural parameters, surface morphology of friction pair, lubricant oil, Axis symmetry and other factors.

This paper comprehensively considers the factors of deformation between cylinder liner and piston ring, the surface roughness of contact surface, and the gas leakage caused by the piston ring, combing all of the friction, lubrication and dynamic behaviors and establishes the frictional model of piston ring-cylinder liner using the software of GT-Suite.

Then, the frictional characteristics of piston ring and cylinder liner under rated condition was studied by using this model. The influence of oil temperature and rotational speed on the friction characteristics was also analyzed.

2 Theoretical analysis

[a] Corresponding author : 63962510@qq.com

2.1 Analysis of Hydrodynamic Lubrication between Piston Ring and Cylinder liner

For the interaction between piston ring and cylinder liner surface, adopt the theory of raised by Patir and Cheng. Using the finite difference method in the length of axial direction in the piston ring, we can get the whixh represents the fluid frictional force.

Fluid dynamic pressure force between piston and cylinder surface is:

$$\frac{\partial}{\partial x}\left(\emptyset_x h^3 \frac{\partial p_h}{\partial x}\right) + \frac{\partial}{\partial y}\left(\emptyset_y h^3 \frac{\partial p_h}{\partial y}\right) = 6\mu U\left(\frac{\partial h_T}{\partial y} + \sigma \frac{\partial \emptyset_s}{\partial y}\right) + 12\mu \frac{\partial h_T}{\partial t} \quad (1)$$

In the equation, p_h is average fluid pressure,σ is the surface roughness between two surfaces, U is axial direction movement velocity of piston, h is nominal thickness of oil film, h_T is the actual thickness of oil film,$\emptyset_x$ and $\emptyset_y$are pressure flow factor, μ is dynamic viscosity, t is time, $\emptyset_s$ is shear flow factor.

The dynamic viscosity of lubrication oil can be defined by equation

$$\mu = \frac{5.6625\times10^4}{(t+40)^{4.5076}}\rho \quad (2)$$

In the equation, μ is the temperature of ubrication oil, ρ is the density of lubrication oil.

Shear force and hydrodynamic friction F caused by hydrodynamic pressure are listed below

$$\tau = -\frac{\mu U}{h_T}\left(\phi_f + \phi_{fs}\right) + \phi_{fp}\frac{h_T}{2}\frac{\partial p_h}{\partial x} \quad (3)$$

$$F_h = R\iint_A \tau(\theta, y)\cos\theta d\theta dy \quad (4)$$

In the equation, ϕ_f, ϕ_{fs} and ϕ_{fp} are shear pressure factor, τ is the function of cut stress in the Axial and circumferential direction of piston.

2.2 Asperity elastic contact analysis

Due to the fact that there is always a thin shear film on the rough contact surface, the rough surface contact theory proposed by Greenwood and Trip is used to obtain the peak friction.

Assuming that the surface height obey the Gaussian, the peak load on the perimeter of the piston ring-cylinder liner under elastic deformation is:

$$W_A = \frac{16\sqrt{2}}{15}\pi(\eta\beta\sigma)^2 E'\left(\frac{\sigma}{\beta}\right)^{1/2} A_a F_{5/2}\left(\frac{d}{\sigma}\right) \quad (5)$$

$$A_r = \pi^2(\eta\beta\sigma)^2 A_a F_2\left(\frac{d}{\sigma}\right) \quad (6)$$

In the equation, η is the peak density in the roughness surface, βis Radius of curvature at the top peak, A_a is the nominal contact area, d is the distance between two surfaces (nominal oil film height), E' is the integrated elastic modulus between two surfaces, $F_{5/2}$ and F_2 is the equation model of F_n.

When the piston ring and cylinder liner surface contact occurs, the peak shear force and peak friction F_A are

$$\tau_\alpha = \tau_0 + \alpha p_\alpha \quad (7)$$

$$F_A = \int_A (\tau_0 + \alpha p_\alpha)\, dA = \tau_0 A + \alpha W_A \quad (8)$$

3 ESTABLISH THE MODEL

This paper analyzed and studied the friction characteristics of piston ring-cylinder liner in diesel engine. Established models mainly include the working process model of diesel engine and the single cylinder piston-cylinder friction model, and the working process model of diesel engine provides the boundary conditions for the friction model.

Piston rings in the diesel have three different rings, in which the first two rings are gas rings, and the third ring is the oil ring. The specific parameters of these rings are showed in Table 1. The GT-Suite software was used to establish a single cylinder piston-cylinder liner friction diesel engine model, which was shown in Fig.1.

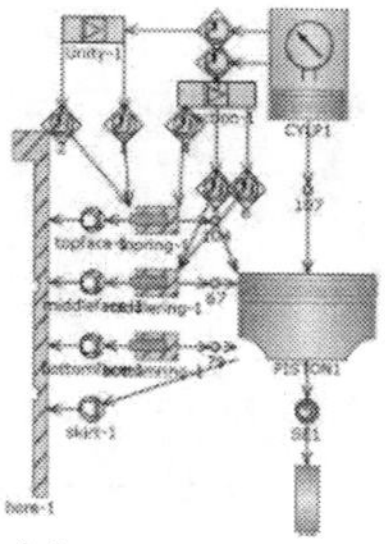

Figure1. Piston ring - cylinder liner friction model

3.1structural parameters

The single cylinder diesel engines piston ring - cylinder liner structural parameters and some performance indicators was shown in Table 1.

Table1 Piston ring - cylinder liner structural parameters and performance indicators

Constructural factor and power indicator of piston ring and cylinder			
Mass of piston	2.63kg	Poissons ratio	0.32
Length of crank	91mm	Diameter of cylinder	150mm
Height of piston	120mm	Mass of connecting rod	7.986kg
Roughness of piston surface	0.4um	Length of connecting rod	320mm
Diameter of first ring	140mm	Height of first ring	3mm
Distance from first ring to top of piston	30mm	Elasticity of piston ring	16.1N
rpm	2000	The elasticity of first ring	24N

3.2 Boundary condition

The boundary conditions of the piston-cylinder liner friction model are the cylinder gas pressure and the cylinder liner inner wall temperature distribution. The GT-power software calculates the cylinder gas pressure and cylinder liner wall temperature distribution under the calibration conditions.

Reference [5] Piston ring group leak analysis method, the first ring gas pressure is the cylinder gas pressure; in the piston ring opening gap throttling effect, the gas pressure between the first ring and the second ring are one fifth of the cylinder gas pressure or so, and the location of the maximum pressure value postponed; after two gas rings seal function, the amount of gas leakage between the second ring and the oil ring are small, so the gas pressure is not considered.

The boundary conditions of piston ring - cylinder liner friction model are shown in Figure 2.

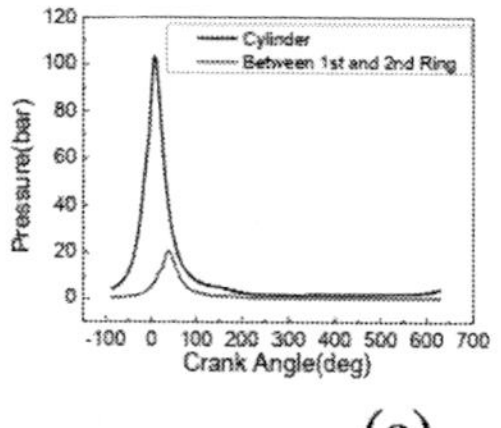

(a)

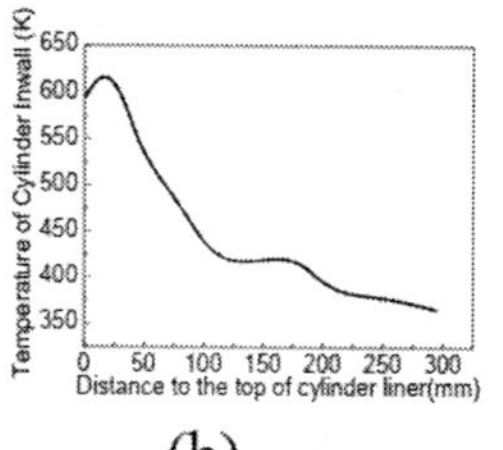

(b)

Figure2(a). Gas pressure at piston ring

Figure2(b). Temperature distribution in the inner cylinder

4 Calculation results and analysis

The maximum film pressure, film thickness at the first piston rings shown in Figure 3.

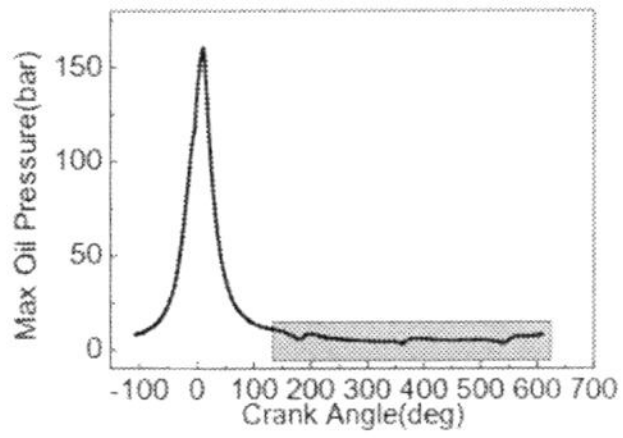

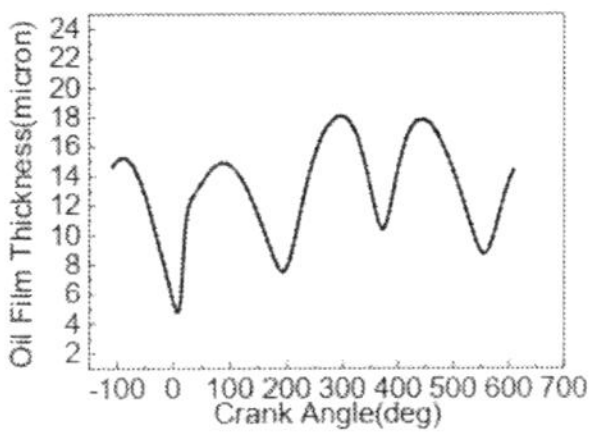

Figure3(a). Change rule of the maximum oil film pressure of the piston ring **Figure3(b).**Change rule of oil film pressure of the piston ring

It can be seen from Fig.3 (a)that the maximum oil film pressure of the piston ring is generally larger than the gas pressure between the piston rings which because of the convergence of the oil film, resulting in an increase of the pressure of the oil film in the convergent zone.

The maximum oil film pressure of the first ring and the change gradient is large, the maximum value appears at the place where piston is near the compression top dead point, and the maximum oil film pressure of the oil ring is small and appears the average periodic changes, this is because the gas pressure between the first ring and the second ring on the oil film plays a major role. After the first two piston rings seal, the gas pressure between the oil rings drastically decreases. The elastic force and motion of the oil ring play a major role on the oil film pressure.

It can be seen from Fig. 3 (b)that the minimum oil film thickness of the first piston rings appears at 0 ℃A, during one stroke of the pistons, which correspond exactly to the upper and lower dead points of piston ring movement, which because that at this time the piston movement speed is very low, resulting in lower viscosity of lubricating oil;

In a piston stroke, the film thickness is at its largest when piston ring is located in the middle of the cylinder, due to dynamic pressure lubrication theory, the higher the velocity, the more viscosity of lubricating oil, then the thicker the thickness of the film, when the piston ring speed reaches its maximum , the film thickness also reaches its maximum;

The thickness of the top oil film is the smallest at the last stage of the working stroke (0 to 180) and the compression stroke (-80 to 0), indicating that the gas pressure greatly affects the oil film thickness.

4.1Friction and friction power consumption

The first piston rings friction and friction power are shown in Figure 5.

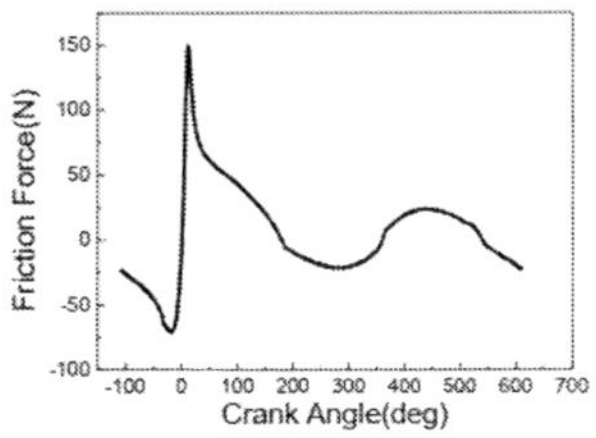

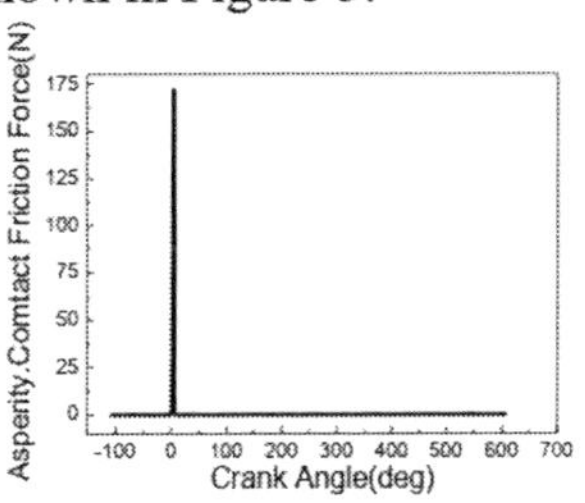

(a) (b)

Figure4(a). The variation of friction force at piston ring **Figure4(b).** Change rule of the peak friction of the piston ring

It can be seen from Fig.4 that the peak friction of the first ring mainly occurs near the top dead center of combustion, and the value of the friction force is obviously higher than that of the other moments the peak element friction of the oil ring is almost zero during the whole cycle, which is due to the lubricating condition of the oil ring is good and the oil film is thicker.

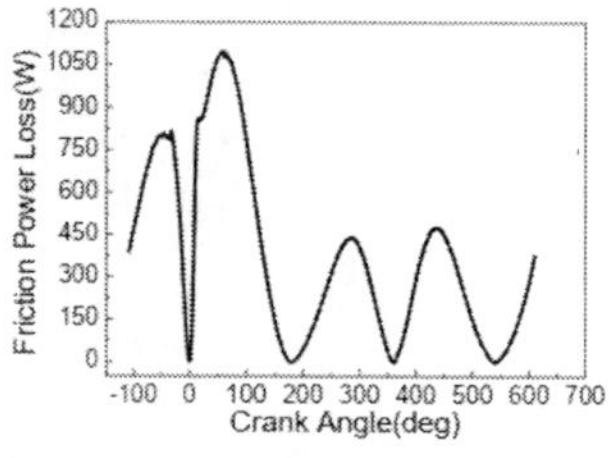

Figure5. Change rule of piston ring friction power consumption

It can be seen from Fig.5 that ,for the first ring, the maximum frictional power dissipation of 1096.86W appears at 69.3°C A with a slight lag from where the maximum friction occurs due to the fact that the frictional power loss is the result of a combination of friction and velocity Compression stroke and work stroke friction power consumption was significantly higher than other strokes, exhaust stroke and intake stroke friction power consumption quite.

Through the simulation results can also be seen: the first ring of the worst working environment Lubrication conditions are poor, high wear and tear, heat load, should be taken into account in the design. The following emphasis on the first ring for research and analysis.

4.2 piston ring torsional deformation and instantaneous elastic deformation

The first piston ring torsional deformation and instantaneous elastic deformation, as shown in Figures 6.

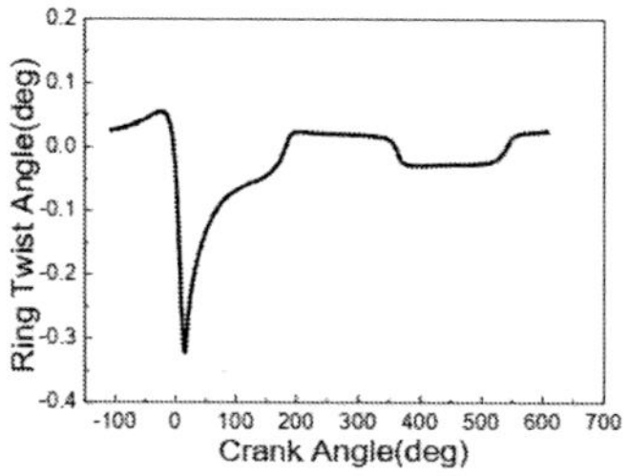

Figure6. Change rule of the distortion of the first ring

It can be seen from Fig. 6 that the distortion of the first ring is obviously large, especially the work stroke, the maximum twist and the maximum gas pressure in the cylinder. This is because the gas pressure inside the cylinder is large Resulting in a larger gas torque resulting in a larger amount of twisting. The amount of twisting of the first ring and the second ring is averaged over the other three strokes. The amount of twisting of the oil ring is periodically changed and averaged throughout the cycle. The amount of torsional deformation of the piston ring The direction opposite to the direction of the piston movement; the larger the amount of torsion of the piston ring, the smaller the corresponding friction and oil film thickness, the less the use of lubrication, which is due to the piston ring torsion lead to reduction of piston ring and cylinder liner contact surface, Large, film thickness decreases.

4.3 Effect of oil temperature on friction characteristics

The front of the diesel engine calibration conditions piston ring - cylinder friction characteristics were analyzed and found that the state of the lubricant has a great influence on the friction characteristics, and in the actual work process oil temperature is also affected by the cooling system. Therefore, the effect of oil temperature on the friction characteristics is further studied. The friction characteristics of the first ring when the oil temperature is 60℃, 70℃, 80℃, 90℃ and 100℃ are calculated respectively, as shown in Fig. 7.

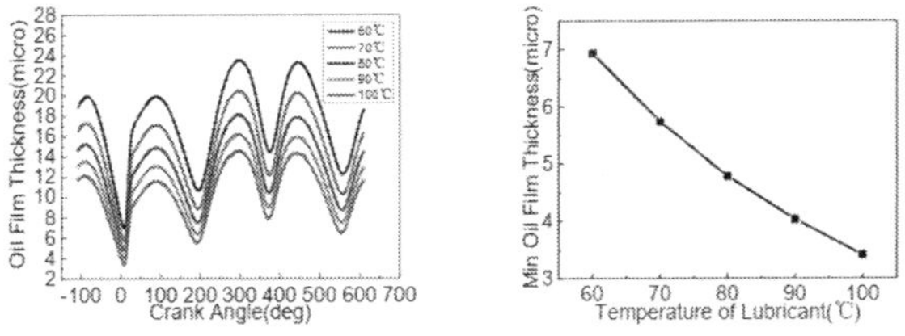

(a) (b)

Figure7(a). The effect of oil temperature on t the thickness of the first ring oil film **Figure7(b).** The trend of the thickness of he minimum oil film thickness

As can be seen from Fig.7, the trend of the thickness of the first ring oil film is similar at different oil temperatures. The higher the oil temperature, the thinner the oil film thickness is. This is due to the significantly lower viscosity of the lubricating oil with increasing temperature. The temperature rise of the minimum film thickness showed a significant decline, but the decline decreased. Oil temperature increased from 80℃to 100℃ , the minimum oil film thickness decreased 28.7%, showing the oil temperature has a greater impact on the thickness of the oil film; at the same time as the oil temperature increases, the minimum oil film thickness decreases, piston ring - cylinder friction form in mixed lubrication The chances of lubrication with the boundary increase, not conducive to lubrication.

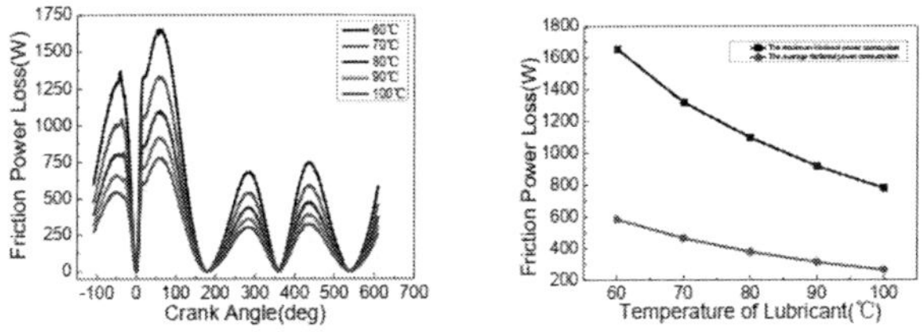

(a) (b)

Figure8(a). The effect of oil temperature on first ring frictional power consumption **Figure8(b).** The effect of lubricating oil temperature on frictional power consumption

From Fig. 8(a) and Fig.8(b), it can be seen that the change tendency of the frictional power consumption of the first ring at different oil temperatures is similar. The higher the oil temperature, the lower the frictional power loss. This is attributed to the friction of the fluid caused by the decrease of the viscosity of the lubricating oil with the increase of the temperature As the oil temperature increases, the maximum frictional power consumption decreases significantly, but the decline decreases gradually. When the oil temperature rose from 80℃ to 100℃ , the maximum friction loss decreased by 41% and the average friction loss decreased by 37.3%. The effect of oil temperature on frictional power consumption was significant.

4.4 Effect of Speed on Frictional Characteristics

Through the above analysis, it is found that the moving speed of the piston ring has a great influence on the frictional characteristics. Therefore, the influence of the moving speed on the frictional characteristics is also studied in this paper. The friction characteristics of the first ring was studied when the engine speed is 2000 r / min, 1800 r / min, 1600 r / min and 1400 r / min.

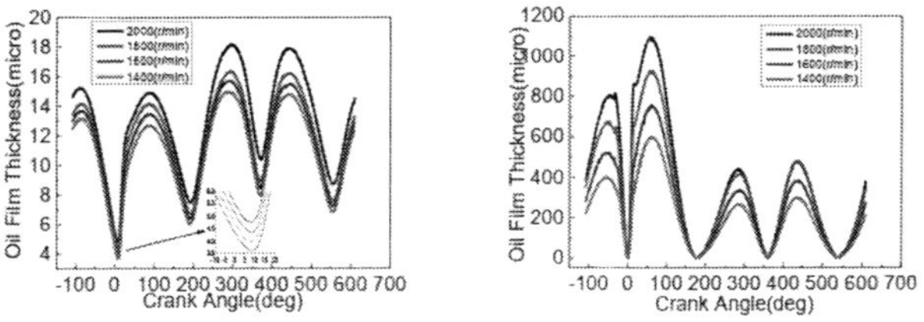

(a) (b)

Figure9(a). Effect of rotational speed on the thickness of the first oil film **Figure9(b).** Effect of Speed on the Frictional Power Consumption of the First Ring

From Fig. 9(a), it can be seen that the changing trend of the thickness of the first oil film at different speeds is similar. The higher the speed is, the thicker the oil film thickness is, but the influence is not big. This is because according to the dynamic pressure lubrication theory, Oil viscosity increased when the speed increased.

From Fig. 9(b), it can be seen that the trend of the frictional power loss of the first ring at different rotational speeds is similar, and the higher the speed is, the higher the frictional power is. This is because the higher the rotational speed per unit time, the longer the piston ring moves. At the same time, the change of speed has a significant effect on the frictional power consumption of compression stroke and work stroke.

5 Conclusion

This article establishes the model of friction between piston ring and cylinder liner using the software of GT-Suite, combining the friction, lubrication and dynamics, and researching the whole frictional characteristic of piston ring. The conclusions are listed below:

The max oil film pressure at the first ring vary greatly and the latitude of vary is big too, the max point occur around the comprocess top dead point. The frictional force and frictional power at the first ring are big and vary greatly at the stage of power and comprocesses. Therefore, the working condition at the first ring is the worst, lubrication condition, friction wear, and the temperature load are worse, we should pay more attention at the stage of design and use. Under different oil temperatures, the variation tendency of the friction characteristics of the piston rings is similar, but have a significant effect on the friction characteristics of the piston rings.

As the oil temperature increases, the oil film thickness is significantly reduced at the same time the frictional power consumption is significantly reduced too. Considering the effect of oil temperature on oil film thickness and frictional power consumption, we can conclude that the minimum oil film thickness and frictional power loss are ideal when the oil temperature is in the range of 80 ~ 90℃ , which provides some references for the matching of lubricating oil cooling system.

The changing tendency of the friction characteristics of piston rings under different speed conditions is similar. With the increase of rotational speed, the thickness of the oil film increases while the frictional power consumption increases too. The rotational speed has little effect on the oil film thickness, but has a significant impact on the frictional power consumption.

References

1. Ye Xiaoming, Jiang Yankun,Hao Xiuli et al.Effect of Radial Deformation of Cylinder Liner on Elastohydrodynamic Lubrication Performance of Piston Ring[J].Vehicle Engine,2007,168(2):22-25
2. Zhou Long, Bai Minli, Lu Jihe et al. Study on heat transfer, lubrication and friction of piston ring-cylinder liner of internal combustion engine by coupling analysis method [J]. Journal of Internal Combustion Engines, 2008, 26(1): 69-75
3. Li Minghai, Li Xiaowei, Yan Shukuan. Effects of oil temperature on lubrication performance of main journal bearings [J]. Diesel Locomotives, 2009,
4. Liu Na. Study on the friction and lubrication performance of key friction pairs of diesel engine [D]. Shandong: Shandong University, 2012.
5. Yang Liping, et al. Analysis of Deformation Friction of Engine Cylinder Tubes and Approaches to Reduce Measures [J]. Automotive Practical Technology, 2017, 03.54-56.

Complex Logic Control Algorithm and Simulation Application Based on Matrix Modeling

Yu Li[1,a], Gang An[2,b] and Xin Li[3,c]

[1] System Integration Center, FAI AVIC,Xi'an, CN
[2] School of management, Xi'an Jiaotong university, Xi'an, CN
[3] Landing Gear System Research Institute, FAI AVIC, Xi'an, CN

Abstract. The complex logic control of discrete signals is a common problem in engineering applications. The traditional program design using multi branch conditional transfer algorithm, program complexity increases with the complexity of the problems, therefore, the ability to solve logic control severely limits the scale and complexity of problems. The complex logic control algorithm based on matrix modeling is based on the analysis and modeling of logical relations, and the implementation of the two-dimensional matrix makes the complexity of the program irrelevant to the problem. The algorithm can well solve the display and control simulation problem of the multi-function display in aircraft cockpit display and control equipment. It has good expansibility and popularization.

1. Introduction

The multi-function display (MFD) is an important component of human-computer interaction system in modern cockpit. In flight and execution tasks, it provides real-time video and graphical information required by pilots. MFD is a controllable display device, providing a rich working mode and a variety of control modes. The pilot can toggle the work mode by function keys as needed. Because the function keys have different control function definitions in different working modes, there is a complex logic control relationship between the conversion of working states and the control of peripheral keys. In the simulation of MFD display control, matrix modeling is used to solve the complex logic control relation transformation problem.

The matrix method is different from the traditional multi branch conditional transfer control algorithm. It uses the two-dimensional array to represent the logic relation[1]. When the program is implemented, only the information in the two-dimensional array is read and correctly interpreted, so that the logic conversion control function can be completed. Compared with the traditional algorithm, the matrix method can simplify the logic in programming complexity, easy to design, modify and expand the function, and can form a standard program algorithm is extended to state control, such as mobile phone equipment and the automatic teller machine. This paper will combine MFD display and control system simulation, from the matrix modeling, program design and performance analysis, in-depth study and description of this algorithm.

[a] Corresponding author :kingliyu@sina.com;
[b]Corresponding author : An_gang@163.com;
[c] Corresponding author :370713464@qq.com

2. System modeling

2.1. Analysis of complex logic control relations

In the design of MFD display and control system of aircraft avionics system, it is an extremely critical functional requirement to realize the accurate and accurate transformation between multi task and multi operating modes.MFD provides 19 main operating modes such as MENU, HSD, DATA, BIT, PDR, and so on.Each of the main operating modes contains several sub modes of operation.As an example, take part of the working mode and its conversion relation as shown in figure 1. The meaning expressed in Figure 1 is:

For example, in MENU mode, pressing the peripheral key L2 enables the MFD to switch to HSD mode, and press the peripheral key R5 in HSD mode to switch the mode to TCN.

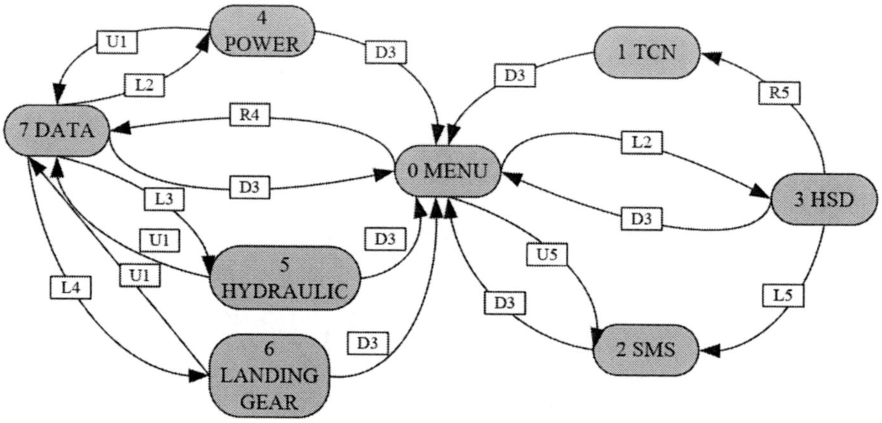

Figure 1. Work mode switching

When dealing with this kind of problem, the traditional multi branch conditional transfer statement can be implemented by the multi layer nesting algorithm[2]. The algorithm is a kind of algorithm without data structure support, and it has many limitations:

- From analysis to design to test the entire development process, it is not suitable for large-scale complex logic processing.
- Requirements change and function expansion are unavoidable problems in software development, and traditional algorithms are difficult to adapt.
- Due to different programming habits and different understanding of the problem, the traditional algorithm is usually designed code differences, it is difficult to form a standard algorithm to packaging, promotion and code reuse.
- For large-scale problems with complex logic relations, programs written by traditional algorithms have large amount of code, complicated nested relationships, and poor readability and maintainability of source programs.

To solve the above problems, an algorithm based on matrix modeling is designed:

- The large scale complex logic relation is represented by the data structure of the two-dimensional matrix.
- The complex logic control program based on matrix method is designed by accessing the program of two-dimensional matrix.

2.2. Matrix modeling of complex logic control relations

When modeling the working mode switch shown in Fig. 1, the operation mode and the press operation of the function keys need to be digitally abstracted so that the computer can recognize and process the operation. The digital process is:

- The 8 states shown in Figure 1 are numbered in accordance with 0~7.

- Press the peripheral keys U1~U5, L1~L5, R1~R5, D1~D5 when 20 key values in turn according to 0~19 number.
- This gives you the MFD state transition table as shown in table 1.
- Line represents the current working state of the MFD. The column represents the key value when pressing the peripheral key.
- The content determined by rows and columns is the number of MFD target states that should be switched to when a peripheral key is pressed in the current working state.
- "-1" stands for the corresponding peripheral key in this work state without the operation definition.

Table 1. State transition table

Working condition	The key value of the function key presses operation																			
	U 1	U 2	U 3	U 4	U 5	L 1	L 2	L 3	L 4	L 5	R 1	R 2	R 3	R 4	R 5	D 1	D 2	D 3	D 4	D 5
0	-1	-1	-1	-1	2	-1	3	-1	-1	-1	-1	-1	-1	7	-1	-1	-1	-1	-1	-1
1	-1	-1	-1	-1	-1	-1	-1	-1	-1	-1	-1	-1	-1	-1	-1	-1	-1	0	-1	-1
2	-1	-1	-1	-1	-1	-1	-1	-1	-1	-1	-1	-1	-1	-1	-1	-1	-1	0	-1	-1
3	-1	-1	-1	-1	-1	-1	-1	-1	-1	-1	-1	-1	-1	-1	1	-1	-1	0	-1	-1
4	7	-1	-1	-1	-1	-1	-1	-1	-1	-1	-1	-1	-1	-1	-1	-1	-1	0	-1	-1
5	7	-1	-1	-1	-1	-1	-1	-1	-1	-1	-1	-1	-1	-1	-1	-1	-1	0	-1	-1
6	7	-1	-1	-1	-1	-1	-1	-1	-1	-1	-1	-1	-1	-1	-1	-1	-1	0	-1	-1
7	-1	-1	-1	-1	-1	-1	4	5	6	-1	-1	-1	-1	-1	-1	-1	-1	0	-1	-1

3. Software design

3.1. Matrix data structure

In programming, you should first solve the data structure problem when the matrix model is represented in the computer[3]. The simplest and direct method is to use a two-dimensional array, and the C language pseudo code is described as:

Int stateTable [total number of work modes] [key number].

In this case, it is represented as stateTable[8][20]. The data in the array is the contents of table 1. Thus, the stateTalbe array represents such a two-dimensional matrix as shown in figure 2.

```
-1 -1 -1 -1 2  -1  3 -1 -1 -1 -1 -1 -1  7 -1 -1 -1 -1 -1 -1
-1 -1 -1 -1 -1 -1 -1 -1 -1 -1 -1 -1 -1 -1 -1 -1 -1 0  -1 -1
-1 -1 -1 -1 -1 -1 -1 -1 -1 -1 -1 -1 -1 -1 -1 -1 -1 0  -1 -1
-1 -1 -1 -1 -1 -1 -1 -1 -1 -1 -1 -1 -1 -1  1 -1 -1 0  -1 -1
7  -1 -1 -1 -1 -1 -1 -1 -1 -1 -1 -1 -1 -1 -1 -1 -1 0  -1 -1
7  -1 -1 -1 -1 -1 -1 -1 -1 -1 -1 -1 -1 -1 -1 -1 -1 0  -1 -1
7  -1 -1 -1 -1 -1 -1 -1 -1 -1 -1 -1 -1 -1 -1 -1 -1 0  -1 -1
-1 -1 -1 -1 -1 -1  4  5  6 -1 -1 -1 -1 -1 -1 -1 -1 0  -1 -1
```

Figure 2. Two-dimensional matrix

The first dimension of each element subscript in the array represents the current working mode number of the MFD, and the second dimension represents the key value for the operation of the peripheral key. For example, the element stateTable[0][6] in an array has the meaning of:

In the 0 mode press the 6 key, MFD will switch to the working mode of stateTable[0][6]=3.

3.2. Design of complex logic control algorithm based on matrix mode

Different from the traditional multi branch program process to solve complex logic control problem, using matrix to deal with the problem, only the two-dimensional array stateTable storage MFD mode switch table release read and correct according to working mode and keys around the current operation is a realization way is very simple and effective. Algorithms using C language pseudo code are described below:

```
int stateAnswer( int curState,  int keyValue)
{
    if ( stateTable [curState][keyValue] != -1 )
    {
        return stateTable[curState][keyValue];
    }
    else
    {
        Return curState;
    }
}
```

Description: stateAnswer function through the operation of the stateTalbe array, the corresponding switch to achieve the working mode. The parameter curState indicates the current working mode number, and keyValue represents the key number for the current push of the peripheral key operation. The return value indicates the target working mode number to be switched to.

```
void stateExplain( int stateID)
{
    switch(stateID)
    {
    case 0:
      MENU Program entry for work mode dependent processing;
      break;
    case 1:
      TCN Program entry for work mode dependent processing;
      break;
    ......;
    }
}
```

Description: stateExplain function, the completion of a variety of work mode number translated into the corresponding specific functions. The parameter stateID indicates the number of working states that need to be converted.

When the MFD display logic handler starts, you should set the default mode to the current mode, and then begin to receive instructions for the peripheral key. The default mode of operation is performed without operation, and the stateAnser function is called to get the number of the next working mode. Finally, call the stateExplain function to perform the specific function of the new working mode.

3.3. Design method

According to the summary and abstraction of the simulation modeling and program design process and method of MFD display and control system, the matrix modeling and programming method is refined, as shown in figure 3.

System requirements analysis and function definition process, focusing on the complex logic transformation of the status and conversion conditions are analyzed in detail, resulting in a state transition diagram[4].

According to the state transition diagram, all States and trigger conditions are digitally abstracted and encoded, and a state transition table is formed, which is the basic work of software design[5].

According to the state transition table, the data structure of the matrix is defined by using the two-dimensional array.

Design control logic software to access the matrix.

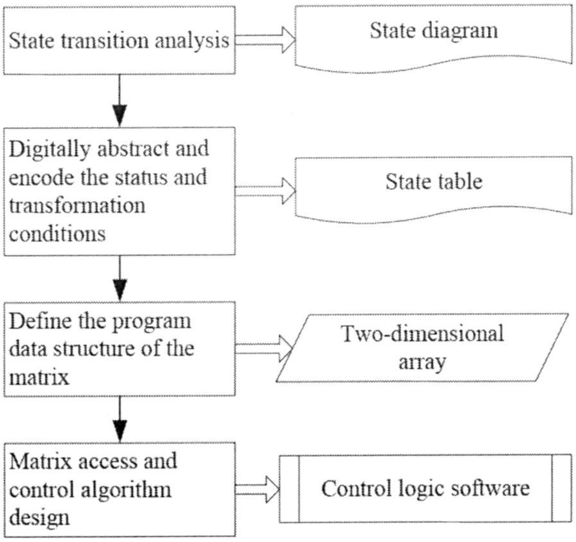

Figure 3. Matrix modeling and programming method

4. Performance analysis

4.1. Simplify the complexity of problem analysis

Complex logic transformation control relationships usually exhibit complex mesh structures (Figure 1), and it is often difficult for a programmer to make a correct judgment because of the limitation of thinking ability and scale of the problem in system analysis[6]. When using a matrix for system analysis, no matter how complex the logical relationship is, the programmer just needs to clear up each state and its trigger conditions, and then fill in the table of status transitions shown in table 1. Without the logical relationship between the depth and breadth of consideration and analysis, always at the same level to make the whole system analysis, the scale and complexity of problems, can greatly simplify the system analysis difficulty for solving large-scale complex logic control problem is particularly advantageous.

4.2. Easy to design, maintain, modify and extend function

Traditional procedural algorithms use complex branch conditional transfer to solve complex logic problems. The condition of judgment, the number of branches, the depth of nested hierarchy, and the length of program vary with the complexity of the problem. Thus, programmers have to write different programs according to different problems, even if the same problem has to be changed massively when the requirements change, the software upgrades or the functions are expanded. At the same time, the state is large and the logic relation is complex, which brings difficulties to the test. Using the matrix, you only need to modify the status response table and the stateExplain function. Thus, it is very easy to design, maintain, modify and extend the program. For example, suppose that MFD display control program was originally designed only shown in Figure 1 of the right part of "original design mode", this state response is shown in Table 1 of the 0~3 state, when the program needs to expand the left part of Figure 1 shows the functional expansion increased working mode "when simply fill in Table 1 can be (in Table 1 with shading to modify the content of the function expansion, then) will be associated with the increase of the working mode of the program can be added to the stateExplain function in the entrance. The test switch just to check "state response table", and the test function can be stateAnswer, because the number of stateAnswer function statement, the algorithm is simple, therefore, both white box and black box testing are very easy.

4.3. Increase execution speed

In order to improve the efficiency of CPU implementation, modern computer architecture is designed using pipelining technology. A pipelining is a task decomposition technique that divides an instruction into several steps that are executed by different execution units in parallel[7]. Due to pipelining processing instruction set is different from the processing conditions for the transfer of the sequential execution of the instruction set, only when the instruction is executed to the "operation", in order to determine the next step should jump to the first few instructions, since this preprocessing work instructions will be adjacent to the waste, but also disrupted the normal flow of pipeline. Using the traditional multi branch conditional transfer nested algorithm, it is easy to waste the preprocessing because the conditional transfer statement is many and the nesting level is deep. The complex logic control algorithm based on matrix modeling is directly positioned to the target state by array index, so it can avoid the occurrence of this situation and improve the execution speed.

4.4. Analysis and solution of the contradiction between execution speed and storage space

In the design of computer algorithms, time and space are always a pair of contradictions. The complex logic control algorithm based on matrix modeling improves the efficiency of program execution. But it must be admitted that due to the use of two-dimensional matrix to represent the logical conversion relationship, the computer storage space has to be increased. Therefore, the algorithm is a programming algorithm designed to sacrifice space for efficiency[8]. However, with the rapid development of computer hardware technology, no matter the speed of CPU execution or memory storage capacity, it is no longer the key to the development of software. Single board computer or mobile phone even as automatic teller machines that, for space utilization is strict, the two-dimensional representation of state transition matrix usually appears in sparse array form, so it can take compression array storage, which fall in the value of -1 compression matrix without operation defined in space and space efficiency in contradiction to seek compromise. Or change the data type, change the plastic number to bit mode, and save space.

5. Conclusions

Through practical application in an aircraft engineering simulator and theoretical analysis of its performance, it is demonstrated that: The complex logic control algorithm based on matrix modeling is very effective for solving large-scale complex logic control problems. It makes every link of analysis, design and test program design simplified and standardized. The algorithm is simple and readable, and can be used as a standard program algorithm for dealing with such problems.

References

1. Soares M D S, Vrancken J. Model-Driven User Requirements Specification Using SysML[J]. Joural of Software, 2008,3(6).
2. J Fisher. Model-Based Systems Engineering: A New Paradigm[J]. INCOSE Insight, 1998,1(3).
3. Thomas Johnson, Christian J. J. Paredis. Integrating Models and Simulations of Continuous Dynamics Into SysML[J]. Journal of Computing and Information Science in Engineering, 2012,12(3).

4. Cai Shubin, Ming Zhong, Li Shixian, Liu Xianming, ontology based model integration. Journal of Electronic Science and technology, 2009 (4).
5. One day, Zhang Yan, Wang Linzhang, Li Xuandong, in science, design pattern modeling and model transformation based on MDA. Journal of software engineering, 2008.19 (9).
6. Zhou Yi, Liu Lei, fan Ren hong. Ontology mapping method based on pattern structure classification [J]. Journal of electronics. 2011 (04).
7. Li Lanying, Wang Feng. Research on scheduling algorithm of reconfigurable system [J]. Computer engineering and application. 2014 (22).
8. Zhang Jiafeng. Modeling and verification of reconfigurable discrete event control systems [D]. Xi'an Electronic and Science University. 2015.

Computer-aided Design for the Route of the Test Waveguides

Wang Runze[1,a], Jian Yabin [1], Yin Xiaofang [1], Hou YaQin [1]and Su XinMing [1]

[1]Beijing Institute of Spacecraft Environment Engineering, Beijing 100094, China

Abstract. With the rapid development of China's space industry, digitization and intelligent is the tendency of the future. The applications of the waveguide are gradually widespread. During the thermal test phase, the routes of the test waveguides are similar for each spacecraft. Although the waveguides are highly standardized, so far it needs engineers to design the particular route of the test waveguidess, then map the engineering drawing for every test. In order to efficiently design the route of waveguide, it needs to design an application to help the engineers. With the help of the MFC(Microsoft Foundation Classes) and the pro/toolkit, it is easily to do the modeling and simulation. After automatic design the particular the route of the waveguide, the API of AutoCAD type library is used to help to modify the engineer drawing. Engineers can supervise every step of this application, and easily to modify the key parameters.

1. Introduction

Waveguides are widely used in spacecraft such as satellites, space station and so on. Just like coaxial cables, the function of the waveguide is to transmit the signal. Compare with the coaxial cable, the advantage of waveguide is lower attenuation and high power [1]. Especially in the high frequency the above two advantages are obvious. The standing wave and the insertion loss of the waveguide are good.

To ensure the spacecraft can work normally, before the launch of the spacecraft, it needs to be test. During the test, the spacecraft is deposited in a sealed container. To receive the signal of the spacecraft, waveguides need to be applied. There are dozens of test every year, the waveguide path needs to be redesigned for each test. It will consume engineers a lot of time and effort to design the test waveguides route and engineer drawings. The waveguide is highly standardized and the routes of the waveguide are similar, so the computer- aided design technology can be used to help engineers design the routes of the test waveguides. This report is present a foundation research about the computer- aided design technology which based on the Pro/E secondary development tool (Pro/toolkit) and AutoCAD secondary development API (AutoCAD type library), with the help such technologies the route of waveguides can be automatic designed easily, in a result of labour saving.

2. Design preparation

Computer-aided design technology will liberate engineers from a lot of repetitive work. It can improve the design efficiency and reduce wasted time greatly. In this paper, engineers set some boundary conditions and parameters. Then the program runs automatically according to the above conditions and interact with the model library. Afterward it will design and assemble the test

[a] Corresponding author :rwangar@connect.ust.hk

waveguides with the help of Pro/Toolkit. Finally, it will mapping the engineer drawings. The system architecture is shown in Figure.1.

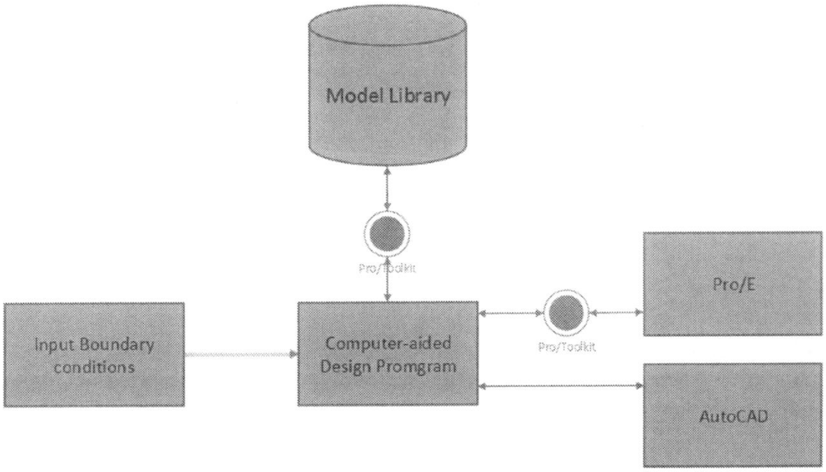

Figure 1. System architecture.

3. Scheme implementation

3.1. Pro/E secondary development

Pro/Engineer provides powerful three-dimensional geometric modelling capabilities that creates complex models. Sometimes these features can't suit the design requirements. For example the standard parts, their structure is the similar, the difference is only their size. The most effective method is to use the similarity of standard parts and Take advantage of its parametric design. Using the secondary development tool such as Pro/toolkit to improve design efficiency [2].

There are several methods to realize Pro/Engineer secondary development, the most popular method is using Pro/toolkit. To interact with Pro/Engineer, the specific version of Visual Studio need to correspond to the specific version of Pro/E, otherwise the built DLL file can't interact with Pro/E. In this paper the programing environment is Visual Studio 2008, programing language is C++. Using the function of MFC DLL to build DLL file and dynamic link with Pro/E. Microsoft Foundation Classes(MFC) is a class libraries, it encapsulate Windows API, it Contains an application framework to reduce the workload for application developers. The version of Pro/E is Creo Element Pro/E 5.0, and the revision is 18.

When doing spacecraft experiments, the spacecraft is fixed on a bracket and deposited in a sealed container. The test waveguidess need to connect to the spacecraft and pass through the sealed flange. As it is shown in the figure 2 that the translucent part is the sealed container, the red colour part is test waveguides, the blue colour part is a bracket, and the grey colour part is sealed flange. The test waveguidess are standard parts, their structure and route are the similar. This paper will present a method to help engineers to design this test waveguides path easily.

Figure 2. Simulating Test waveguides Assembly

To fulfil this goal, some factors need to be sure. Firstly, the coordinate relationship of sealed flange, the bracket and the test waveguides connection point. Secondly the number and type of waveguides. Lastly, the route of the test waveguides.

The structure of Pro/toolkit program begin with 'int user_initialize()', it is the initializing program of the Pro/toolkit program, users can create the interactive interface in this function such as creating menu. The function 'int user_terminate()' invocated when users terminate synchronized mode[3].

To create a friendly human-machine interface (HMI), the menus are essential. The menu is created by the function ProMenubarMenuAdd () in the Pro/toolkit. The action is defined in the function ProCmdActionAdd () [3]. With the help of such functions, users can create a friend HMI and interact with Pro/E.

To generate the test waveguides path, engineer needs to know the coordinates of the sealed flange and the test waveguides connection point. The function ProSelect() is an interface function when user select a part this function return a mark to the program. Then the function ProAsmcomppathTrfGet() will be used, the capabilities of this function is get the correspondence relation between part coordinate system and component coordinate system, the return value of this function is a matrix (ProMatrix()),a four-order matrix, The first three lines of the matrix describes the relative orientation of the two coordinate systems, the last line describe the translation of the two coordinate systems. ProMatrix()41, ProMatrix()42, ProMatrix()43 represents the translation of the x,y,z axis, respectively [4].

To calculating the number of the test waveguides, the function ProAsmcompMdlGet() can be used. The main program will use this function to traverse the whole assembly model, then output the name of each part in the assembly model. Then the number of test waveguidess can be calculated by looping method [5].

To realize the goal of designing the test waveguides path, parts assembly is necessary. The assembling matrix ProMatrix() is also needs to be used to define the boundary condition such as translation distance , rotation angle and so on. The function ProAsmcompconstraintAsmreferenceSet() is used to define the reference assembly. The constrain type can be defined by the function ProAsmcompconstraintTypeSet() such as point constrain, linear constrain, plane constrain, axis constrain and coordinate constrain and so on [6]. With help of the functions above, the test waveguides can be assembled according the route design algorithm.

3.2. Route design

Usually the test waveguidess is divided into five sections, each section is a strict line, so the route of the waveguides can be simplified by a mathematical model. The functions of the route design algorithm is described below.

$$S_{1X}= W_X \quad (1)$$

$$S_{1Y}= W_Y \quad (2)$$

$$S_{1Z}= B_Z + 100 \quad (3)$$

$$L_1= B_Z + 100 - W_Z \quad (4)$$

In the above functions S1X is the first section X coordinate position, S1Y is the first section Y coordinate position, S1Z is the first section Z coordinate position. WX is the X coordinate position of test waveguides connection point, WY is the Y coordinate position of test waveguides connection point, WZ is the Z coordinate position of test waveguides connection point. BZ is the Z-direction length of the bracket. L1 is the length of the first section of the test waveguides.

$$S_{5X}= |F_X| \quad (5)$$

$$S_{5Y}= S_{4Y} \quad (6)$$

$$S_{5Z}= S_{4Z} \quad (7)$$

$$L_5= |F_X| - |S_{4X}| \quad (8)$$

In the above functions S5X is the fifth section X coordinate position, S5Y is the fifth section Y coordinate position, S5Z is the fifth section Z coordinate position, S4X is the fourth section X coordinate position, S4Y is the fourth section Y coordinate position, S4Z is the fourth section Z coordinate position. FX is the length of the bracket X-direction. L5 is the length of the first section test waveguides.

The other three sections are similar to the two sections presented above.

3.3. AutoCAD development

As it is shown in the figure 3 that the engineer drawing of the waveguide isn't complicated. Because the waveguides are standard parts, so if the type of the waveguide can be definite. Then the engineer drawing template can be chosen according to the specific type of waveguide. The geometric tolerance, surface roughness, constraint, the size of flange and waveguide wall are same for different engineer drawing. The different parts are length of the waveguide, the quantity text and the length text in the table text.

There are many methods to do AutoCAD secondary development, such as VB, ObjectARX For AutoCAD R14, Auto Lisp, ADS and .Net. Because of its simplicity, high fault-tolerant rate and friendly HMI, VB is widely used.

When the windows form application is designed completive then it needs to connect to the AutoCAD. Adding the reference of AtuoCAD 2014 Type Library and AutoCAD/ObjectDBX Common 19.0 Type Library in the COM menu. Using the function Acaddoc.ModelSpace.AddText() to add the text of quantity and length in the template form[7].

 https://doi.org/10.1051/matecconf/201817901022

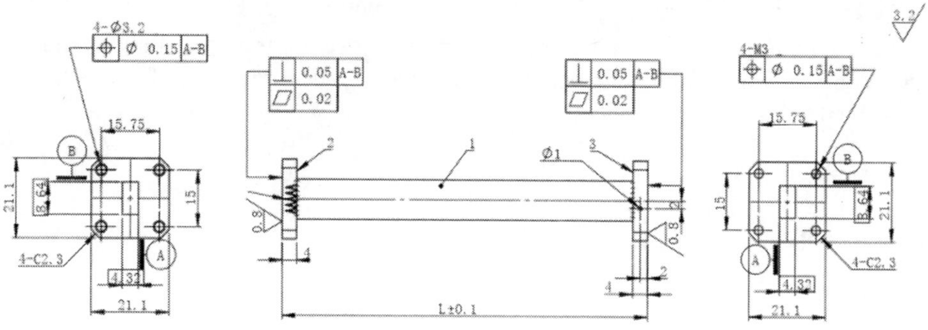

Type	Number	L
BJ260ZF1		
BJ260ZF1		
BJ260ZF1		

Figure 3. Engineer Drawing template of Waveguides

4. Conclusion and improvement

During the study of this project, the basis of computer-aided design for the route of the test waveguides has been outlined. With the help of this research the design efficiency of engineers can be improved greatly and the repetitive works can be reduce greatly. Some basic concept about the Pro/E secondary development has been elaborated. Also, there a lot of things can be improved about this research. For example, the path of the test waveguides which connected to the hydro-cooling load, such path needs to install closely to the spacecraft surface, there are a lot of protuberance on the spacecraft surface, so the obstacle avoidance algorithm can to be added. The hydro-cooling load automatic assemble algorithm can be add. These will be improved in the later works.

References

1. Dong J.M., Microwave Technology, China Machine Press, Beijing, 50(2014).
2. Wu L.J., Foundation of Pro/Engineer Secondary Development Technology, Publishing House of Electronics Industry, Beijing, 1(2006).
3. Wang W.B., Pro/E Wildfire 4.0 Secondary Development Cases resolution, Tsinghua university Press., Beijing, 10-20(2010).
4. Wang W.B., Pro/E Wildfire 4.0 Secondary Development Cases resolution, Tsinghua university Press., Beijing, 153-155(2010).
5. Wang W.B., Pro/E Wildfire 4.0 Secondary Development Cases resolution, Tsinghua university Press., Beijing, 150-153(2010).
6. Wu L.J., Pro/Engineer Wildfire Foundation of Pro/Engineer Secondary Development Technology, Publishing House of Electronics Industry, Beijing, 250-276(2006).
7. Zhang J.Z., Visual Basic and AutoCAD Secondary Development Cases resolution, Tsinghua university Press., Beijing, 130-180(2002).

Investigation on the Thermal Control Algorithm of the Heat Sink Temperature Regulating System for Thermal Vacuum Tests

Jian Yabin[a], Zhang Chunyuan, Wang jing, Lin Boyin, Wang Runze and Yin Xiaofang

Beijing Institute of Spacecraft Environment Engineering, Beijing 100094, China

Abstract. The heat sink temperature is often set in a semi-empirical way based on steady-state temperatures . As a result, the entrance parameters of the heat sink working fluid are often too conservative, and the rate of temperature rise is difficult to control. In this paper, transient thermal models for each component of the heat sink temperature regulation system are established. Then they are programmed into a dynamic simulation model by using Matlab/Simulink as the thermal control algorithm.It is shown that the model can accurately reflect the dynamic and steady state characteristics of the heat sink temperature regulation system, and can provide guidance for the selection of control strategies and working fluid parameters in the thermal vacuum test.

1 Introduction

The thermal vacuum test is to assess the performances and to expose the design and process problems of spacecraft components, subsystems or the whole spacecraft under extreme hot and cold vacuum conditions. In thermal vacuum tests, the use of heat sink temperature regulation has the advantages of economy, convenient operation, fast cooling, and strong adaptability to complex shape and moving parts. It is an important development direction of spacecraft component level thermal vacuum test[1-3]. At present, in the process of heat sink temperature regulation, the initial temperature, the final temperature and the cooling time usually are taken to be the known conditions. The temperature of the heat sink is set by a semi-empirical method based on steady-state model. The parameters of the inlet working fluid are regulated automatically by PID. Due to the large lag caused by the heat capacity of the heat sink, the inlet parameters obtained by steady state calculation are usually much more conservative than the actual conditions during the test. The heat sink temperature regulation time is too long, and the rate of temperature rise and fall of the test article is difficult to control [4].

In the past, the research of heat sink temperature control system mainly focused on heat sink structure and external cryogenic system. There was less research on the dynamic temperature characteristics of the whole system, and the thermal model of the test article was relatively simple [5-6]. Here, based on transient thermal model, the heat transfer relationship of the whole heat sink temperature regulating system is analyzed, and a dynamic simulation model is established by Matlab/Simulink. The dynamic response relationship among the entrance working fluid parameters ,the heat sink and the test article is obtained. The model is programmed into the

[a] Corresponding author : cast5118@163.com

temperature control algorithm for the test article, and applied in the thermal vacuum tests. The results show that the established dynamic model and simulation program can reflect the steady and dynamic characteristics of the test article accurately, which is of great significance for the formulation of the test strategy, the selection of test parameters, and the prediction of the test process.

2 Model of heat sink temperature regulating system

The heat transfer of heat sink can be divided into the inner part and the outer part. There are two paths for the outer heat transfer, one is to the test article, the other is to the wall of the chamber (including radiation shield). In these two paths heat is exchanged by thermal radiation. For the inner part, heat is exchanged with the working material by convection heat transfer. Based on the idea of object-oriented programming and parametric modeling, the simulation system is composed of 3 separate modules: heat sink heat exchanger module, test article module and radiation heat transfer module [5].

2.1 Heat sink heat exchanger module

The heat sink is a pipe network system composed of finned tubes. For the convenience of modeling, one of the single branch pipes was analyzed. The branch was divided into n nodes, and the length of the microelement dl was taken as the control body. Based on energy conservation, the change of energy in the control body is equal to the algebraic sum of heat flux in and out. The temperature of each control body in the tube wall at the τ moment can be expressed as follow:

$$\Delta m \cdot c \cdot \frac{dT_i}{d\tau} = \delta q + h_i S_1 (T_i' - T_i) \tag{1}$$

$$\delta q = q / k / n \tag{2}$$

$$q = q_1 + q_2 \tag{3}$$

Where q_1 is radiation heat transfer between heat sink and test article; q_2 is radiation heat transfer between heat sink and chamber; k is the number of branch pipes; n is the number of nodes of one branch pipe; $\Delta m = M / k / n$ is mass of the control body; M is mass of heat sink; T_i' is temperature of the fluid at the node i; T_i is temperature of the heat sink at the node i; c is the average heat capacity of heat sink; S_1 is convective heat transfer area between heat sink and working fluid.

Using the fitted equations of refrigerant properties[6] and the heat transfer correlations[7-9], The convective heat transfer program can be written, and then it can be executed to dll files using the mex command. Through the S-function, the convective heat transfer calculation module can be given as shown in figure 1. The input is external heat flux, and the output is the average temperature of the heat sink .

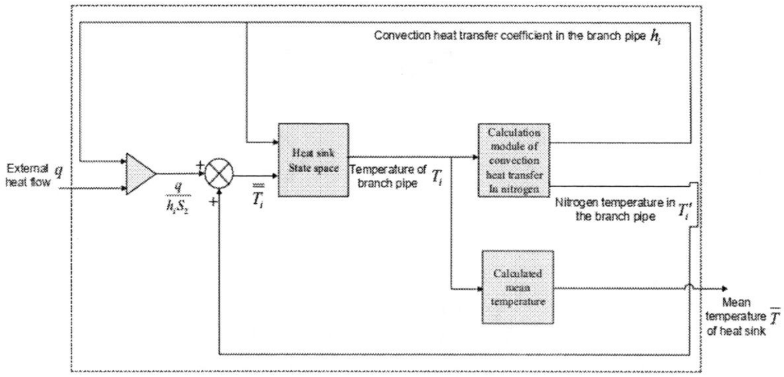

Figure 1. Heat exchange module of heat sink-working fluid

The parameter needed by the simulation can be a single variable or variables in a form of a matrix.

2.2 test article module

The test article is simplified as a solid cylinder with the same material, mass and surface area as the actual test article. It is treated as one-dimensional heat conduction. This simplification is feasible for most engineering applications .

The structure of the test article is divided into m layers, as shown in Figure 2.

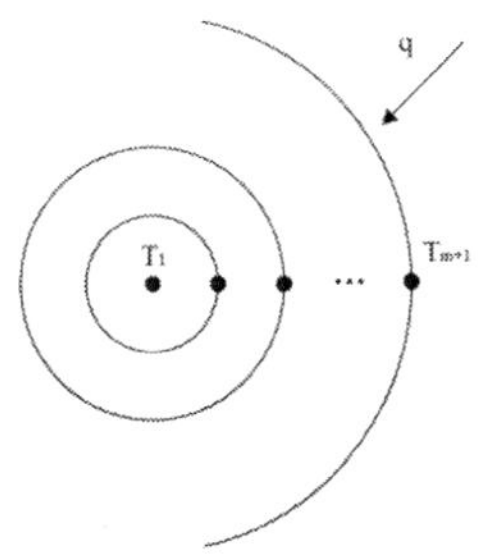

Figure2. Simplified heat transfer model of test article

The one-dimensional heat conduction model of the cylindrical object is discretized by the two order formulation [8].

$$C\frac{dT_i}{d\tau}=\frac{T_{i-1}-2T_i+T_{i+1}}{R}-\frac{T_{i-1}-T_{i+1}}{2(i-1)R} \tag{4}$$

where C is specific heat capacity of test article; $R=\Delta r/\lambda$ is the thermal resistance.

For the $m+1$ point, which is on the outer boundary, the energy change in the control body is equal to the heat transferred from the m point added the heat transferred from the outside, which can be simplified to:

$$\frac{dT_{m+1}}{d\tau}=\frac{2(m-1/2)}{RC(m-1/4)}(T_m-T_{m+1})+\frac{\rho c_p}{C^2L\pi(m-1/4)}q \tag{5}$$

When the difference between the average temperature of the test article and the outside surface is small, it is equivalent to the lumped parameter method ignoring the internal thermal resistance of the test article..

The module take the radiative heat flux as the input, and output the temperature of the test article.

It can be selected whether or not to treat the test article by lumped parameter method. When the lumped parameter method is selected, only the correlation parameters and initial temperature of the test article are required for input. When the lumped parameter method is not selected, the parameters such as the test article 's density and thermal conductivity coefficient should also be input.

2.3 Calculation module of radiation heat transfer

The heat transfer between the heat sink and other surfaces (the chamber and the test article) can be regarded as the typical radiation heat transfer: thermal radiation among isothermal grey surfaces.

The mathematical model of radiation heat transfer is provided by Simulink, it is built and encapsulated into a radiation heat transfer subsystem to form a radiation heat exchange module.

2.4 Simulation model of heat sink and temperature regulating system

According to the modules established above, the schematic diagram of the system simulation model can be built,as shown in Figure3.

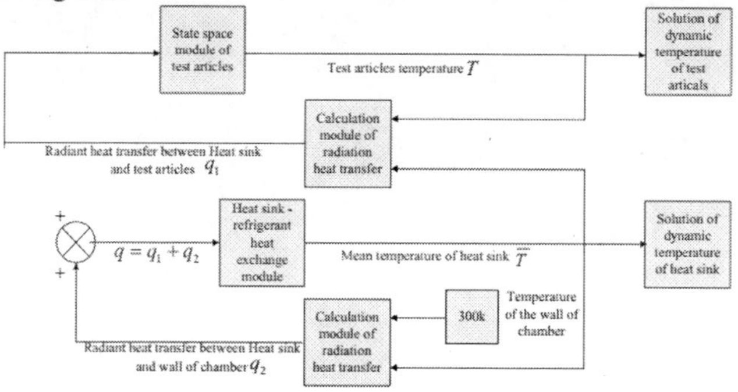

Figure3. Schematic diagram of heat sink temperature regulating system simulation model

For a small vacuum chamber, the entrance pressure of the working fluid (nitrogen pressure) is 0.4Mpa, the entrance temperature is 150K, the total flow rate is $1.5\, m^3/\mathrm{min}$. The temperature variation of the heat sink branch pipe nodes (k=50) in 3600 seconds are shown in Figure 4.

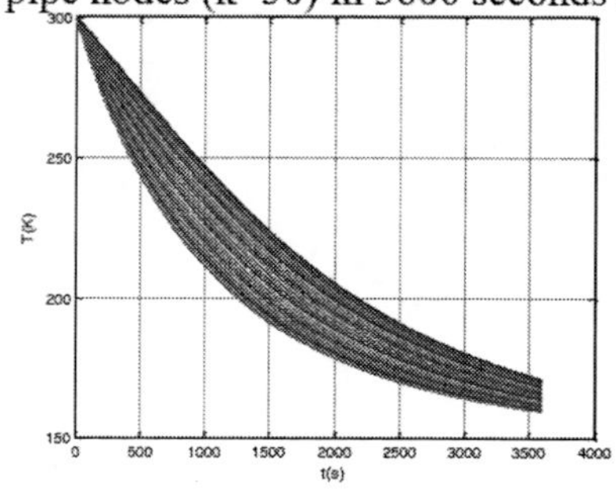

Figure4. Response curve of heat sink temperature

The test article is a electro-thermal component with mass 10kg, surface area $5X10^{-3}m^2$, internal heat source 20 W, and surface emissivity 0.9. The temperature variation during heat sink cooling is shown in Figure 5.

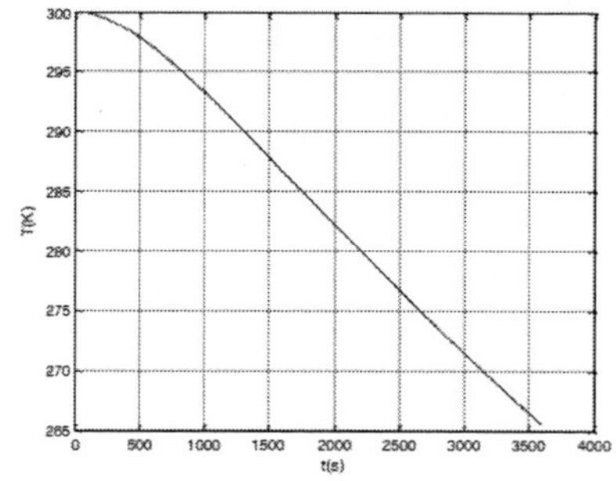

Figure5. Response curve of test article temperature

This dynamic simulation model is superior to the traditional model in that it predicts a temperature variation agreed well with the actual cooling process.

3 Temperature control algorithm based on transient thermal model

Using the Matlab external commands, the simulation model can calculate the steady and dynamic parameters of the heat sink, the test article and the entrance working fluid parameters. It can also do the reverse operation to calculate the entrance working fluid parameters which satisfy the requirement of the target temperature and the temperature change rate of heat sink or the test article. In the following, the dynamic calculation is performed, and it is illustrated how to guide the setting of the entrance working fluid parameters of the thermal vacuum test.

During the thermal vacuum test, the temperature change rate of the test article is often specified. For the simulation model established, the temperature change rate is solved in two ways: "Positive calculation" and "Inverse calculation". "Positive calculation" is to calculate the temperature change rate of the heat sink or the test article given the parameters of the heat sink entry. "Inverse calculation" is to calculate the inlet variables of the heat sink to satisfy the temperature rate. In practice, the "Inverse calculation" is of more practical value.

Using SIM command, provided by Matlab ,we can get the output of the system at any time under various input conditions. In the thermal vacuum test, temperature variation of the heat sink and test article can be obtained through the SIM command effectively.

The key to calculating the temperature change rate("Positive calculation") is to get the time duration required for the system to reach the target temperature. The result of the SIM command can be seen as a "function" of the input time variable. The time required is regarded as the solution equation of the function. A secant method is used to calculate the time required to reach a certain temperature. The program is saved as searcht.m to be called during the "Inverse calculation" process of the temperature change rate.

For a single input system, there is only one demarcation point that meets the requirements of the temperature change rate.The solution of the point T_0 can be carried out according to the following algorithm: First, the trim command provided by Matlab is used to calculate the heat sink inlet temperature T_s, which can make the heat sink stable at the target temperature T_2. Then the average temperature change rate κ' of the heat sink from the initial temperature T_1 to the target temperature T_2 is calculated at the inlet working fluid temperature $T_0 = T_s$. If $\kappa' < \kappa$, then $T_0 = T_0 - 1K$, if $\kappa' > \kappa$,then $T_0 = T_0 + 1K$, until the boundary point of the condition is satisfied.

Generally, The inlet parameters calculated by the target stable temperature are difficult to meet the requirement of temperature change rate. The program diagram for solving the parameters of the heat sink entrance to meet the temperature change rate of the test article is shown in Figure 6.

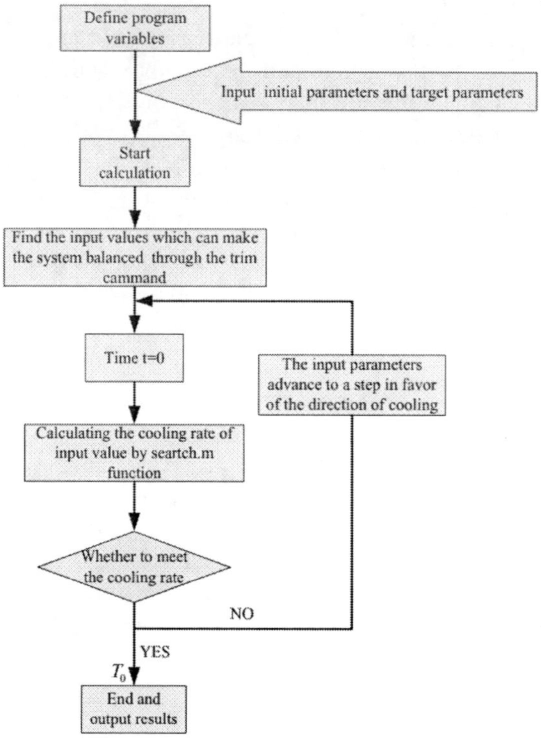

Figure6. Block diagram of cooling rate

The above entry parameters can be the flow or temperature of the heat sink inlet working fluid.

Combined with a test equipment, the average cooling rate of the test article from 30℃ to -60℃ is not less than 0.5℃/min, and the temperature is eventually stabilized at -60℃. The compressor outlet pressure is 0.4Mpa, the heat sink inlet flow rate is 5m3/min(standard condition), In this case, the heat sink inlet temperature can be calculated to meet the cooling rate and the target temperature of the test article.

Input the known parameters and initial parameters in the interface, select "control test article" in the drop-down menu of "select control target" ("control test article" or "control heat sink"). "-0.5" stands for the cooling rate is 0.5 ℃/min. In the choice of " heat sink entrance conditions "(" flow "or" temperature ") pull-down menu ,select" flow ", standing for the flow of the inlet working fluid is known. The simulation curve calculated by the program is shown in Figure7.

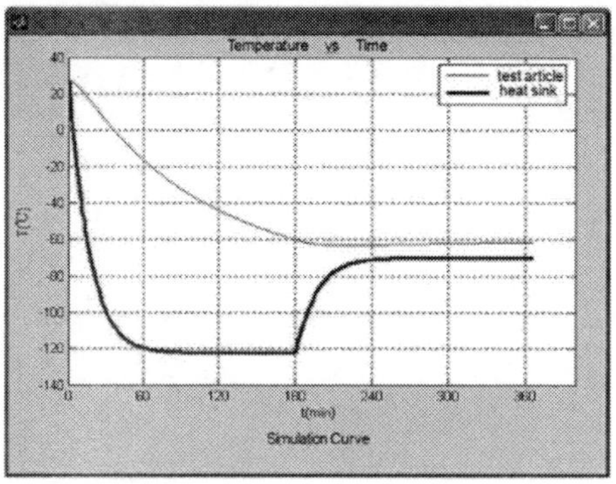

Figure7. The simulation curve interface of model

The inlet flow rate of the heat sink Qv is 5m^3/min, the inlet temperature T_0 is -124.9℃. Then the temperature of the test article is reduced from the initial temperature to -60℃at t=180min ,Meeting the requirement of cooling rate. At the end of the cooling stage, the inlet flow rate is constant, the temperature is changed to =-71.8 ℃, The final temperature of the test article is finally stabilized at -60 ℃. the final temperature of the heat sink is -70.3 ℃.

4 Conclusion

In this paper, mathematical models for the heat transfer of the working fluid, heat sink and test article are established. They are integrated into the heat sink temperature regulating system. Then the Simulink model of the overall heat sink temperature control simulation system are built, and the temperature control algorithm based on the transient thermal model is realized. It can not only simulate the steady and dynamic simulation with the various input parameters, but also can carry out the "Inverse calculation" under certain test conditions to provide guidance for the parameter setting of the heat sink temperature regulating system in the thermal vacuum test.

References

1. HUANG Bencheng, *Environment test technology of Spacecraft[M]*.,2002:60-98
2. SHAO Xingguo, LIU Zijun, TAN Changhai,et al. Research on heat flux measurement and simulation in thermal balance test of spacecraft[J]. Spacecraft Engineering,2007,**16**(4): 94-98
3. Andrina G, Cane L, Panelh B. Solar Simulation with Infrared Techniques Applicability to Mercury/ Bepicolombo Probe System Thermal Test. Proceedings of 4th International Symposium on Environmental Testing for Space Program, ESTEC, Noordwijk Netherlands. 2001:327-332
4. Michael T. Fournier. Controller for an environment test chamber. SCI, United States Patent NO. US6,023,985, 2000,1
5. JIAN Yabin , ZHANG Chunyuan, DING Wenjing, et al. Simulation design based on state-pace for regulating heat sink[J] Spacecraft Environment Engineering,2008,**25**(4):373-376
6. JIANG Chansheng. Dynamic Simulation Technology for Cryogenic Shrouds in Large Space Simulators [D].Beihang University,1999
7. Yakhot V, Orzag S A. Renormalization group analysis of turbulence:basic theory[J].Journal of Scientific Computing,1996(1):3-11
8. LI Qingfang, ZHANG Guozhong, LIN Yuanyuan,et al.Study of heattransfer and pressure drop in tubes with rectangle-like fins[J].Journal of Engineering Thermophysics, 2009,**30**(12):2122-2124
9. LUO Baojun, WANG Zhaoli, YAN Tao,et al. Design and Implementation of a Space Sub-KelVin Sorption Cooler[J]. Journal of Astronautics,2015,**36**(7):855-859

MATEC Web of Conferences **179**, 01024 (2018) https://doi.org/10.1051/matecconf/201817901024
2MAE 2018

The IR Characteristics Modeling and Simulation of the HEO Target

Tao Dongxing[1,a], Lin Boying[1], Du Peng[1], Bi Yanqiang[1], Shang Yonghong[1], Li Xiyuan[1] and Wang Jing[1]

[1] *Beijing Institute of Spacecraft Environment Engineering, Beijing 100094, China*

Abstract. Space target infrared (IR) characteristics model can be used to design space target detection sensor, generating simulation data to validate the data processing algorithms, such as target detection and tracking. In the work, a HEO target IR characteristics model is built. The model consists of the geometry module and the IR radiometric module. Unlike the traditional space IR model, the temperature of cabin inner is used as the simulation origin. Using the model, the irradiance spectra of HEO target are calculated for different target temperature. With the calculation results, the main wavelength range for HEO detection is analyzed. Using the equivalent temperature, the work also designs a simulative target, which has the similar IR characteristics of a HEO target. The simulator can be used in the ground test for the imaging sensor or target decoy.

1 Introduction

As a passive detection manner, IR detection has many advantages, such as the excellent concealment, adapting to day and night work, and the high sensitivity for the temperature difference between target and background [1]. So that, IR detection has been widely developed and applied for the space target detection. Now, there has abundant work dedicated to the modeling and digital simulation of the space target IR characteristics [2, 3]. The digital simulators can provide the IR radiation of the target for the given orbit and attitude, and it could be used to assist the detection sensor design and provide simulation data for processing method development. The traditional modeling methods use the target surface temperature as the simulation origin to calculate the IR radiation [4]. However, the target surface temperature is unknown in the normal condition. The temperature message can be gotten is the inner cabin and the instrument. Usually, the cabin is covered by the thermal control unit and coating. Therefore, the model should use the inner cabin temperature as the simulation origin, and the heat transmission of the space target should be considered. In order to evaluate the detection ability of the developed IR imaging sensor in the ground or design decoy target for the important space target, a simulative target with the similar IR characteristics is needed. In the work, the IR characteristics modeling of the HEO target is built with the target orbit, surface material, inner cabin temperature, and attitude considered. Moreover, a simulative target with the given IR radiation is designed for the ground test.

[a] Corresponding author : taodx2016@163.com

2 The IR characteristic modeling of the HEO target

2.1 Geometry module

The body surface of space target can be expressed by a large number of facets based on computer graphics theory. All facets are convex polygons. Their shapes, sizes, and positions can be uniquely defined by their 3D coordinates and exterior normal directions in a coordinate system. The professional geometry modeling software 3DS MAX is used to build the geometry model [5]. Then the material characteristic of each facet is defined.

Figure 1. The sketch of the geometry configuration.

As the special equipment, each space target has the different configuration. The complex degree of the configuration determines the calculation quantity on the digital simulator. Therefore, a trade-off should be made between the simulation precise and the calculation quantity. In the work, a satellite geometry module, which consists of satellite body, two solar wings, and two antennas, is built. Figure **1** shows the sketch of the space target geometry configuration.

2.2 IR Radiometric Module

The IR radiation of the space target, which reaches the imaging sensor, comprises the reflected radiation and the facet-self emission radiation:

$$L(\lambda) = L_e(\lambda)+L_r(\lambda) \tag{1}$$

Where L is the radiance the imaging sensor received, L_e is the emission part, L_r is the reflection part, and λ represents the wavelength.

2.2.1 Reflected irradiance of the target facet

The radiation received by the target facet includes three parts: solar irradiance, lunar irradiance, and earth irradiance. The lunar irradiance reaches the earth satellite is very low because of the 3.8×10^5 kilometers of the lunar-earth distance. Therefore, the irradiance reaches the target (E_{tr}) is:

$$E_{tr}(\lambda) = E_s(\lambda)+[\,E_{es}(\lambda)+ E_s(\lambda)\cos\theta_1 r_e(\lambda)]\times\tau_a(\lambda) \tag{2}$$

Where E_s is the TOA (Top of Atmosphere) solar irradiance, E_{es} is the emission irradiance of the earth, θ_1 is the solar altitude angle relative to nadir point, r_e is the equivalent reflectance of the earth, τ_a is the transmittance of the atmosphere. Based on the Planck's Law [6], E_s and E_{es} are represented by:

$$E_s(\lambda) = \frac{2hc^2}{\lambda^5 d_{se}^2\left\{\exp\left[hc/(\lambda k_B T_s)\right]-1\right\}}\varepsilon_s(\lambda) \tag{3}$$

$$E_{es}(\lambda)=\frac{2hc^2}{\lambda^5 d_{et}^2\left\{\exp\left[hc/(\lambda k_B T_e)\right]-1\right\}}\varepsilon_e(\lambda) \tag{4}$$

Where h is the Planck's constant, k_B is the Boltzmann's constant, c is the velocity of light, T_s is the temperature of the solar surface, T_e is the temperature of the earth surface, ε_s is the emissivity of the solar surface, ε_e is the emissivity of the earth surface, d_{se} is the solar-earth distance, and d_{et} is the earth-target distance.

The received irradiance from the earth is very low for the HEO target, and it could be ignored in the work. Then, the L_T is represented by [7]:

$$L_T(\lambda) = E_s(\lambda) r_{BRDF}(\lambda)\cos\theta \tag{5}$$

Where r_{BRDF} is the BRDF (Bidirectional Reflectance Distribution Function) of the target surface, and θ is the solar altitude angle relative to surface facet.

2.2.2 Emission irradiance of the target facet

In order to get the emission irradiance of the target, the surface temperature should be calculated firstly. For the HEO target, the environment heat flux is only the solar radiation with the earth radiation ignored. Considering the conversation of energy in the heat balance condition, which is showed in the Figure 2, the surface temperature can be represented by:

$$\alpha q = -\sigma\frac{T_i - T_o}{\delta} \tag{6}$$

Where α is the absorptivity of the facet, q is the environment heat flux, σ is the thermal conductivity of the facet material, δ is the depth of the thermal control unit and coating, Ti and T0 are the temperature of the cabin inner and the surface separately.

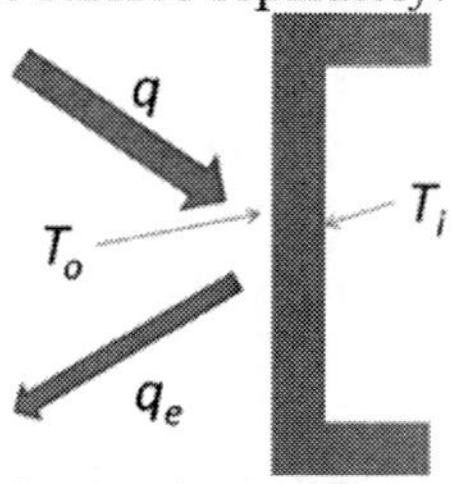

Figure 2. The sketch of the heat transmission.

The target received heat flux can be calculated when the 3D-configuration and attitude of the target is provided. Then the target surface temperature is represented by:

$$T_o = \frac{\alpha q \delta}{\sigma} + T_i \tag{7}$$

In order to improve the thermal control ability of the satellite, isothermal temperature design of the cabin is used widely. Therefore, each cabin and solar wing can be defined a uniform temperature separately, and the target emission irradiance can be calculate.

In the work, the emission characteristic of the target surface is supposed isotropic. Therefore, the Planck's Law showed in Equation (3) is used to calculate the Emission irradiance of the target E_e with the emissivity of the facet ε_t considered.

$$E_e(\lambda)=\frac{2hc^2}{\lambda^5\left\{\exp\left[hc/(\lambda k_B T_t)\right]-1\right\}}\varepsilon_t(\lambda) \tag{8}$$

Where T_t is the temperature of the target facet.

2.3. HEO Target radiation characteristic

Using the Equation (3) and (6), the irradiance target received and emission is showed in Figure 3 for different temperature.

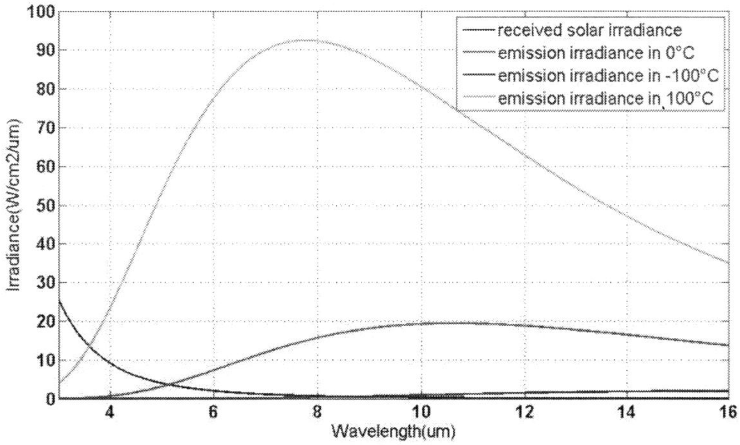

Figure 3. The received solar irradiance and emission irradiance.

It can be shown in Figure 3 that the emission irradiance is mainly distributed in the wavelength range of 3~14 μm. For the HEO target, the filter is usually used as the wavelength chosen device. With the sensor SRF (Spectral Responsible Function) R_s considered, the IR radiation be detected is represented by [8]:

$$E_d = \frac{\int_{\lambda_1}^{\lambda_2} L(\lambda)\times\Omega\times R_s(\lambda)d\lambda}{\int_{\lambda_1}^{\lambda_2} R_s(\lambda)d\lambda} \tag{9}$$

Where λ_1 and λ_2 is the response wavelength of the sensor, and Ω represents the solid angle of the sensor. With the Equation (9) considered, the ratio between received solar irradiance and the emission irradiance in Figure 3 is calculated for different temperature and wavelength band. The calculation results are showed in the Table 1.

Table 1. The ratio between received irradiance and emission irradiance.

Calculation Parameter	-100°C	0°C	100°C
3~14 μm	0.3	4.9	25.0
8~14 μm	4.6	72.7	283.5

The target temperature will not decrease to -100 °C, so that the main radiation of the target is the emission part. Moreover, the reflected radiation will not enter the FOV (Field of View) sometimes for the HEO target. Therefore, the emission irradiance of the target is more important for the detection sensor. The temperature of the target will not rise to 100 °C, so that wavelength range of 8~14 μm is the concerned only.

3 HEO target simulator

In the work, a HEO target simulator, which has the similar IR characteristic as the truly target, is designed. The target is in the geostationary orbit, and the attitude is earth orientation. The 3D-configuration is the same as the Figure 1. Two working cases of the target, which are the worst cold case and the worst hot case, are considered in the work. The worst cold case appears when the target crosses the equinox point in its early life, and the worst hot case appears when the target crosses solstice point in its end life. For the HEO target, the underside cabin cannot be observed all the time. And the temperature of other cabin and solar wing is showed in the Table 2. The surface materials are the MLI (Multi-Layer Insulation), the thermal control paint, the OSR (Optical Solar Reflector), and the carbon fibre envelope.

Table 2. The temperature of each cabin and the solar wing.

position	The worst cold case (°C)	The worst hot case (°C)
Earth orientation cabin	0	13.0
Antenna	passive	passive
East cabin	6.0	20.0
West cabin	6.0	20.0
South cabin	4.0	30.0
North cabin	-4.0	0
Solar wing	40.0	42.0

Based on the Equation (7) and the material parameters, the surface temperature is calculated and showed in the Table 3. Then the IR irradiance is calculated using the IR radiometric module and showed in the Figure 4.

Table 3. The temperature of each cabin surface and the solar wing.

position	The worst cold case (°C)	The worst hot case (°C)
Earth orientation cabin	4.4	11.3
Antenna	-11.2	-7.8
East cabin	2.6	16.5
West cabin	-46	-35.1
South cabin	4.4	30.3
North cabin	-3.8	0.6
Solar wing	40	42

In order to simplify the simulator structure, only the thermal control paint is used as the surface material. And the equivalent surface temperature is designed to get the same IR irradiance as the Figure 4 showed. Then the surface temperature of the simulator is calculated using the reversing progress of the IR irradiance calculation and showed in the Table 4.

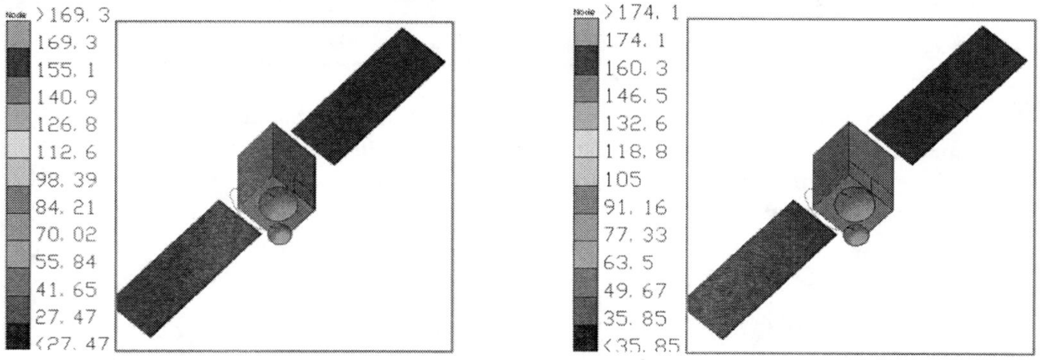

(a) The worst cold case (b) The worst hot case

Figure 4. The surface temperature.

The IR characteristic of the simulator is measured in the cryogenic vacuum chamber, and the heater band and thermocouple are used to control the temperature of the simulator. Liquid nitrogen is used to provide the cryogenic environment in the chamber.

Table 4. The temperature of the simulator surface.

position	The worst cold case (°C)	The worst hot case (°C)
Earth orientation cabin	-9.7	-3.46
Antenna	-11.2	-7.8
East cabin	-11.4	1.25
West cabin	-55.8	-45.8
South cabin	-0.97	24.0
North cabin	-8.9	-4.6
Solar wing	37.8	37.0

4 Discussion and conclusion

In the work, an IR characteristic model for HEO target is represented. Unlike the traditional model, the temperature of the cabin inner is used as the origin of the simulation. In this way, the input of the modelling could be achieved more practical. 3DS MAX software is used to build the 3D-configuration, so that the model can fit for different target demand. Moreover, the coordinate of each facet is used to calculate the shelter relation among the facets. In the model, some radiation contribution to the target is simplified to decrease the calculation quantity. The irradiance for different temperature implies that the simplified action is valid. Moreover, the simplified action brings in other advantage that the model could be used to calculate the IR radiation of the moving target real-time. Using the model, a HEO target simulator is designed, and the equivalent surface temperature is used to simplify the simulator structure. In the future work, the simulator IR characteristic will be measured in the cryogenic vacuum chamber to evaluate the capability. And the results are also used to evaluate the accuracy of the model further.

References

1. Y. Zheng, Z. Wei, X. Huizhong, Infrared and Laser Engineering, **37** (2008)
2. W. Jun, H. Yu, D. Na, S. Kefeng, S. Dandan, X. Junhu, W. Wenli, G. Jiaobo, Proceedings of SPIE, **9449** (2015)
3. G. Sili, T. Xinyi, Y. Yang, X. Fengting, Proceedings of SPIE, **7383** (2009)
4. W. Ying, H. Jianming, W. Xiangquan, Infrared and Laser Engineering, **44** 9 (2015)
5. T. Donging, J. Guorui, Z. Huijie, International Journal of Remote Sensing, **35** 20 (2014)
6. F. Jianying, W. Yang, G. Guanglei, Journal of Harbin University of Science and Technology, **14** 6 (2009)
7. S. Chengming, Y. Yan, Z. Xiubao. Proceedings of SPIE, **7383** (2009)
8. M. J. Duggin, C. J. Robinove, International Journal of Remote Sensing, **11** (1990)

Research on Hot-Wire Low Pressure Wind Speed Measurement System

Yin Xiaofang[1,a], Li Xiyuan[1], Gao Qinghua[1] and Wang Jing[1]

[1]*Beijing Institute of Spacecraft Environment Engineering, Beijing 100094, China*

Abstract. In order to verify the correctness of the thermal control design, the Mars rover needs to conduct a thermal test on the ground during the construction of the Mars exploration mission. In the course of the thermal test, real-time measurement of the wind speed at low pressure is required. Therefore, it is needed to establish a low-pressure wind speed measurement system. This article describes the composition of the test system, the choice of measurement sensors, in which the impact of wire resistance, electromagnetic interference and other factors are taken into account, and the design of the test process is completed.

1 Introduction

Mars also enveloped the atmosphere, similar to the Earth's atmosphere. Mars's main component of the atmosphere is carbon dioxide, followed by nitrogen, argon, in addition to a small amount of oxygen and water vapor. There is nitrogen in both the Martian atmosphere and the Earth's atmosphere, which is the biggest similarity between Mars and the Earth. The density of the atmosphere in Mars is less than one percent of the Earth's atmosphere, the surface atmospheric pressure 500-700pa[1-2].

According to previous tests, the convective heat transfer coefficient at 1000Pa low pressure can be measured by the wind speed sensor even though the relative atmospheric pressure is reduced. After re-calibration of the hot-wire wind speed sensor, the convection heat transfer coefficient can be used for low pressure[3].

2 Hot-wire low pressure wind speed measurement system[4-5]

2.1 Hot-wire anemometer measurement principle

Hot air anemometer measurement circuit diagram is as follows:

[a] Corresponding author : 190064602@qq.com

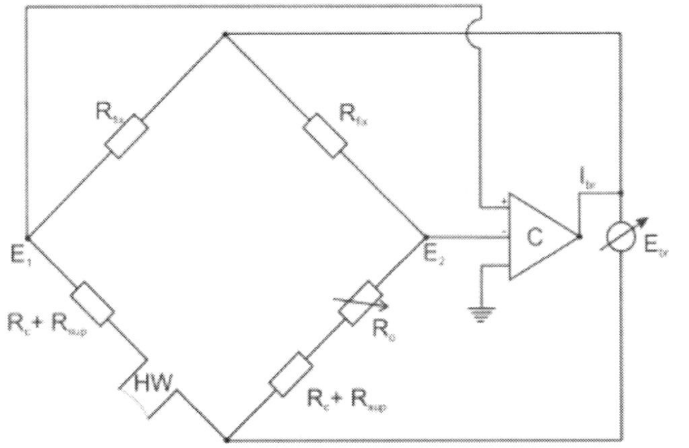

Figure 1 bridge measurement circuit

As shown in the figure 1, where Rfix is the internal resistance of the bridge, Rc + Rsup is the resistance of the wire + the resistance of the bracket. In order to accurately measure the hot wire HW, it is necessary to ensure that Rc + Rsup does not change during the test. Can not change too much. The system limits are as follows:

a) The cable resistance changes little during the test. The wire resistance and calibration state are the same when measuring.

b) The system needs a brush slip ring;

c) Since the wind speed sensor is a high-frequency analog signal, electromagnetic interference needs to be considered.

Aiming at these constraints, the system diagram of wind speed measurement under low pressure is as shown in Figure 2:

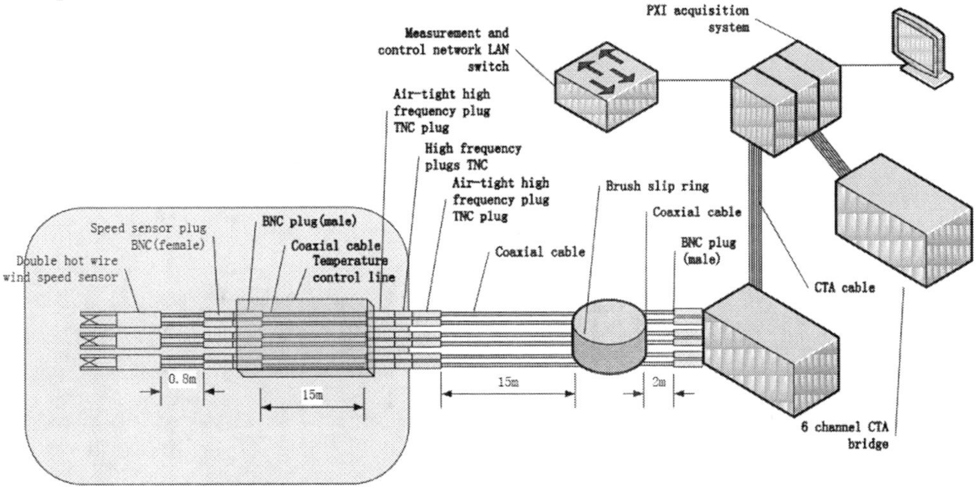

Figure 2 low pressure wind speed measurement system

The following figure shows the wind speed measurement system in calibration mode:

As can be seen from the figure 3, the system includes two parts of the cable inside the container and the cable outside the container, wherein the cable inside the container is a slip ring in the calibration mode, and the 15m coaxial cable is outside the container. Wind speed measurement mode, the exchange of the cable inside and outside the container, to ensure that the calibration and use of the state of the same resistance. At the same time, in the calibration and use mode, all through the aluminum temperature control temperature of the cable temperature, reduce the temperature changes on the cable resistance.

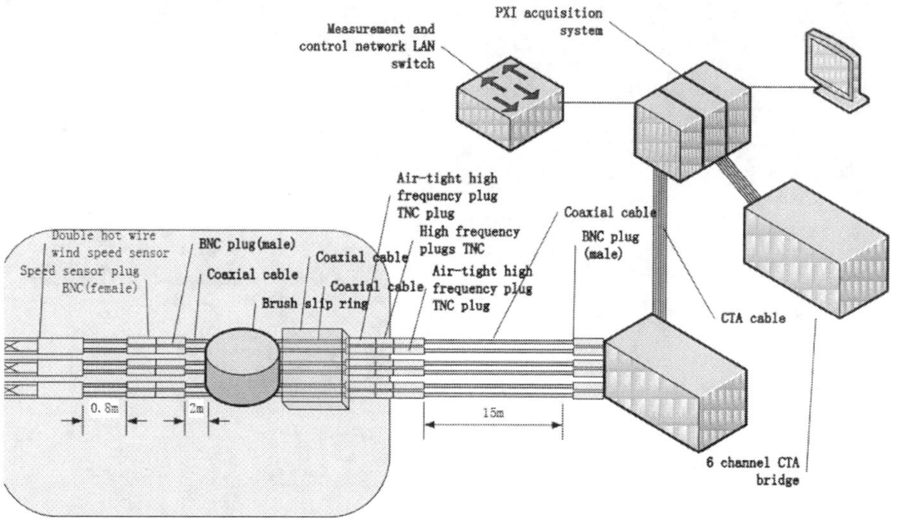

Figure 3 low pressure wind speed measurement system (calibration mode)

2.2 System Configuration

According to preliminary tests, the use of 55R01 series coated hot wire sensor can be under low pressure on the wind speed measurement accuracy, select the same type of x-type probe. In order to meet the wind speed measurement requirements of different directions, select two types of wind speed probe, model 55R52 and 55R51 respectively, as shown in the Figure 4:

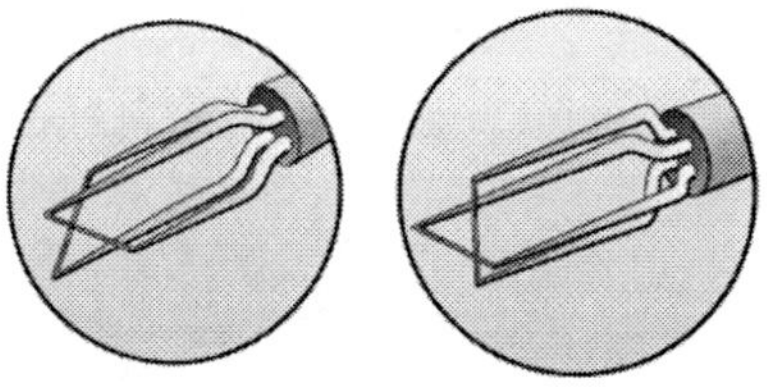

Figure 4 55R51 and 55R52 probe

One probe are 0.5um coating, each resistance is about 6Ω , the temperature coefficient of 0.44% / ℃, the maximum temperature of 300 ℃. According to the test at low pressure, the sensitivity of the probe is about 4-8mv / (m / s), as shown in Figure 5:

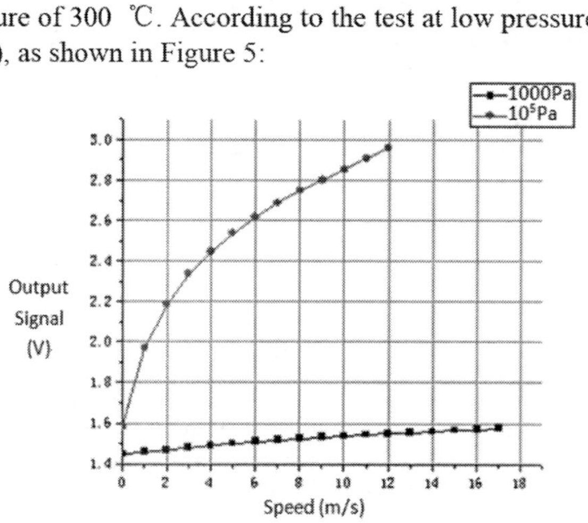

Figure 5 probe output at low pressure

Measurement accuracy to meet the requirements of the wind speed can be accurately measured.

Probe support played a role in the wind speed probe support and signal transmission, in order to take into account the various test point of view, select the two probe holder, respectively, straight probe and 90 ° probe, as shown in Figure 6 and Figure 7:

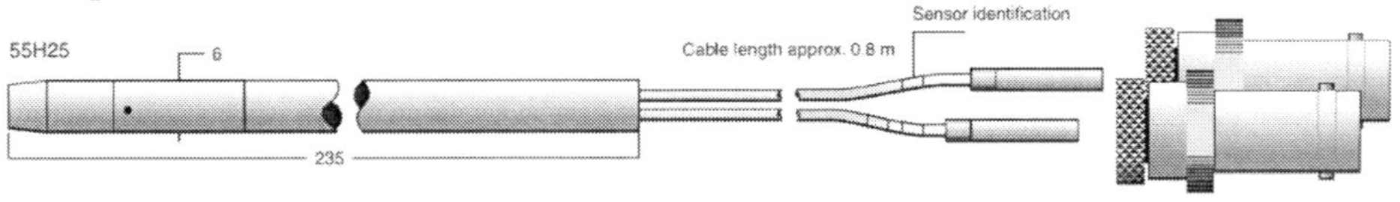

Figure 6 9055H025 Probe Holder

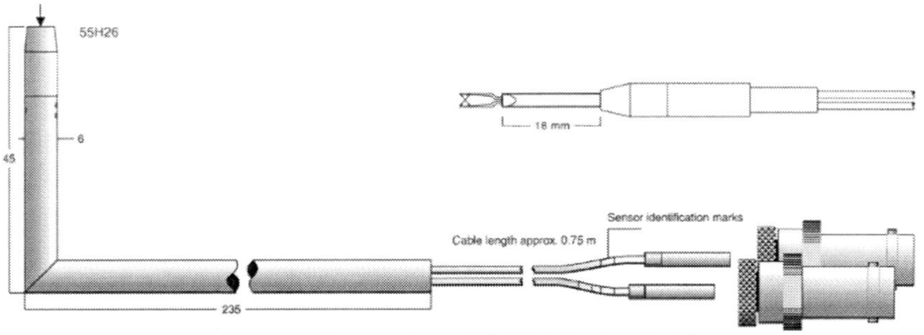

Figure 7 9055H026 Probe Holder

Each two-dimensional wind speed probe occupies two bridges of the bridge and a total of 12 probes are required for the six probes. Two Dantec 6-channel bridges with six wind speed measurement channels and one platinum resistance temperature measurement channel are available Consider the influence of temperature on wind speed measurement. Dantec 6-channel wind speed measurement system is shown in Figure 8.

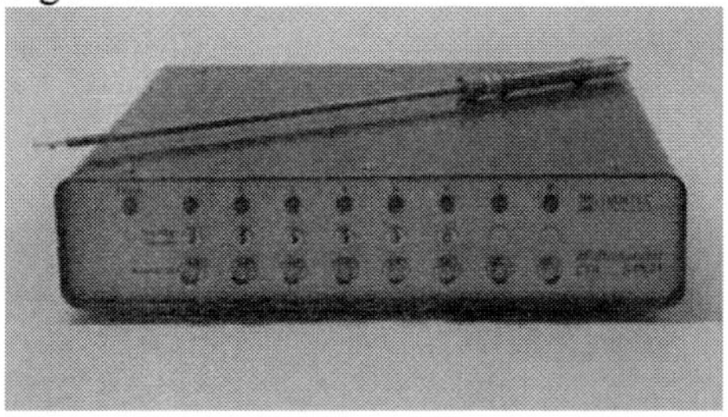

Figure 8 Dantec 6-channel wind speed measurement system

The data acquisition system chooses NI Corporation based on the PXI measuring instrument, The program can measure the 0-5V signal output by the wind speed probe, of which 24-bit accuracy of up to 1.192uV, to meet the measurement requirements.

There are two types of measurement cables: Slip-free and Slip-ring, respectively defined as A and B cables, A cable used to measure the status of the in-container cable and the calibrated out-of-container cable, B cable used to measure the status of Container outside the cable and calibration of the container cable, A, B each requires two group.

The use of constant heat flow to drive the wind speed probe, the probe temperature is constant temperature, the total heat loss through the probe:

$$Q = sh\Delta t = \frac{sNu\lambda}{l} \; t_w - t_e \tag{1}$$

Which Nu can be expressed as a function of Re, due to the cylinder, sphere approximation, and the ball has been under low pressure for many times to verify the ball, for example, can be expressed as:

$$Nu = 2 + (0.4Re^{0.5} + 0.06Re^{2/3})Rr^{0.4} \tag{2}$$

According to the test at low pressure, the output of the wind speed sensor calculated by this formula at 40000Pa is completely consistent with the atmospheric pressure, but the deviation is about

20% at about 1000Pa. At present, according to the literature, the Nu-Re number caused by low atmospheric pressure is usually correlated According to the test, at about 700Pa and 0-15m / s, the equation can be modified by the formula : h * (P1 / P0) 2

$$h = h_0 \times (\frac{P_1}{P_0})^{0.07} \tag{3}$$

Where P1 and P2 are the operating pressure and atmospheric pressure, respectively, and h0 is the convective heat transfer coefficient calculated according to the criterion. The corrected h can be used to evaluate the output of the wind speed sensor under low pressure.

The picture of turntable calibration system is in Figure 9:

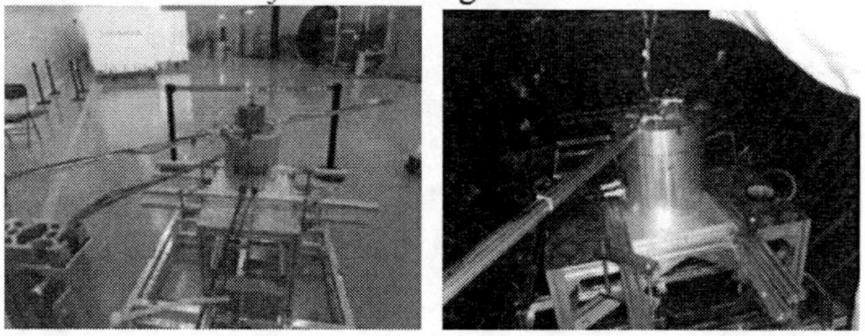

Figure9 turntable calibration system

2.3 Systematic Error Analysis

According to the previous simulation of the flow field near the cantilever using the CFD simulation method, the influence of the cantilever itself on the flow field can be reduced beyond 0.1 m outside the cantilever, reducing the error level to about 0.2 m / s.

Turntable speed can be adjusted between 0 ~ 100RPM, the diameter of the cantilever 2.6m, according to the formula :

$$V = RPM \cdot \frac{\pi R}{60} = \frac{\pi}{60} \cdot RPM \cdot R \tag{4}$$

The error for speed is

$$dV = \frac{\pi}{60} \cdot \left[d(RPM) \cdot R + RPM \cdot d(R) \right] \tag{5}$$

The RPM error is less than 0.5RPM. The error of radius R is measured with ruler and the error is less than 0.5mm. The error of wind speed calibration can be calculated for each working condition. In actual calibration, the combination of pressure, temperature and output needs to be calibrated with the wind speed. Under each combination of pressure and temperature, the calibrated wind speed condition is initially determined as shown in Table 1:

Table 1 calibration conditions

No.	Wind speed m/s	Rotating speed(RPM)	error	No.	Wind speed m/s	Rotating speed(RPM)	error
1	0.5	3.67	0.03	21	10.5	77.13	0.04
2	1	7.35	0.03	22	11	80.8	0.04
3	1.5	11.02	0.03	23	11.5	84.47	0.04
4	2	14.69	0.03	24	12	88.15	0.04
5	2.5	18.36	0.03	25	12.5	91.82	0.04
6	3	22.04	0.03	26	13	95.49	0.04

7	3.5	25.71	0.03	27	13.5	99.17	0.04
8	4	29.38	0.03	28	14	102.84	0.04
9	4.5	33.06	0.03	29	14.5	106.51	0.04
10	5	36.73	0.03	30	15	110.18	0.04
11	5.5	40.4	0.04	31	15.5	113.86	0.04
12	6	44.07	0.04	32	16	117.53	0.04
13	6.5	47.75	0.04	33	16.5	121.2	0.04
14	7	51.42	0.04	34	17	124.88	0.04
15	7.5	55.09	0.04	35	17.5	128.55	0.04
16	8	58.76	0.04	36	18	132.22	0.04
17	8.5	62.44	0.04	37	18.5	135.89	0.04
18	9	66.11	0.04	38	19	139.57	0.04
19	9.5	69.78	0.04	39	19.5	143.24	0.04
20	10	73.46	0.04	40	20	146.91	0.04

As can be seen in the figure, the error can be less than 0.5 m / s.

Due to the influence of pressure on the wind speed calibration, and the temperature effect is relatively small, for the pressure calibration and temperature calibration, to be calibrated 1300, 1350, 1400, 1450, 1500Pa three pressure conditions, temperature calibration -120, -100, 80 three working conditions.

Calibration process is: the container is first pumped to a vacuum, pressurized to the lowest calibration pressure by nitrogen, adjusted to the highest temperature to be calibrated, calibrated at each speed, then repressurized to a second pressure, calibrated in sequence until the current pressure and velocity are calibrated, Heat sink to adjust the temperature to the next condition, until all conditions calibration completed.

3 Conclusion

This paper introduces the design method of low-pressure air velocity measurement system in detail, introduces the structure of the test system, selects the measuring sensor, considers the influence factors such as wire resistance and electromagnetic interference, and completes the design of the test flow to obtain the wind speed data under low pressure. Build a good foundation.

References

1. Johnson K. Simulation of Mars Surface conditions for Characterization of the Mars Rover Thermal Response[J].
2. Numata D, Anyoji M, Sugino Y, et al. Characteristics of Thermal Anemometers at Low-Pressure Condition in a Mars Wind Tunnel[C]// Aiaa Aerospace Sciences Meeting Including the New Horizons Forum and Aerospace Exposition. 2013.
3. Wilson C F. Measurement of wind on the surface of Mars[J]. University of Oxford, 2003.
4. Ransome T, Pesket S, Toplis G. Thermal Balance Testing of the Beagle2 Mars Lander[C]. International Symposium Environmental Testing for Space Programmes, 2001.
5. Quix H, Quest J, Brzek C. Hot-Wire Measurements in Cryogenic Environment[C]// Aiaa Aerospace Sciences Meeting Including the New Horizons Forum and Aerospace Exposition. 2013.

Research on Thermal Insulation Performance of Lightweight Thermal Protection Materials for High Speed Aircraft under Different Boundary Conditions

Feng Wang, Dafang Wu[a] and Haoyuan Ren

School of Aeronautic Science and Engineering, Beihang University, Beijing 100191, China

Abstract. The determination of thermal insulation performance of thermal protection materials or structures is an indispensable and important step in the safety design of high speed flight vehicles. To obtain the temperature difference of the radiating surface for plate specimens under three different boundary conditions in heat insulation experiments (the specimens were placed either vertically or horizontally with the radiating surface facing down or horizontally with the radiating surface facing up), three thermal test setups were established to test the thermal insulation performance of light-weight ceramic specimens at different temperatures. The results show that the radiating surface temperature was the highest when the specimen was placed horizontally with the radiating surface facing down, while it was the lowest when the specimen was placed horizontally with the radiating surface facing up.The numerical calculation results agreed very well with the experimental ones, confirming the credibility and accuracy of the experimental results. The different thermal insulation performances of the plate specimens obtained under three different boundary conditions will provide important guidance for designers in the design of thermal protection systems for large cabins of high speed flight vehicles.

1 Introduction

With the continuous improvement of design speed for hypersonic vehicles, the high temperature environment resulting from aerodynamic heating becomes more and more harsh. Therefore, it is imperative to implement thermal protection measures to protect equipment inside the vehicles and ensure their safe operation. Because of the strict restriction on the weight and thickness of the thermal insulation materials, light-weight and high-efficiency thermal insulation materials are required to construct a thermal protection system. The determination of thermal insulation performance of thermal protection materials or structures is an indispensable and critical step in the safety design of high-speed flight vehicles. Therefore, the experimental investigation of thermal protection materials in high-temperature environments is receiving increasing attention. Based on the experimental results of the thermal insulation performance of ZrO_2 ceramic fiber materials, J.J. Sun and Z.J. Hu et al.[1] analyzed the effect of sintering temperature, binder contents and other factors on the microstructure, linear shrinkage and thermodynamic properties of highly porous ceramic fiber materials and, thus, optimized the chemical composition and preparation process for the thermal insulation material. By

[a] Corresponding author : wdf1950@163.com

measuring the thermal conductivity of Al_2O_3 nanomaterials at different temperatures using the transient plane source method, J.N. Li and Z.J. Hu et al.[2]. investigated the high-temperature thermal insulation performance of Al_2O_3 nanomaterial sheets and their thermal stability. Based on the results of the thermal insulation experiments of porous wet-lay plate materials in high temperature environments, R. Figueiredo and J.J. Costa et al.[3] explored the effect of water content on the thermal protection performance of the porous material. Based on the aerodynamic heating numerical simulation and experimental test of a metallic honeycomb plate structure and by taking into account the coupling of heat conduction and radiation, Chen and Gao et al.[4] established a finite element model for the heat conduction of a metallic honeycomb plate and obtained the temperature distribution of the metallic honeycomb plate structure along the thickness direction in high temperature environments. Qing and Jiang et al.[5] built a quartz lamp-based infrared heating device to investigate the thermal insulation performance of plate-type thermal protection components in high speed flight vehicles. Because there are no experimental setups for the thermal insulation performance of high-temperature plate materials or structures with uniform regulations on the market, researchers normally need to customize experimental systems for the simulation of aerothermodynamic experiments[6-11]. In the cabin structure of high-speed flight vehicles as shown in Figure 1, the directions of the radiating surfaces of the internal walls of the cabin are different. Thermal insulation performance experiments of plate materials or structures need to be simplified to the following three different experimental modes: (1) the specimen is placed vertically; (2) the specimen is placed horizontally with the radiating surface facing down; and (3) the specimen is placed horizontally with the radiating surface facing up. For the different thermal test modes, the boundary conditions of the specimens are also different. In fact, for the heat insulation experiments at high temperatures, researchers often only use one of the three modes. Among the three experimental modes, when the temperature of the radiating surface is the same, which mode has the highest radiating surface temperature? Moreover, which mode has a lower radiating surface temperature? What is the difference? These questions are very important to designers in the design of thermal protection systems for high-speed flight vehicles and the test scheme.

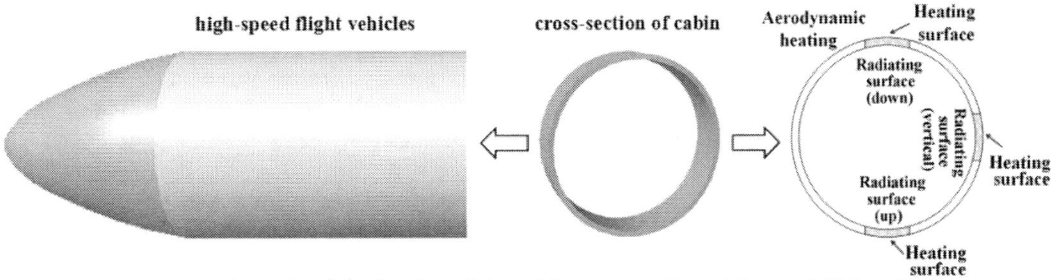

Figure 1. Schematic of the heating of the cabin structure in a high speed flight vehicle

Based on the three different thermal boundary conditions in the thermal insulation experiments of the plate materials, the corresponding experimental setups were established: the specimen was placed vertically; the specimen was placed horizontally with the radiating surface facing down; the specimen was placed horizontally with the radiating surface facing up. Test equipment with different heating directions were used for testing the thermal insulation performance of light-weight ceramic specimens to obtain the radiating surface temperatures of the plate specimens at different boundary conditions. In addition, numerical simulation was also performed to calculate the thermal insulation properties of light-weight ceramic materials. The calculation results were compared to experimental ones to confirm the credibility and accuracy of the experimental results.

2 Thermal insulation test under different boundary conditions

2.1 Specimen

A light-weight porous ceramic material specimen is shown in Figure 2. It is mainly composed of Al_2O_3 and SiO_2 particles. This material has characteristics such as high-temperature resistance, low thermal conductivity, corrosion resistance, high strength and high hardness. The X & Y dimensions of the specimen are 150 mm × 150 mm, and the thickness is 20 mm. The density and porosity of the material are 330 kg/m^3 and 85%, respectively.

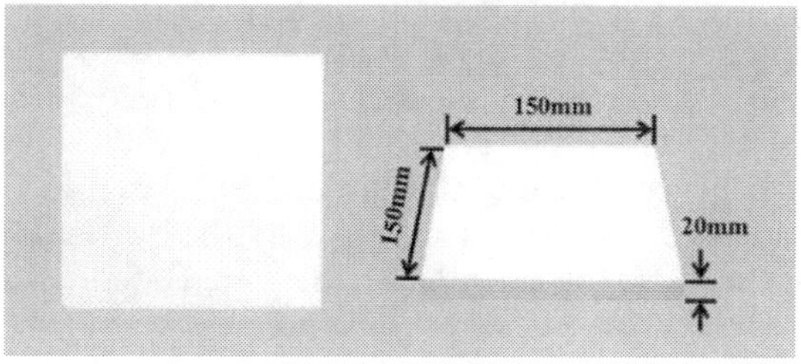

Figure 2. A specimen of the light-weight ceramic material

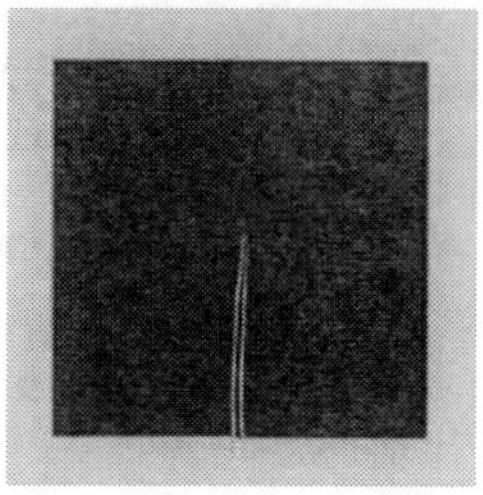

Figure 3. Thin metal plate

2.2 Measurement of the radiating surface temperature

For some light-weight thermal insulation materials such as ceramic fiber blanket, nanomaterials, and porous ceramic materials because of their high porosity, it is very difficult to blacken the heating surfaces. Because the surfaces of these materials are soft, it is difficult to keep the installed temperature sensors in place in high temperature environments. To accurately measure and control the radiating surface temperature of the light-weight fiber blanket, nanomaterials or porous ceramic materials, a high-temperature thin metal plate (1 mm thickness) as shown in Figure 3 can be firmly pressed on the heating surface of the specimen. The surface of the metal plate was blackened to reduce the undesired impact of local grayscale variation on the surface in high temperature environments on the uniformity of the temperature field of the heating surface. In this paper, to understand the impact of the addition of 1-mm-thick thin metal plate on the heating surface temperature of the ceramic specimen, a verification test was performed. First, temperature sensors were installed in the center of the front and back surfaces of the thin metal plate. A 20 mm thick soft ceramic heat insulation material was installed on the outside of the back surface. A computer automatic control system was used to control the front-surface temperature of the metal plate[12] in real-time and measure the back-surface temperature of the metal plate. Figure 4 and Table 1 show the set temperature and the actual control temperature of the front surface of the 1-mm-thick thin metal plate, and the back-surface temperature of the metal plate. It can be seen from Figure 4 that the difference in the temperatures of the front and back surfaces of the metal plate was very small. It can be seen from the data in Table 1 that in the temperature ascending stage, the relative error of the front and back surface temperatures of the metal plate was less than 1.6%. When the temperature entered the stable stage, the relative error of the front and back surface temperatures of the metal plate was within 0.2%. Therefore, the addition of the 1-mm-thick thin metal plate has very little impact on the heating surface temperature of the light-weight ceramic specimen.

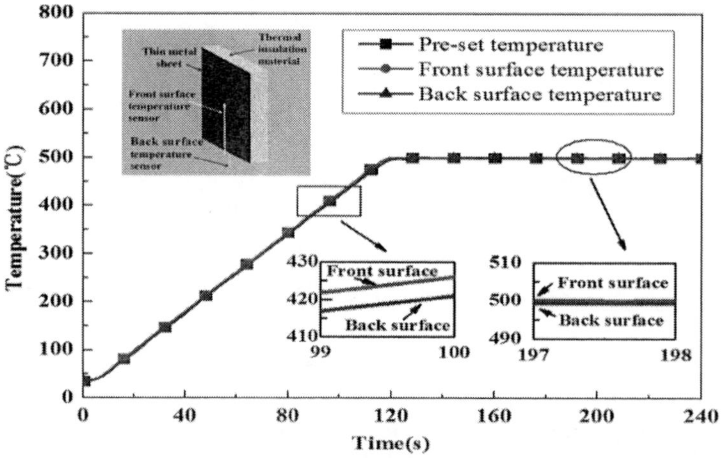

Figure 4. Measured front and back surface temperatures of the 1-mm-thick metal plate

Table 1. Front and back surface temperatures, and the relative error of the 1-mm-thick metal plate

Time /s	0	60	90	120	150	180	210	240
Pre-set temperature /℃	35.0	262.3	385.2	500.0	500.0	500.0	500.0	500.0
Front-surface /℃	35.3	261.8	386.0	499.4	500.9	500.2	500.5	500.9
Back-surface /℃	35.0	257.6	381.2	496.2	500.1	499.4	499.7	500.0
Difference /℃	0.3	4.2	4.8	3.2	0.8	0.8	0.8	0.9
Relative error /%	0.9	1.6	1.2	0.6	0.2	0.2	0.2	0.2

2.3 Thermal boundaries of the specimen

Figure 5 shows the schematic of the natural convection heat transfer of the radiating surface under three different boundary conditions in the heat insulation performance tests for the plates, and the picture of the quartz lamp-based radiation heating tests. Figure 5(a) shows the vertically placed specimen; Figure 5(b) shows the horizontally placed specimen with the radiating surface facing down; Figure 5(c) shows the horizontally placed specimen with the radiating surface facing up. To obtain a uniform temperature field, the quartz lamp heating arrays were densely arranged with an inter-lamp spacing of approximately 3-5 mm. The distance between the heating source and the plate specimen was approximately 100 mm. The radiating surface of the specimen has an open style.

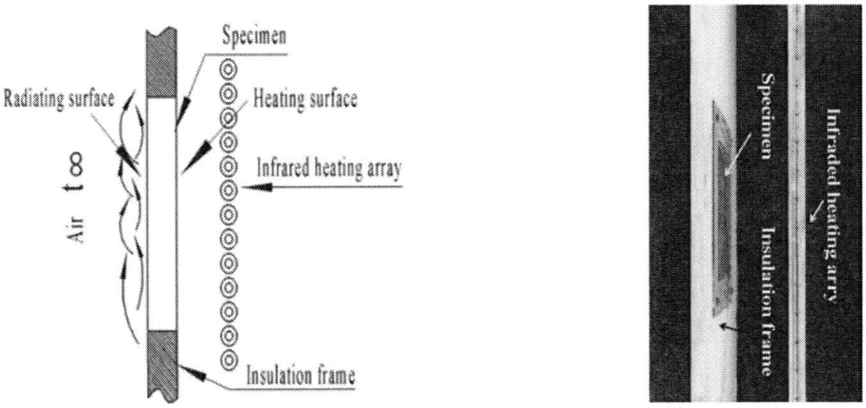

(a) Vertically placed

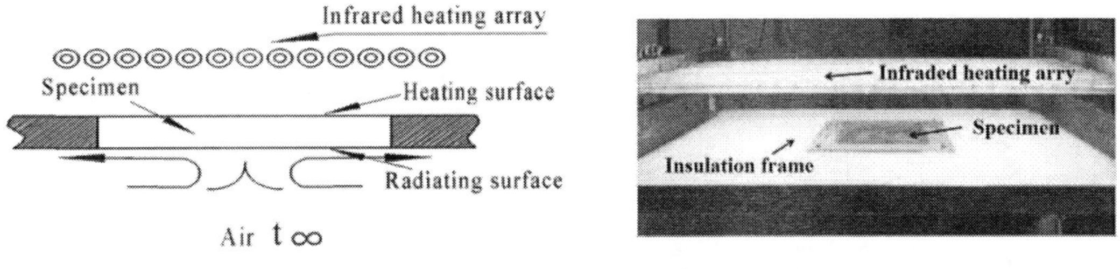

(b) Horizontally placed with the radiating surface facing down

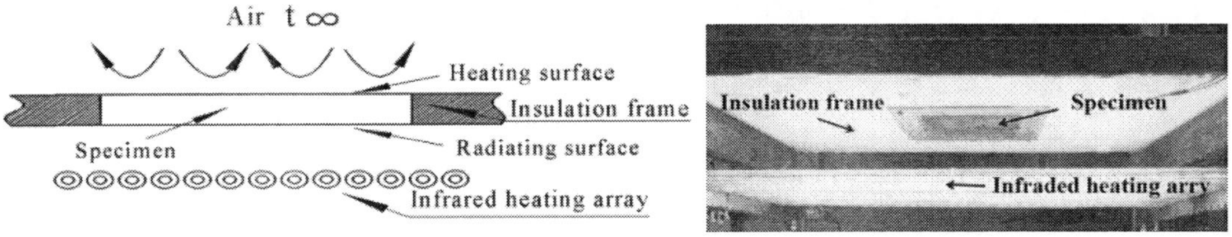

(c) Horizontally placed with the radiating surface facing up

Figure 5. The three installation modes for the plate tests

2.4 Test temperature

In testing the thermal insulation of the light-weight ceramic materials, different temperature conditions were set as shown in Figure 6 (1,000 °C, 800 °C, 600 °C, 400 °C and 200 °C). The heating process is as follows: the thermal control system heated the heating surface of the specimen to the set temperature in 60 seconds and then maintained the heating surface temperature constant; the duration of the whole heating process was 1,800 seconds. The temperature curves in Figure 6 show the excellent agreement between the measured temperature and the set temperature. Table 2 lists the set temperature and measured temperature values of the specimen when the steady-state temperature was 1,000 °C. It can be seen from Table 2 that in the whole heating process, the relative error between the set temperature and actual control temperature was less than 1%, indicating that this test system can very well simulate steady-state or dynamic thermal environments.

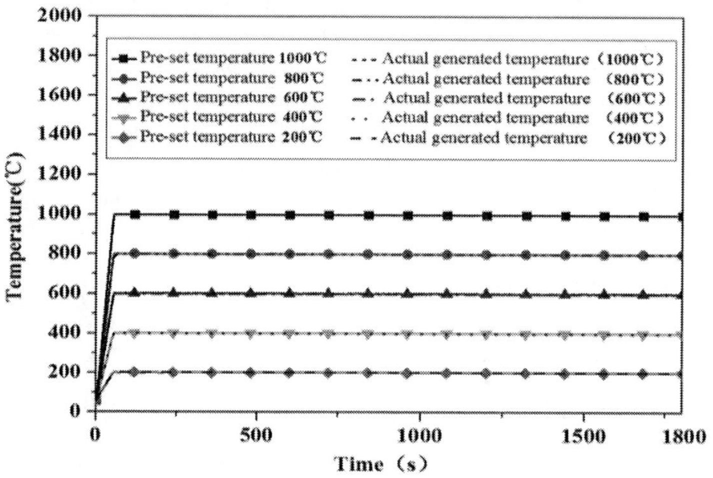

Figure 6. Set temperature and actual control temperature curves of the specimen (vertically placed)

Table 2. Set and actual values of the heating surface temperatures for the light-weight ceramic materials

Time /s	25	50	300	600	900	1200	1500	1800
Pre-set temperature /℃	389.6	921.1	1000.0	1000.0	1000.0	1000.0	1000.0	1000.0
Controlled temperature /℃	391.3	917.5	1002.4	1001.2	1000.2	1000.2	1000.4	1000.2
Difference /℃	1.7	3.6	2.4	1.2	0.2	0.2	0.4	0.2
Relative error /%	0.44	0.39	0.24	0.12	0.02	0.02	0.04	0.02

2.5 Test results

Figure 7 shows the heat insulation results for the light-weight ceramic materials under three different boundary conditions when the heating surface temperature was 1,000 °C, 800 °C, 600 °C and 400 °C, respectively (because of overlap, the test results for the heating surface temperature of 200 °C were omitted). It can be seen from Figure 7 that for the different heating modes, even when the heating surface temperatures were the same, the radiating surface temperatures would be different. The radiating surface temperature of the Horizontally placed specimen with the radiating surfacefacing up (Figure (a)) was lower than that of the vertically placed specimen (Figure (b)). The radiating surface temperatures of the modes in Figure (b) and (c) were close.

To further compare the heat insulation results of the light-weight ceramic materials under three different boundary conditions, Figure 8 shows the heating surface temperature curves of the specimens when the radiating surface temperature was 1,000°C, 800°C or 200°C. In Figure 8, the radiating surface temperature of the specimen for the three different modes was labeled as Tv, Td, and Tu, where Tv is the radiating surface temperature of the vertically placed specimen, Td is the temperature of the horizontally placed specimen with the radiating surface facing down, and Tu is the temperature of the horizontally placed specimen with the radiating surface facing up.

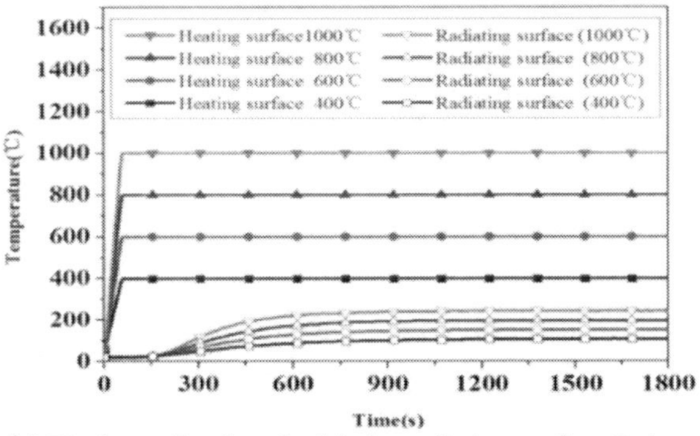

(a) Horizontally placed with the radiating surface facing up

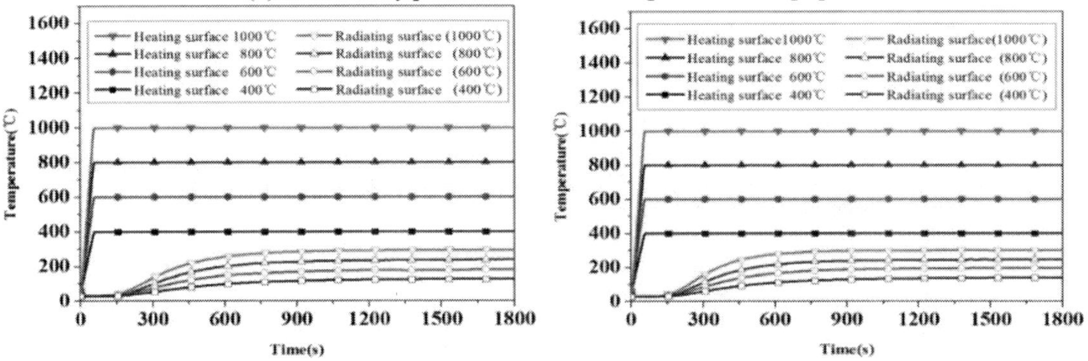

(b) Vertically placed (c) Horizontally placed with the radiating surface facing down

Figure 7. Thermal insulation test results for the ceramic materials under three different boundary conditions

It can be seen from Figure 8 that among the three different thermal boundary conditions, Tu was significantly lower than Td, and Tv was close to Td.Table 3 lists the radiating surface temperatures of the light-weight ceramic materials under three different boundary conditions when the heating surface temperature was 1,000°C. It can be seen from Table 3 that at 1800 s, Td was the highest (301.6°C), and Tu was the lowest (242.2°C with a temperature difference of 59.4°C and a relative error of 19.7%. This is because when the air near the radiating surface of the specimen was heated, its density decreased, and its volume increased. As a result, the hot air generated a lifting force. When the radiating surface of the plate specimen faced down, the ascending motion produced by the hot air was hindered, leading to the increase of heat accumulation on the specimen surface. As a result, the lost energy was less than the energy carried away by the hot air when the radiating surface of the specimen facedup.It can be seen from Table 3 that Tv is not much different from Td in that Tv is 2.3% lower than Td.

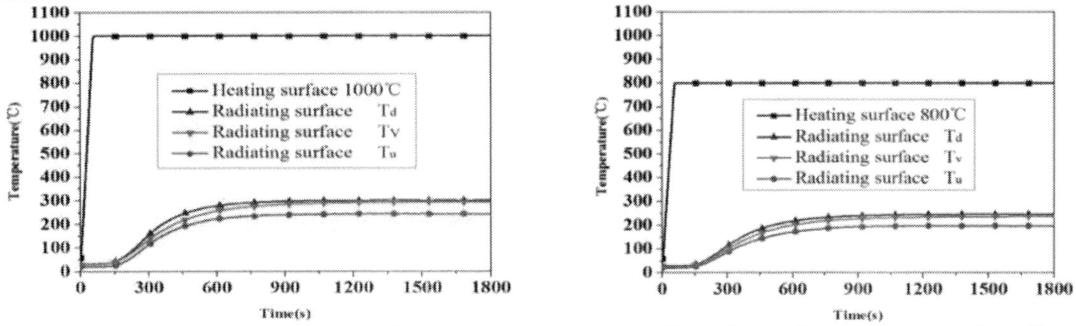

(a) Heating surface temperature of 1,000 °C (b)Heating surface temperature of 800℃

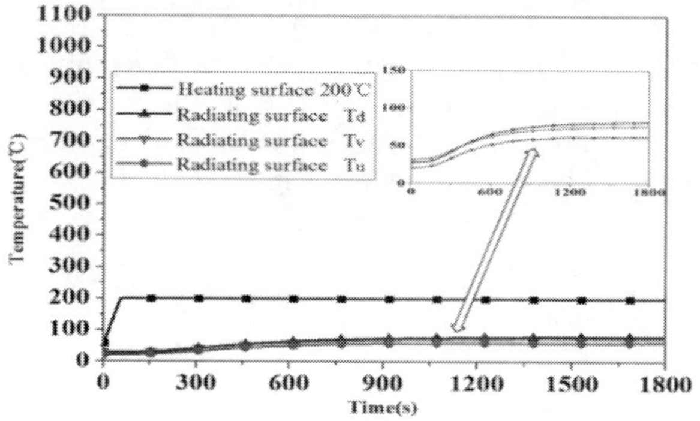

(c) Heating surface temperature of 200℃

Figure 8. Thermal insulation effect of the ceramic specimen under three different boundary conditions

Table 3. Radiating surface temperatures of the ceramic plates under three different boundary conditions (heating surface temperature of 1,000 °C)

Time /s	200	400	600	800	1000	1200	1400	1600	1800
Heating surface temperature /℃	1000.0	1000.0	1000.0	1000.0	1000.0	1000.0	1000.0	1000.0	1000.0
Radiating surface temperature Td /℃	71.1	222.1	278.2	294.2	299.2	300.6	301.3	301.3	301.6
Radiating surface temperature Tv /℃	67.6	195.6	256.6	279.2	288.5	292.1	293.6	294.4	294.6
Radiating surface temperature Tu /℃	46.5	170.2	222.1	237.1	242.1	243.2	244.0	243.7	242.2

3 Numerical calculation

3.1 Finite element model

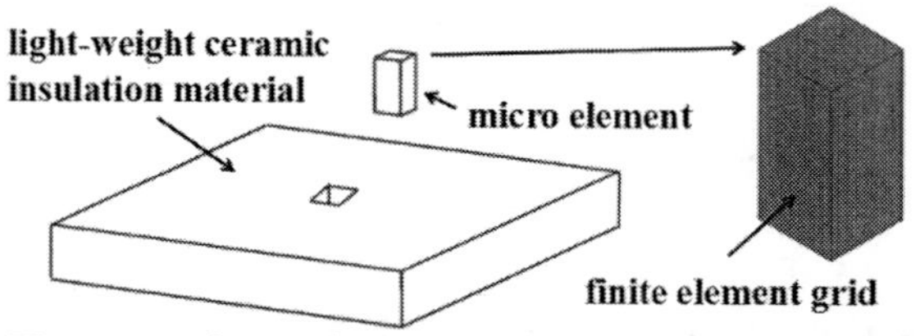

Figure 9. Element cutting method and finite element mesh generation

The finite element model as shown in Figure 9 was established for the numerical calculation of the thermal insulation performance of the plate materials at three different boundary conditions.A rectangular-shaped block with a size of 10 mm × 10 mm × 20 mm was cut from the center of the plate. The cut block was subjected to mesh partition using the 6-plane, 8-node and Solid70 element to generate 16,000 elements and 18,081 nodes. Assuming that the temperature variation of the model in the plane was small, there was no need to consider heat exchange within the plane. The heat was mainly transferred along the thickness direction and can be considered as one-dimensional, transient heat transfer. The temperature T(x, t) along the thickness direction inside the material satisfies the following heat conduction equation in the Cartesian coordinate system[13]

$$\rho c\frac{\partial T}{\partial t}=\frac{\partial}{\partial x}(k\frac{\partial T}{\partial x}) \qquad (1)$$

where ρ is the density of the thermal insulation material, k is the thermal conductivity of the material, and c is the specific heat capacity of the material.The material parameters are the same as the light-weight ceramic specimen. This experiment only used one-side heating. Because the temperature along the thickness of the material was different, thermal conductivity and specific heat capacity at different temperatures were used in the calculation. Table 4 lists the specific heat capacity and thermal conductivity of the light-weight ceramic materials in the temperature range of 30-1200 °C.

Table 4. Physical parameters for the light-weight ceramic materials

Temperature /℃	30	100	200	300	400	500	600	700	900	1200
Thermal conductivity W/(m·℃)	0.05	0.055	0.061	0.069	0.076	0.082	0.09	0.098	0.113	0.138
specific heat J/(kg·℃)	850	870	900	930	1050	1190	1230	1270	1370	1430

3.2 Boundary conditions

Because the circumference of the light-weight ceramic specimen was installed with an insulation border, the circumferential boundary can be approximated as adiabatic. The boundary conditions of the rectangular-shaped block were mainly determined by the states of the front and back surfaces. Because the front surface temperature of the specimen was generated by the thermal control system based on pre-set values, the front surface temperature was a known value. Therefore, the boundary condition of the front surface is the Dirichlet boundary condition, i.e., the predefined temperature boundary condition. The heat transfer between the radiating surface of the specimen and outer space is mainly consisted of radiative heat transfer and natural convection. For the three different modes, there is little difference in the energy transfer due to radiation on the radiating surface. In contrast, the geometric distribution of the radiating surface has a decisive impact on the motion state of the fluid. The coefficient of the natural convection heat transfer for the radiating surface, h, can be calculated from the following equation

$$h = \frac{Nu\,\lambda}{H} \tag{2}$$

where λ is the thermal conductivity of air, H is the characteristic length of the light-weight ceramic specimen, and Nu is Nusselt number. Nu is determined by the experimental correlation of the natural convection heat transfer in the large space under the uniform wall temperature boundary condition:

$$Nu = C\left(Gr\,Pr\right)^n \tag{3}$$

where Pr is the Prandtl number, and C and n are the constants determined by the air flow. Gr is the Grashof number and can be calculated from the equation below:

$$Gr = \frac{g\,\alpha\,(T_w - T_\alpha)\,H^3}{\nu^2} \tag{4}$$

where α is the expansion coefficient of air, Tw is the back surface temperature, Tα is the temperature of the external space, and ν is the kinematic viscosity coefficient of air.

For the three corresponding modes in Equation (3), their constant C values that represent air flow states are different and related to Gr. When the specimen was vertically placed, the Grashof number (Gr) for the radiating surface of the specimen calculated at different temperatures ranged from 2.6× 106~2.4×107. The gas flow state in the natural convection of the radiating surface belongs to laminar flow. Therefore, C is 0.59[13]. When the specimen was horizontally placed, the product of Gr and Pr calculated from Equation (4) at different temperatures ranged from 2.8×104~2.6×105. For the vertically placed specimen, when its radiating surface faced up, C is 0.54[14]; when its radiating surface faced down, C is 0.27[15].

4 Comparison of the calculated and experimental results

Figure 10 shows the comparison curves of the heat insulation performance from the calculated and experimental results for the light-weight ceramic plate under three different boundary conditions. It can be seen from Figure 10 that the calculated and experimental results agreed very well with each other under the three different boundary conditions. Table 3 lists the calculated value, experimental value and relative error of the radiating surface temperature when the heating surface temperature was 1,000 °C, 800 °C, 600 °C, 400 °C or 200 °C.The results in Table 5 show that the relative error between the calculated and experimental values was less than 5%. Therefore, the two had excellent agreement. The calculated results provide support to the credibility and accuracy of the test method and results.

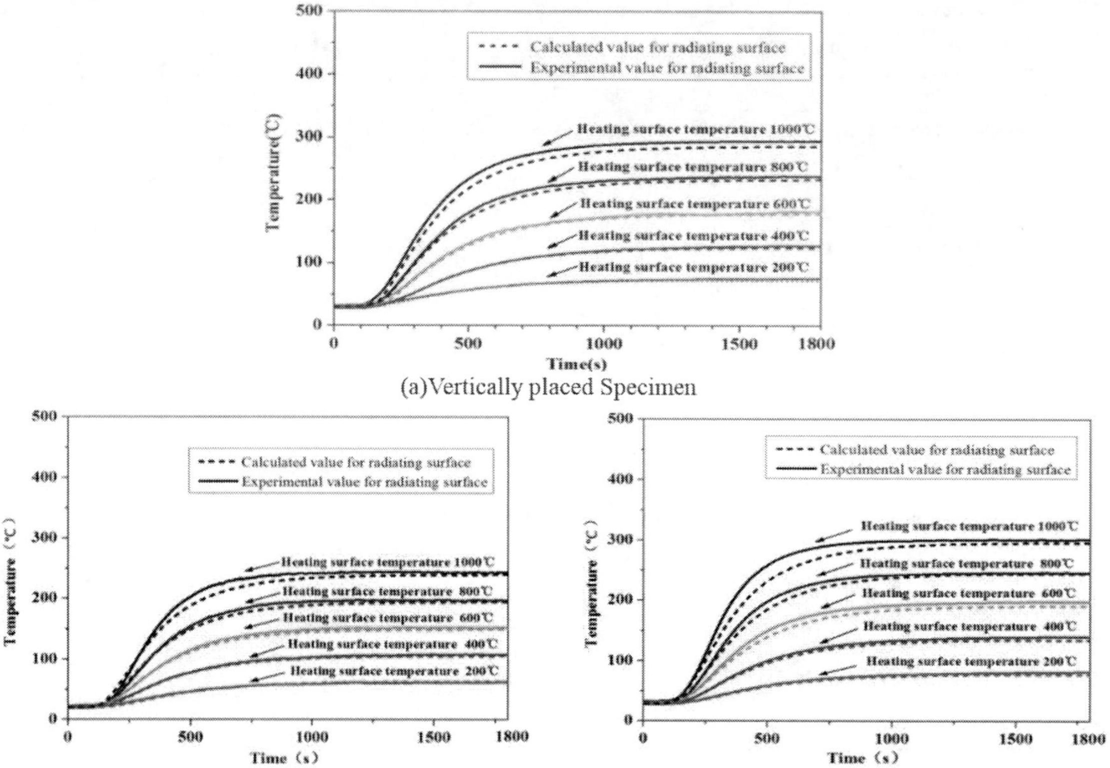

(a)Vertically placed Specimen

(b)Horizontally placed specimen (radiating surface facing up)

(c) Horizontally placed specimen (radiating surface facing down)

Figure 10. Calculated and experimental radiating surface temperature curves

Table 5. Calculated value, experimental value and relative error of the radiating surface temperature when the light-weight ceramic plate was vertically placed

Time /s		0	300	600	900	1200	1500	1800
Heating	Experimental value /℃	30.5	137.3	256.6	284.7	292.1	294.1	294.6
surface	Calculated value /℃	30.0	133.6	246.1	276.0	284.1	286.0	286.2
1000℃	Relative error /%	1.6	2.7	4.1	3.1	2.7	2.8	2.9
Heating	Experimental value /℃	27.8	100.1	200.8	226.7	233.7	236.7	238.1
surface	Calculated value /℃	29.0	97.6	192.5	220.6	229.1	231.7	232.5
800℃	Relative error /%	4.3	2.5	4.1	2.7	2.0	2.1	2.4
Heating	Experimental value /℃	31.5	75.6	148.2	169	178.5	178.9	181.7
surface	Calculated value /℃	30.0	73.9	144.2	167.5	174.9	177.3	178.1
600℃	Relative error /%	4.8	2.2	2.7	0.9	2.0	1.2	0.9
Heating	Experimental value /℃	27.8	52.8	98.6	116.7	123.5	126.4	127.7
surface	Calculated value /℃	29.0	52.3	98.5	115.8	121.8	123.7	124.4
400℃	Relative error /%	4.3	0.9	0.1	0.8	1.4	2.1	2.6
Heating	Experimental value /℃	31.7	43.7	62.2	70.4	73.8	75.3	76.3
surface	Calculated value /℃	31.0	42.0	61.6	69.5	72.2	72.8	72.7
200℃	Relative error /%	2.2	3.9	1.0	1.3	2.2	3.3	4.7

5 Conclusion

Based on the three different thermal boundary conditions in the thermal insulation experiments of plate materials, the corresponding experimental setups were established. The heat insulation test and numerical simulation were performed for the light-weight ceramic plates at different temperatures. The following conclusions were drawn:

1) In the finite-plane radiation heating tests of thermal insulation materials, under the same heating surface temperature, the radiating surface temperature was the highest when the specimen was placed horizontally with the radiating surface facing down, while the radiating surface temperature was the lowest when the specimen was placed horizontally with the radiating surface facing down When the heating surface temperature was 1,000°Cand the time was 1,800 s, the radiating surface temperature of the horizontally placed specimen with the radiating surface facing down was 301.6°C, while that of the horizontally placed specimen with the radiating surface facing up was 242.2°C; the relative difference was 19.7%.

2) The radiating surface temperature of the horizontally placed specimen was lower than that of the horizontally placed specimen with the radiating surface facing down. In this experiment, when the heating surface temperature was 1,000°C and the time was 1,800s, the radiating surface temperature of the vertically placed specimen was 2.3% lower than that of the horizontally placed specimen with the radiating surface facing down.

3) The numerical calculation results agreed very well with experimental results, confirming the credibility and accuracy of the experimental results.

The heat insulation results of the light-weight ceramic materials obtained under different boundary conditions in the present paper will provide valuable guidance for designers in their analysis of the temperature distribution inside the large cabins of high-speed flight vehicles and the final thermal test plan.

Acknowledgment

The authors appreciate the support from the National Natural Science Foundation of China (No. 11427802).

References

1. J. Sun, Z. Hu, J. Li, H. Zhang, C. Sun, Ceram. Int. **40**, 11787 (2014)
2. J. Li, Z. Hu, X. Wang, J. Yang, C. Sun, Ceram. Int. **43**, 8343 (2017)
3. A .R. Figueiredo, J.J. Costa, Int. J. Heat Mass Transfer. **47**, 11(2003)
4. Y. Chen, D. Gao, X. Wang, PTCA. A **39**, 234 (2003)
5. Q. Qin, J. Jiang, Z. Cheng, E. Jia, Q. He, Aircraft. D. **36**, 41 (2016)
6. M. Hanao, J. Japan Society powder powder Metall. **55**, 732 (2008)
7. D. Wu, L. Zheng, B. Pan, Y. Wang, B. Sun, M. Mu, Chin. J. Theor. Mech. **44**, 297 (2017)
8. W. Yuen, G. Cunnington, Aiaa Aero. Sci. Meeting Exhibit, 737 (2006)
9. D. Wu, L. Shang, Y. Pu, H. Wang, Z. Gao, Spacecraft Environ. Eng. **33**, 7 (2016)
10. D. Wu, Y. Wang, Z. Gao, J. Yang, J. Mater. Eng. Perf. **24**, 3373 (2015)
11. W. Wang, Z. He, M. Huang, Y. Hu, Z. Wang, Ceram. Int. **43**, 38 (2016)
12. D. Wu, B. Pan, Z. Gao, M. Mu, L. Zhu, Y. Wang, J. Exp.Mech. **27**, 255 (2012)
13. S. Yang, W. Tao, Heat Transfer 3th ed, 269 (1998)
14. F. Incropera, D. Dewitt, Fundamen. Heat Mass Transfer. 430 (2002)
15. J. Holman, Heat transfer, 285 (2002)

Cutting force analysis in machining of titanium alloy with solid carbide cutters of different geometriy

Liu Jianyong, Qiao Lihong and Chen Wuyi[a]

School of Mechanical Engineering and Automation, BeiHang University, Beijing 100191, China

Abstract. In order to study the cutting force in machining of titanium alloy with solid carbide cutters of different geometriy, variable helix (VH), variable pitch (VP) and standard (SD) milling cutter were used to machine titanium alloy TB6 via dry milling. The influence of cutting parameters and geometric structure parameters of milling cutters on the cutting force was investigated by the analysis of radial and tangential forces in the cutting process with three kinds of milling cutters. The experiment results showed that cutting parameters had the same influence on the radial force and tangential force of the three milling cutters, and the influence of geometric structure parameters on radial force and tangential force was different. Compared with the SD milling cutter, the change of pitch and helix resulted in the increase of radial and tangential forces, the pitch change led to the radial force being larger than the tangential force, and the helix change led to the radial force being smaller than the tangential force.

1 Introduction

Titanium alloys have been widely used in the aerospace, automotive, biomedical and other manufacturing industries because of their good strength-to-weight ratio, high strength and toughness, excellent corrosion resistance etc. [1-4]. However, some material characteristics of titanium alloy such as low thermal conductivity, low elastic modulus and high chemical activity etc., can increase the machining difficulty, and impair the parts qualities [5].

Milling process is widely used in machining processes for aeronautical, aerospace, die, and automobile components etc. Cutting force has a great influence on workpiece precision, surface quality, cutting system vibration, cutting power and tool life, especially when the milling cutter has a complex geometry in machining processes [6]. The cutting forces are influenced by several factors, such as tool geometry, properties of the workpiece material, cutting conditions, etc. [7], cause deflections in the cutting system by tool-workpiece-machine [8], which cause significant geometrical errors in the machined workpiece [9]. These errors are particularly important when a high tool length/diameter ratio is used, when the inclination of the machined surface is high and when tool wear is significant [10]. Cutting forces cause vibrations in machining process [11], the vibrations occur when the damping capacity of the tool-workpiece-machine system is insufficient to absorb the energy transmitted by the cutting [12]. Several authors argued that the workpiece surface quality could be improved by preventing chattering [13,14]. Many scholars studied on reducing vibration by changing the geometry of milling cutters. Huang et al. [15,16] investigated the cutting forces by the analysis of

[a] Corresponding author : wychen@buaa.edu.cn

time domain and frequency domain in side milling of titanium alloy TC4, the results showed that vibration reduction effect of the VP milling cutter was more obvious than SD milling cutter, the radial forces of VP milling cutter were the minimum compared with SD and VP milling cutter. Wang et al. [17] studied the cutting stability of VP and VH milling cutter, it was found that the VP and VH milling cutter could obtain larger stable cutting area than uniform pitch and uniform helix milling cutter, and the effect of suppressing chatter of VP and VH milling cutter was obviously.

In summary, cutting forces can cause the vibration of cutting system, changing the geometric structure parameters of milling cutter is an effective method to reduce the vibration and improve the quality of machined surface. In fact, the published literatures are rarely in relation to machining high strength titanium alloy TB6 with vibration-free milling cutter.

This study will investigate the cutting forces in cutting titanium alloy TB6 with vibration-free milling cutter (VP and VH milling cutter) and the SD milling cutter, analyze the influence of cutting parameters and geometric structure parameters of milling cutter on cutting forces.

2 Experimental design

Titanium alloy TB6 was machined with VH, VP and SD milling cutter in the experiment. The nominal chemical composition of TB6 is shown in Table 1, and the mechanical properties at room temperature of TB6 were shown in Table 2. The geometric structure parameters of the three milling cutters were shown in Table 3.

Table 1. Chemical composition of TB6 (mass fraction, %)

Element	Al	V	Fe	C	O	N	H	Ti
Wt%	3.15	10.5	2.1	0.02	0.10	0.03	0.002	Balance

Table 2. Mechanical properties of TB6 at room temperature

Tensile strength (MPa)	Yield Strength (MPa)	Elastic modulus (GPa)	Elongating (%)	Hardness (HV)	Fracture toughness (MPa·$\sqrt{m}$)
1105	1035	106	4~10	>375	≥60

Table 3. Geometry structure parameters of three kinds of milling cutter

Milling cutter	Blade number	Helix angle (deg)	Tooth pitch (deg)	Blade length (mm)	Diameter (mm)	Tool length (mm)
SD	4	45	90	25	10	72
VH	4	41, 45	85, 95	25	10	72
VP	4	45	86, 94	25	10	72

The three types of milling cutters had the same geometric structure parameters except the helix angle and tooth pitch. The cross section of three types of milling cutters was shown in Figure 1.

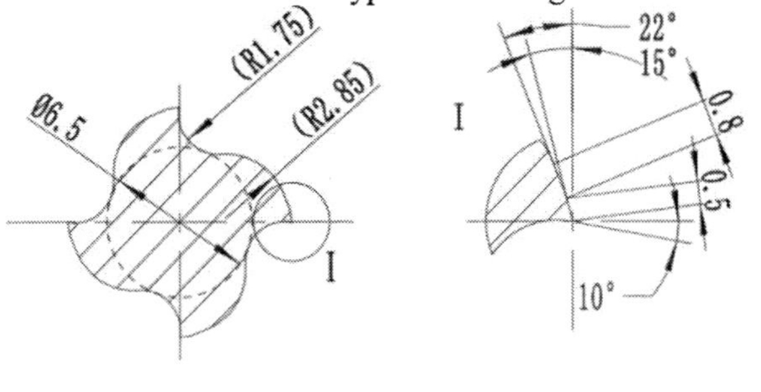

Figure 1. Cross section of three types of milling cutters (The unit of length is mm)

The three types of mill cutter were all carbide cutters. The VH and VP end mills had no deposited coating and SD end mill was coated with TiAlN.

The cutting parameters were listed in Table 4 and the cutting system coordinates used were shown in figure 2. The workpiece was solid block of 115mm × 70mm × 70mm and the milling cutters' hang out was all 40mm. The cutting experiments were carried out on a five-axis CNC machine center (XKR50A) under down milling and dry machining condition. The cutting forces were measured with dynamometer (KISTLER 9123C).

Table 4. Cutting parameters of milling experiment

Test number	Cutting speed (m/min)	Feed per tooth (mm/z)	Axial depth (mm)	Radial depth (mm)	Milling cutter
1	20	0.04	5	1	VH
2	20	0.06	5	1	
3	30	0.04	5	1	
4	30	0.06	5	1	
5	20	0.04	5	1	VP
6	20	0.06	5	1	
7	30	0.04	5	1	
8	30	0.06	5	1	
9	20	0.04	5	1	SD
10	20	0.06	5	1	
11	30	0.04	5	1	
12	30	0.06	5	1	

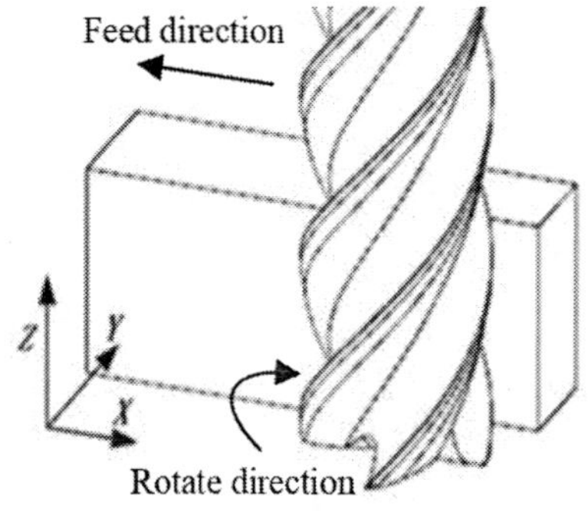

Figure 2. Cutting system coordinates

3 Test results and analysis

Figure 3(a) showed the change of average radial cutting forces with cutting parameters within 2 s of stable cutting time of three kinds of milling cutter. As shown in this figure, the radial forces of three kinds of milling cutter increase with the increase of feed, the radial forces of VP and SD milling cutters increase with the cutting speed increasing, but the radial forces of the VH milling cutter decrease with the cutting speed increasing. Under the same cutting parameters, compared with the radial forces of three kinds of milling cutter, it was found that the radial force of the VP milling cutter was the largest, the radial force of the VH milling cutter was the second, and the radial force of the SD milling cutter was the smallest. This phenomenon was different from that in milling titanium alloy TC4. Vibration-free milling cutter and standard milling cutter were used to machine titanium alloy TC4 under dry machining, side milling and down milling conditions. It was found that the radial force of the VH milling cutter was the largest, the radial force of the SD milling cutter was the second, and the radial force of the VP milling cutter was the smallest [16]. Different titanium alloy materials used to machine may be the cause of these differences.

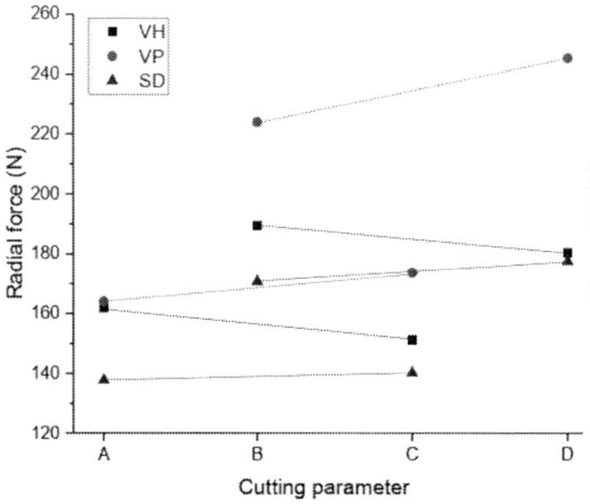

(a) Average radial cutting forces

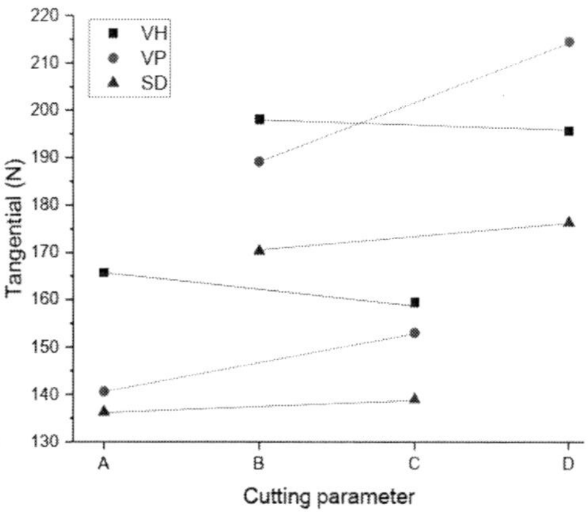

(b) Average tangential cutting forces

Figure 3. Average radial cutting and tangential cutting forces of three kinds of milling cutter with the change of cutting parameters A: V=20 m/min, f=0.04 mm/z; B: V=20 m/min, f=0.06 mm/z; C: V=30 m/min, f=0.04 mm/z; D: V=30 m/min, f=0.06 mm/z.

Figure 3(b) showed the change of average tangential cutting forces with cutting parameters within 2 s of stable cutting time of three kinds of milling cutter. It could be seen from this figure that the tangential forces of three kinds of milling cutter increased with the increase of feed, the tangential forces of VP and SD milling cutters increased with the cutting speed increasing, but the tangential forces of the VH milling cutter decreased with the cutting speed increasing. Under the same cutting parameters, compared with the tangential forces of three kinds of milling cutter, it was found that the tangential force of the VH milling cutter was the largest, the tangential force of the VP milling cutter was the second, and the tangential force of the SD milling cutter was the smallest.

According to the result of analysis based on above mentioned radial and tangential forces of the three kinds of milling cutter, it could be seen that cutting parameters had the same influence on the radial force and tangential force of the three milling cutters, and the influence of geometric structure parameters on radial force and tangential force was different. Compared with the SD milling cutter, the changes of pitch and helix resulted in the increase of radial and tangential forces, the pitch change led to the radial force being larger than the tangential force, and the helix change led to the radial force being smaller than the tangential force.

Figure 4 showed the contrast between the radial force and the tangential force. It could be seen from this figure that for the VH milling cutter, the radial force was less than the tangential force; for the VP milling cutter, the radial force was greater than the tangential force; for the SD milling cutter, there was little difference between the radial force and the tangential force. Compared with the SD milling cutter, the radial force was larger than the tangential force because of the change of pitch, and the radial force is smaller than the tangential force because of the change of helix.

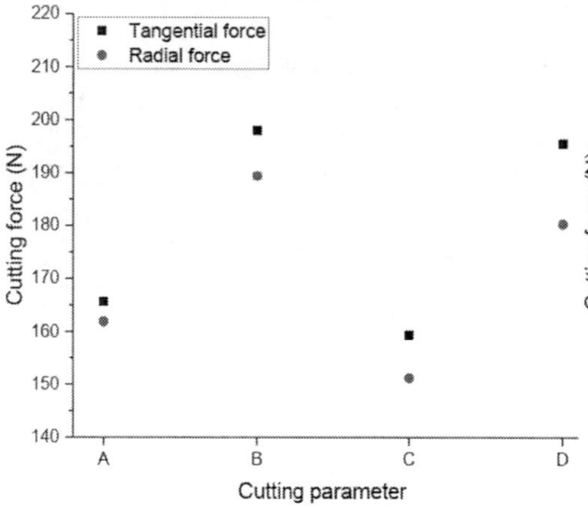

(a) VH end mill

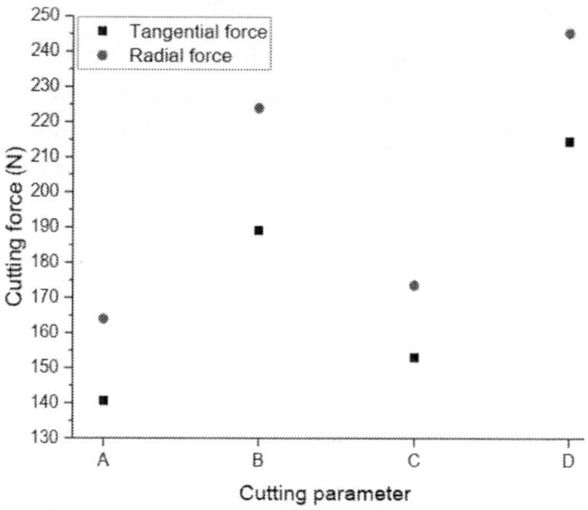

(b) VP end mill

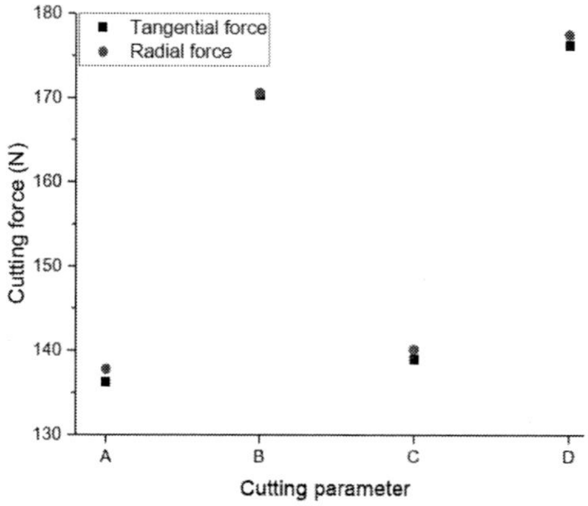

(c) SD end mill

Figure 4. Contrast between radial and tangential cutting force of three kinds of milling cutter with the change of cutting parameters A: V=20 m/min, f=0.04 mm/z; B: V=20 m/min, f=0.06 mm/z; C: V=30 m/min, f=0.04 mm/z; D: V=30 m/min, f=0.06 mm/z.

4 Conclusions

In this research, titanium alloy TB6 was dry machined using the three types of milling cutters to investigate the cutting forces. The major results are summarized as follows:

1) The influence of cutting parameters on cutting Force

The radial and tangential forces of the three kinds of milling cutter increase with the feed increasing, the radial and tangential forces of the VP and SD milling cutters increase with the increase of cutting speed, but the radial and tangential forces of the VH milling cutter decrease with the cutting speed increasing.

2) The influence of geometric structure parameters on cutting Force

Under the same cutting parameters, the radial force of the VP milling cutter is the largest, the radial force of the VH milling cutter is the second, and the radial force of the SD milling cutter is the smallest. The tangential force of the VH milling cutter is the largest, the tangential force of the VP milling cutter is the second, and the tangential force of the SD milling cutter is the smallest. Compared with the SD milling cutter, the radial force is larger than the tangential force because of the change of pitch, and the radial force is smaller than the tangential force because of the change of helix.

References

1. E.Ezugwu, Z.Wang, Journal of Materials Processing Technology, **68**, 262-274 (1997)
2. R.Boyer, R.Briggs, Journal of Materials Engineering & Performance, **14**, 681-685 (2005)
3. M.Rahman, Z.Wang, Y.Wong, Jsme International Journal, **49**,11-20 (2006)
4. D.Ulutan, T.Ozel, International Journal of Machine Tools and Manufacture, **51**, 250-280 (2011)
5. A.Daymi, M.Boujelbene, A.Amara, Materials Science & Technology, **27**,387-394 (2010)
6. B.Ikua, H.Tanaka, F.Obata, Precision Engineering, **25**, 266-273 (2001)
7. C.Toh, Journal of Materials Processing Technology, **152**,346-356 (2004)
8. G.Kim, B.Kim, C.Chu, International Journal of Machine Tools & Manufacture, **43**, 917-924 (2003)
9. H.Erdim, I.Lazoglu, B.Ozturk, International Journal of Machine Tools and Manufacture, **46**, 747-757 (2006)
10. L.Lacalle, A.Lamikiz, J.Sánchez, Journal of Materials Processing Technology, **123**, 292-302 (2002)
11. M.Polli, *Análise da estabilidade dinâmica do processo de fresamento a altas velocidades de corte* (Ph.D. Thesis, Federal University of Santa Catarina, 2005)
12. A.David, A.gapiou, S.John, Journal of Petroleum Science and Engineering, **16**, 315-316 (1996)
13. H.Zhou, C.Wang, Z.Zhao, Journal of Materials Processing Technology, **207**, 154-162 (2008)
14. H.Paris, G.Peigne, R.Mayer, International Journal of Machine Tools & Manufacture, **44**, 1567-1576 (2004)
15. P.Huang, J.Li, J.Sun, Journal of Cleaner Production, **67**, 258-264 (2014)
16. P.Huang, J.Li, J.Sun, International Journal of Advanced Manufacturing Technology, **67**, 1385-1391 (2013)
17. Y.Wang, T.Wang, Z.Yu, Shock and Vibration, **2015**, 1-9 (2015)

COMPRESSIVE PROPERTIES OF GEOPOLYMER MATRIX COMPOSITES

Robin Hron[1,a], František Martaus[1] and Martin Kadlec[1]

[1] *VZLU - Czech Aerospace Research Centre, Czech Republic*

Abstract. Polymer based resin is presently the most used resin for preparing of composites constructions, because of its undisputable benefits; however, there are some limits. The aircraft industry has especially strict requirements for fire, smoke and toxicity (FST) properties which are limited when using organic polymers. Conventional polymer resins resist to temperatures usually up to 120 °C and then they lose stiffness and strength. However, geopolymer matrix is a new type of resin with high potential for cost-efficient applications dealing with temperatures up to 1 200 °C.

This paper presents compressive properties of a new geopolymer resin and a fibre reinforced composite with the geopolymer matrix (geocomposite). The effect of a harsh environment exposition on the strength was also evaluated, specifically the impact of the exposure in hot-wet and salt mist conditions. Samples were tested in accordance with ASTM D695 in case of pure resin and in accordance with ASTM D6641 in case of the geocomposite. All tests were performed at room temperature and additionally, pure geopolymer resin was tested at 400 °C. The high temperature caused 35 % decrease of the compressive strength in comparison with the room temperature. Geopolymers behaves like a ceramic and have some unique properties such as high thermal stability, non-flammability and do not generate toxic smoke and fumes.

1. Introduction

At present, there is an increasing need for materials eliminating fatal results of fire in case of aircraft accident [1]. Currently used organic matrix composites formally meet requirements such as FAR 25, appendix F; nevertheless, in the event of real fire they quickly deteriorate at temperatures above 300 °C emitting toxic fumes and gases. True non-flammability of present-day composites is limited to minutes [2]. The material, known as geopolymer (first geopolymer resin was described in France by J. Davidovits in 1979 [3]) withstands temperatures of over 1000 °C and can be utilized as matrix in fibre reinforced composites. The use of geopolymer materials looks like a good choice for constructions with increased requirements on fire, smoke and toxicity (FST) properties. The main advantages of geopolymers are the excellent temperature stability; fire resistance with no generation of toxic fumes and smokes, low thermal conductivity and good specific strength [4-7]. The material behaves like a ceramic, but does not require high processing temperatures or pressures, so it can be worked up like common organic resins [8]. On the other hand, the geopolymer materials have also disadvantages

[a] Corresponding author : hron@vzlu.cz

when compared with commonly used polymer resins. The most significant are lower mechanical properties.

2. Experiment

Compressive properties of a new geopolymer resin and carbon fibre reinforced geopolymer composite (geocomposite) were examined. The influence of the temperature and environment (hot-wet and salt mist conditions) on these properties was also investigated in the experiment.

Samples were tested in accordance with the ASTM D695 – Standard Test Method for Compressive Properties of Rigid Plastics [9], in case of pure resin and in accordance with ASTM D6641 – Standard Test Method for Compressive Properties of Polymer Matrix Composite Materials Using a Combined Loading Compression (CLC) Test Fixture [10], in case of geocomposite samples.

2.1. Material

Geopolymer resin was prepared by mixing of the components listed in Table 1. The table also shows the weight and weight ratio of the individual components. In the first step, pure geopolymer resin was prepared (by mixing components 1-5) and then short ceramic and carbon fibres were added.

Table 1. Geopolymer resin composition.

	Component	m (g)	wt. %
1	SiO_2 (Thermal silica)	513.24	35.24
2	$Al(OH)_3$	453.32	28.74
3	H_2O	158.48	10.05
4	NaOH	111.16	7.05
5	KOH	163.80	10.39
6	Ceramic fiber LYTX-311-L	84	5.33
7	Carbon chopped fiber Tenax® – A HT C124 3 mm	93.24	5.91
Σ		1577	100

Unidirectional carbon fibre tape, type TCU175, 3K [11] was used for geocomposite samples.

2.2. Test samples

Resin samples were prepared by casting the geopolymer mixture into the form. The filling of the geopolymer resin into the forms was carried out on a vibration table. After casting, the test samples were cured at 80 °C/ 20 hour. The test samples were then extruded from the form and tempered at 80 °C/ 9 hours + 100 °C/ 12 hours at the chamber. Final tempering was done in a cascade mode from 105 to 170 °C.

Carbon fibre-reinforced polymer (CFRP) samples were prepared by cutting of the test panel. Test panel were prepared by contact laminating (standardly 400 g of resin / m^2 of fabric). Parameters of the curing cycle were as follows: heat up rate 2-8 °C/ min, cured temperature 85 ±3°C, dwell time 18 hours, vacuum min. -80 kPa, cooling rate max. 8 °C/ min. Test samples were cut from the test panel by a diamond blade.

The surface protection of all test samples to improve the resistance to external influence (especially by water) was performed by immersing test samples into a mixture of ethyl silicates of polysilic acids, concurrently with Dynasylan Silbond 40 [12]. Test matrix is showed in Table 2.

Table 2. Test matrix

Sample type	Exposure	Test temperature	No. of samples	Test method
Geopolymer resin	-	RT	5	ASTM D695
		400 °C	5	
	hot-wet	RT	5	
		400 °C	5	
	salt mist	RT	5	
		400 °C	5	
Geocomposite samples	-	RT	6	ASTM D6641
	hot-wet	RT	5	
	salt mist	RT	5	

2.3. Exposure

Exposure in salt mist (SM) was performed according to standard EN ISO 9227 with concentration of the NaCl solution 50 g/l, temperature 35 °C and relative humidity 100 %. Hot/Wet (HW) condition at 70 °C and 85 % relative humidity until saturation per EN 60068-2-78, was applied.

2.4. Tests

All tests were performed in the testing laboratory of Strength of structure department of the Czech aerospace research centre according to valid standards on the mechanical loading machine Instron 55R1185.

2.4.1. Pure resin compression test

Test was performed according to ASTM D695. Specimen was placed between compressive plates parallel to the specimen surface (platen surfaces parallel was within 0.03 mm across the sample contact area), and compressed along its major axis at constant rate of displacement (1.0 mm/min) until the specimen fracture occurred. The test assembly is shown in Figure 1a and Figure 1b for RT and elevated temperature, respectively.

2.4.2. Coupon compression test

Test was performed according to ASTM D6641. Specimen was fixed in the combined loading compression (CLC) test fixture so that the end of the specimen was in flush with the ends of the CLC test fixture. All screws on the CLC fixture were screwed to 2.5 Nm. The assembled fixture was placed between well-aligned, fixed flat platens (platen surfaces parallel was within 0.03 mm across the fixture base) and compressed along its longitudinal axis at constant rate of displacement (1.3 mm/min) until the failure. To determine the compression modulus of the geopolymer, extensometer Instron 2620-601 with gauge length 12.5 mm was attached on the test sample. The test assembly is shown in Figure 1c.

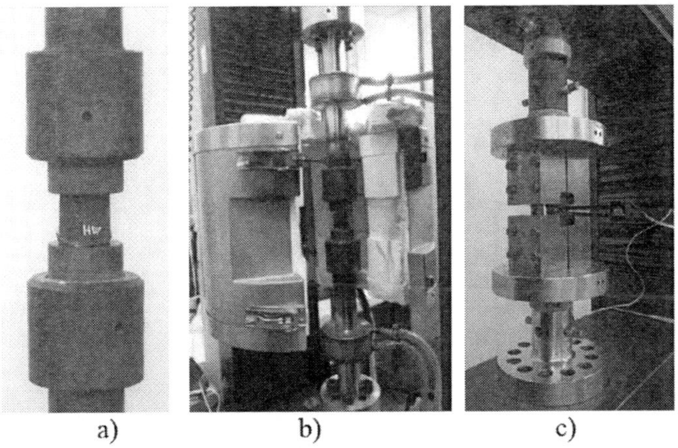

Figure 1. Test assemblies: a) pure resin at RT, b) pure resin at temperature chamber, c) geocomposite sample in CLC test fixture

3. Results

All measured results were tested by Dixon's Q test on presence of outliers. Test was performed for significance level 0.05 and outlying values were not included into the test evaluation. A T-test was performed for determination whether the individual sets are significantly different from each other. Statistical significance 0.05 was chosen.

3.1. Compressive strength of the pure resin.

The highest average compressive strength of 43.6 MPa was measured on samples tested at room temperature without previous exposition in the harsh environment. Mean values with standard deviations are shown in Figure 2. The measured data showed that the elevated temperature of 400 °C decreased in all cases the compressive strength compared to RT sets. The decrease was evaluated as statistically significant for all the sets. In the case samples measured at RT after the environmental exposition, the hot-wet condition had not statistically significant influence on compressive strength. However, the effect of salt mist was statistically significant. For the tests at 400 °C, the influence of the environment was statistically insignificant. Typical failure modes of the geopolymer resin samples are showed in Figure 3a.

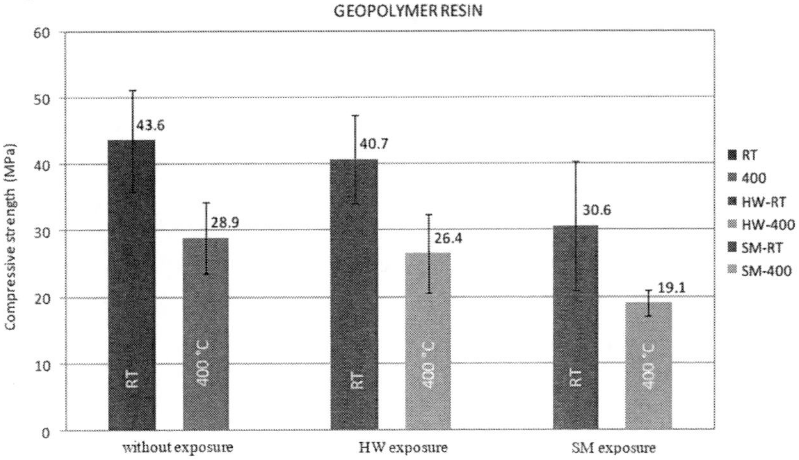

Figure 2. Compressive strength – mean values with sample standard deviation bars, geopolymer resin.

a) b)

Figure 3. Specimens after the test: a) pure geopolymer resin, b) geocomposite.

3.2. Compressive strength and modulus of the geocomposite

The highest average compressive strength of 225.5 MPa was measured for set that was not exposed in the environment. Average compressive strength measured on the samples exposed at hot-wet conditions was approximately by 26 % lower (168.1 MPa) compared to the baseline and average strength of salt mist exposed sample was by 21 % lower (178.5 MPa) than for the not exposed set. Typical failure mode is shown in Figure 3b and mean values with standard deviations are shown in Figure 4a.

Differences between the measured compressive modulus of elasticity were evaluated as statistically insignificant. The difference between the highest and lowest measured average compressive modulus of elasticity was less than 15 % (96.8 GPa for not exposed set, 83.0 GPa for hot-wet set and 86.4 for salt mist set). Mean values of the compressive modulus with standard deviations are shown in Figure 4b.

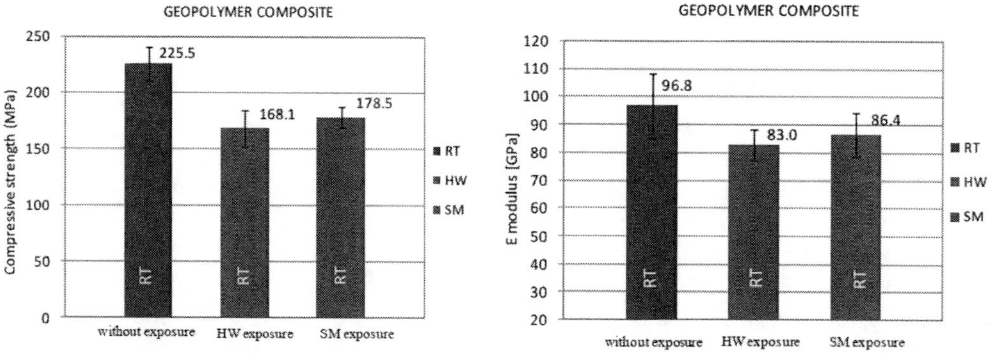

a) b)

Figure 4. Compressive strength and modulus – mean values with sample standard deviation bars, geocomposite.

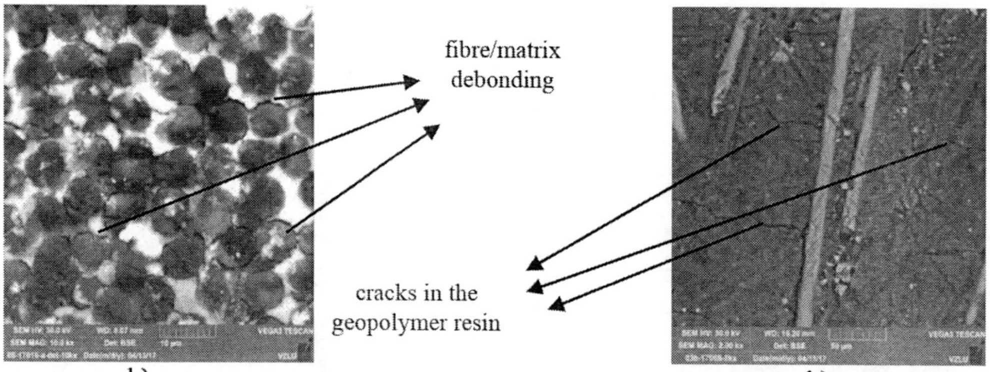

b) b)

Figure 5. Microscopy analysis before testing: a) matrix debonding from fibres, b) cracks in the resin.

3.3. Microstructure analysis

Microstructure analysis using scanning electron microscopy was performed on selected specimens before the testing; see Figure 5. The analysis showed that in some places the resin is debonded from the carbon fibres. Cavities and cracks in the pure resin were observed. The origin of these defects was probably in the shrinkage of the geopolymer resin. These defects were probably the cause of a large scatter of the measured values.

4. Conclusion

The research of geopolymer based materials showed that compressive strength of pure geopolymer resin at 400 °C was decreased by approx. 35 % in comparison with the room temperature values. Furthermore, the exposure of salt mist was significantly deteriorating on the strength of geopolymer resin for both test temperatures. On the other hand, the influence of the hot-wet environment had insignificant influence on measured compressive strength.

For geocomposite samples, the effect of the harsh environment on values of compressive strength was significant for both the cases reaching up to 25 % decrease. However, the influence of the environment on compressive modulus of the geocomposite samples was statistically insignificant.

The measured values clearly demonstrated that the geopolymer based materials still retain approximately 65 % of the strength at 400 °C. Compared to that, conventional polymeric materials would be disintegrated at that temperature.

Geopolymeric materials can be recommended for the structures in airplanes interiors where strict requirements for fire, smoke and toxicity are defined according to FAR 25 regulations.

Acknowledgement

This research was funded by the European Union in the frame of "Future Sky Safety" project, the Joint Research Initiative Safety P7, project "Mitigate risk of fire, smoke and fumes".

References

1. R. E. Lyon, "Fire Safe Aircraft Cabin Materials," in Fire and Polymers, ACS Symposium Series Number 599, G.L. Nelson, ed., American Chemical Society, Washington, D.C., p. 618, 1995.
2. R. E. Lyon, Fire Response of Geopolymer Structural Composites, DOT/FAA/AR-TN95/22, 1996
3. J. Davidovits, J. L. Sawyer. Early high-strength mineral polymer, U. S. Patent 4,509,985, United States, 1985.
4. J. Davidovits, Geopolymer Chemistry & Applications, Institute Géopolymèr, France, 2008.
5. R. E. Lyon, et. all. Fire resistant aluminosilicate composites. Fire and materials **21(2)** pp. 67-73, 1997.
6. Duxson, P. et al. (2007). Geopolymer technology: the current state of the art. Journal of Materials Science **42(9)**: pp. 2917-2933.
7. T. D. Hung, et all. New Generation of Geopolymer Composite for Fire-Resistance, Advanced in Composite Materials – Analysis of natural and Man-Made Materials, InTech, 2011
8. J. Davidovits, 30 Years of Successes and Failures in Geopolymer Applications - Market trends and Potential breakthroughs. Geopolymer 2002 Conference, Melbourne, Australia, Geopolymer Institute, pp. 1-16, 2002.
9. ASTM D695-15 - Standard Test Method for Compressive Properties of Rigid Plastics.
10. ASTM D6641-16 – Standard Test Method for Compressive Properties of Polymer matrix Composite materials Using a Combined loading Compression (CLC) Test Fixture.
11. Fabrics for composites, Multiaxial knitted fabric MC-45C-0200, Product Data Sheet, 2014.
12. Evonik Industries AG, Dynasylan 40, Product Data Sheet, 2011.

Development of Gypsum Block of Energy Conservation and Waste Utilization

Yafei Hao[1], Lingxiao Li[1], Xiongmu Chen[1] and *Fengqing Zhao*[1,2,a]

[1]*Department of Chemical Engineering, Hebei University of Science & Technology, Shijiazhuang, 050018, China*
[2]*Hebei Engineering Research Center of Solid Waste Utilization, Shijiazhuang, 050000, China*

Abstract. Waste, parings and scraps (WPS) in shoemaking industry are difficult to deal with by traditional approaches, which will pollute the environment and results in waste of resources if not handled properly. In this paper, WPS, after modification, is used as reinforcing material in gypsum block, in which livestock-waste based retarder is used to control setting time. The result showed that the proper mass ratio of hemi-hydrate gypsum to WPS is 95:5. The optimal dosage of retarder is 0.23% (on the basis of total solid). The addition of modified WPS remarkably improved the characteristics of anti-cracking, water absorption, dry shrinkage and thermal conductivity of the gypsum block, with little decrease in mechanical strength values.

1 Introduction

The acceleration of urbanization has promoted the vigorous development of the construction and building industry, and the corresponding building energy consumption is also increasing. According to statistics, 99 % of existing 40 billion square meters of buildings and more than 95 % of new buildings in China belong to high energy consumption buildings. In recent years, as a lightweight material, gypsum block gradually replace the traditional clay brick as non-load bearing interior walls, partition walls, cavity walls, skin walls and pillar casing indoors, which not only can utilize waste gypsum, but also has the advantages of heat preservation, comfortable, and fire-resistant, etc.

In order to further improve the performance of gypsum, some admixtures were added [1,2]. Considering the cost, solid waste is the first choice. At present, a large number of shoe-making plants produce large amount of waste, parings and scraps which will pollute the environment and results in waste of resources if not handled properly. Analysis shows that the main components of them are rubber, adhesives and fiber etc. The combined use of these wastes with gypsum to produce construction block may be a possible approach considering the properties.

The aim of this work is to use WPS in gypsum block, giving full play the synergistic effects between different materials[3,4]. In order to reduce water requirement and increase the compatibility with gypsum, WPS is modified beforehand[5].

2 Experimental

[a] Corresponding author : zhaofq3366@126.com

2.1 Raw materials

The construction gypsum was provided by shijiazhuang changan jiayi building materials Co. Ltd, which is obtained by the desulfurization gypsum calcined at 150-170℃. See Fig. 1. Table 1 gives the composition. Shoe-making industry waste is from the Xiong'an District, see Fig. 2. Gypsum retarder was prepared from livestock-waste [6]. Citric acid, sodium hexametaphosphate and other chemical reagents were bought on the market.

Table 1 Principal composition of desulfurization gypsum, wt%

Composition	CaO	SiO_2	FeO	Fe_2O_3	MgO	Al_2O_3	MnO	f-CaO	SO_3
Desulfurization gypsum	37.76	1.38	-	0.32	0.16	0.27	-	-	50.34

Figure 1. Construction gypsum

Figure 2. WPS in shoe-making industry

2.2 Methods

(1)Modification of WPS. Potassium methyl silicate, VAE emulsion and water are mixed to form modification liquid, in which small WPS fragments are soaked, air dried.

(2)Preparation of gypsum block sample[7]. Construction gypsum, modified WPS fragment, retarder and water are mixed to form gypsum slurry. The slurry is put into mold, then de-mold, curried and air dried to obtain the final product. Fig. 3 gives the flow-sheet of the preparation of gypsum block.

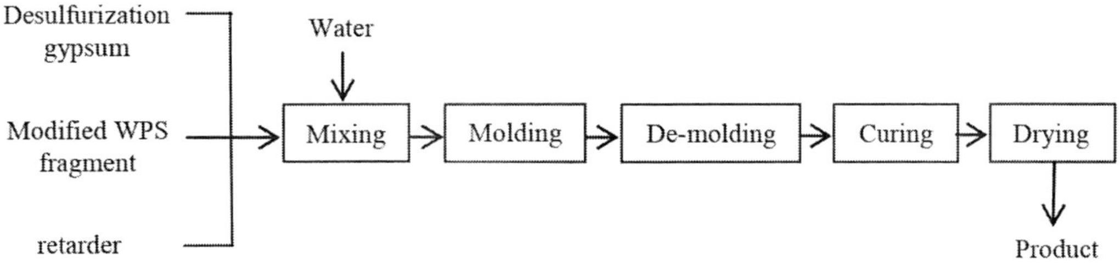

Figure. 3 Preparation of gypsum block

(3)Performance test. Specimens with dimension of 40 mm×40 mm×160 mm were molded in lab to test the performance of the product. The specimen is cured under standard curing conditions and then dried in thermostatic oven at (40±5)°C to constant weight. The standard consistency, setting time and mechanical strength of the gypsum were tested according to GB/T17669.4-1999. The flexural strength value was the average of every three tests, whereas the compressive strength value was the average of every six tests. The dry shrinkage and water absorption of the specimen is tested according to GB/

T2542-2512. The thermal conductivity of the specimen was determined on DRM thermal conductivity tester[8,9].

3 Results and discussion

3.1 Effect of retarder on setting time and mechanical strength

In this work, gypsum retarder was prepared using slaughter plant waste, by hydrolysis and neutralization process [1]. Based on the same initial setting time, the effects of four retarders on the bending and compressive strength of desulphurization gypsum were observed, See table 2.

Table 2 Effects of different retarders on the strength of construction gypsum

Retarder	Construction gypsum/%	Retarder /%	Initial setting time/min	Absolute dry strength/MPa	
				Bending	Compression
Blank	100	0	6	11.5	24.1
Citric acid	100	0.03	23	7.6	12.4
Sodium citrate	100	0.16	25	8.2	13.2
Sodium hexametaphosphate	100	0.16	20	7.1	12.0
Slaughter plant waste based retarder	100	0.11	23	12.5	17.3

It can be seen from table 2 that, the slaughter plant waste based retarder has less negative effect on the mechanical strength of gypsum, compared to the other three retarders.

3.2 Effect of modified WPS on the properties of gypsum block

The first choice is the larger size of the scraps, through the exploration of different dosage, testing the density, water absorption, bending and compressive strength.

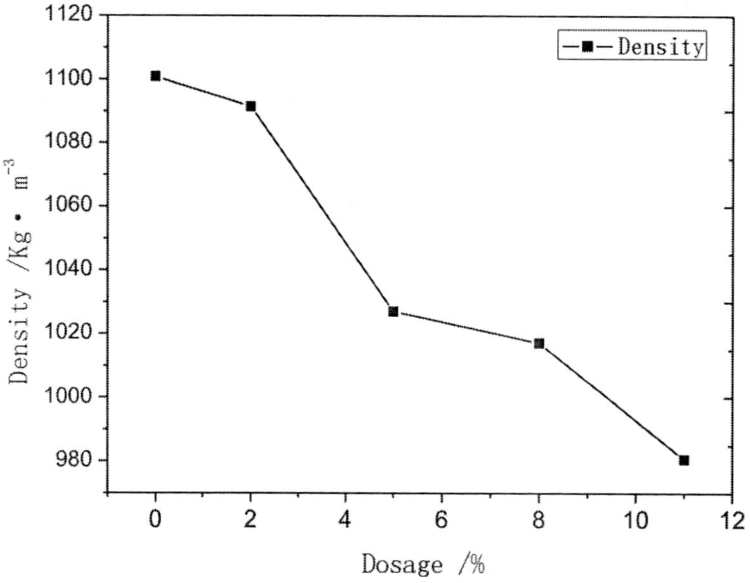

Figure. 4 Effect of different dosage on the density of gypsum

As we can see from Figure 4, with the increasing in WPS scraps content, the density of the block shows a downward trend.

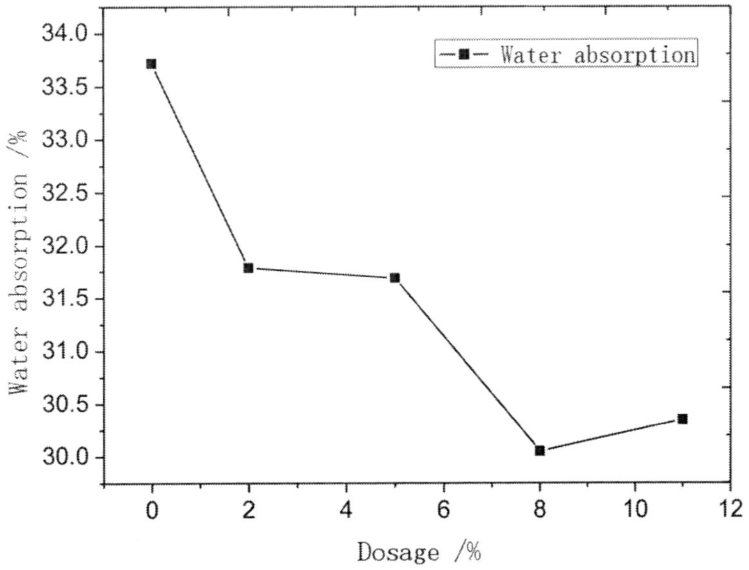

Figure. 5 Effect of different dosage on the water absorption of gypsum

As can be seen from Fig. 5, with the increase in WPS scraps content, the water absorption of block gradually decreased, indicating that WPS has a significant effect on gypsum system after modification, and the water absorption rate is reduced by 10% compared with the blank experiment.

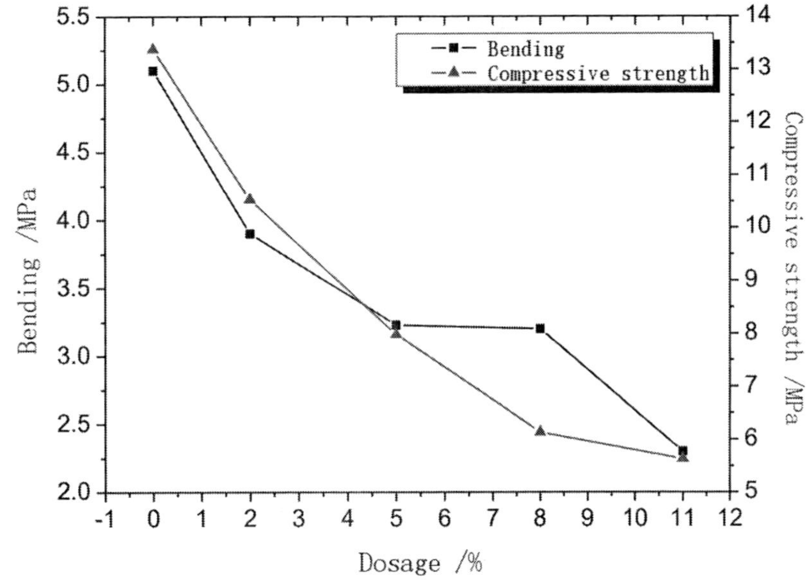

Figure. 6 Effect of different dosage on the bending and compression strength of gypsum

Fig 6 shows that the mechanical strength decreases with the increase of the dosage WPS. The flexural strength decreases sharply with the increase of the content of the WPS, and the compressive strength is slower.

Figure 7 and Figure 8 show the cross section morphology and profile of the block. It can be predicted that the addition of WPS can improve the crack resistance of the test block.

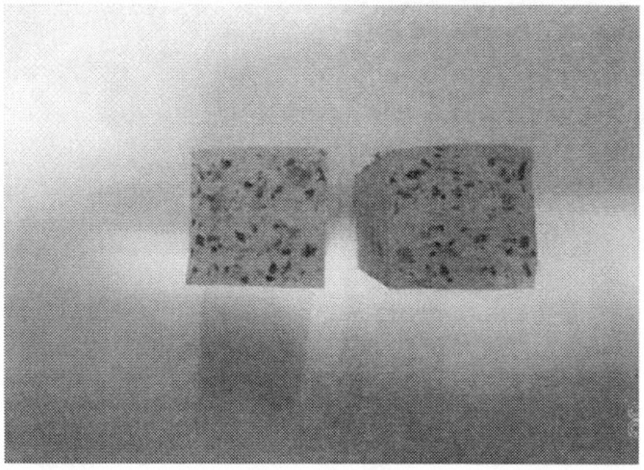

Figure 7 Distribution of WPS scraps in gypsum

Figure 8 Broken test block

3.3 Effect of scraps on dry shrinkage value and thermal conductivity

The dry shrinkage value and thermal conductivity were tested for addition of WPS scraps in the gypsum. See table 3.

Table 3 Effect of the scraps on the thermal insulation and deformation of gypsum

Dosage/%	Dry shrinkage value	Thermal conductivity
0	0.69	0.1864
5	0.39	0.1688

As can be seen from table 3, compared with the blank, the dry shrinkage and thermal conductivity of the gypsum system decrease by 43 % and 9.4 % respectively.

That is to say, the addition of WPS scraps can effectively improve the thermal insulation effect of gypsum blocks and reduce the dry shrinkage value[10, 11].

4 Summary

(1)A kind of environment-friendly gypsum insulation block of energy conservation is prepared by using desulphurization gypsum, WPS in shoemaking industry and slaughter plant waste based retarder, realizing solid waste utilization.
(2)The slaughter plant waste based retarder has good retarding effect on construction gypsum and has less negative effect on the mechanical strength of gypsum, compared to the other three retarders.
(3)The addition of modified WPS improved the characteristics of anti-cracking, water absorption, dry shrinkage and thermal conductivity of the gypsum block. In this case, the mechanical strength values of the block decrease[12-14].

References

1. M. M. Zhang, M. M. Chen,etal. Improving Water proof Property of Gypsum Block with Organic-Inorganic Compound Materials, Proceedings of 2nd International Conference on Material Engineering and Application, 707-711 (2015)
2. M. M. Chen, F. Q. Zhao, T. L. Fan, M. M. Zhang, S. J. Liu, H. J. Liu. A waterproof and impermeable storage gypsum block and its preparation method. CN 105924120A [P]. (2016)
3. Y. W. Wang, Y. H. Wu, Y. Yao, F.Nan. Study on the preparation and influencing factors of gypsum-based expanded perlite insulation block [J]. C&CP.(03):83-85(2014)

4. P. Zhang, B. Wang. Research on the preparation of functional materials using construction waste gypsum-waste mineral wool [J]. FJBM.(2016)
5. B. G. Ma, J. Liu. Building functional materials [M]. Wuhan: wuhan university of technology press, (2004)
6. X. M. Chen, M. M. Zhang, L. X. Li, F. Q. Zhao. Utilization of slaughter plant waste proteins in gypsum retarder: A case study. ICMAA, **108**:53-56 (2017)
7. D. Y. Li. Building energy saving technology [M]. Beijing: mechanical industry press, (2006)
8. B. Jin, X. Wang, L. S. Yang. Study on the preparation of lightweight thermal insulation block by aluminum coal ash residue [J]. C&CP.**2**:64-67(2017)
9. J. S. Yuan. Application of waste polystyrene foam and phosphogypsum in external wall insulation block [J]. BES. **34**(189):38-39(2006)
10. Z. P. Zhang, Z. Li, Y. L. Dong. The development status and prospect of building insulation and energy-saving wall [J]. ENG PLAST APPL, **24**(4):121-127(2007)
11. J. M. Li, Z. W. Wang, Y. P. Ai, et al. Study on the material of EPS- gypsum - slag insulating wall [J]. BCHINCERAMSOC, **32**(3):533-536(2013)
12. Z. M. He, C. W. Yan, J. F. Zhang. Experimental study on the preparation technology of thermal insulation mortar for expansion glass microbeads [J]. NEWBUILDMATER, (8):73-76(2007)
13. C. C. Ruan, L. M. Liu, Y. P. Huang, et al. Effects of water-repellent agent on the performance of the α-type hemihydrate [J].NEWBUILDMATER,**08**:42-46(2014)
14. F. Q. Zhang, G. Li, J. Li, et al. Study on the improvement of water resistance of gypsum with composite Portland cement [J]. MON MET MINES, **02**:17-19(2016)

Charge transport mechanism of polyaniline/ZnO nanocomposites based on inorganic/organic heterojunctions

Lei Han[1,a], Wen Li[1], Chao Meng[1], Yan Chen[1] and Shan Fan[1]

[1]*Department of Mechanics, Wenhua College, No. 8 Wenhua Yuan Road, Wuhan 430074, P.R. China*

Abstract. The key to improve the photocurrent is how to promote the generation rate and reduce the recombination rate of the photo-hole pair. In this paper, a p-n heterjunction structure in Polyaniline (PANI)/ZnO nanocomposites was formed *via* mechanical mixing process. Compared with ZnO, PANI doped ZnO nanocomposites exhibit obviously enhanced response. It might be attributed to the p-n heterojunctions formed between p-type and n-type enhanced response, enhancing the charge separation and in turn promoting the photoconductivity of the film. The 5at-%PANI doped ZnO nanocomposites demonstrates the highest response.

1.Introduction

Heterojunction composites of conjugated polymers and semiconductor nanoparticles (NPs) have attracted a great deal of attention in photocatalysis [1], solar cells [2], gas sensors [3] and photodiodes [4]. The hybrid organic–inorganic semiconductor junctions are envisaged to complement each others' functional advantages while overcoming their respective drawbacks. The high quality schottky junction or p-n junctions formed the composites interface have improved charge separation. [5]

Among the inorganic nanoparticles, ZnO nanomaterials have received enormous interest in exclusive electrical, catalytic, electronic and optical properties due to their wide band gap (3.3eV), low power threshold sand high exciton binding energy (~60 meV) as well as their low cost. [6] The conducting polymers, which have substantial π-electrons through the polymeric chain, are the excellent hosts for the trapping of the carriers from the semiconducting nanomaterias.[7] PANI has received much attention as a promising material for microelectronic devices due to due to unique electrical, optical, photoelectric properties and good environmental stability. [8] PANI is extensively used as p-type material which could be deposited by various techniques such as wet, chemical, and electrophoretic methods on n-type semiconducting electrodes for p-n heterostructure devices. [9] Various methods have been widely used for the synthesis of the nanocomposites of PANI with the inorganic nanoparticles ZnO.

The results indicate that ZnO together with PANI, an inorganic/organic hybrid structure shows a promising application in the UV photodetection. [10-11]Under the light illumination, ZnO and PANI produce photo electric-hole pair. The p-n heterointerface between ZnO and PANI can promote to reduce the recombination rate of the photo-generated electron-hole pair. Although there are many methods to prepare the composite material, the main purpose is to improve the transfer of the photo-generated electrons, which the PANI can act as a bridge to transfer the photo-generated electrons from

[a] Corresponding author : xiaohan0522@163.com

the ZnO. However, the interfacial area in the structure may accelerate recombination, suggesting the need to manage the content of ZnO or PANI.

In this work, we designed porous PANI/ZnO nanocomposites with the different mass ratios of PANI:ZnO from 0:100 to 100:0 to evaluate photoconductivity response. The photoconductivity measurement is made to test the photoelectrical response of each sample, and the best content of PANI is confirmed based on the photoconductivity. The aim of this paper is to evoke the importance of a subject that has been overlooked, as well as to establish a method to calculate the percolation threshold of p-n heterojunction nanocomposites, which is based on photoconductivity measurement. The PANI/ZnO nanocomposites have been characterized by FT-IR, UV visible spectroscopy and SEM.

2. Experimental Section

2.1 Materials

Commercial PANI (average size 1 μm) and ZnO (average size 80nm) were used. Other chemicals used in the experiment were of analtytically pure grade. All of them were used as received without any pretreatment. Fixed quantity of ZnO nanoparticles were mixed with different weight of doped polyaniline powder. Six different ratios of ZnO(Z) and Polyaniline(P) were studied labeled as ZP100:0(ZP000), ZP99:1(ZP001), ZP95:5(ZP005), ZP90:10(ZP010), ZP50:50(ZP050), and ZP0:100(ZP100) were prepared with different weight ratio as indicated above.

The two pure powders and a certain amount of organic solvent (a 55:30:10:4:1 weight-% combination of terpineol, butyl carbitol, di-n-butyl phthalate, span 85, andethyl-cellulose) were mixed in a 7:3 weight ratio. Organic solvent was used mainly as thickening agent and rheological agent for screen printing. The mass of powder in each combination was 10 g. Each sample was processed in a QM-3SP04 ball miller (Nanjing University Instrument Factory, China) for 3 hours at 250 rpm to obtain a paste suitable for screen printing. The paste was printed onto an Au interdigital electrode which was preprinted on an alumina substrate. The spacing of the fingers in the Au interdigitated electrode is 0.4mm. The dimensions of each sample were 8 mm by 6 mm by 10 μm thickness. The latter value was controlled by the screen printing machine (WJ-LCD4050CHL Shenzhen Screen-Star Instrument Co. China). These samples were heat-treated for 30 min at 200 °C to eliminate the organic solvent. After the above procedures, a thick porous film was obtained in each case. [12]

2.2 Photoconductivity Measurement

The photoconductivity of each device in the material library was measured by a homemade photoconductivity measurement platform. All the samples were located in the stainless steel test chamber and tested under different light sources (UV LED 365 nm, 36 W/m^2) with constant specific bias voltage and ambient air conditions (20 °C and below 20% relative humidity) which was controlled by the dry air generator and the mass flow controller.

A typical conductivity-time curve (in this film) is shown in Figure 4. The bias voltage was added at 10s, and the dark conductivity was obtained. Illumination by LED was initiated at 20s and conductivity was monitored, reaching a maximum (at which the generation and recombination of electron-hole pairs achieves dynamic equilibrium) by 160 seconds for all samples. The LED was turned off at 180s. Finally the bias voltage was unloaded at 350s. All the photoconductivity measurements were lasted for 360s.

To obtain the I–V characteristics, current was continuously measured as function of voltage. In the light irradiation experiments, we used UV LED array (365nm, 36W/m^2) and the lamp-to-sample distance was 1cm.

2.3 Characterzation

The morphologies of the prepared samples were observed on field-emission scanning microscopy (FSEM, Sirion 200 FEI). Room-temperature Fourier transform infrared (FT-IR) spectra were recorded in the range of 4000 - 500 cm^{-1} on VERTEX 70 (made by Bruker Co., Germany) using the KBr pellet technique. The UV-vis DRS spectra (Lambda 35, PerkinElmer) was used to obtain the reflectance spectra of the samples over 300-800 nm.

3. Results and Disscuccion

3.1 Characterizations

The FT-IR spectra of the different composites are showed in Figure 1. The band at 490 cm^{-1} observed in the spectra of Figure 1 is assigned to ZnO.[13] All the characteristic bands of PANI [14], such as the C=C stretching deformation of the quinoid (1556 cm^{-1}) and benzenoid (1472 cm^{-1}) rings, the C-N stretch of secondary aromatic amines (1300 cm^{-1}), the aromatic C-H in-plane bending (1140 cm^{-1}), and the N-H stretching (3447 cm^{-1}), can be observed in Figure 1.

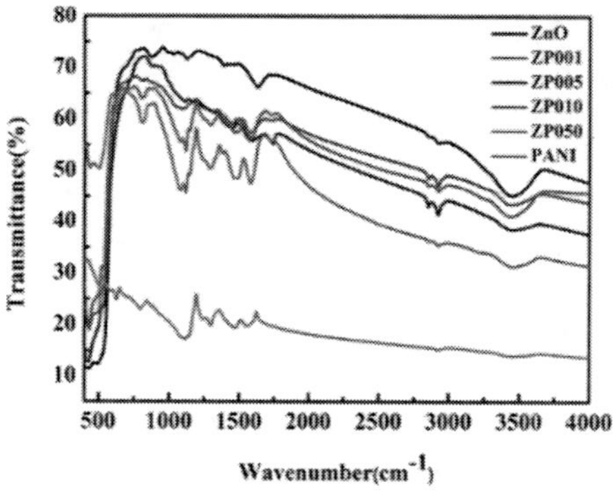

Figure 1 FT-IR spectra of the samples

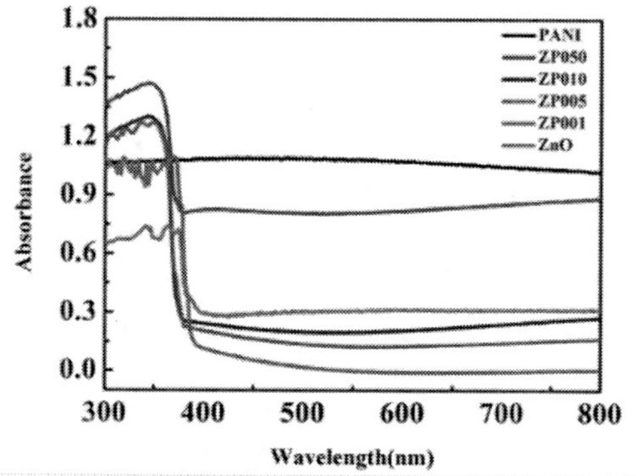

Figure 2 UV-visible adsorption spectra of the samples

Figure 2 shows the UV-visible adsorption spectra of the PANI/ZnO nanocomposites and the PANI. The PANI shows sharp and a broader bands at 350 and 650 nm which are ascribed to $\pi \rightarrow \pi^*$ transitions of the benzenoid groups and $n \rightarrow \pi^*$ transitions due to polaron formation into the conducting PANI, respectively.[14] It can be clearly seen that the absorption edge of the ZnO/PANI nanocomposites could be shifted into visible spectrum. This red shift might occur due to the interaction between the hydroxyl groups of ZnO and the quinoid ring of emeraldine salt (ES) which facilitates the charge transfer from highly reactive imines groups at quinoid unit of ES to ZnO nanomaterials via partial hydrogen bonding. [15] Additionally, the relative intensity of the peaks increases with the increased weight percent ratio of PANI nanomaterials in PANI/ZnO nanocomposites. This increased intensities of the peaks might at tribute to the enhanced hydrogen bonding in the form of NH– (PANI) ·····O—Zn [16].

The SEM images of the samples are shown in Figure 3. The composite is composed of the ZnO nanoparticles with the diameter of 40~60 nm and the PANI with500 nm. From the SEM image, the size of PANI particles is larger than that of the ZnO particles. When the content of ZnO is larger than 50%, the PANI particles are surrounded by ZnO, which like as islands in ZnO.

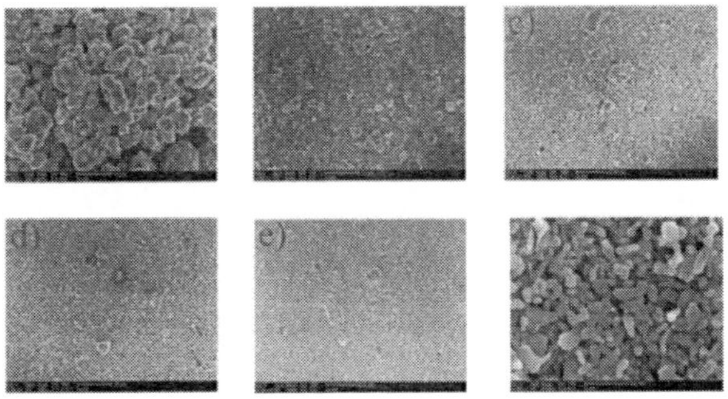

Figure 3 SEM images of the samples

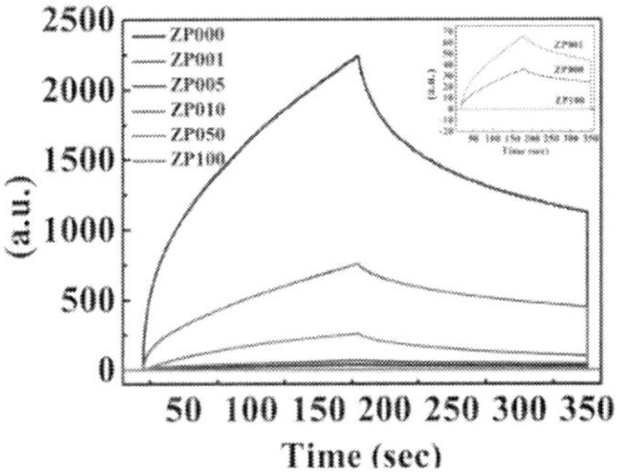

Figure 4 Photoconductivity responses of the samples

3.2 Photoconductivity response of the PANI/ZnO nanocomposites

Figure 4 shows the photoconductivity responses of composites with the mass ratios of PANI:ZnO from 0:100 to 100:0. It is shown that the conducting polymer PANI performs the less response to UV because of the high recombination of electron and hole [16-17]. In the whole experimental range, the PANI/ZnO nanocomposites demonstrate much higher response than that of the pure ZnO, indicating that the addition of PANI is beneficial to the UV responsing of the ZnO nanoparticles. Especially, the 5 at-%PANI doped ZnO nanocomposites (ZP005) exhibits markedly higher response than pure, 1, 10, 50 and 100at-%PANI-ZnO nanocomposites. It also suggests obviously the increasing photoconductivity amplitude can be seen with the increase of the PANI to ZnO from 0/100 to 5/100. When the initial ratios of PANI to ZnO larger than 5/100, the photoconductivity amplitude begins to decrease.

3.3 I-V characteristics of the PANI/ZnO nanocomposites

Figure 5 illustrates the *I-V* characteristics of the n-ZnO/p-PANI (ZP050 and ZP010) nanocomposites heterojunction in the dark and UV irradiation conditions at room temperature, respectively. The linear curve in the Figure 5 (a) can indicate that the ohmic contact is fromed between the ZnO and PANI (ZP010). The inset on upper side in Figure 5 (a) shows no current flow in the measured voltage range, indicative the resistance of the film being very high. The two non-linear curves in Figure 5 (b) exhibit almost symmetric characteristic with an extended plateau suggestive of tunneling conduction through the PANI/ZnO nanocomposites. This result confirms that a p-n junction was successfully formed.

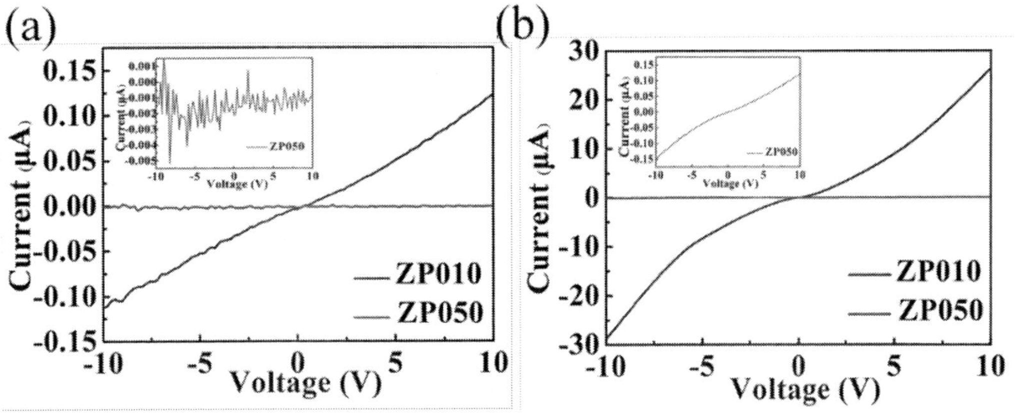

Figure 5 *I-V* characteristics of the PANI/ZnO nanocomposites in the dark (a) and in the light (b)

3.4 Mechanism analysis

The conducting polymer PANI is p-type semiconductor and can form the holes-accumulation layer in the grain boundary. [14] The metal oxide ZnO is typical n-type semiconductor and the ZnO surface can easily adsorb oxygen molecules. [18-19] The chemical adsorb oxygen molecules, which trap the free electrons from the conduction band of ZnO, can transform to negative oxygen ions. [20] So the electrons-depletion layer is formed at the grain boundary of ZnO. Based on the research of Lin et al. [10], both of the energy bands will bend up nearby the surface of the p-n heterjunction (as shown in Figure 6 (a)).

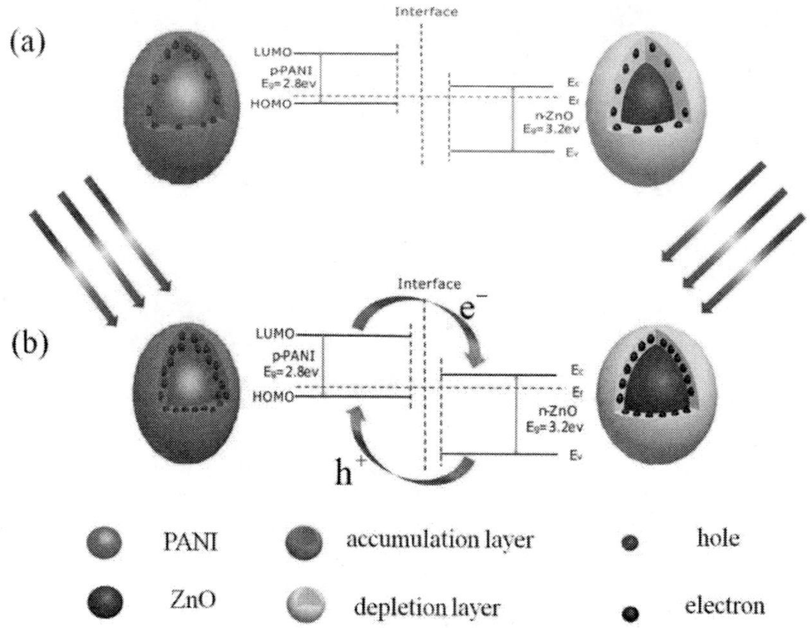

Figure 6 Schematic illustration of the energy bands for the grain boundary of the PANI/ZnO nanocomposites (a) in the dark (b) in the light

Under UV illumination, the interface of ZnO and PANI will be changed by photo-generated electrons and holes. The photo-generated holes migrate to the ZnO surface and react with the negative oxygen ions. The photo-generated electrons transfer between the LUMO of PANI and the CB of ZnO. The photo-generated holes transfer between the VB of ZnO and the HOMO of PANI. These lead to the decrease of holes-accumulation layer of PANI and the increase of the electrons-depletion layer of ZnO, which is as shown in Figure 6 (b).

Here e^- and h^+ represent photogenerated electrons and holes, respectively. The subscript Z and P mean the related subjects stay in ZnO and PANI, respectively. During the on and off illumination test, several processes might have happened:

$$ZnO + photo(UV) \rightarrow e^-_{(Z)} + h^+_{(Z)} \tag{1}$$

$$PANI + photo(UV) \rightarrow e^-_{(P)} + h^+_{(P)} \tag{2}$$

$$e^-_{(P)} \rightarrow e^-_{(Z)} \tag{3}$$

$$h^+_{(Z)} \rightarrow h^+_{(P)} \tag{4}$$

$$e^-_{(Z)} + O_2 \rightarrow O_2^- \tag{5}$$

$$O_2^- + h^+_{(Z)} \rightarrow O_2 \uparrow \tag{6}$$

$$e^-_{(Z)} + h^+_{(Z)} \rightarrow recombination(heat) \tag{7}$$

$$e^-_{(P)} + h^+_{(P)} \rightarrow recombination(heat) \tag{8}$$

During illumination, O_2^- reacts with photo induced holes. [20] The adsorbed oxygen molecules are released with the help of this reaction, which will lower the barrier height as proposed by Muraoka.[20]

4. Conclusion

A facile, simple and environment friendly mechanical mixing process method was used to synthesise PANI doped ZnO p-n heterojunction nanocomposites. Compared with pure ZnO nanoparticles, PANI doped ZnO p-n heterojunction samples demonstrate enhanced UV responding performances.It might be ascribed to the formation of p–n heterojunctions between p-type PANI and n-type ZnO interfaces, further promoting the generation rate and reducing the recombination rate of the photo-generated electron-hole pair. Among all fabricated samples, the 5at-% PANI doped ZnO nanocomposites demonstrate the highest responses in the UV light. All experiments indicate the 5at-% PANI doped ZnO p-n heterojunction nanocomposites as a promising sensing candidate for UV detection in practical application. The PANI is like some islands that storage the holes from the VB of the ZnO. This island-structure is helpful to separate electron-hole pairs.

Reference

1. Q.L. Li, C.R. Zhang, J.Q. Li, Appl. Surf. Sci. **257**, 944-948 (2010).
2. A.A. Khan, M. Khalid, J. Appl. Polym. Sci. **117**, 1601-1607 (2010).
3. J. Huang, T.L. Yang, Y.F. Yang, Y. Wang, S. Wang, J. Nat. Gas Chem. **20**, 515-519 (2011).
4. O. Game, U. Singh, T.Y. Kumari, A. Banpurkar, S. Ogale, Nanoscale **6**, 503-513 (2014).
5. W.W. Wu, S. Bai, N.Y. Cui, F. Ma, Z.Y. Wei, Y. Qin, E. Xie, Sci. Adv. Mater. **2**, 402-406 (2010).
6. L. Irimpan, V.P. N. Nampoori, P. Radhakrishnan, Sci. Adv. Mater. **2**, 578-582 (2010).
7. Z.H. Mbhele, M.G. Salemane, C.G.C.E. Sittert, J.M. Nedeljkovic, V. Djokovic, A.S. Luyt, Chem. Mater. **15**, 5019-5024 (2003).
8. S. Ameen, M.S. Akhtar, Y.S. Kim, O.B. Yang, H.S. Shin, Microchim. Acta **172**,471-478 (2011).
9. F. Ahmed, S. Kumar, N. Arshi, M.S. Anwar, S.Y. Lee, G.S. Kil, D.W. Park, B.H. Koo, C.G. Lee, Thin Solid Films **519**, 8375-8378 (2011).
10. Y.J. Lin, T.H. Su, J.C. Lin, Y.C. Su, Synthetic Met. **162**, 406-409 (2012).
11. J. Wu, H.Y. Li, Y. Liu, C.S. Xie, J. Appl. Phys. **110**, 123513-123516 (2011).
12. Z.J. Zou, Y. Liu, H.Y. Li, Y.C. Liao, C.S. Xie, J. Comb. Chem. **12**, 363-369 (2010).
13. T. Guo, L.S. Wang, D.G. Evans, W. S. Yang, J. Phys. Chem. C **114**, 4765-4772 (2010).
14. S. Ameen, M.S. Akhtar, Y.S. Kim, O.B. Yang, H.S. Shin, Colloid & Polym. Sci. **289**, 415-421 (2011).
15. D.X. Han, Y. Chu, L.K. Yang, Y. Liu, Z.X. Lv, Colloid. & Surf. A: Physicochemical & Eng. Asp. **259**, 179-187 (2005).
16. Q.W. Tang, L. Lin, X. Zhao, K. Huang, J.H. Wu, Langmuir **28**, 3972-3978 (2012).
17. Y.M. Xiao, G.Y. Han, Y.Z. Chang, H.H. Zhou, M.Y. Li, Y.P. Li, J. Power Sources **267**, 1-8 (2012).
18. C.S. Lao, M.C. Park, Q. Kuang, Y.L. Deng, A.K. Sood, D. L. Polla, Z. L. Wang, J. Am. Chem. Soc.**129**, 12096-12097 (2007).
19. Y.Q. Bie, Z.M. Liao, P.W. Wang, Y.B. Zhou, X.B. Han,Y. Ye, Q. Zhao, X.S. Wu, L. Dai, J. Xu, Adv. Mater. **22**, 4284-4287 (2010).
20. Y. Muraoka, N. Takubo, Z. Hiroi, J. Appl. Phys. **105**, 103702-103707(2009).

A New Characterization Parameter of out-of-plane Constraint for C(T) Specimens

Zhao Lingyan [1,a], Cui Yinghao [2] and Yang Fuqiang [1]

[1]*School of Science, Xi'an University of Science and Technology, Xi'an 710054, China*
[2]*School of Mechanical Engineering, Xi'an University of Science and Technology, Xi'an 710054, China*

Abstract. As the resistance of structures against the crack-tip plastic deformation, constraints has attracted much attention in the research of material fracture behaviour. In order to increase the accuracy of structural integrity assessment, many characterization parameters of in-plane and out-of-plane constraints have been considered and proposed in the last few decades. Three-dimensional finite element analyses have been conducted for five C(T)50 specimens with different out-of-plane constraint. The distributions of four constraint parameters (σ_{22}/σ_0, T_z, h and D_p) along crack fronts were calculated. To characterize the out-of-plane constraint, the capability of constraint parameters such as σ_{22}/σ_0, Tz, h and D_p were analyzed and compared. Results show that, compared with the three constraint parameters (σ_{22}/σ_0, T_z and h) based on crack-tip stress fields, the parameter D_p based on crack-tip equivalent plastic strain is more sensitive to out-of-plane constraint, and may effectively characterize the out-of-plane constraint.

1 Introduction

As the resistance of structures against the crack-tip plastic deformation, constraints has attracted much attention in the research of material fracture behaviour. Constraints can be divided into in-plane constraint and out-of-plane constraint. The in-plane constraints such as the crack length relate to the specimen dimension in the direction of growing crack. The out-of-plane constraints such as the specimen thickness relate to the specimen dimension parallel to crack front. In order to increase the accuracy of structural integrity assessment, many characterization parameters of in-plane and out-of-plane constraints have been considered and proposed, such as T-stress[1], Q[2], A_2[3], T_z[4, 5], stress triaxiality factor h[6] and A_p[7, 8] etc.

In this work, σ_{22} is the stress component at crack tip in the direction parallel to the crack propagation direction, and it is the most common parameter for charactering the crack-tip driving force. Usually, the stress σ_{22} is normalized by the yield strength σ_0, then a constraint parameter can be described by σ_{22}/σ_0. And the numerical solutions of σ_{22}/σ_0 were obtained through detailed 3D finite-element (FE) calculations.

The constraint parameter T_z introduced by Guo[4, 5] is defined as the ratio of out-of-plane stress (σ_{zz}) to sum of the in-plane stresses (σ_{xx} and σ_{yy}), and it can characterize the 3D crack-tip stress fields:

$$T_Z = \frac{\sigma_{zz}}{\sigma_{xx} + \sigma_{yy}} \tag{1}$$

[a] Corresponding author : gloomy2@foxmail.com

The stress triaxiality factor h is defined as the ratio of the hydrostatic stress σ_m to the von Mises stress σ_e:

$$h = \frac{\sigma_m}{\sigma_e} = \frac{\sigma_1 + \sigma_2 + \sigma_3}{3\sqrt{\frac{1}{2}[(\sigma_1 - \sigma_2)^2 + (\sigma_2 - \sigma_3)^2 + (\sigma_3 - \sigma_1)^2]}} \tag{2}$$

where σ_i (i=1, 2 and 3) are the principal stresses. The factor h incorporates both in-plane and out-of-plane principal stresses. Thus, it may have a capability to describe both in-plane and out-of-plane constraint effects.

Based on crack-tip equivalent plastic strain, a new characterization parameter D_p of out-of-plane constraint was defined as follows:

$$D_p = \frac{D_{PEEQ}}{D_{ref}} \tag{3}$$

where D_{PEEQ} is the gradient of equivalent plastic strain (ε_p) in the middle-plane area surrounded by a specific ε_p isoline along a evaluated specimen crack front, and D_{ref} is the ε_p gradient in the middle-plane area surrounded by a specific ε_p isoline along the crack front of a standard reference specimen with high constraint level at fracture.

2 Finite element modelling

2.1. Specimen model

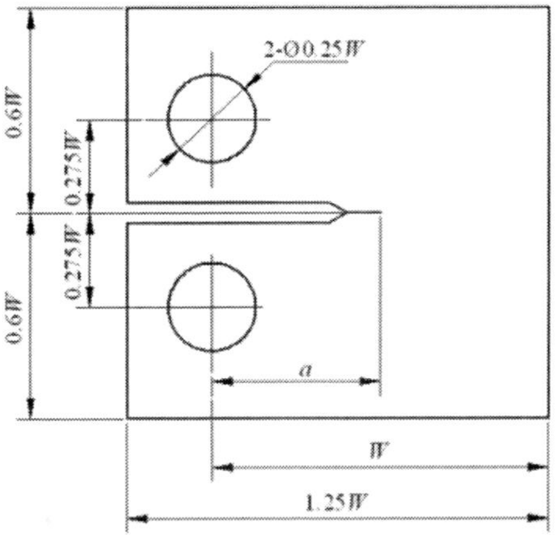

Figure1. Geometry of C(T) specimen (W=50 mm)

A compact tension (CT) specimen with a constant load was used to measure SCC growth rate in high temperature water environment [9]. Five types of C(T) specimens (specimen width W=50 mm) with thickness-to-width ratio B/W=0.0625, 0.125, 0.25, 0.5 and 1 were chosen, and their relative crack length a/W=0.5 after pre-cracking. The geometry and size configurations of a C(T)50 specimen is shown in figure 1. It can be considered that the out-of-plane constraints were changed by the changing specimen thickness B, and the specimen sizes are shown in table 1.

The calculated results of fracture toughness are also listed in table 1. The stress-strain field distributions along crack fronts were calculated at the same average J-integral value of J_{ave}=20 kJ/m^2 along the specimen crack fronts with different out-of-plane constraints, and the maximum values of J-internal J_{max} at the crack fronts mid-point were also listed. The values of fracture toughness K_{max} in table 1 were calculated from J_{max} by equation 4.

$$K_{max} = \sqrt{J_{max} \frac{E}{1 - \nu^2}} \tag{4}$$

where, E is the Young's modulus and v is the Poisson's ratio.

Table 1. C(T) specimen sizes and the calculated results of fracture toughness

Specimen	a/W	B (mm)	B/W	P (kN)	K_{max} (MPa·m$^{1/2}$)	J_{max} (kJ·m^{-2})
		3.125	0.0625	1.83	74.78	23.67
		6.25	0.125	3.7	74.56	23.53
C(T)50	0.5	12.5	0.25	7.57	74.27	23.35
		25	0.5	15.6	74.05	23.21
		50	1	32.4	74.02	23.19

2.2. Material model

The pressure vessel steel 22NiMoCr3-7 is investigated in this paper, and its non-linear relationship between stress and strain beyond yield can be described by Ramberg-Osgood equation [10]:

$$\frac{\varepsilon}{\varepsilon_0}=\frac{\sigma}{\sigma_0}+\alpha\left(\frac{\sigma}{\sigma_0}\right)^n \tag{5}$$

where ε is the strain, including elastic and plastic strain. σ is the total stress; ε_0 is the yield strain of the material, σ_0 is the yield stress of the material, and n is the strain hardening exponent of the material, α is the material offset coefficient.

The mechanical properties data and the Ramberg-Osgood constitutive parameters of 22NiMoCr3-7 steel at -60℃ are given in table 2.

Table 2. Mechanical properties of 22NiMoCr3-7 steel at -60℃[11]

Material	Young's modulus E (MPa)	Poison ratio v	Yield stress σ_0 (MPa)	Hardening exponent n	Hardening parameter α
22NiMoCr3-7	215 000	0.3	517	3.7	6.2

2.3 FE Model

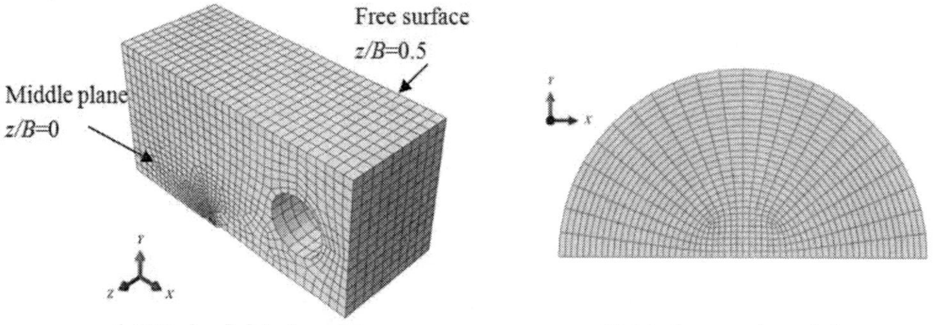

(a) Mesh of global model (b) Mesh around crack front

Figure 2. The global model mesh and mesh around crack front of C(T)50 specimen with B=25 mm and a/W=0.5

Five 3D FE models in table 1 were built by the commercial finite element code ABAQUS [12]. The eight-node isoperimetric elements with reduced integration (C3D8R) were used for all models. To reduce the computation time, considering the loading and geometry symmetry, only 1/4th of the specimen was modeled. The symmetry boundary conditions were applied on the un-cracked ligaments. The FE model meshes of C(T)50 specimen with B=25 mm and a/W=0.5 are shown in figure 2.

3 Results and Discussion

To characterize the out-of-plane crack-tip constraints, the capability of constraint parameters such as σ_{22}/σ_0, *Tz*, *h* (based on crack-tip stress fields) and D_p (based on crack-tip equivalent plastic strain) were analyzed and compared. The parameters σ_{22}/σ_0, T_z and *h* were calculated and evaluated at $r=2J/\sigma_0$ and $\theta=0$ by using equations 1 and 2, respectively.

The reference ε_p gradient D_{ref} in equation 3 was calculated at the area surrounded by the $\varepsilon p=0.002$ isoline in the middle plane of a standard reference C(T)50 specimen with $B/W=0.5$ and $a/W=0.5$ in Table 1, which is with high constraint.

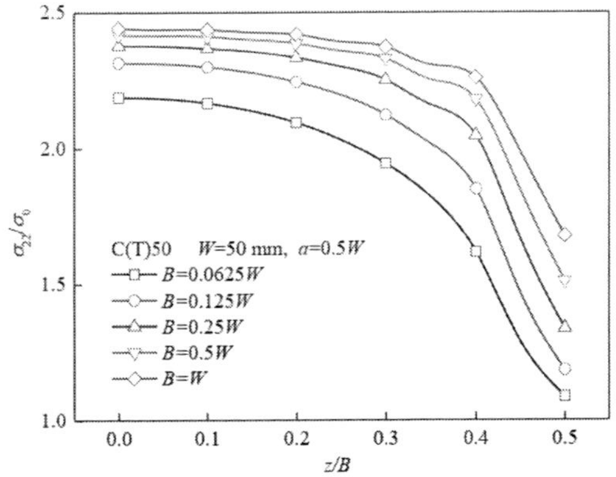

Figure3. Distributions of constraint parameter σ_{22}/σ_0 ahead of crack fronts

Figures 3-6 show the distributions of the four constraint parameters for C(T) specimens with the same $a/W=0.5$ (same in-plane constraint) and different B (different out-of-plane constraint). The constraint parameters are plotted against the normalized specimen thickness *z/B*, and the $z/B=0$ is the middle plane and the $z/B=0.5$ is the free surface of the specimens.

It can be seen from figures 3 and 4 that the centre region (from $z/B=0$ to 0.4) of all specimens has higher constraint, and the surface region (near $z/B=0.5$) has lower constraint. The normalized constraint parameter σ_{22}/σ_0 and T_z in the centre region have very small variation with increasing specimen thickness B. This implies that the parameters σ_{22}/σ_0 and T_z are insensitive to out-of-plane constraint and can not effectively characterize it.

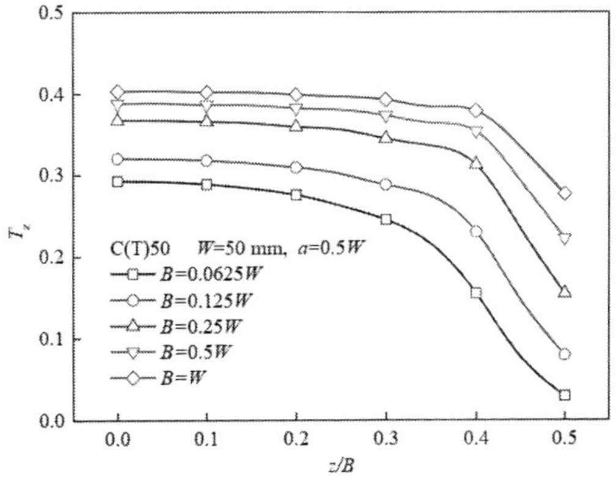

Figure 4. Distributions of constraint parameter T_z ahead of crack fronts

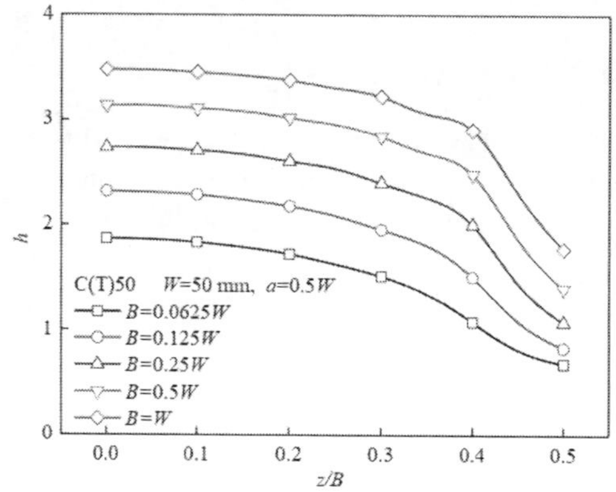

Figure 5. Distributions of constraint parameter h ahead of crack fronts

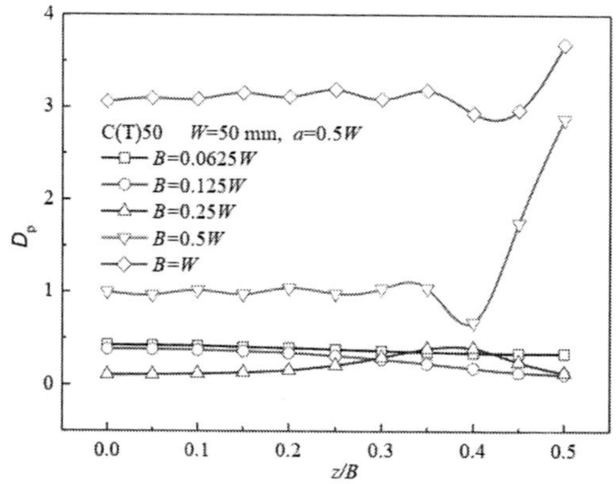

Figure 6. Distributions of constraint parameter D_p ahead of crack fronts

In figure 5, the parameters h increases with the increasing thickness B, which indicates that the parameters h is sensitive to out-of-plane constraint effect induced by thickness B. Figure 6 shows that the parameter D_p significantly increases with the increasing thickness B (the increase of D_p implies the decrease of constraint). The greater values of D_p mean that the thicker specimens surface regions (near z/B=0.5) have lower constraints and lead to more deformations. This implies that the parameter D_p may more efficiently capture the out-of-plane constraint effect than the three parameters σ_{22}/σ_0, T_z and h.

4 Conclusion

Three-dimensional finite element analyses have been conducted for five C(T)50 specimens with different out-of-plane constraint. The distributions of four constraint parameters (σ_{22}/σ_0, T_z, h and D_p) along crack fronts were calculated. Results show that, compared with the three constraint parameters (σ_{22}/σ_0, T_z and h) based on crack-tip stress fields, the parameter D_p based on crack-tip equivalent plastic strain is sensitive to out-of-plane constraint, and may effectively characterize the out-of-plane constraint.

Acknowledgements

This work is financially supported by the National Natural Science Foundation of China (Grant no. is 11502195).

References

1. S. G. Larsson, A. J. Carlsson, J. Mech. Phys. Solids., **21**, 263 (1973)
2. N. P. O'Dowd, C. F. Shih, J. Mech. Phys. Solids., **40**, 939 (1992)
3. Y. J. Chao, S. Yang, M. A. Sutton, J. Mech. Phys. Solids., **42**, 629 (1994)
4. W. L. Guo, Engng. Fract. Mech., **46**, 93 (1993)
5. W. L. Guo, Engng. Fract. Mech., **46**, 105 (1993)
6. H. Clausmeyer, K. Kussmaul, E. Roos, Appl. Mech. Rev., **44**, 77 (1991)
7. J. Yang, G. Z. Wang, F. Z. Xuan, S. T. Tu, Eng. Fract. Mech., **115**, 296 (2014)
8. J. Yang, G. Z. Wang, F. Z. Xuan, S. T. Tu, Fatigue Fract. Eng. Mater. Struct., **37**, 132 (2014)
9. Y. J. Kim, J. S. Kim, S. M. Cho, Y. J. Kim, Eng. Fract. Mech., **71**, 1203 (2004)
10. L. Y. Zhao, H. Xue, F. Q. Yang, Y. H. Suo, Frat. Integrita Strutr., **29**, 410 (2014)
11. J. Hebel, J. Hohe, V. Friedmann, D. Siegele, Int. J. Fract., **146**, 173 (2007)
12. ABAQUS Inc, ABAQUS/Standard User's Manual Version 6.14 (2015)

The Vortex Aircooler System for Cutting Materials in Aerospace Industry

G.O. Belov[1], S.S Dostovalova[2], E. O. Barbonov[2], D.A. Uglanov[2,a], P.A. Chertykovtsev[2] and R.A. Panshin[2]

1 – JSC «Salut», 20 km, Moskovskoeshosse, Samara, 443028, Russian Federation
2 – Samara National Research University34, Moskovskoeshosse, Samara, 443086, Russian Federation

Abstract. In aerospace industry there is a situation, when it is not possible to use liquid cooling of cutting zone, because we have special material or special construction. How often in that case used air fluent, but it has some problems with life cycle tools, quality and cutting time. For solving these problems was use the vortex effect, it was open less than one hundred years ago. The unit for making vortex effect is could vortex tube.

It was done the design the vortex tube of air cooler system for cutting using some researching [1-5]. Some experimental dates gave verification. The construction was made by metal materials and now is using on JSC «Salut», Samara..

1 INTRODUCTION

Modern machining takes place in the presence of a standard lubricating coolant (coolant), which is fed directly into the cutting zone through the channels in the tool or external tubes. However, a large group of materials due to low corrosion resistance has to be treated with the use of special coolant, but it should be noted that in real mechanical production it is not always possible to stop the machine to replace the coolant. Taking into account the fact that all aerospace production is small-scale, frequent replacement of coolant is an economically inappropriate action. There is also a whole group of structural materials, the mechanical processing of which is carried out at all without the use of a lubricating cooling liquid. There are cases when the use of coolant must be abandoned not because of the properties of the material of the part, but because of the design itself. For example, drilling large-sized sheet metal parts is carried out without the use of coolant, because in view of the large dimensions of the component, the liquid will not circulate in the machine as it should, but it will drain over the sheet metal into the non-working space, which is unacceptable. It is also worth noting that not all machines are equipped with a coolant supply system.

Traditionally, in these cases a jet of compressed air is used for cooling, the efficiency of which is several times smaller, which negatively affects the tool's durability, the quality of the surface being treated and the laboriousness of the machining. In order to partially remove these negative phenomena, it is proposed to use the vortex effect, which was discovered less than a hundred years ago. This phenomenon, realized in a special device called a vortex tube, allows the compressed gas flow to be

[a] Corresponding author: dmitry.uglanov@mail.ru

divided into hot and cold [6]. We will feed the cold stream into the cutting zone, and we will transfer the hot stream to the periphery. Use in the design of the vortex tube is due to its reliability and simplicity of the device.

2 CALCULATION OF THE MAIN PARAMETERS OF THE DEVICE

The device for the countercurrent vortex tube is shown in Fig. 1. It shows the following elements - a tube 2 with a tangential supply of compressed air through the cochlea 1; one end of the tube is closed by a lid with a diaphragm 4 - a cold flow through it, and through the other end of the tube with the throttle 3 a hot stream emerges.

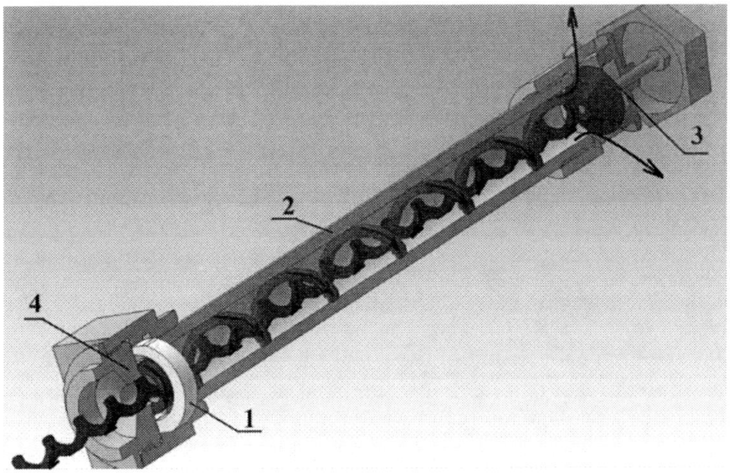

Figure 1. Arrangement of a countercurrent vortex tube.

Despite the relative simplicity of the tube device, the vortex effect is not fully understood. At the moment, the most accurate and logical description of the vortex effect for the countercurrent tube is provided by the "vortex interaction hypothesis". There are two vortices in the tube. The first "free", which displaces air through the throttle, the second "forced", acting in the axial direction and forcing air through the diaphragm. As a result of the intense turbulent interaction of these vortices, a cooled gas exits through the diaphragm, and a hot gas through the choke. [1]

There are a large number of papers, including [1-5,7], in which differential equations describing the processes of turbulent energy-mass exchange are presented. The numerical solution of these equations explains the processes qualitatively but does not show them quantitatively. It is also worth noting that the results of the turbulent energy-mass exchange simulation largely depend not only on the boundary conditions, but also on the methods for their solution. Because of the lack of a common opinion, not only the initial system of equations, but also the methods of solutions, we will have to abandon numerical calculations and turn to empirical calculation methods. Here it is worth noting that different researchers of the vortex effect proposed their own techniques, which, unfortunately, with the same initial data give different values of the geometric parameters of the device.

At the moment there is a large number of not only theoretical, but also experimental studies of the vortex effect. Using the volume of experimental data representing the influence of various factors on ΔT_{x} (the difference in temperatures at the entrance to the tube and the temperature of the cold flow), the following experimental dependences of ΔT_{x} on [2] were found:

- the area of the inlet nozzle F_n;
- vortex tube diameter D_T;
- the relative diameter of the diaphragm $\overline{D_{\text{D}}} = \frac{D_{\text{D}}}{D_T}$;
- air pressure in front of the vortex tube p_{in};

- the ratio of the flow of the cold flow to the flow at the inlet to the tube μ;
- tube length in calibers$\overline{L} = \frac{L}{D_T}$.

Having considered these characteristics, we can conclude that ΔT_x is a function of 6 variables:

$$\Delta T_x = f\ (\overline{L}; \mu; p_{in}; D_T;\ \overline{D_D};\ F_n) \tag{1}$$

Since the considered construction was tested only with a spiral inlet, and different nozzle areas were investigated on a tube of the same diameter, it is advisable to introduce instead of Fn the relative equivalent diameter of the nozzle inlet of the vortex tube:

$$\overline{d_{eq}} = \frac{\sqrt{\frac{4 \cdot F_n}{\pi}}}{D_T} \tag{2}$$

Taking this into account, we rewrite the function ΔT_x:

$$\Delta T_x = f\ \left(\overline{L};\ \mu;\ p_{in}; D_T;\ \overline{D_D};\ \overline{d_{eq}}\right) \tag{3}$$

Since the graphical dependencies considered are continuous functions (on the ranges studied), we expand the above function in a Taylor series. The Taylor series for a function of six variables takes the form:

$$f(x_1; \dots x_6) = f(x_{01}; \dots x_{06}) + \lim_{n\to\infty} \sum_{n=1}^{n} \frac{1}{n!} * [d^n(f)]_0, \tag{4}$$

where$[d^n(f)]_0$ –the total differential of the n-th order in a neighborhood $(x_{01}; \dots x_{06})$.

We draw attention to the fact that from the graphical dependencies given in [2], we can define only partial derivatives of the nth order with respect to one variable. A complete differential of the first order contains only partial derivatives of the first order and can be easily determined, while the differentials, beginning with the second order, contain derivatives of several variables. By the property of the Taylor series, each subsequent order of the differential is less "weight" than the previous one, therefore, for approximate calculations, it is often limited to the first order. The last remark enables us to conclude that the effect of individual variables on a function is significantly higher than their mutual influence. Neglecting, therefore, partial derivatives with respect to several variables, we can rewrite (4) as:

$$f(x_1; \dots x_6) \approx f(x_{01}; \dots x_{06}) + \sum_{j=1}^{6} \lim_{n\to\infty} \sum_{n=1}^{n} \frac{(x_j - x_{0j})^n}{n!} * \frac{\partial^n f}{\partial x_j^n}(x_{01}; \dots x_{06}) \tag{5}$$

where$x_1 \dots x_6$ – variables, $x_{01} \dots x_{06}$ – some meaning of these variables. For the expansion (3) in a series, according to the formula given for six variables, we choose well-traced values of the variables on the considered graphical dependencies (common point):

$$\overline{L}_0 = 15; p_{in\_0} = 0{,}5\ \text{МПа}; D_{T\_0} = 28\ \text{мм}; \overline{D_{D\_0}} = 0{,}5; \mu_0 = 0{,}5; \overline{d_{eq\_0}} = 0{,}25 \tag{6}$$

Function Value ΔT_xat these values is 38K.

Considering the graphical dependences$\Delta T_x = f\ (\overline{L})$; $\Delta T_x = f\ (p_{in})$; $\Delta T_x = f\ (D_T)$it can be noted that the approximating curves are linear in the range under consideration, and therefore all partial derivatives of these functions starting with the second order can be equated to zero. The functions $\Delta T_x = f\ \left(\overline{D_D}\right)$,$\Delta T_x = f\ (\mu)$are reliably approximated by a second-order polynomial function, so the derivatives of these functions starting with n=3 equate to zero. The graphical dependence $\Delta T_x = f\ \left(\overline{d_{eq}}\right)$ is more complex, for its reliable approximation in the range of values used a third-order polynomial function is used, thus the fourth and subsequent orders of this function are equated to zero. In view of the foregoing, expanding (3) in a series (5) we obtain:

$$\Delta \mathrm{Tx} \approx a + b_1(\overline{L} - c_1) + b_2(p_{in} - c_2) + b_3(D_{\mathrm{T}} - c_3) + b_4(\overline{D_{\mathrm{D}}} - c_4) + b_5(\overline{D_{\mathrm{D}}} - c_4)^2 + b_6(\mu - c_5) + b_7(\mu - c_5)^2 + b_8(\overline{d_{\mathrm{eq}}} - c_6) + b_9(\overline{d_{\mathrm{eq}}} - c_6)^2 + b_{10}(\overline{d_{\mathrm{eq}}} - c_6)^3, \quad (7)$$

where$a = \Delta T_{x_0} = 38K$;
$b_1 \ldots b_{10}$ – values of advisory partial derivatives taking factorial into account;
$c_1 \ldots c_6$ – the corresponding values of the variables (6).

Values $b_1 \ldots b_{10}$ were obtained from the experimental data [2] by means of a PC and are summarized in Table 1.

Table 1. Values of the coefficients a, $b_1 \ldots b_{10}$, $c_1 \ldots c_6$.

b_1		b_2	b_3	b_4	b_5	b_6	b_7	b_8	b_9	b_{10}
0,1		5,7	0,52	4,81	-239,8	4,17	-41,7	281,5	4915	-313314

	a	c_1	c_2	c_3	c_4	c_5	c_6
	38	15	5	28	0,5	0,5	0,25

Like any other dependence, the obtained "cooling" function of air$\Delta T_x = f\ (\overline{L};\ \mu;\ p_{in};\ D_T;\ \overline{D_{\mathrm{D}}};\ \overline{d_{\mathrm{eq}}})$can be reliably used in a certain range of values. The range of values of the function variables is given in Table 2.

Table 2. Range of application of the formula (7) for calculationΔT_x.

$\overline{L}$	p_{in}, МПа	D_T, мм	$\overline{D_D}$	μ	$\overline{d_{eq}}$
10...25	0,3...0,6	10...30	0,413...0,715	0,3...0,8	0,228...0,28

The dependence (7) showed the greatest convergence with the empirical technique given in [1]. By this method, the basic geometric parameters of the vortex tube for cooling the object with heat release up to 1 kW:

- diameter of the vortex tube 17.9 mm;
- the height and width of the nozzle should be 3.5 mm and 7 mm;
- diameter of the aperture 9.6 mm;
- length of the vortex tube is not less than 161,1мм

3 DEVELOPMENTS OF CONSTRUCTION AND INTRODUCTION TO PRODUCTION

Based on the obtained parameters, the design of the product was developed in the form of a 3D model, shown in Fig. 2. In this figure, we can see a vortex tube 1 with a tangential gas supply through a collet with a throttle 3, adjusting the temperature and flow of cold and hot flow with Using the handle 2 and the poly-hinge tube 4 to feed the jet into the cutting zone.

The construction comprises the following features:

1. Reliability due to the absence of moving elements during operation;
2. Ease of connection to a pneumatic system through standard factory fitting compound;
3. Convenient supply of cold air to the cutting area through a poly-hose tube;
4. Wideadjustmentrange.

According to the developed 3D models of individual parts and assembly of the vortex tube, drawings were made. In the metal, this design was implemented in the tool shop of JSC «Salut», since the choke 3 (Figure 2) was equipped with a fitting clamp of a standard pneumatic tube, the connection to the factory compressor station did not cause any difficulties. The temperature and the ratio of the hot and cold flow were adjusted freely with the help of the handle 2. Moreover, with a considerable opening of the throttle, the tube operated as a vacuum pump, that is, air was sucked through the polycharne tube 4. When the choke 2 was compressed, a flow separation was observed, and through the poly- "Cooled" air, and through the throttle 2 heated.

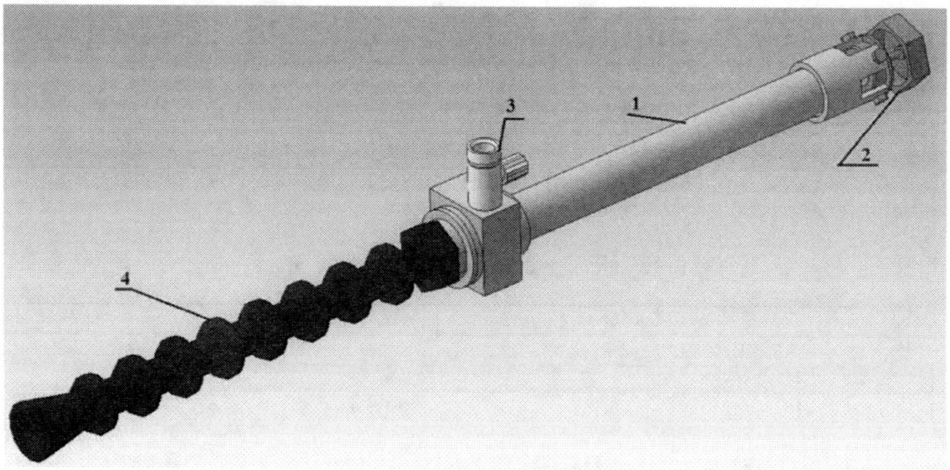

Figure 2. Designed construction.

The air temperature and pressure in the compressor station were measured in automatic mode and displayed. During the experiment, the pressure was 0.67 MPa, the temperature was 98°C (T_{in}). Thus, the product was tested in hot enough air, which contained moisture. Temperature measurements were made in the range of μ from 0.4 to 0.8 using a chromel-alumel thermocouple (TXA). The experimental dependence of the cooling effect on the flow coefficient is shown in Fig. 3.

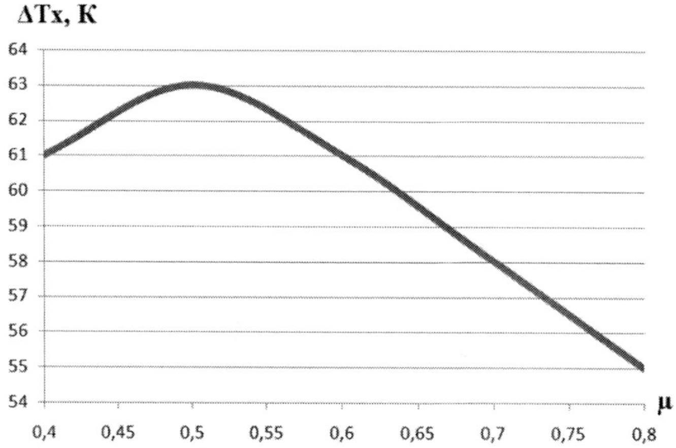

Figure 3. Dependence of the cooling effect on the flow coefficient.

The introduction into production was carried out on the turning of a very whimsical to the thermal processes of the structural material - aluminum alloy AL4. When machined, this material tends to stick to the chip. This process limits the cutting modes. After installing the vortex tube on the machine, the machine operator was able to increase the feed of the cutter from 0.6 mm/rev to 1 mm/rev for the same depth of cut and spindle speed. The morphology of the chips did not change at the same time. Thus, the introduction of a vortex cooler made it possible to reduce the labor intensity of the operation by a factor of 1.7.

4 CONCLUSION

A formula for determining the effect of air cooling by a vortex tube is proposed and justified, depending on the geometric parameters of the device and the parameters of the input flow. With its help, the design of an air cooling system for the cutting zone for aerospace materials has been developed and implemented. This device was introduced into production at JSC "Salut.

REFERENCES

1. Merkulov A.P. *Vortex effect and its application in engineering Izd*. 2nd revision. and additional. Samara, Optima. (1997)
2. Martynov A.V. Borodyansky V.M. *What is a vortex tube* (Energy,1976)
3. Tarnapolsky A.V. *Vortex heat and power devices*. Penza: Publishing house of the state. University, (2007)
4. PiralashviliSh.A., Polyaev VM and others. *The vortex effect. Experiment, theory, technical solutions*(UNPTs "Energomash", 2000)
5. Suslov A.D., Ivanov S.V. and others. *Vortex devices*.(Mechanical Engineering, 1985)
6. GOST 22616-77. *Pipes are swirling. Terms and Definitions.*
7. Barbonov, EO, Biryuk, VV, Gaynullin, MN, Sotova, VA, Chertykovtsev, *PA Hybrid cooling system for on-board IR receivers based on vortex and thermoelectric effects*(Bulletin of the International Academy of Refrigeration№ 2, 2017)

Parameter Research of UAV Vertical Rope-type Recovery System

Chengjing Ma[1], Xing'an Liu[2,a] and Caiwen Zhang[3]

[1] *College of Mechanical and Electrical Engineering, Nanjing University of Aeronautics and Astronautics, Nanjing 210016, China*
[2] *School of Aeronautic Science and Engineering, Beihang University, Beijing 100191, China*
[3] *Research Institute of Unmanned Aerial Vehicle, Nanjing University of Aeronautics and Astronautics, Nanjing 210016, China*

Abstract. Vertical rope-type recovery system is a new accurate recovery method for small fixed-wing unmanned aerial vehicles (UAV). This paper models the recovery system and finishes the simulation of recovery process. The recovery stand model of the recovery system is built by analyzing the physical essence, and the aircraft model is based on dynamics and kinematics equations as well as the mechanical characteristics of arresting rope. Finally the simulation analysis of the recovery process is completed. On the basis of the modeling simulation, the system parameters' effect on recovery performance such as maximum resistance force, maximum overload and maximum radius has been discussed by varying each key parameter. Eventually, the pattern of how initial conditions when hitting the rope as well as the arresting rope's stiffness and damping coefficient influence recovery performance has been researched, and the result could provide theoretical reference for the design of UAV using vertical rope-type recovery and the system in the future.

1 Introduction

The development of UAV technology has been mature so far, and it could undertake a large range of tasks [1]. Because of its convenient operation, flexible deployment and excellent performance-cost ratio, the deployment area has been extended to ships, islands and other locations [2]. Vertical rope-type recovery is a way to achieve accurate recovery in these limited areas. It's especially favourable for small fixed-wing UAV, due to the large recovery window during the recycle process, low demand for control accuracy in the height direction, and low cost in terms of mass and volume [3]. The US military has successfully developed UAV vertical rope-type recovery system and deployed it, for instance, the Scan Eagle UAV and RQ-21 UAV, built by INSITU corporation, have used sky-hook recovery system [4,5].

So far, many research institutes and schools have explored this technology. For example, Dai Chuan et al. of Chengdu Aircraft Design Institute proposed a vertical rope-type recovery device of UAV [6]; Lu Wei from Northwestern Polytechnical University simulated the UAV dynamic characteristics, buffer and energy-absorbing capability during the rope-hook recovery process, and researched the feasibility of rope-hook recovery mode [7]. However, the recovery stand model in

[a] Xing'an Liu : liuxingan94@163.com

current research differs from the actual recycling devices in use, and the influence pattern of the system parameters on the recovery performance is not studied in depth.

This paper improves the recovery system model based on the actual physical characteristics of the vertical rope-type recovery system, and uses it for simulation. On the basis of it, the influence of initial conditions when hitting the rope and the arresting rope's stiffness and damping coefficient on recovery performance has been researched, and it could provide reference for design of UAV and recovery system.

2 Recovery system modelling

Analysis of the recycling device shows that the vertical rope-type recovery system is mainly composed of a ground recovery stand, arresting rope device and the intercepting hook on wing tip [8]. The recovery system can be simplified as **Figure 1**. In the figure, the fixed point of the recovery system on the ground is point Q, QH is the vertical bar, QB and AD are cross bars, composing the main body of the recovery stand. AD could rotate around QH, and is constrained by two damping rope EF and FG. The arresting rope is represented by PAB in the figure, the damper which is simplified as the stiffness and damping coefficient of arresting rope in subsequent research. As for the aircraft, the center of gravity is point K, the point of wing root is point I, the point of wing tip is point J, and the point which the leading edge of UAV wing contacts the arresting rope is point C.

Figure 1. Schematic diagram of vertical rope-type recovery system

2.1 Recovery stand modelling

The recovery stand will deform due to the force transferred by the arresting rope during the recovery, and the process of its deformation and recovery plays a role of buffer and energy absorption. Therefore, it can't be considered as rigid body during simulation. According to the actual dynamic and kinetic characteristics of the recovery stand, it can be simplified to two parts which are connected together. Among them, bracket QH is fixed on the ground, therefore only the bending and torsion deformation is considered. Strut AD is hinged on bracket QH, and could only rotate around the QH axis. Hence, during the dynamic analysis, the recovery stand is assumed balanced by overall external forces, and only the change of attitude angle caused by the external moment is considered, so the dynamic equation is:

$$\boldsymbol{M}_S = \boldsymbol{I}_S \boldsymbol{\omega}_S + \dot{\boldsymbol{\omega}}_S \times \left(\boldsymbol{I}_S \boldsymbol{\omega}_S \right) \tag{1}$$

In the formula, $\boldsymbol{\omega}_S$ is the angular velocity of the recovery stand in the local coordinate system, $\boldsymbol{I}_S$ is the inertia tensor of the recovery stand.

For strut AD, there are 3 external forces on it, the reaction force F_{CA} of the arresting rope, the resistance force F_{FG} and F_{EG} caused by damping rope, as shown in **Figure 2**. Its resultant force equation is not considered as the assumption that the force is balanced, so these 3 forces only cause it rotating around axis QH.

The moment for the whole recovery stand caused by F_{CA} can be calculated as:

$$\boldsymbol{M}_{CA} = \boldsymbol{r}_{CA} \times \boldsymbol{F}_{CA} \tag{2}$$

The component of moment $\boldsymbol{M}_{CA}$ in plane ADF causes strut AD rotating around axis QH, the other components transmit to strut QH through the hinge, and are finally supported by QH.

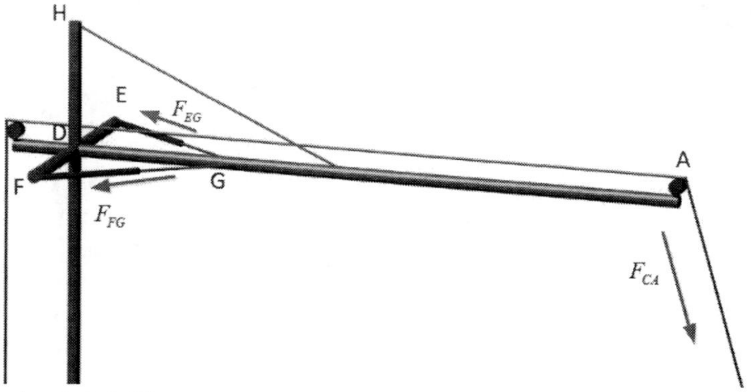

Figure 2. Force analysis of strut *AD*

The resistance force F_{FG} and F_{EG} caused by damping rope can be calculated as:

$$F = k\Delta l + c\Delta \dot{l} \tag{3}$$

In the formula, k and c denote stiffness and damping coefficient of the damping rope respectively. The change of length Δl and the rate of the change $\Delta \dot{l}$ are related to the position change of point E, F and G, which can be obtained by the coordinate of these points. The resultant moment caused by the resistance force can be calculated as:

$$\boldsymbol{M}_{EF} = \boldsymbol{r}_{EG} \times \boldsymbol{F}_{EG} + \boldsymbol{r}_{FG} \times \boldsymbol{F}_{FG} \tag{4}$$

As for strut QH, the external moments it takes are the resulting moment caused by the counter force of $\boldsymbol{F}_{FG}$ and $\boldsymbol{F}_{EG}$, and the moment generated by the force transmitted by the hinge point connected with strut AD. Its own material stiffness and damping are simplified into elastic stiffness and damping coefficient in the direction of three altitude angle, which can provide recovery moment and damping moment.

From the foregoing analysis, the position and rate of change of the position of the feature points on recovery stand need to be determined in the modeling process, which can be solved by determining the coordinates of these points in the same coordinate system. This paper uses the vertical ground coordinate system to determine the coordinates, and establishes local coordinate system to assist the determination. Set the origin of vertical ground coordinate system V as point O, and establish local coordinate system respectively with point Q and point D as the origin. So r_{OA} and $\dot{r}_{OA}$ can be expressed as follow [9]:

$$\boldsymbol{r}_{OA} = \boldsymbol{r}_{OD}^{V} + \boldsymbol{T}_{VD} \cdot \boldsymbol{r}_{DA}^{D} = \boldsymbol{r}_{OQ}^{V} + \boldsymbol{T}_{VQ} \cdot \boldsymbol{r}_{QD}^{Q} + \boldsymbol{T}_{VD} \cdot \boldsymbol{r}_{DA}^{D} \tag{5}$$

$$\begin{aligned}\dot{\boldsymbol{r}}_{OA} &= \dot{\boldsymbol{r}}_{OQ}^{V} + \dot{\boldsymbol{T}}_{VQ} \cdot \boldsymbol{r}_{QD}^{Q} + \boldsymbol{T}_{VQ} \cdot \dot{\boldsymbol{r}}_{QD}^{Q} + \dot{\boldsymbol{T}}_{VD} \cdot \boldsymbol{r}_{DA}^{D} + \boldsymbol{T}_{VD} \cdot \dot{\boldsymbol{r}}_{DA}^{D} = \boldsymbol{\omega}_{Q}^{V} \times \boldsymbol{T}_{VQ} \cdot \boldsymbol{r}_{QD}^{Q} + \boldsymbol{\omega}_{D}^{V} \times \boldsymbol{T}_{VD} \cdot \boldsymbol{r}_{DA}^{D} \\ &= \boldsymbol{\omega}_{Q}^{V} \cdot \boldsymbol{r}_{QD}^{V} + \boldsymbol{\omega}_{D}^{V} \cdot \boldsymbol{r}_{DA}^{V}\end{aligned} \tag{6}$$

In the formula, $\boldsymbol{r}_{OD}^{V}$ means the vector $\boldsymbol{r}_{OD}$ expressed in vertical ground coordinate system V, $\boldsymbol{r}_{DA}^{D}$ means the vector $\boldsymbol{r}_{DA}$ expressed in the local coordinate system with point D as the origin, $\boldsymbol{T}_{VD}$ represents the transformation matrix from local coordinate system D to vertical ground coordinate system V. Other similar expressions in this paper have the same meaning, and if no special reference is made, the default coordinate system is vertical ground coordinate system V. r_{OG}, $\dot{r}_{OG}$, r_{OE}, $\dot{r}_{OE}$, r_{OF} and $\dot{r}_{OF}$ can also be obtained using similar method in appropriate coordinate system. Then we can use the results to get relevant forces and moments, and the recovery stand can be modeled based on the dynamics analysis above.

2.2 Aircraft modelling

The aircraft six-degree-of-freedom dynamic equations are as follow [10]:

$$\begin{cases} X - mg\sin\theta = m(\dot{u} + qw - rv) \\ Y + mg\cos\theta\sin\phi = m(\dot{v} + ru - pw) \\ Z + mg\cos\theta\cos\phi = m(\dot{w} + pv - qu) \end{cases} \begin{cases} L = I_x\dot{p} - I_{zx}(\dot{r} + pq) - (I_y - I_z)qr \\ M = I_y\dot{q} - I_{zx}(r^2 - p^2) - (I_z - I_x)rp \\ N = I_z\dot{r} - I_{zx}(\dot{p} - qr) - (I_x - I_y)pq \end{cases} \quad (7)$$

Compared with conventional aircrafts, additional resistance force and resulting moment are added due to the arresting rope during the vertical rope-type recovery. Therefore, the modelling of the aircraft needs to increase the force $\boldsymbol{F}_l$ and moment $\boldsymbol{M}_l$ caused by the arresting rope on the basis of the dynamic equations of the aircraft.

Each force caused by the arresting rope PAC and PBC is along its own deformation direction, and the magnitude is as follows:

$$F_{AC} = -\left(k_l \Delta l_{PAC} + c_l \Delta \dot{l}_{PAC}\right), \quad F_{BC} = -\left(k_l \Delta l_{PBC} + c_l \Delta \dot{l}_{PBC}\right) \quad (8)$$

The resistance force caused by the arresting rope is the resultant force of $\boldsymbol{F}_{AC}$ and $\boldsymbol{F}_{BC}$:

$$\boldsymbol{F}_l = \boldsymbol{F}_{AC} + \boldsymbol{F}_{BC} \quad (9)$$

In order to calculate the force, the recovery process is divided into two phases: in the first phase, the arresting rope hit the wing leading edge, and slide along the leading edge until the intercepting hook at wing tip; in the second phase, the arresting rope resistance effects on wing tip. Therefore, in view of the different physical characteristics of the two stages, the resistance of the arresting rope at each stage should be considered separately.

In the first stage, it can be assumed that the leading edge of the wing is smooth and there is no friction on the arresting rope, thus the rope is balanced itself and the constraint condition is:

$$F_{AC} = F_{BC} \quad (10)$$

Besides, we can consider the arresting rope as entirety because the rope can slide around the leading edge, which means:

$$\dot{r}_{AC} + \dot{r}_{BC} = \Delta\dot{l}_{PAC} + \Delta\dot{l}_{PBC} \quad (11)$$

According to the geometric relationship, we got:

$$\Delta l_{PAC} = \left|\boldsymbol{r}_{AC}\right| - h_A, \quad \Delta l_{PBC} = \left|\boldsymbol{r}_{BC}\right| - h_B \quad (12)$$

h_A and h_B respectively means the height between the initial hit-rope position and point A or point B. Thus, in order to solve the constraint equations (10) and (11), we need to get r_{AC}, r_{BC}, $\dot{r}_{AC}$ and $\dot{r}_{BC}$, which can be determined by the coordinates of end point:

$$\boldsymbol{r}_{AC} = \boldsymbol{r}_{OC} - \boldsymbol{r}_{OA} \quad (13)$$

$$\dot{r}_{AC} = \dot{\boldsymbol{r}}_{AC} \cdot \frac{\boldsymbol{r}_{AC}}{\left|\boldsymbol{r}_{AC}\right|} = \left(\dot{\boldsymbol{r}}_{OC} - \dot{\boldsymbol{r}}_{OA}\right) \cdot \frac{\boldsymbol{r}_{AC}}{\left|\boldsymbol{r}_{AC}\right|} \quad (14)$$

Because of the initial pre-tightening force, the position of the contact point C should satisfy the condition that it's on line IJ, the leading edge of the wing, and it minimizes the change of the length of the arresting rope (the distance to AB is the shortest) .So point C is the intersection of the common perpendicular of line IJ and AB and line IJ. Solve the equations of common perpendicular and line IJ, we can get point C's coordinate $r_{OC} = \left(x_C, y_C, z_C\right)$.

In order to get $\dot{r}_{OC}$, Body Axis system with the centre of gravity point K as origin is established, and the coordinate of point C in Body Axis system can be expressed as: $\boldsymbol{r}_{KC}^{K}=\left[l_{IK}-l_{C}\tan\Lambda,-(b/2-l_{C}),0\right]^{T}$, in which l_{IK} is the distance between wing root point I and point K in the direction of body, l_C is the distance between point C and wing tip in the direction of wingspan, b is length of the wingspan, and Λ is the sweep angle of wing leading edge.

Point C's coordinate in vertical ground ordinated system can be expressed as $\boldsymbol{r}_{OC}^{V}=\boldsymbol{r}_{OK}^{V}+\boldsymbol{T}_{VK}\cdot\boldsymbol{r}_{KC}^{K}$, and take the derivative with respect to time, we can get:

$$\dot{\boldsymbol{r}}_{OC}^{V}=\dot{\boldsymbol{r}}_{OK}^{V}+\dot{\boldsymbol{T}}_{VK}\cdot\boldsymbol{r}_{KC}^{K}+\boldsymbol{T}_{VK}\cdot\dot{\boldsymbol{r}}_{KC}^{K}=\dot{\boldsymbol{r}}_{OK}^{V}+\boldsymbol{\omega}_{K}^{V}\times\boldsymbol{T}_{VK}\cdot\boldsymbol{r}_{KC}^{K}+\boldsymbol{T}_{VK}\cdot\dot{\boldsymbol{r}}_{KC}^{K}=\dot{\boldsymbol{r}}_{OK}^{V}+\boldsymbol{\omega}_{K}^{V}\times\boldsymbol{r}_{KC}^{V}+\boldsymbol{T}_{VK}\cdot\dot{\boldsymbol{r}}_{KC}^{K} \quad (15)$$

$\boldsymbol{\omega}_{K}^{V}=\left[p,q,r\right]^{T}$ means the angular velocity vector of UAV in vertical ground coordinate system. From the former modelling, r_{OA}, r_{OB} and $\dot{r}_{OA}$, $\dot{r}_{OB}$ have been obtained, and therefore we can get $\boldsymbol{F}_{AC}$ and $\boldsymbol{F}_{BC}$, and finally we can get $\boldsymbol{F}_l$ and its resultant moment $\boldsymbol{M}_{l}=\boldsymbol{r}_{KC}\times\boldsymbol{F}_{l}$.

In the second stage, when the arresting rope gets locked by the intercepting hook, it couldn't slide in the direction of rope due to the friction between the hook and itself. Therefore the magnitude of two arresting rope's resistance force differs from each other and should be considered respectively. At this moment, point C coincides with point J, and the constraint condition becomes as follow:

$$\dot{r}_{AJ}=\Delta\dot{l}_{PAC},\quad \dot{r}_{BJ}=\Delta\dot{l}_{PBC} \quad (16)$$

Similar to the first stage, we should determine the coordinate of end points. The wing tip point J's coordinate in Body Axis system can be expressed as $\boldsymbol{r}_{KJ}^{K}=\left[l_{IK}-\frac{b}{2}\tan\Lambda,-\frac{b}{2},0\right]^{T}$ .Point J's coordinate in vertical ground ordinated system can be expressed as $\boldsymbol{r}_{OJ}^{V}=\boldsymbol{r}_{OK}^{V}+\boldsymbol{T}_{VK}\bullet\boldsymbol{r}_{KJ}^{K}$, and take the derivative with respect to time, we can get:

$$\dot{\boldsymbol{r}}_{OJ}^{V}=\dot{\boldsymbol{r}}_{OK}^{V}+\boldsymbol{\omega}_{K}^{V}\times\boldsymbol{r}_{KJ}^{V}\left(\dot{\boldsymbol{r}}_{KJ}^{B}=0\right) \quad (17)$$

In the same way, we can get $\boldsymbol{F}_l$, using r_{OA}, r_{OB} and $\dot{r}_{OA}$, $\dot{r}_{OB}$,and its resultant moment $\boldsymbol{M}_{l}=\boldsymbol{r}_{KC}\times\boldsymbol{F}_{l}$.

So far, the resistance force and moment in the recovery process can be determined. Combined with six-degree-of-freedom dynamic equations of conventional aircraft, the aircraft modelling is completed.

3 Simulation of recovery process

3.1 Initial conditions

Taking a small fixed-wing UAV as an example, we simulated the recovery process in the environment of Matlab/Simulink [11]. The UAV's parameters are as follow: the weight of UAV is m=60 kg, the length of wingspan is b=5 m, the leading edge sweep angle is Λ=11°. The initial kinematics conditions are as follow: the vertical distance between the rope-contacting point and the upper end of the recovery stand point A and the vertical distance to point B is h_A=h_B=8 m, the distance between the rope-contacting point and wing tip in the direction of wingspan is l_C=0.5 m. The starting time of the simulation is when UAV hits the rope.

3.2 Motion trajectory and change of velocity

Figure 3 shows the motion trajectory of the center of gravity of the aircraft in horizontal plane. It shows that when the arresting rope slides along the leading edge of the wing at first, the attitude of the aircraft is almost unchanged and the original speed direction is maintained. After that, the arresting rope hangs on the wingtip. When the arresting rope was stretched long enough to provide centripetal force (x_K=10 m in the figure), the UAV enters the circular motion under the centripetal force provided by the resistance of the arresting rope, and the motion radius reduces continuously due to the energy absorption effect of recovery system. After about 2 laps, the aircraft speed decreased to the point where the circular motion could not be maintained, and then swung slightly around the equilibrium position until the speed was reduced to 2.5 m/s below, which is thought the end of the recovery process.

The trajectory of the rope-contacting point in the horizontal plane is shown in **Figure 4**. It can be seen that when the arresting rope slides along the leading edge toward the wingtip, the motion trajectory of the rope-contacting point can be approximated as a straight line (as indicated by the dotted red line); then the trajectory of the rope-contacting point coincides with the wingtip trajectory, forming the fluctuation on the basis of the circular motion of aircraft as shown in the figure.

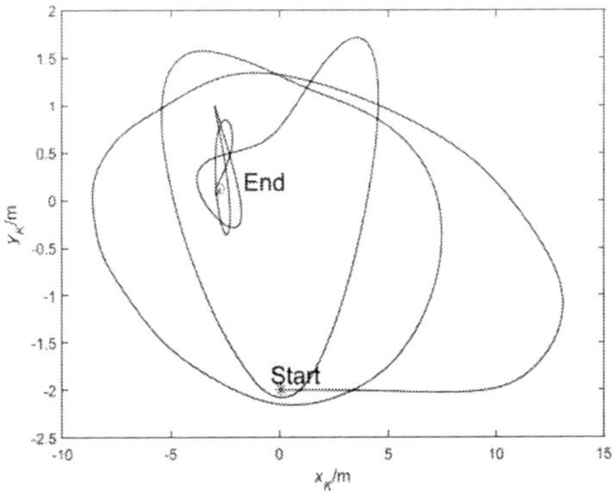

Figure 3. Trajectory of UAV's center of gravity in horizontal plane

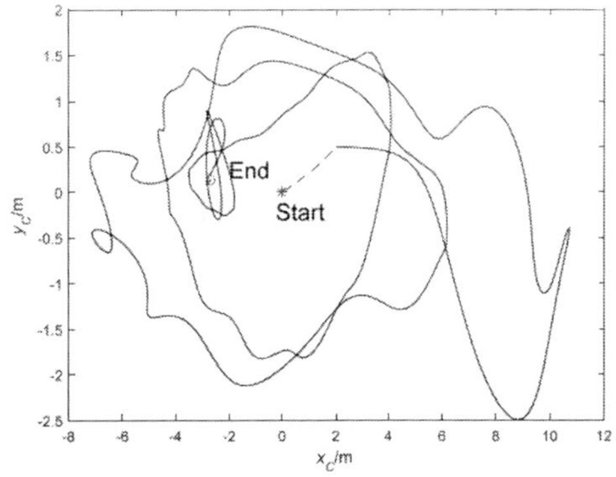

Figure 4. Trajectory of rope-contacting point in horizontal plane

The response curve of the UAV velocity V_a with time t in the recovery process is shown in **Figure 5**. It can be seen that the velocity of UAV dropped rapidly to less than 5 m/s after two fluctuations within 5 s, and then stabilized to below 2.5 m / s after about 10 s, when the UAV can be took down and recovered. Due to the energy absorption effect of the recovery stand, the peak of each fluctuation declines continuously.

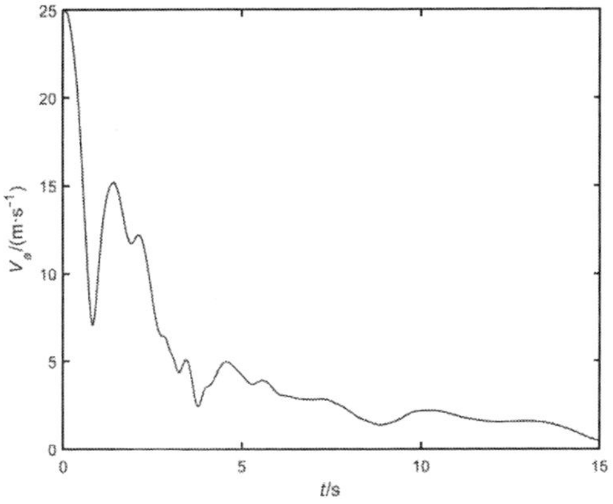

Figure 5. Curve of velocity of UAV during recovery

3.3 Characteristics of resistance force

When the parameters of arresting rope are set as k_l=500 N/m, c_l=40 N/(m • s^{-1}), the curve of magnitude of resistance force $\left[F_x, F_y, F_z\right]$ and the resultant force F_R is as **Figure 6**:

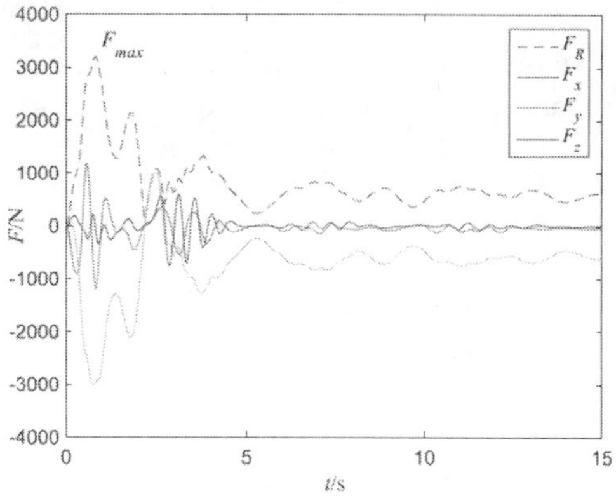

Figure 6. Curve of resistance force

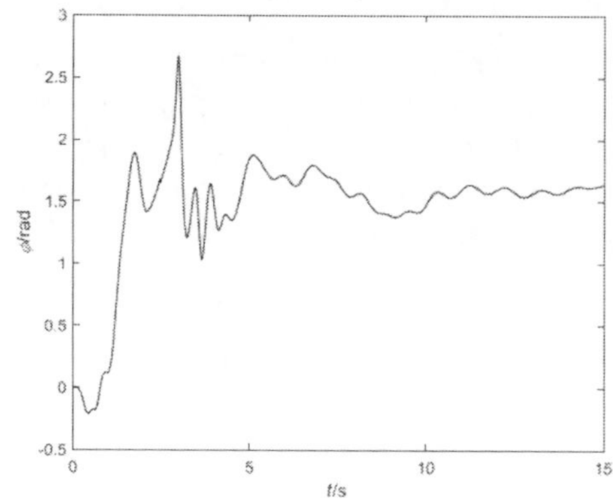

Figure 7. Curve of roll angle

It shows that the resistance force of arresting rope increases rapidly after the aircraft hit the rope and reaches a peak at about 0.8 s, and the overload is about 5.41. At this moment, the aircraft moves to the farthest end from the balance position and begins to enter the circular motion. As for the change tendency of the resistance force in all directions, it shows that the force in the direction of fuselage (x-axis of the Body Axis) increases rapidly at first; however, in order to balance the centrifugal force of the aircraft after entering the circular motion, the force in the direction of wingspan(y-axis of the Body Axis) rapidly increases and become the main part of the total resistance; at last, when the aircraft reaches equilibrium, the arresting rope resistance almost exists only in the direction of wingspan, and is approximately equal to gravity, which is because at the equilibrium stage, the fuselage has a roll angle of about 90° relative to the initial state (as shown in **Figure 7**), and the force of the arresting rope is in the direction of wingspan to balance gravity.

4 Research on system parameters' effect on recovery performance

4.1 Effect of initial conditions when hitting the rope

For a UAV with a definite appearance, the initial recovery velocity and the hit-rope position are uncertain factors and difficult to achieve accurate control. Therefore, the influence of the above factors on the recovery performance needs to be analysed.

During the recovery process, the initial recovery velocity of UAV determines the maximum kinetic energy, and thus affects the maximum resistance and maximum overload. But limited by maximum lift coefficient and the effect of the wind field, the recovery speed is usually higher than the stall speed. The influence of the recovery velocity V_0 on the maximum overload $n_{\max}$ is analysed under the setting of the system parameters of k_l=500 N/m, c_l=40 N/(m • s^{-1}), the simulation result is as **Figure 8**:

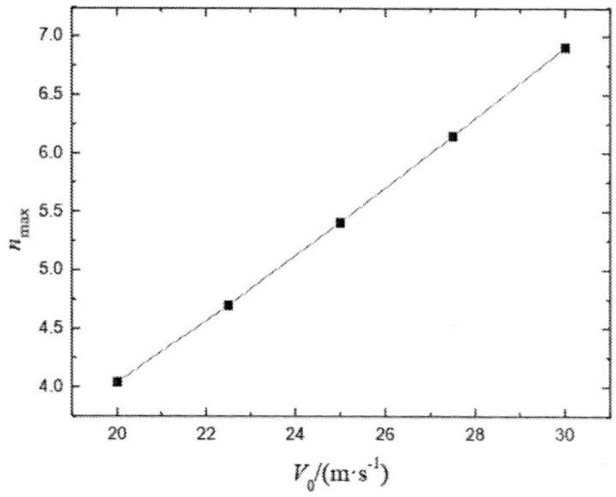

Figure 8. Influence of recovery velocity on maximum overload

The figure shows that the effect of recovery velocity on the maximum overload is almost linear. It can be obtained that the maximum recovery speed should not exceed 33 m/s through fitting, in order to ensure the maximum overload lower than 10.

The hit-rope position may vary in both the wingspan direction and the height direction. Therefore, how it impacts the recovery performance need to be analysed. Maintain the same parameters of arresting ropes, analysing the impact of the location of the arresting ropes on the maximum resistance F_{max} are shown in **Figure 9** and **Figure 10**:

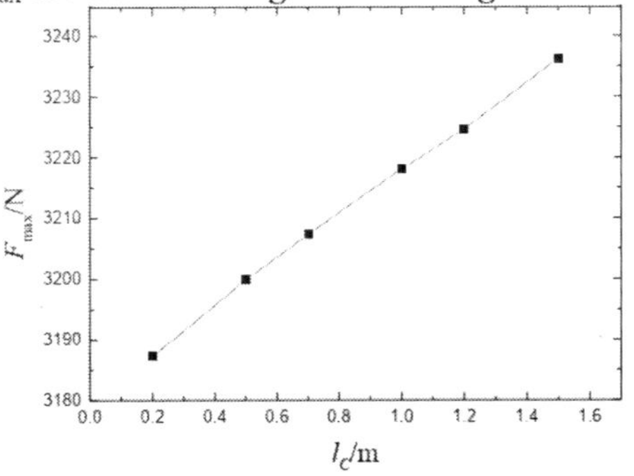

Figure 9. Influence of the hit-rope position along wingspan on maximum resistance force

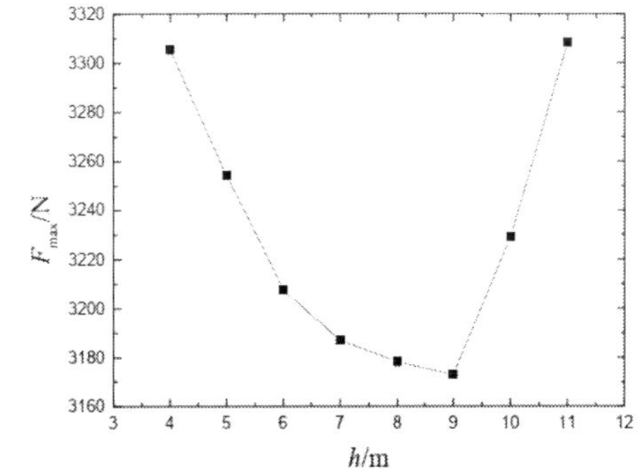

Figure 10. Influence of the height of the hit-rope position on maximum resistance force

It can be seen from the figure that the influence of the position change along the wingspan direction on the maximum resistance is monotonous. The closer the hit-rope position to wing tip is, the smaller the maximum resistance is. However, due to the short span of the wing, the range of maximum resistance is limited. Influence of the height change is non-monotonic. The resistance is the minimum when it's near the midpoint of the recovery stand. When the height of the hit-rope position is above the midpoint, the resistance is more sensitive to altitude changes.

4.2 Effect of the stiffness and damping coefficient of arresting rope

The recovery stand's stiffness and damping coefficient are related to its material, and are inconvenient to change, while the stiffness and damping coefficient of arresting rope PAB are related to damper settings. Therefore, this article mainly analyses how the stiffness and damping coefficient of arresting rope affects the maximum resistance. In order to avoid unreasonable set of parameters so that the range of motion of UAV during recovery is too large, the influence of the parameters on the maximum radius during recovery is also analysed.

Influence of the rope's stiffness k_l on maximum resistance force F_{max} is researched. And when c_l=40 N/(m • s^{-1}), influence of the rope's stiffness k_l on maximum overload and maximum radius is considered. The results are as **Figure 11** and **Figure 12**:

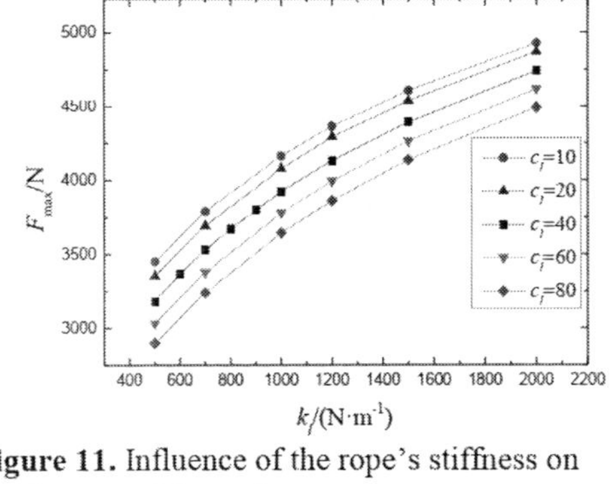

Figure 11. Influence of the rope's stiffness on maximum resistance force

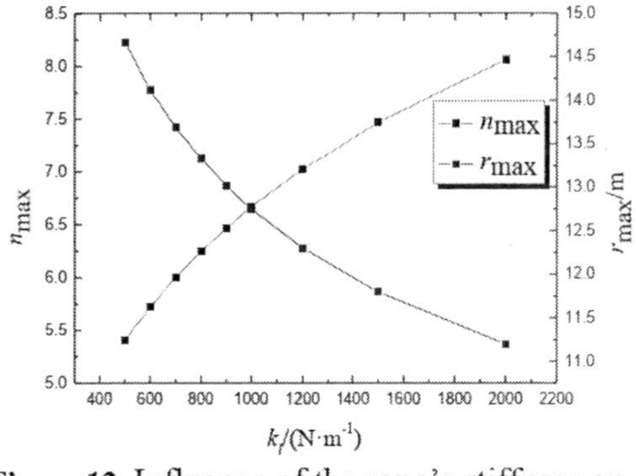

Figure 12. Influence of the rope's stiffness on maximum overload and radius

The above two figures shows that the recovery performance with the changing of the stiffness of the arresting rope is consistent under different damping coefficients: the greater the stiffness is, the greater the maximum resistance and the maximum overload are, and the smaller maximum radius of the UAV's trajectory is. Besides, the slope of the curve is greater when stiffness is smaller.

Influence of the rope's damping coefficient c_l on maximum resistance force F_{max} is studied. And when k_l=500 N/m, influence of the rope's damping coefficient c_l on maximum overload and maximum radius is considered. The results are as **Figure 13** and **Figure 14**:

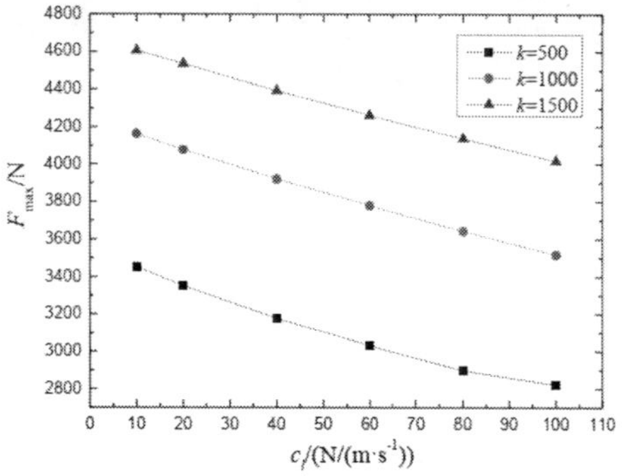

Figure 13. Influence of the rope's damping coefficient on maximum resistance force

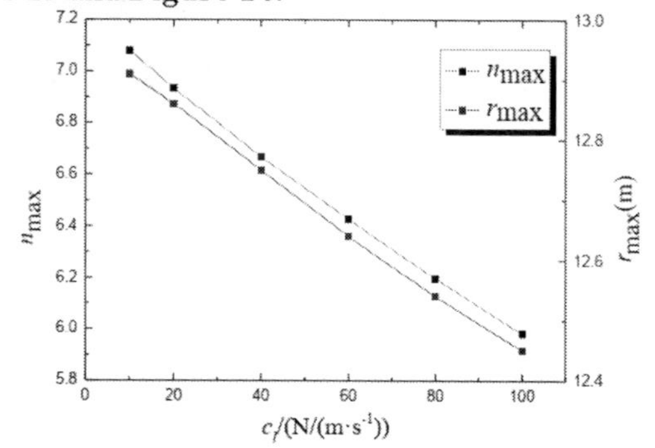

Figure 14. Influence of the rope's damping coefficient on maximum overload and radius

The above two figures shows that the recovery performance with the changing of the damping coefficient of the arresting rope is consistent under different stiffness: the greater the stiffness is, the lower the maximum resistance and the maximum overload are, and the smaller maximum radius of the UAV's trajectory is. And the trend is basically linear.

In addition, it can also be seen from the velocity curves of the aircraft during the recovery with different damping coefficients (as shown in **Figure 15**) that increasing the damping coefficient can reduce the peak velocity of each fluctuation and increase energy absorption. However, the influence mainly works in the initial stage of the recovery process, especially the change of the peak velocity of the first fluctuation after the collision.

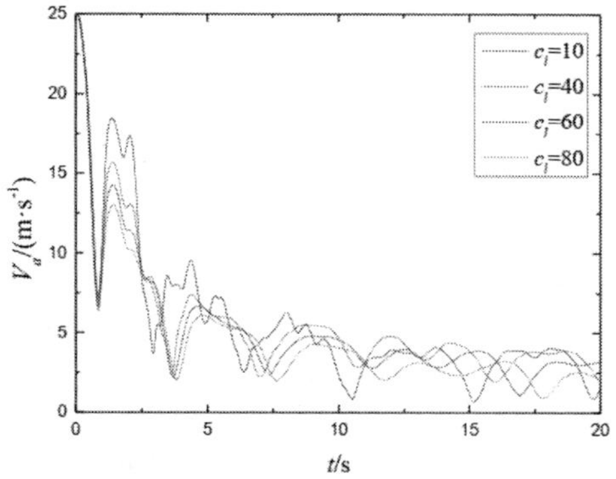

Figure 15. Velocity curves of UAV with different arresting rope's damping coefficient

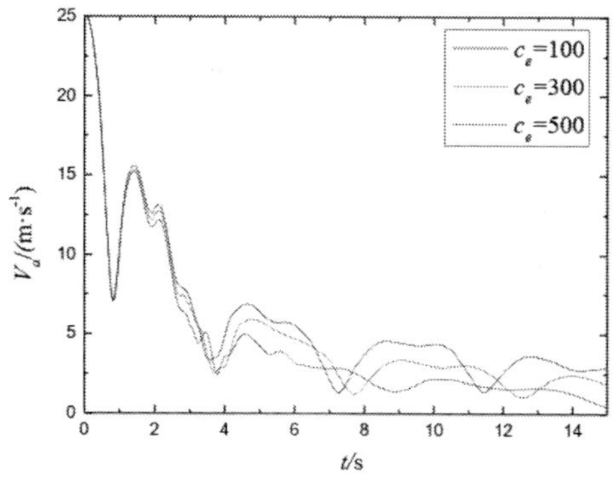

Figure 16. Velocity curves of UAV with different damping rope's damping coefficient

4.3 Effect of the damping coefficient of damping rope

The damping ropes EG and FG on the upper support of the recovery stand also contribute greatly to the cushioning and energy absorption effects of the recovery process. Therefore, its influence on recovery performance must also be considered. Different curves of velocity V over time t when damping coefficient of damping rope varies are compared as **Figure 16**.

It shows that the change of the damping coefficient of the damping ropes EG and FG has little effect on the initial buffering and energy absorption performance of the recovery process. There is no obvious difference of the peak velocity of the first fluctuation in curve, under the three different damping coefficients. However, as time goes on, increasing the damping coefficient of the damping rope can obviously improve the energy absorption effect. Using a larger damping coefficient, the fluctuation weakens, and the convergence tendency gets obviously faster.

Compared with the former velocity curves when damping coefficient of the arresting rope PAB varies, as shown in **Figure 15**, it can be seen that the absorbing and buffering effect of the arresting rope PAB is mainly reflected in the initial stage of the recovery process, while the absorbing effect of the damping rope EG and FG responses later, mainly reflects in the convergence stage. The reason is that in the initial stage, deformation of arresting rope PAB caused by the aircraft is the main factor, the recovery stand scarcely rotates or deforms, and thus energy absorption mainly lies on the arresting rope; while the motion of UAV turns into rotating around the recovery stand, deformation of the arresting rope PAB reduces and the deformation of the damping rope EG, FG caused by the rotation of the recovery stand AD becomes dominant energy absorption factor.

5 Conclusions

In this paper, a model of UAV vertical rope-type recovery system was established, and the UAV's position, attitude, velocity and resistance response curve during the recovery process were simulated. The effects of recovery speed of UAV, hit-rope position and stiffness as well as damping coefficient of arresting ropes on recovery performance were studied. These are the conclusions:

1) The effect of recovery velocity on the maximum overload is approximately linear and the lighter the UAV's weight is, the more sensitive the maximum overload is to the change of recovery velocity.

2) The maximum resistance force tends to decrease, when the hit-rope position is close to wing tip in wingspan direction, or close to the midpoint of recovery stand in height direction.

3) Increasing the stiffness of the arresting rope can reduce the maximum radius of the UAV's trajectory in the recovery process, but will increase the maximum overload. Increasing the damping coefficient can improve the buffering characteristics of the recovery process, which can reduce the maximum overload and the maximum radius of the UAV's trajectory during recovery.

4) Increasing the damping coefficient of the arresting rope PAB mainly improves the buffering and energy absorption performance in the initial stage of the recovery process, and reduces the maximum overload; while increasing the damping coefficient of the damping rope EG and FG can improve the concussion characteristics and speed up the convergence of UAV's velocity.

References

1. QIN B, WANG L. Summary of development of UAV. *Winged Missiles Journal*, **8**:4-10 (2002).
2. ZHAO T. Development of shipborne UAV. *Ship Electronic Engineering*, **30(4)**:21-24 (2010).
3. HONG D, ZHOU L, ZHENG Z S. Research of the development of foreign small shipborne fixed-wing UAV recovery technology. *Winged Missiles Journal*, **4**:50-54 (2014).
4. Dennis B D. Methods and apparatuses for capturing and recovering unmanned aircraft, including a cleat for capturing aircraft on a line: United States, 7059564B2.2006-06-13.
5. ZANG C X, CHU P J, HU Q, et al. The status and development trend of UAV recovery technology. *Winged Missiles Journal*, **3**:39-42 (2016).
6. DAI C, ZHANG Y, XIA X Y, et al. Vertical rope-type recovery device of small UAV: China,201761644U. 2011-03-16.
7. LU W, MA X P, ZHOU M, et al. Simulation analysis of dynamic characteristic of UAV rope-hook recovery system. *Acta Aeronautica et Astronautica Sinica*. **36(10)**:3295-3304 (2015).
8. WANG H X, LIU C L, CHENG J. UAV recovery technology and its development. *Winged Missiles Journal*, **1**:27-31 (2016).
9. B. Etkin. *Dynamics of Atmospheric Flight* (Beijing: Science press) pp 112-120 (1979).
10. FANG Z P, CHEN W C, ZHANG S G. *Aeronautic Aircraft flight dynamics* (Beijing: Beihang University Press) pp 174-180 (2005).
11. WANG Y L. Unmanned Aerial Vehicle Simulation in Flight envelope based on Matlab/Simulink. Nanjing: Nanjing University of Aeronautics and Astronautics, 2006:6-27.

Integrated Guidance System of a Commercial Launch Vehicle

Nickolay Zosimovych[1,a], Niu Xiaodong[1], Alex Noel Joseph Raj[1], Alessandro Simeone[1] and Xiongbin Peng[1]

[1]*Intelligent Manufacturing Key Laboratory of Ministry of Education, Shantou University, 243, Daxue Road, Shantou, Guangdong, 515063, China*

Abstract. This paper proposes a concept of integrated on-board navigation systems for commercial launch vehicles in the context of the current task, and provided mathematical models of its elements for different variants of designing structure and composition. Has been set and simulated the technical problem of the conceptual design of an integrated navigation system for the space launch vehicle qualified to insert small artificial Earth satellites into low and medium circular orbits with application of GPS technologies.

1 Introduction

A key tendency in the development of affordable modern navigation systems is exposed by the use of integrated GPS/INS navigation systems consisting of a gimballed inertial navigation system (GINS) and a multichannel GPS receiver [1, 2]. The investigations show [3, 4], that such systems of navigation sensors with their relatively low cost are able to provide the required accuracy of navigation for a wide class of highly manoeuvrable objects, such as airplanes, helicopters, airborne precision-guided weapons, spacecraft, launch vehicles and recoverable orbital carriers.
The study of applications of GPS navigation technologies for highly dynamic objects ultimately comes to solving the next problems [1, 2, 5-7].

Within the framework of this study we shall consider a light launch vehicle which has been jointly developed by the European Space Agency (ESA) and the Italian Space Agency (ASI) since 1998 [8, 9]. It is qualified to launch satellites ranging from 300 kg to 2000 kg into low circular polar orbits. As a rule, these are low cost projects conducted by research organizations and universities monitoring the Earth in scientific missions as well as spy satellites, scientific and amateur satellites. The launch vehicle Vega [9-11] is the prototype of the vehicle under development.

The planned payload delivered by the launch vehicle to a polar orbit at an altitude of ~700 km shall be 1500 kg [12]. The launch vehicle is adapting for missions to low Earth and Sun-synchronous orbits. During the first mission, the light class launch vehicle is to launch the main payload, a satellite weighing 400 kg, to an altitude of 1450 km with an inclination of the orbit 71.500. Unlike most single-body launchers, this vehicle is to launch several spacecraft.

The launch vehicle under consideration is the smallest one developed by ESA [13, 14]. The injection is presenting according to the most popular and simplest (and the cheapest) scenario, more specifically: the instrument unit and the navigation system ride atop the third stage of the launch vehicle. Thus, launching until separation of the fourth stage-carrying payload is piloting in accordance

[a] Corresponding author : nzosimovych@stu.edu.cn

with the data provided by the navigation system, which estimates 12 components of the launcher state vector, including position, velocity, orientation angles and angular velocities. Basically, launching may be done upon implementation of any of the possible algorithms, for example, a terminal one, that provides accuracy of the 3rd stage launching to the calculated point of separation of the 4th stage or the traditional algorithm which minimizes the deviation of the centre of mass of the launcher from the preselected programmed trajectory [2, 6, 10, 12].

From the standpoint of the problem concerned, namely the synthesis of the navigational algorithm of the space launcher in the proposed injection sequence we are interested only in the first factor, i.e. accuracy of lifting of the 3rd stage to the point of separation 4th stage [15].

Thus, we may determine the main criterion of the accuracy of the navigation task in relation to the integrated inertial navigation system of the space launch vehicle: we need to ensure maximum accuracy in determining the position and velocity vectors of the 3rd stage of the launch vehicle in the exo-atmospheric phase of the mission in the selected for navigation coordinate system. Consequently, in the case of the proposed injection sequence the simplest and most obvious criterion for evaluation of the accuracy of the synthesized system should be adopting. It is required to ensure maximum accuracy in determining the vectors of position and the centre -of- mass velocity of the launcher during the flight of the1st-3rd stages, i.e. in atmospheric and exo-atmospheric phases of the mission. This accuracy can be characterizing by the value of the dispersions posteriori of the corresponding components of the mentioned vectors [12, 16].

2 Problem Formulation

Now let us consider the possible integration schemes for GINS and GPS receiver with respect to this technical problem. As it has been aforementioned, currently we can think of three possible integration schemes as follows [2, 6, 7]:

- uncoupled (separated subsystems);
- loosely coupled;
- tightly coupled (ultra-tightly coupled).

The peculiarities of these schemes are discussing below. Uncoupled systems are the simplest option for simultaneous use of INS and GPS receiver [2, 6]. Both systems operate independently. However, as INS errors constantly accumulate, it is necessary eventually to make correction of INS according to data provided by the GPS receiver. In loosely coupled systems, GINS and GPS also generate separate solutions, but there is a binding unit in which GPS-based measurements and GINS readouts assess the status vector and make corrections of data provided by GINS [6].

A loosely coupled complex envisages an independent identification of navigation parameters both by GINS and by Self-Guided System (SGS). Different navigation parameters (coordinates, velocities) are provided by GINS and SGS. They are then using in the Kalman filter to determine errors occurring in GINS with a purpose of their subsequent compensation [14, 15]. The drawback is in correlation of errors, arriving from SGS to the input of the second Kalman filter and the need of strict synchronization of measurements provided by INS and SGS [6, 12].

In sources, loosely coupled systems are separating into three following types [2, 6]: the standard, "aggressive" and the so-called MAGR schemes. The difference between "aggressive" scheme and the standard one is that the former one uses the information on acceleration for extrapolation of navigation sighting executed by SGS provided by GINS in the period between measurements (Fig. 1). The Rockwell MAGR scheme uses inertial measurement from the SGS receiver made in carrier tracking loop (Fig. 2) [8].

In tightly coupled systems, the role of the INS is decreasing only to the measurement of the primary parameters of translational and rotational motions. For this reason, in such systems INS are only inertial measurement units, and the GPS receiver is without own Kalman filter. In such a structure both INS and SGS provide a series of measurements for a common computing unit [15].

Tightly coupled systems use the only "evaluator" (as a rule, the Kalman filter) that uses differences between pseudoranges and/or pseudovelocities, calculated (predicted) by INS and measured by Self-Guided System. Advantages of such a scheme described in next publications [6, 8, 12].
The disadvantages of closely coupled systems are the following [6, 8]:

- the need for special equipment for Self-Guided Systems;
- use of complex equations for measurements;
- low reliability because INS failure may result if failure of the whole system.

The later drawback can be excluding by introducing a parallel Kalman filter only for Self-Guided System [16]. Thus, the main differences between a tightly coupled system and a loosely coupled system are as follows [6, 12]:

- use of the INN output information on acceleration in the code and carrier frequency tracking loop. This allows narrowing the loop bandwidth and improving performance and tuning accuracy;
- use pseudoranges and pseudovelocities (instead of coordinates and velocities) to estimate errors in INS.

A separate embodiment of the tightly coupled systems are the so-called ultra-tightly coupled systems [12]. In such systems, estimations are undertaken in the integrated Kalman filter, and the GPS receiver is further simplified [6]. In this case, the Kalman filter is of order 40 and its implementation requires a computer with a very high speed.

Proceeding from the above information, we may conclude that an integrated navigation system of future launchers should have a structure that, depending on the functionability of SGS receiver, shall allow operating in accordance with the algorithms both as an uncoupled and tightly coupled system. It should be capable of processing coordinates and velocities as well as pseudoranges and pseudovelocities.

The structure of the complex is to be open to information from other on-board navigation tools and external consumers of navigation information. This may be by introducing the corresponding input/output ports.

Clearly, the first of the above schemes using both GINS and GPS receiver is not acceptable for our task, because here the receiver is not consuming for calibration (adjustment) of GINS during the mission by evaluating the drift component. As a result, in the absence of GPS-data errors of GINS grow at the same rate as in the absence of the receiver [8].

Thus, we conclude that in the present study it is appropriate to examine both schemes of interconnection: tightly and loosely coupled, and based on the results of simulation, draw conclusions in favour of one of the possible solutions.

3 Simulation of the Operation of the Onboard Integrated System for a Commercial Launch Vehicle

3.1 Mathematical Model of Integrated Onboard System

The Integrated onboard guidance and navigation systems used in launch vehicles allow applying modern information technologies most appropriately to ensure the required quality (accuracy and reliability) of navigation [6, 8]. The analysis shows that the onboard-integrated control systems (OICS) have a number of features, main among them being unification of respective functional groups on the level of technical solutions [14]. For example, all processor sections of a computer system are the same, irrespective of the problem they solve: navigation, guidance or stabilization. This fact makes away with one of the main disadvantages of the traditional (composite) on-board control system – excess range of schematic and technical solutions [16].

The architecture of OICS is influencing to solve a particular set of tasks on the board and is adapted for a class of problems for solving on the board with about 30 % redundancy. At the same time, it must support communications to ensure survivability of the system, and must be able to reprogram it from outside, etc. [7].

We must note that the mentioned requirements are typical for any modern computer system and meeting them does not cause any essential difficulties. Thus, OICS have the following characteristic properties [15]:

- functional flexibility and the ability to reprogram its functions;
- high performance characteristics, i.e. a possibility of creating compact control and start-up equipment with high probability of fulfilment of the task set;
- survivability, as in the case of failure of the equipment the system will be regenerated or switch to one of the particular algorithms, or elects self-destruction.

Finally, orientation of computer architecture to solving a particular class of specific problems provides a gain in sizes and power consumption. Since, as it was already specifying above, all the logic tasks on the board are solving within the computer system, all information flows pass there through. By writing all the input information of each processor into the correspondent register it is possible (provided this element is preserved) to reproduce all states of each processor after the experiment in the laboratory conditions. This makes debugging and testing of onboard algorithms and programs less complex.

3.2 Model of Inertial Navigation System

In the INS under consideration, the main components are the gyrostabilized platform and the block of integrating accelerometers. The gyrostabilized platform (GSP) embodies the on-board navigation coordinate system (ONCS). A functional circuit for INS using GSP is expressive in Fig. 1.

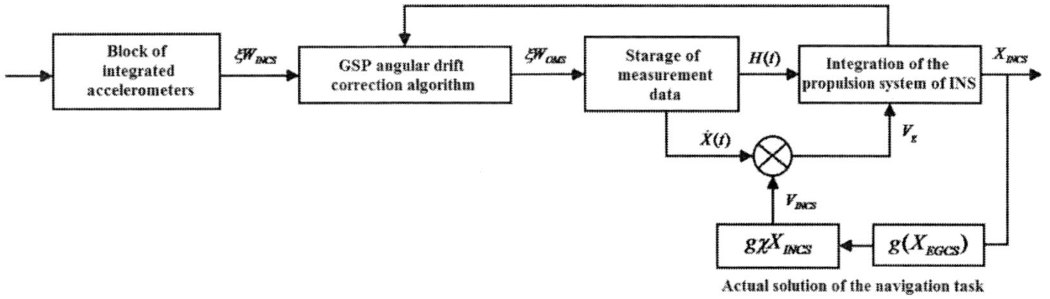

Figure 1. A functional circuit for INS using GSP

Functionally INS may be seeing composed of two blocks: a pre-processor and a navigational block.

3.3 Source Data for Modelling

According to mathematical models and algorithms described above, we created mathematical software for mathematical modelling of the process of nautical positions of the launch vehicle using an inertial navigation system, GPS receiver and integration algorithm for system data [7].

We shall point out that such formulation of a guidance task is naturally simplified as much as possible, and doesn't allow, in particular, to study the process of injecting a payload into the Earth orbit, because this process supposes manoeuvre of the final stage, and consequently solving on board a respective boundary task in one form or another. However, as it has been many times emphasized [1, 2, 6, 12], the purpose of the present research is to formulate efficient and precise navigation algorithms, based on the use of SINS and the GPS - receiver. In connection with the foregoing, it is obvious that high precision solution of a navigation task will allow implementing manoeuvres needed to launch a payload, all other conditions being equal vectors [12, 16].

4 Simulation Results

According to mathematical models and algorithms described above, we created mathematical software for mathematical modelling of the process of nautical positions of the launch vehicle using an inertial navigation system, GPS receiver and integration algorithm for system data [6].

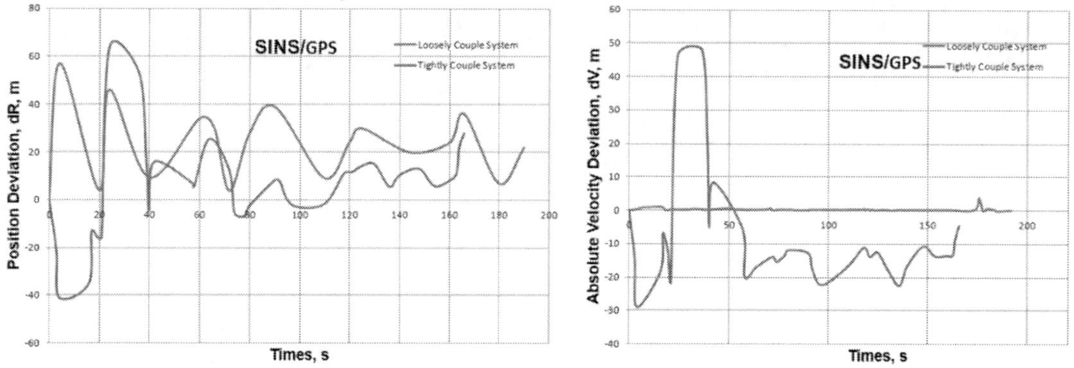

Figure 2. Deviation of the radius-vector dR and absolute velocity dV for Loosely Couple and Tightly Couple Systems with SINS/GPS

Based on the modelling conducted [2, 6-8, 12], we obtained dependencies, which are shown in Fig. 2.

Fig. 3 is shows average in implementation values of errors in calculation of pitch and yaw angles of the launch vehicle.

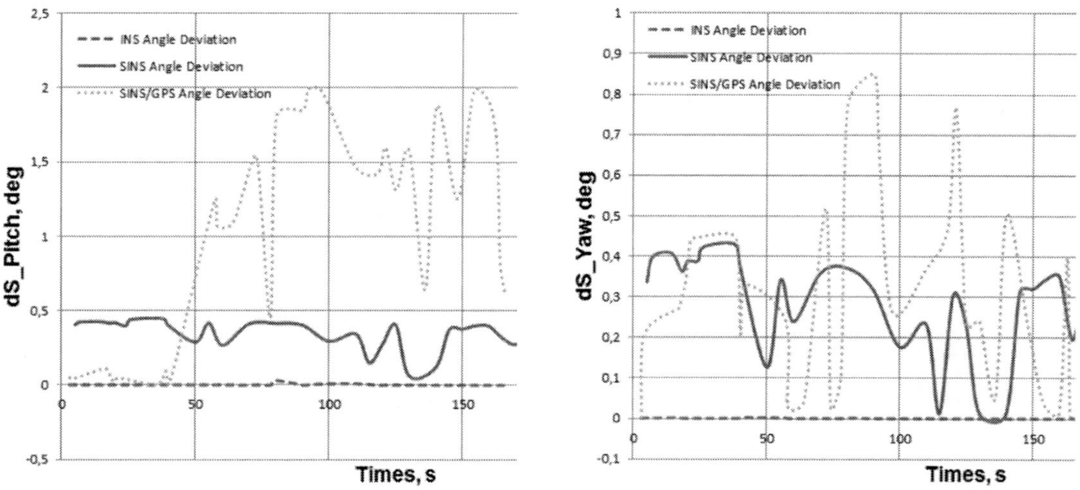

Figure 3. Errors in calculations of pitch and yaw

5 Conclusion

1. Have offered a concept of the integrated navigation system of a commercial launch vehicle, based on GPS technology.
2. Have developed a complex of mathematical models and algorithms, which help model both the process of functioning of the integrated navigation system and operation of the onboard navigation system of the launch vehicle itself.
3. We have developed an object-oriented computer soft implementing all the above models and algorithms for modelling of flight of the launch vehicle with different onboard systems: a) only GSP; b) GSP+GPS receiver; c) only SINS; d) SINS+GPS receiver for: uncoupled scheme; loosely coupled scheme; tightly coupled scheme.

4. Performed mathematical simulation of the operation of the navigation system of the LV in all the above implementations confirming the efficiency of all models and algorithms.
5. It has been shown that the use of the tightly coupled system does not result in significant improvement of navigation (deviations of the radius- vector 16.2 m and of the absolute velocity m/s) attended by an increase in the cost of its elements and amplification of integration schemes.

References

1. N. Zosimovych, Commercial Launch Vehicle Design, LAP LAMBERT Acad. Publ., **184** (2016)
2. N. Zosimovych, Conceptual Design of Navigation System for a Commercial Launch Vehicle, The Int. Journ. of Eng. and Sc. (IJES), Vol. 5, **31-42**, 10 (2015)
3. Flight instruments and navigation systems: Aircraft systems (lecture notes), Politecnico di Milano, Dipartimento di Ingegneria Aerospaziale, (2004)
4. G. M. Siouris, Aerospace Avionics Systems, Acad. Press INC (1993)
5. Kyu Sung Choi, Development of the commercial launcher integrated navigation system the mathematical model, using GPS/GLONASS technique, 52nd Int. Astronautical Congr., France, Toulouse, (2001)
6. N. Zosimovych, Integrated Guidance System of a Commercial Launch Vehicle, Modern European Science – 2014, Proc. X Int. Sc. and Pract. Conf., Sheffield, **10-22**, 18 (2014)
7. N. Zosimovych, Studying Guidance System of a Launch Vehicle with Application of GPS Technologies, Бъдещето Въпроси от Света на Науката - 2015, Proc. XI Int. Sc. and Pract. Conf., Sofia, Vol. 20, **24-38** (2015)
8. N. Zosimovych, Guidance System of a Launch Vehicle with Application of GPS Technologies, Cambridge Journ. of Edu. and Sc., Vol. 3, Cambridge University Press, **647-674**, 1(15) (2016),
9. The European Small Launcher: BR-257, ESA Publ., the Netherlands (2007)
10. Vega (raketa-nositel), http://ecoruspace.me/Вега.html
11. Vega: User's Manual. Issue 3/Revision 0, France: Arianespace, **188** (2006)
12. N. Zosimovych, Modeling the Integrated Guidance System of a Commercial Launch Vehicle, Int. Ref. Journ. of Eng. and Sc. (IRJES), Vol. 3, **39-54**, 6 (2014)
13. R.F. Appazov, O.G. Sytin, Metody proektirovaniya traektoriy nositeley i sputnikov Zemli, M., Nauka, **440** (1987)
14. Yu.G. Siharulidze, Ballistika letetelnyh apparatov, M., Nauka, **351** (1982)
15. A.A. Lebedev, G.G. Adjimamudov, V.N. Baranov, V.T. Bobronnikov, Osnovy sinteza system letatelnyh apparatov, M., MAI, **224** (1996)
16. D.J. Bayley, Design Optimization of Space Launch Vehicles Using a genetic Algorithm, Auburn University, Alabama, **196** (2007)

A Fixed Wing UAV Path Planning Algorithm Based On Genetic Algorithm and Dubins Curve Theory

Peng Yan[1], Zhuo Yan[2], Hongxing Zheng[1] and Jifeng Guo [1,a]

[1]*School of Astronautics, Harbin Institute of Technology, Harbin 150001, China*
[2]*Aerospace Propulsion Test Technology Institute, Xi•an 710000, China*

Abstract. Path planning is an essential problem in the autonomy research of UAVs. This paper proposes a new path planning algorithm for fixed wing UAVs based on genetic algorithm (GA) and Dubins curve theory. Path planning is treated as a global optimization problem under certain constraints, including the velocity vectors of initial and goal points and the minimum turning radius of UAVs. Dubins curve theory is utilized to satisfy the velocity vector constraints. GA is utilized to generate the shortest threats avoidance path in a 2D environment. A new encoding scheme is proposed, taking into account initial circles, goal circles, threats circles and the positional relationship between the path and these circles. The 2D Dubins path is converted to Dubins airplane path by adding a flight-path angle to it. The algorithm was tested in a complex flight environment and the planned path was tracked by a 6DOF Simulink model of a fixed wing UAV. Results show that the algorithm can generate shortest threats avoidance path in a complicated 3D environment and meanwhile satisfy constraints mentioned above.

1 Introduction

Path planning plays an important role in enhancing the ability of autonomous flight of unmanned aerial vehicles (UAVs). In general, there are two different approaches to construct this path: 1) on-line path planning and 2) off-line path planning [1]. On-line path planning is real-time local planning and it requires real-time access to environmental information. Off-line path planning is global planning based on the known environment information, and its goal is to generate an optimal flyable path from an initial point to a goal point under certain constraints. This paper gives an algorithm to solve the global path planning problem of fixed wing UAVs.

There already exist different kinds of path planning algorithms. The most widely used algorithms are graph searching algorithms such as A* and Dijkstra, which are applied to grids and visibility graph [2]. In addition, sampling based algorithms like the Probabilistic Roadmap Method (PRM) and the Rapidly-exploring Random Tree (RRT) are commonly used in path planning problems [3]. The Potential Fields method, called also Artificial Potential Fields, is usually used with other algorithms due to the good results [2, 4]. In [5], a two-level path planning algorithm was proposed, which deals map uncertainty and utilizes Voronoi diagrams and A* to search the global path and Potential Fields to plan local path. Besides, heuristic-based algorithms are widely used in path planning problems due

[a] Corresponding author : guojifeng@hit.edu.cn

to their high robustness compared other existing search methods under a relatively high number of constraints [4]. In [6], Li et al. proposed a Genetic-PSO (Particle Swarm Optimization) algorithm to optimize the flight paths for a group of multiple agricultural UAVs. In [7], Liu et al. presented a 3D path planning algorithm based on adaptive sensitivity decision operator combined with PSO technique.

Compared with other heuristic-based algorithms, GA is commonly used in path planning problems. For example, Li and Huang et al. [8] proposed a hierarchical genetic algorithm with optimized search region (OSR-HGA) to solve detours. Lin et al. [9] proposed a hybrid solution for UAV multi-target path planning, which integrates GA with DP (Dynamic Programming) for UAV to guarantee the safety requirement and feasibility for path planning. Sonmez et al. [10] used GA to find the optimal flyable path for the UAVs in a 3D environment.

Each aforementioned method has its own strength over others in certain aspects. However, for fixed wing UAVs, the motion constraints are not taken into account, which means they usually cannot be used in practical situations. The first constraint is the maximum curvature of the UAV under certain speed, and this limits the minimum turning radius of the UAV. The second constraint is the velocity vectors of initial and goal points, and this determines what kind of direction the UAV leaves the initial point and reaches the goal point. To meet these constraints, the Dubins curve theory [11] is utilized.

Since Dubins found that the paths of minimum length and bounded curvature can be easily constructed, the results have been widely used in path planning problems [12-14]. In [15], Shanmugavel et al. used CLC paths for path planning to solve the problem of the simultaneous arrival of UAVs and the shortest path of each UAV was found out from the available set of CLC paths. In [16], Yang et al. proved that the shortest path consists of no more than five parts, which is either an arc or a straight line under the assumption that there is only one circular obstacle. In [17, 18], Savla et al. proposed methods to solve the traveling salesperson problems (TSP) for Dubins' vehicle. In [19], Gao et al. described an algorithm that can find the shortest path derived from the theory of the Dubins curve during maneuvers and improves the performance of the UAV to follow the waypoints. The Dubins curve is also used in 3D environment. In [20], Wilburn et al. proposed a complete algorithm for producing a Dubins path in 3D environment through a set of 2 poses. In [21], Beard et al. proposed a method that extended the Dubins car model into a Dubins airplane model by adding a flight-path angle to the Dubins car path. The Dubins airplane path between two waypoints consists of helical and straight-line paths. However, these methods are used in simple environments and are difficult to obtain results in complicated environments which have more than one threat. To solve the problem of UAVs path planning in complicated environments, this paper proposes a new algorithm based on GA and Dubins curve theory.

This paper treats path planning as a global optimization problem with mission, environment and UAVs physical constraints. It has three main contributions. First, the minimum turning radius of UAVs is considered. Second, the velocity vector constraints of initial and goal points are considered. Third, the proposed algorithm can be used in a complicated 3D environment and obtain the shortest threats avoidance path. Simulation results demonstrate that the algorithm is capable of generating high quality path.

2 Problem formulation and Dubins paths

This section briefly describes the UAVs path planning problem to be solved. Path planning is defined as generating the shortest path from initial point to goal point under certain constraints, including the minimum turning radius of UAVs and the velocity vector constrains of initial and goal points.

2.1 Assumptions

Some assumptions are used to simplify the problem:

- Threats are treated as circles in a 2D environment and cylinders in a 3D environment and the radius of circles and cylinders denote the scope of threats.

- The UAV is assumed to fly at a constant speed and the influence of the wind is ignored.
- The velocity vector constraints are only acting on a horizon plane and have no influence on the vertical direction.

The flight environment in a horizon plane can be represented in Figure 1, where the red dotted circles represent threats, the small circle and the star point represent initial point and goal point, respectively, $P_s(x_s, y_s, \vec{v}_s)$ and $P_f(x_f, y_f, \vec{v}_f)$ denote the initial configuration and goal configuration in a 2D environment, respectively.

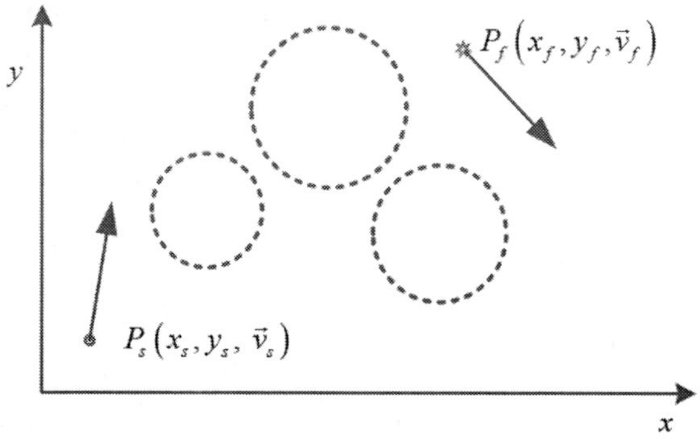

Figure 1. Flight environment in a horizontal plane.

2.2 The minimum turning radius of the UAV

The minimum turning radius of the UAV depends on the maximum roll angle of the UAV and the speed of the UAV, which can be derived from equations

$$\begin{aligned} F_{lift} \sin \gamma_{max} &= m \frac{v^2}{r_{min}} \\ F_{lift} \cos \gamma_{max} &= mg \end{aligned} \tag{1}$$

where F_{lift} represents the lift force on the UAV, γ_{max} represents the maximum roll angle of the UAV, m represents the mass of the UAV, v represents the speed of the UAV, r_{min} represents the minimum turning radius of the UAV, g represents the acceleration of gravity.

From equation (1), the minimum turning radius r_{min} can be calculated by equation

$$r_{\min} = \frac{v^2}{g \tan \gamma_{\max}} \tag{2}$$

2.3 Dubins paths between two configurations

The Dubins curve theory shows that the shortest path between two configurations in a 2D plane is the concatenation of circular arcs and their tangent lines. Given a certain configuration in a 2D plane, the UAV can turn right or left. Thus a certain configuration has two tangent circles as shown in Figure 2. One turns right and is marked as R, another turns left and is marked as L.

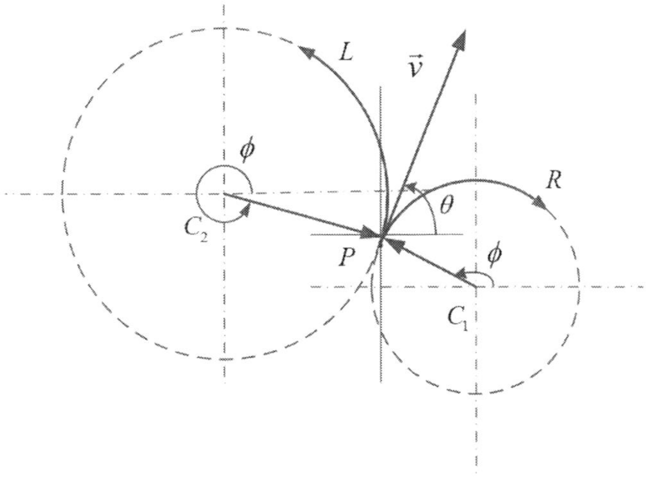

Figure 2. Tangent circles.

In Figure 2, P and $\vec{v}$ represent the position and the velocity vector of the UAV, respectively. ϕ represents the position of the UAV in the tangent circles and can be calculated by equation

$$\phi = \mathrm{mod}(\theta \pm \frac{\pi}{2} + 2\pi, 2\pi) \tag{3}$$

where θ represents the orientation of the velocity $\vec{v}$. The choice of a positive sign is associated with a right turn and the choice of a negative sign is associated with the left turn.

The centers of the tangent circles can be calculated once given the configuration of the UAV. The equations of the circles are:

$$\begin{aligned} x_{cs} &= x_s - r_s \cos\phi_s \\ y_{cs} &= y_s - r_s \sin\phi_s \\ x_{cf} &= x_f - r_f \cos\phi_f \\ y_{cf} &= y_f - r_f \sin\phi_f \end{aligned} \tag{4}$$

where (x_{cs}, y_{cs}) and (x_{cf}, y_{cf}) are the coordinates of centers of C_s and C_f, respectively. The angles ϕ_s and ϕ_f are the positions of the UAV in the initial circle C_s and the goal circle C_f, respectively. r_s and r_f are the radius of the circles C_s and C_f, respectively.

The initial configuration and goal configuration both have two circles. There are four paths between these four circles: RSR, RSL, LSR and LSL [13, 21]. They form the Dubins paths between two configurations.

2.4 Implementation steps

To generate a flyable and shortest path in a 3D environment, our work is divided into two steps:

(1) Generate the shortest path and meanwhile satisfy the velocity vector constraints in a 2D plane. This step generates the shortest Dubins path in a complicated environment by using GA. The path in a 2D environment consists of circular and straight-line paths.

(2) Convert the 2D path to the 3D path by adding a flight-path angle to the 2D path. The path in a 3D environment consists of helical and straight-line paths.

3 GA for Dubins path planning in a 2D environment

This section presents how to use GA to generate the shortest threats avoidance Dubins path in a complicated 2D environment. The main stages in GA include selection, crossover, and mutation. In addition, to determine whether the path intersects threats, the detection operator is proposed. The path is denoted by a chromosome. The chromosome encoding method is taken detailed consideration to make better use of GA search capability.

3.1 Chromosome encoding scheme

The chromosome encoding scheme has a great effect on the design of genetic operators, such as crossover and mutation. We propose a new binary encoding scheme which takes into account the initial circles, goal circles, threats circles, and the positional relationship between the path and these circles.

Given the initial configuration $P_s(x_s, y_s, \vec{v}_s)$, goal configuration $P_f(x_f, y_f, \vec{v}_f)$ and the positions and scopes of threats, which are denoted by T_i (i=1, 2, …, n) (where n is the total number of threats), the shortest path is at least tangent to k ($k\in[0,n]$) threats. Thus, the problem transfers into solving the shortest path (C_s, W, C_f). Where W is a sequence of threats that are tangent to the path, C_s and C_f are initial and goal circles, respectively. The encoding scheme contains all situations in (C_s, W, C_f). Take the positional relationship between the circle and the path into consideration, each threat has three encoding schemes: the threat is tangent to the right of the path, the threat is tangent to the left of the path, and the threat isn't tangent to the path. Because one encoding scheme can represent that the threat is not tangent to the path, the number of encoding schemes to represent n threats is at least $2n+1$. Due to the path is tangent to the C_s and C_f certainly, the initial circle C_s and the goal circle C_f need 4 encoding schemes. Thus, to represent n threats, the number of binary digit is $ceil(\log_2(2n+1))n$. To represent C_s and C_f, the number of binary digits is 2. A chromosome has $ceil(\log_2(2n+1))n + 2$ binary digits. A chromosome is shown in Figure 3.

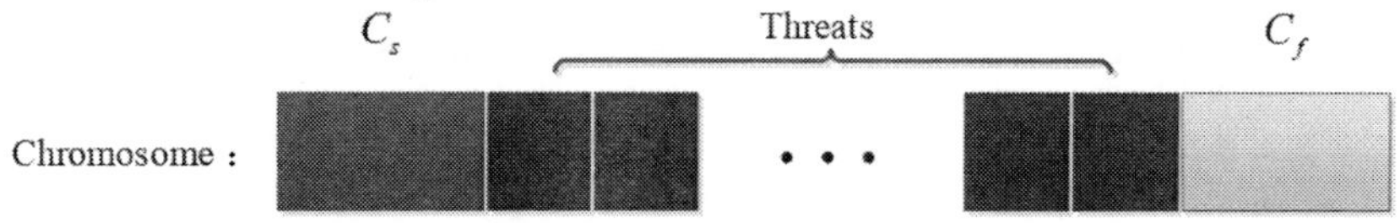

Figure 3. Chromosome composed by C_s, threats and C_f.

3.2 Fitness function

In order to generate the shortest path, the fitness function is designed as

$$f = \frac{d}{L+C} \tag{5}$$

where L is the length of path, C is a binary variable which punishes f when the path intersects with threats, d is the Euclidean distance from the initial point (x_s, y_s) to the goal point (x_f, y_f) which is calculated by

$$d = \sqrt{(x_s - x_f)^2 + (y_s - y_f)^2} \tag{6}$$

C is defined as

$$\begin{cases} C = 0 & \textit{if path doesn't intersect with the threats} \\ C = 10d & \textit{if path intersects with the threats} \end{cases} \tag{7}$$

3.3 Genetic operator

Conventional genetic operators include selection, crossover and mutation. On the basis of these operators, detection operator is proposed to determine whether the path intersects with the threats.

Selection: Two parents are randomly selected based on their fitness by using the roulette wheel selection method.

Crossover: One-point crossover is used. Two crossover sites are chosen randomly and then the genes between the two sites in the two parent paths are exchanged.

Mutation: The parental chromosome is chosen according to the selection method. The mutation operation is done by randomly select a gene in the parental chromosome and replace it with an opposite gene with a certain probability.

Detection: The detection operator is used to determine whether the path generated by the above three operators intersects with the threats. If the path intersects with the threats, the variable C is assigned a value of $10d$. If not, the variable C is assigned a value of zero.

3.4 Control parameters

A number of parameters affect the performance of GA implementation. These parameters include population size and the probability of crossover and mutation. Usually, the population size ranges from 50 to 200. The larger the population size, the more diverse it will be. The probability of crossover determines how often the crossover operator is applied to an individual. The mutation operator is an effective way to maintain population diversity. Commonly the value of the crossover probability ranges from 0.60 to 1.00 and the value of the mutation probability ranges from 0.05 to 0.15 [9]. The GA terminates when the maximal number of iterations is reached.

4 The Dubins airplane path in a 3D environment

In the previous section, the Dubins path in a 2D environment is proposed. In this section, the 2D path is converted to the Dubins airplane path by adding a flight-path angle to the 2D path. Figure 4 depicts a UAV flying with velocity $\vec{V}$ and flight-path angle ϑ in a vertical plane.

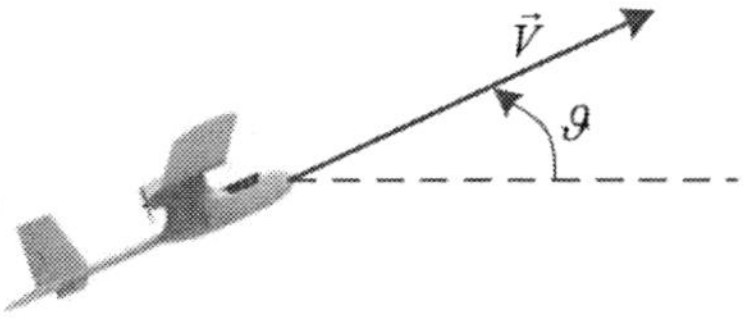

Figure 4. A UAV flies in a vertical plane.

The physical limit on the UAV in the vertical plane is represented by

$$|\vartheta| \leq \bar{\vartheta} \tag{8}$$

where $\bar{\vartheta}$ is the maximum flight-path angle subject to the physical limit of UAV.

The 2D Dubins path consists of circular and straight-line paths. The circular paths are turned into the helical paths when the Dubins path is converted to the Dubins airplane path.

4.1 Minimum distance Dubins airplane path

This subsection describes how to concatenate straight-line and helix paths to produce the minimum-distance path between two configurations for the UAV. Initial and goal configurations are defined as $P_s(x_s, y_s, z_s, \vec{v}_s)$ and $P_f(x_f, y_f, z_f, \vec{v}_f)$ which are formed by adding altitude information to the 2D configurations. As described in[14, 21], there are three different cases for Dubins airplane path that depend on the altitude difference between the initial and goal configuration, the length of 2D Dubins path, and the flight-path angle limit $\bar{\vartheta}$. The three cases are defined to be low altitude, medium altitude, and high altitude [14].

In this study, the low altitude is only taken into consideration. The altitude difference between the initial and goal configuration is said to be the low altitude if

$$|z_f - z_s| \leq L\tan(\bar{\vartheta}) \tag{9}$$

where the term on the right is the maximum altitude difference that can be obtained by flying at the flight-path angle $\pm\bar{\vartheta}$ for a distance L which is the length of a 2D Dubins path.

In the low altitude case, the altitude difference between the initial and goal configurations can be achieved by flying the 2D Dubins path with a flight-path angle satisfying constraint (8). Therefore, the optimal flight-path angle can be calculated by

$$\vartheta^* = \arctan\left(\frac{z_f - z_s}{L}\right) \tag{10}$$

The complete path planning algorithm is summarized in Algorithm 1.

Algorithm 1 Fixed Wing UAV Path Planning Algorithm

```
Calculate the minimum turning radius using equation (2)
Calculate initial and goal circles using equation (4)
Encode initial circles, goal circles, threats circles
Initialize the population
while iteration < the maximum number of iteration
    for individual in population
        Construct Dubins path among circles
        Detect threats
        Calculate fitness
        Perform selection operation
        Perform crossover operation
        Perform mutation operation
    end
    iteration = iteration +1
end
Select the best individual from the population as the shortest Dubins path
Calculate the optimal flight-path angle using equation (10)
Convert the Dubins path to Dubins airplane path by adding the optimal flight-
path angle to the Dubins path.
```

5 Simulation and discussions

The new proposed algorithm was tested in a complex flight environment, which has ten threats. The planned path was tracked using a 6DOF Simulink model of a fixed wing UAV. The trajectory tracking algorithm is provided by [22]. The proposed algorithm was tested in MATLAB environment and ran on a server with a 1.8 GHz CPU and 4.0 GB RAM.

5.1 Parameter settings

The mission of the UAV is to depart from an initial point and arrive at a goal point under certain constraints which include the velocity vectors of initial and goal points in a horizontal plane, the maximum roll angle and the maximum flight-path angle. The flying space is a cube with side length 10000*m*. The positions and radius of the threats are randomly generated with the radius ranging from 800*m* to 1000*m*. The physical constraints of the UAV are listed in Table 1. The parameters of GA are listed in Table 2. The configurations of the UAV are shown in Table 3.

Table 1. UAV physical constraints.

Constraint	Value	Unit
$\gamma_{\max}$	20	*deg*
$\bar{\vartheta}$	15	*deg*

Table 2. GA parameters.

Parameter	Value
Population size	200
Selection rate	0.9
Mutation rate	0.1
Maximum iterations	20

Table 3. Values of the UAV configurations.

Configuration	Value	Unit
Initial coordinates	(0, 0, 3000)	*m*
Goal coordinates	(10000, 10000, 500)	*m*
Initial velocity	(30, 40)	*m/s*
Goal velocity	(0, 50)	*m/s*

5.2 Results

The algorithm runs 10 times and the best results are shown in Figure 5.

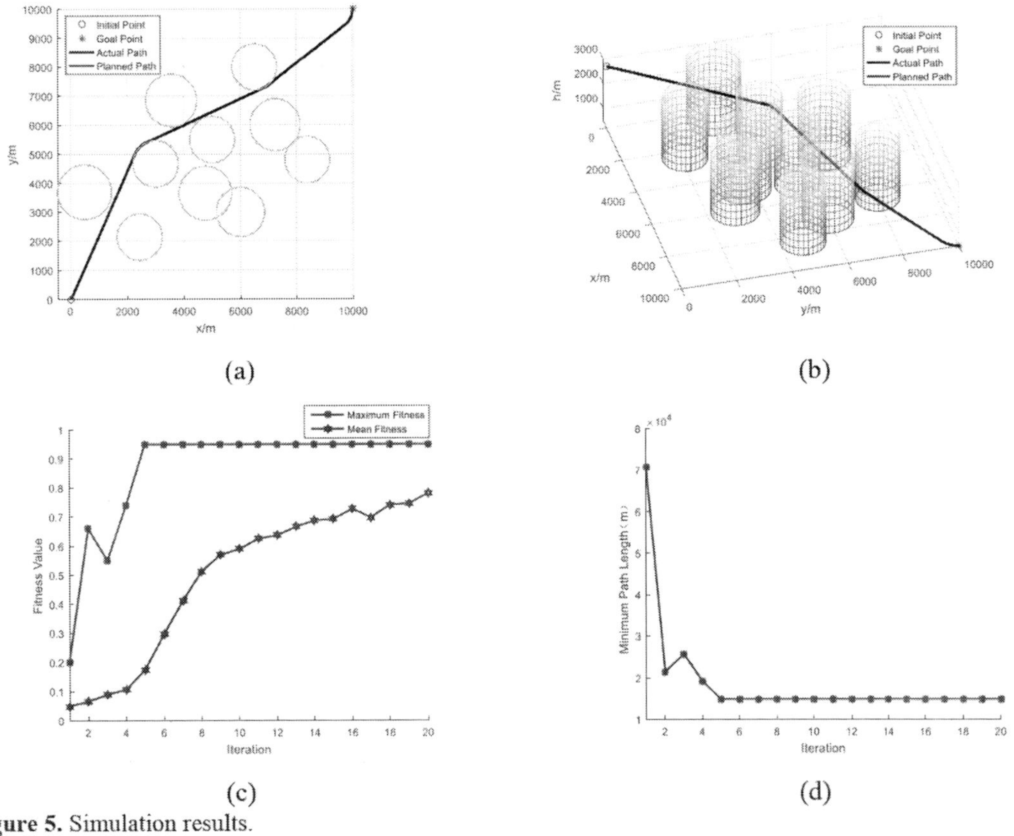

Figure 5. Simulation results.

Figure 5 (a) shows the planned path and the actual path in 2D view. The planned path is tangent to two threat circles. The minimum turning radius is 700.9*m*. The velocity vector constraints of initial and goal points are satisfied by turning left at the initial and goal points at the minimum turning radius, respectively. The path length is 14894.5 *m*.

Figure 5 (b) shows the planned path and the actual path in 3D view. The planned path is from (0, 0, 3000) to (10000, 10000, 500), the altitude difference is -2500*m*. The optimal flight-path angle is -9.53 degrees which satisfies the UAV physical constraint. The length of the Dubins airplane path is 15102.8 *m*. The Dubins airplane path successfully avoids the threats and satisfies the constraints. The actual path and the planned path are almost identical which indicates that the planned path is a flyable path.

Figure 5 (c) shows the fitness value for each iteration. The max fitness value of the population reaches the maximum value after fifth iteration. The mean fitness value continues to increase with the iteration, and it gradually approaches the maximum fitness, which indicates that the algorithm converges smoothly.

Figure 5 (d) shows the minimum path length for each iteration. The min path length reaches the minimum value after the fifth iteration.

5.3 Discussions

As demonstrated by the results, the proposed algorithm can generate the shortest threats avoidance path in a complicated 3D environment and satisfy the constraints mentioned above. However, there are three aspects worth noting. One is the randomness of the algorithm which makes the optimal result need to run the algorithm multiple times. Another is that there are two cases may result that the final path isn't the shortest path. First, for a non-cylindrical threat, in order to generate the path, the threat is treated as a large cylinder. Second, if the path is tangent to a threat whose radius is smaller than the minimum turning radius of the UAV, the radius of the threat will be increased to the minimum turning radius of the UAV and rerun the algorithm to calculate the path until the radius of the threat which is tangent to the path is greater or equal to the minimum turning radius of the UAV. The third aspect is that we only consider the low altitude situation in a 3D environment. For medium altitude and high altitude situations, the solution process is more complex and will be studied in the future work.

6 Conclusions

In this paper, a new path planning algorithm for fixed wing UAVs is proposed which generates the shortest threats avoidance path in a complicated 3D environment and meanwhile satisfies the constraints including velocity vectors of initial and goal points and the minimum turning radius of the UAV. First, we have a simple description of the problem to be solved and make some appropriate assumptions. Second, the Dubins curve theory is utilized to satisfy the velocity vector constraints. Then, in order to generate the shortest path, GA is used and a new encoding scheme is proposed which takes into account initial circles, goal circles, threats circles, and the positional relationship between the path and these circles. At last, the 2D Dubins path is converted to the Dubins airplane path by adding a flight-path angle to it. The simulation results show our algorithm is capable of generating high quality path in a complicated 3D environment.

References

1. H. Ergezer and M. K. Leblebicioğlu, *International Conference on Unmanned Aircraft Systems*, 79-88(2013)
2. O. Souissi, R. Benatitallah, D. Duvivier, and A. H. Artiba, *International Conference on Industrial Engineering and Systems Management*, 1-8(2013)
3. D. González, J. Pérez, V. Milanés, and F. Nashashibi, IEEE Trans. Intell. Transp. Syst. **17**, 1-11(2015)
4. T. T. Mac, C. Copot, D. T. Tran, and R. D. Keyser, Robot. Auton. Syst. **86**, 13-28(2016)
5. K. Ok, S. Ansari, B. Gallagher, and W. Sica, *IEEE International Conference on Robotics and Automation*, 4596-4601(2013)
6. X. Li, Y. Zhao, J. Zhang, and Y. Dong, *IEEE International Conference on TOOLS with Artificial Intelligence*, 691-697(2017)
7. Y. Liu, X. Zhang, X. Guan, and D. Delahaye, Aerosp. Sci. Technol. **58**, 92-102 (2016)
8. J. Li, Y. Huang, Z. Xu, J. Wang, and M. Chen, *IEEE International Conference on Control &*

Automation, 1033-1038(2017).
9. C. E. Lin and Y. M. Syu, JoAAA, **48**, 203-220(2016)
10. A. Sonmez, E. Kocyigit and E. Kugu, *International Conference on Unmanned Aircraft Systems*, 50-55(2015)
11. L. E. Dubins, Am. J. Math. **79**, 497-516(1957)
12. X. Sun, C. Cai and X. Shen, J. Intell. Robot. Syst. **79**, 3-19(2015)
13. I. Lugo-Cardenas, G. Flores, S. Salazar, and R. Lozano, *International Conference on Unmanned Aircraft Systems*, 339 - 346(2014)
14. H. Chitsaz and S. M. Lavalle, *IEEE Conference on Decision & Control*, 2379-2384(2008)
15. M. Shanmugavel, A. Tsourdos, R. Zbikowski, and B. White, *AIAA Guidance, Navigation, and Control Conference and Exhibit*, (2006)
16. D. Yang, D. Li and H. Sun, Math. Probl. Eng, **2013**, 1-6(2013)
17. K. Savla, E. Frazzoli and F. Bullo, IEEE Trans. Autom. Control. **53**, 1378-1391(2008)
18. K. Savla, E. Frazzoli and F. Bullo, *American Control Conference*, 786-791(2005)
19. X. Z. Gao, Z. X. Hou, X. F. Zhu, J. T. Zhang, and X. Q. Chen, Acat. Polytech. Hun. **10**, 221-239(2013)
20. J. N. Wilburn, M. G. Perhinschi and B. K. Wilburn, *AIAA Guidance, Navigation, and Control (GNC) Conference*, 5232 (2013)
21. M. Owen, R. W. Beard and T. W. Mclain, *Handbook of Unmanned Aerial Vehicles*, 1677-1701(2015)
22. D. R. Nelson, D. B. Barber, T. W. Mclain, and R. W. Beard, IEEE Trans. Robot. **23**, 519-529(2007)

The design service life of wide-body commercial aircraft research based on airlines data

Yue Wang[1,a], Dan Gao[1] and Jiangtao Si[1]

[1]Shanghai Aircraft Design and Research Institute, No. 5188, Jinke Road, Pudong New Area, Shanghai, China

Abstract. To reasonably define the design service life of Chinese double aisles wide-body commercial aircraft, the exact market region and design condition are researched. The base market of Chinese double aisles wide-body commercial aircraft is China, Asia, Europe and the Pacific. In these zones, customers should be analyzed carefully, and the customers who are not available but play an important role in these zones should be picked off decisively. Besides, the flight length definition is also studied, which affects the design service life of aircraft obviously in small market region. After research, the results could reduce engineering difficulties, and also increase airlines adaptability and competitiveness in the specific market zone.

1 Introduction

Commercial aircraft is a special commodity, whose internal attribute asks it to meet the market requirements, which turns to be an important condition for success. Market is the source of product, so is in the aviation industry.

Design service life is a limited period confirmed during aircraft certification, displaying in the form of flight cycles and flight hours [1]. Design service life is an important fatigue evaluation parameter affecting the aircraft structure weight directly. So a reasonable design service life is critical to the market competitiveness and economy of the aircraft.

Currently, the method of minimum design service life is commonly used by most of aircraft manufacturers. Three typical flight lengths are provided with the method, including the short, medium and long characteristic flight length [2]. These flight lengths are calculated in general from the airlines data of market. Airlines are scattered all around the world, with different geographical location, natural environment, cities distribution, population distribution, geopolitical position, customers habits, etc., which leads to the unique regional features of different airline markets. So the aircraft manufacturers need to be sensitive to the market and produce the appropriate product to meet exactly requirement of the market.

2 Analysis process of airlines data

To analyze the airlines data effectively and find out the airlines feature reasonably, a bank of airlines data should be established firstly, then the constraints should be defined exactly, and finally the data

[a] Corresponding author : wangyue@comac.cc

of object market could be picked out as a special data collection, which is the basis of analysis and research. The analysis process is shown in figure 1 as below.

In the whole process, the most important step is constraints definition, which defines the object market of product development [3]. For indicating the design service life of double aisles wide body commercial aircraft clearly, three constraints are defined and applied in operation route analysis. The first one is the aircraft types, which classifies the aircraft by flight range and seats. With this constraint, a global object market data for special competitors is collected. The second one is geographical division, which reflects the detailed geographical region of object market. The third one is the customers, who will really operate the aircraft in the future. These constraints make preparation for the analysis to the exact object market data.

Following the process, the analysis results with different constraints would be shown in the paper.

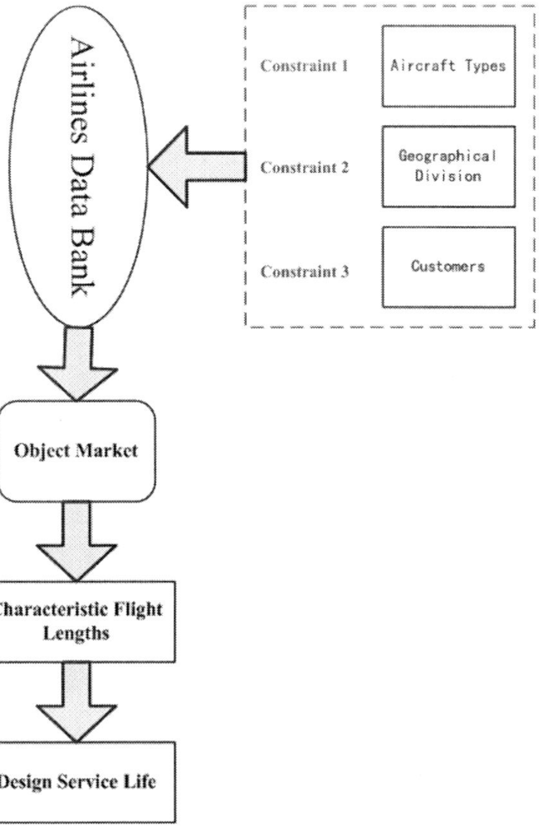

Figure 1. Analysis process of airlines data

3 The method of aircraft design service life

Aircraft design service life is a predefined life in aircraft design, which is achieved by design, maintenance and so on [4]. Generally the method of minimum design service life is used by most aircraft manufacturers, by which the design service life of aircraft could be analyzed rapidly and effectively. In the paper, this method would be used in the analysis.

Generally speaking, minimum design service life is the aircraft's serving life, 20 years at least, cumulated under the harshest operating conditions. During this period, there shall not be obvious cracks in main load-bearing structures of aircraft [2]. With this method, three special flight lengths are provided for defined the flight cycles, see figure 2 for details.

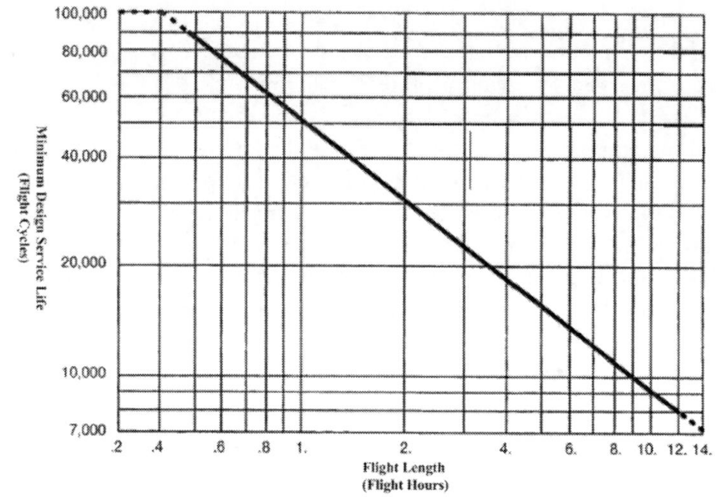

Figure 2. Minimum design service life curve [2]

With the development of commercial airlines, minimum design service life curve would be modified a little. In the paper, the fitting curve equation is approximate to

$$MDSG = 10^{(4.7-0.741\lg h)} \tag{1}$$

Where h is flight hour of a flight length [2].

4 The airlines data analysis of wide-body aircrafts

According to the analysis process of airlines data, the characteristics of different markets will be analyzed and shown.

4.1 Global airlines data analysis of wide-body aircrafts

In general, there are double aisles with 200 to 400 seats in commercial wide-body aircrafts, whose range is no less than 10000 km, such as A330, A350, B787 and B777. But a special aircraft type of B767 should not be neglected, which has been excluded from the market. The airlines operated by B767 aircrafts have been replaced by other wide-body aircrafts gradually. So these airlines should be accounted in the global airlines data of wide-body aircrafts [5].

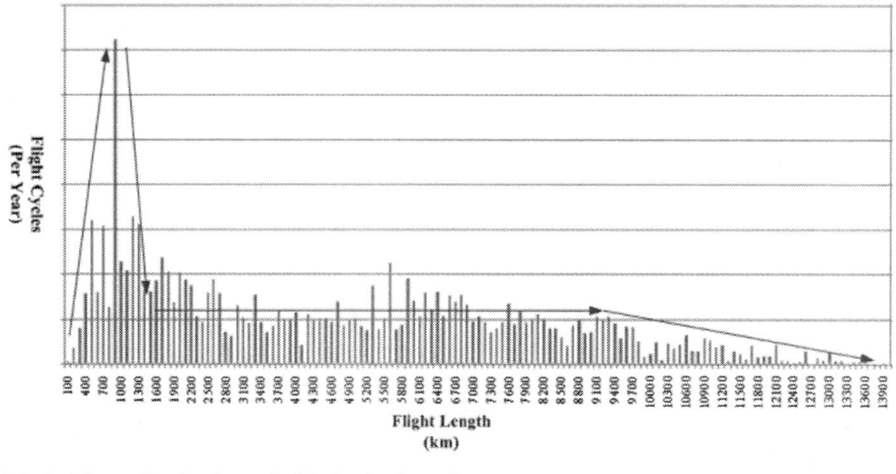

Figure 3. Global airlines distribution of wide-body aircrafts

Figure 3 shows that there is an operating peak value at about 900 km flight length, which is much more than other flight lengths. This phenomenon reflects an explosive market in this flight length, which will be focused and analyzed in chapter 3.3. After the peak value till about 9000 km flight length, the operating frequencies become smooth. And after 9000 km flight length, they reduce gradually.

4.2 Airlines data analysis of wide-body aircrafts in Asia-Europe-Pacific region

The commercial aircrafts of China developed later, so the development goal is different from the western aircraft companies, who face the global market. For developing commercial wide-body aircraft of China, an appropriate and exact market should be focused and chosen, where the product would be competitive and get success easier.

There is a large advantage for commercial aircraft development of China, which is the huge inner market. According to the prediction, during 2025 to 2045, about 1300 new wide-body aircrafts are needed, which is about 13% of global new increasing wide-body aircrafts, so the inner market is big enough for the primary market of commercial wide-body aircrafts development of China. Otherwise, considering the belt and road strategy, the market of commercial wide-body aircrafts development of China is located in the region of Asia-Europe-Pacific.

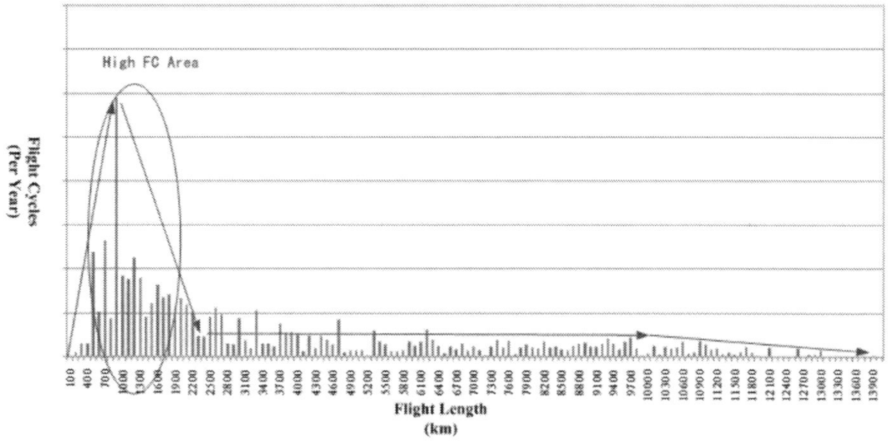

Figure 4. Asia-Europe-Pacific airlines of wide-body aircrafts distribution

Figure 4 is the similar as figure 3 in airlines distribution. In short flight length, there is a peak value at about 900 km. The medium flight length frequencies are smooth. The long flight length frequencies are less than others.

There are large quantities of airlines operated with wide-body aircrafts in Asia-Europe-Pacific region. Because of the large population distribution and rapid economy development in these regions, this market will become more and more important.

For reaching the requirements of this important market, the potential customers should be analyzed carefully, and the interference information should be eliminated.

4.3 Customers airlines data analysis of wide-body aircrafts in Asia-Europe-Pacific region

Through analyzing and classifying the potential customers in object region, especially considering the wide-body aircrafts flight quantities per year, the market of Japan should be concentrated intensively. Because of its population and cities distribution, there are large quantities short-thick airlines operated by wide-body aircrafts in this market. Therefore, Boeing provides a special short range wide-body B787 aircraft according to the characteristics in Japanese airlines market. [4]

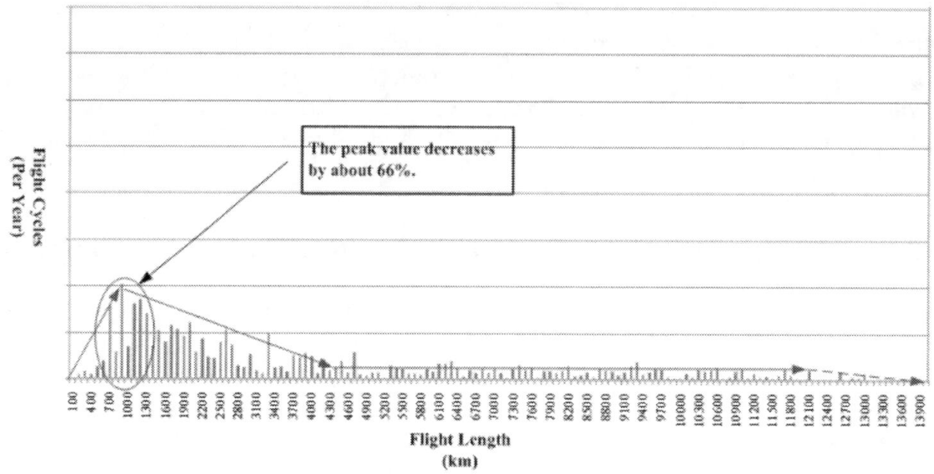

Figure 5. Customers airlines of wide-body aircrafts in AEP region distribution

There is a big different between figure 5 and figure 4. In figure 5, the peak value of flight frequencies decreases by about 66%, which comes from Japan airlines market of wide-body aircrafts. Considering the conventional operating habits of Japan and the difficulties in products design, eliminating the Japanese market from the customers airlines of Chinese wide-body aircrafts in AEP region is proposed.

4.4 The characteristic flight lengths for design service life

Based these markets mentioned above, the characteristic flight lengths have been calculated with a fixed proportion of flight frequencies.

Table 1. The characteristic flight lengths in different object markets of wide-body aircrafts

Market Region	Short length km	Medium length km	Long length km
Globe	1290	5850	10425
Asia-Europe-Pacific(AEP)	880	3830	10165
Customers in AEP	1165	4400	10510

Table 1 shows that the characteristic flight lengths are different in different object markets, especially in short lengths and medium lengths. Primarily because the special market of wide-body aircrafts of Japan exists. For long lengths, they are almost the same, which reflects the long flight length airlines distribute evenly globally.

5 The design service lives of wide-body aircrafts in different object markets

Table 2. The design service lives of wide-body aircrafts in different object markets

Market Region	Flight cycles	Flight hours
Globe	32000	100000
Asia-Europe-Pacific(AEP)	42000	100000
Customers in AEP	34000	100000

In design service life calculation of wide-body aircraft, the short length basically reflects the flight cycles. In table 1, the short length of Asia-Europe-Pacific increases by about 32.3%, from 880 km to 1164 km, after eliminating Japan and Korea market data. And in table 2, the flight cycles decrease about 19%. This is a significant result, which means the more appropriately the market is chosen, the more competitive the product would be.

6 Flight length divisions effecting on the design service life of wide-body aircraft

In above analysis, the characteristic flight lengths for design service life are defined with a fixed proportion of flight frequencies, which means separately fixed percentages of flight times indicating the short, medium and long flight length. This definition method shows advantages in calculation stability in large sample space. But in small sample space, because of discontinuity of data, the calculation result turns out to be not precise enough.

To analyze the special market exactly, it is necessary to redefine the short, medium and long flight lengths. There are two elements related directly with the commercial aircrafts, which are the airline and the passenger, so these two elements are considered primarily in this paper.

At first, short flight length is redefined. For airline companies, the airlines operated with wide-body aircrafts are almost short-thick airlines, which carry large quantities of passengers. When conditions are matched, it is easy to initiate the airline. So for the short flight length definition, more attention should be paid to the requirements of passengers. For business passengers attending business meetings in different cities would be hoped in one day. And for short distance tourists, a shorter time spent in route the better, just like self-driving travel in weekend. Through a limited sample inquiry, the result is that the short flight length is no more than 3 flight hours. This inquiry will be researched in detail in the future.

Long flight length definition is opposite with short flight length definition. The focus of the former is the aircraft operator. For spending long time in air, the pilot weary of flight has the possibility to make mistakes. So the flight hours of pilot is restricted to 10 hours in both CCAR91 and CCAR21.

Therefore, the medium flight length is redefined between 3 to 10 flight hours.

According to the new flight lengths divisions, the characteristic flight lengths are re-analyzed.

Table 3. The characteristic flight lengths of wide-body aircrafts based on new flight lengths definition

Market Region	Short length km	Medium length km	Long Length km
Globe	1225	5470	10225
Asia-Europe-Pacific(AEP)	1215	4830	10325
Customers in AEP	1355	4765	10450

Three characteristic flight lengths in Table 3 are almost the same with Table 1 in global market region, which reflects the new flight lengths definition is reliable. But more attention should be paid to the smaller market region, the result is different obviously, especially in short flight length.

Table 4. The design service life of wide-body aircrafts based on new flight lengths definition

Market region	Flight Cycles	Flight Hours
Globe	32000	100000
Asia-Europe-Pacific(AEP)	32000	100000
Customers in AEP	30500	100000

In Table 4, the design service lives of wide-body aircrafts are different in smaller market region. In the customers' market region, the design service life could decrease by 10.3%, which is a significant benefit for aircraft design. Therefore, to increase the competitiveness in the special market of wide-body aircrafts, the exact design condition should be researched and defined.

7 Conclusions

The airlines data of wide-body aircrafts has been analyzed carefully, and the following results have been achieved:

a). The airlines distribution characteristics in different market regions are different.

b). Japanese wide-body airlines market operating wide-body aircrafts is a special market, because of its population and cities distribution, there are large quantities short-thick airlines operated by wide-body aircrafts in this market. In the flight frequencies distribution chart, this market could make a peak value in a short length region, which affects design service life of wide-body aircrafts a lot.

c). The flight length division should be researched. In analysis of airlines data in special market regions, this effect element plays an important role to increase the competitiveness.

d). The base market of Chinese double aisles wide-body commercial aircraft is China, Asia and the Pacific. In this zone, customers should be analyzed carefully, and the customers who are not available but play an important role in this zone should be picked off decisively. The exact market definition for Chinese double aisles wide-body commercial aircraft could reduce engineering difficulties, and also increase airlines adaptability and competition in the specific market zone.

References

1. CAAC. CCAR PART 25, AIRWORTHINESS STANDARDS: TRANSPORT CATEGORY AIRPLANES[S]. 2011.11. (in Chinese)
2. The damage-tolerance and durability of aircraft structure design manual, volume 2, the fatigue analysis of aircraft structure[M]. Beijing: Aviation Industry Press, 1989. (in Chinese)
3. Chen Bushi, Li Yunjun, Wang Zhiyu. Aircraft Design Manual, volume 5, Civil Aircraft General Design[M]. Beijing: Aviation Industry Press, 2005.8, 16～34. (in Chinese)
4. Egbert Torenbeek. Synthesis of subsonic airplane design[M]. Netherlands: Delft University Press, 1976.
5. The market research center of COMAC. The market prediction annual report of COMAC(2016-2035)[R]. 2016.11. (in Chinese)

Research on Interior Ballistic of a Gas Multi-stage Cylinder Ejection

Mei Huang[1,a], Yi Jiang[2] and Boman Wang[1]

[1]*School of Aerospace Engineering, Beijing Institute of Technology, 100081, Beijing, China*

Abstract. Aimed at increasing aircraft's leaving speed based on a determinate length of aircraft container and ensuring the aircraft's acceleration below one certain value, a new gas multi-stage cylinder ejection is proposed in this paper, and at the same time the primary design and the interior ballistic evaluating are processed. The simulation results can be used for the design of gas-catapult device and provide theoretical support for parameter optimizations in trajector.

1 Introduction

According to the launching power, the launching method for aircraft can be divided into self-launch and ejection. Ejection is a way that uses external power to catapult the aircraft out of the launching tube. When the aircraft reaches to a predetermined height, the engine of the aircraft is ignited and it flies as programmed[1]. Compared with the other launching technology, ejection has its unique advantages. Ejection can make the aircraft get a certain initial speed, thus achieving a longer range. Besides, the aircraft is installed in a sealed launching tube. And thus it can provide a stable storage conditions with constant temperature and pressure. So it has been widely used and developed since the late 1950s. Ejection can be classified into compressed air ejection, gas ejection and gas-steam ejection by working medium. In form of launching device, it can be divided into tube ejection and piston cylinder ejection too[2].

The gas multi-stage cylinder ejection proposed in this paper is built on the basis of piston cylinder ejection. Because of the latter's structure limitations, its actual stroke is less than half the length of the launching tube.If we want to increase the aircraft's leaving velocity, it can be solved by lengthening the launching tube or by increasing the mass of propellants. However, the former will affect the maneuverability and traffic-ability characteristic of the launching vehicle. And the latter will lead to the increase of aircraft's overload[3]. In view of the above contradiction, a new gas multi-stage cylinder ejection is put forward in this paper. The gas generator (also called HPC which is short for high-pressure chamber) is used as the power source of the ejection to gurantee the output power. The design of multi-stage cylinder can not only increase the ejection stroke to improve the maximum ejection speed, but also prevent the aircraft from ablation[4-5]. At last, the the interior ballistic performances of this ejection are calculated with numerical method and verified by dynamic analysis.

2 Physical model

[a] Corresponding author : 18612786701@163.com

In the design of the ejection device, the gas generator and the multi-stage cylinders are integrated to simplify the structure and ensure the balance of multi-stage actuating cylinders. The ejection device, as shown in the Fig. 1, is mainly composed of four parts: the gas generator and the LPC(which is short for low-pressure chamber), four groups of multi-stage cylinders and the shell of the device. And each group is composed of ten stage cylinders, each stage has an effective stroke of 1200mm.

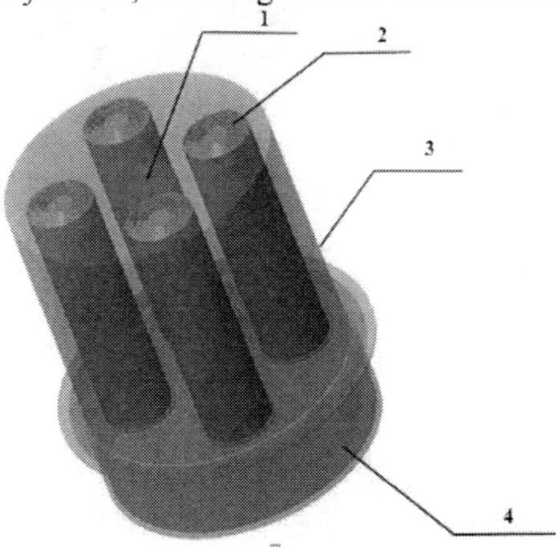

Figure 1. Structure of the gas multi-stage cylinder ejection:

1-the gas generator; 2-multi-stage cylinder; 3-shell; 4-the LPC

3 Mathematical model

3.1 The motion rule of missile

During the ejection process, the missile moves upward along the catapult. It has four forces in the axial direction, namely, ejection force, gravity, and friction. The acceleration expression of missile motion can be obtained from Newton's second law.

$$m\frac{dv_m}{dt}=F_e-F_f-F_g \tag{1}$$

where, F_e is the ejection force, F_f is the friction force, F_g is the gravity.

3.2 Zero dimensional interior ballistic hypothesis

• The gas in high-pressure chamber is treated as perfect gas. And it ingredients, physical and chemical properties will not vary.

• The gas in the low-pressure chamber is treated as perfect gas. And it follows the ideal gas equation.

• The spatial distribution of pressure and temperature in low-pressure chamber is not considered. And suppose they are the same at the time.

• The heat exchange from the chamber wall of the high-pressure chamber to the gas in the low-pressure chamber is neglected.

3.3 Zero dimensional interior ballistics Equations

$$\begin{cases} \dfrac{dP_1}{dt} = \dfrac{RT_1}{V_1}\left(\dot{m}_b - \dot{m}_{t1}\right) \\ \dfrac{de}{dt} = u \\ \dfrac{dv_m}{dt} = \dfrac{S_2 P_2}{\varphi m} \\ \dfrac{dl_m}{dt} = v_m \\ \dot{m}_b = \rho_p S_b u \\ \dot{m}_{t1} = \varphi_{21} K_0 \dfrac{P_1 S_{t1}}{\sqrt{RT_1}} \\ u = ap_1^n \\ P_2 = \dfrac{m_{t1} RT_2}{V_{20} + S_2 l_m} \\ T_2 = \chi_2 k T_1 - \dfrac{(k-1) S_2 P_2 v_m}{R\dot{m}_{t1}} \end{cases} \tag{2}$$

Where, P_1 is the pressure in HPC; T_1 is the temperature in HPC; $\dot{m}_b$ is the burning rate of propellant; $\dot{m}_{t1}$ is the mass flow; v_m is the velocity of missile; ρ_p is the density of the propellant; S_b is the area of the burning face; φ_{21} is the flux modificative coefficient of HPC; S_{t1} is the area of the throat at the nozzle; n is the gas index of pressure; k is the specific heat ratio, and K_0 is a constant connecting with k , $K_0 = \sqrt{k}\left(\dfrac{2}{k+1}\right)^{\frac{k+1}{2(k-1)}}$ [6].

3.4 Algorithm realization of interior ballistic equations

Matlab is used to solve the above equations. It has the characteristics of small code, high programming efficiency, powerful function, and easy to use. On the other hand, it has a unique advantage especially in the numerical calculation and visualization.

The algorithm realization of interior ballistic equations is as follows: Firstly, the above equations can be written into a set of ordinary differential equations. Then, code the program adopting the Runge-Kutta method. At last, Matlab is used to solve the initial value problem of ordinary differential equations.

4 Simulation and analysis

4.1 Calculation conditions

Before the simulation analysis of the multi-stage ejection device, some calculation conditions should be given first.

According to the reference[7-8], the $P_1 - t$ curve and the nozzle are designed shown in Fig. 2-3. Besides, the design performance index is given in Table. 1.

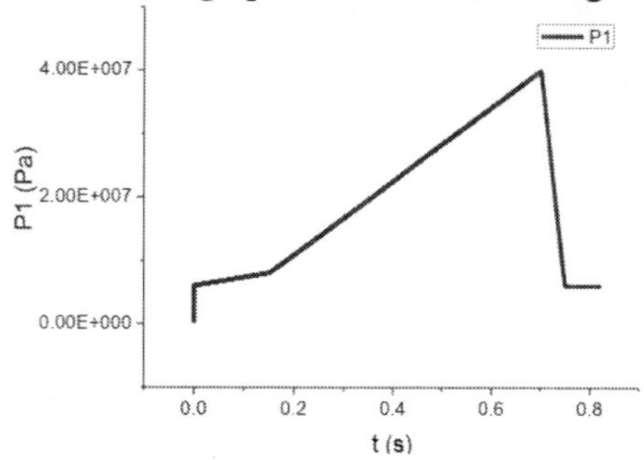

Figure 2. Pressure curve of the HPC .

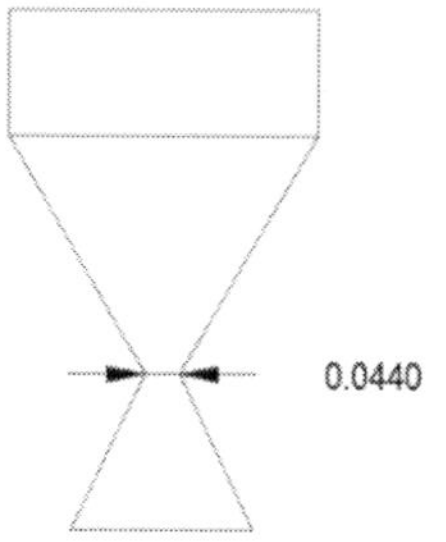

Figure 3. Structure of the nozzle.

Table 1. Design performance index.

Performance index	Value
Leaving velocity	$\geq 25m/s$
Overload	$\leq 8g$
Leaving time	$\leq 1s$

4.2 Internal ballistic simulation

Using the four order Runge-Kutta method to solve the ballistic equations combined with the above scheme design, the corresponding curves of the interior ballistic parameters varying with time can be obtained. And they are illustrated in Fig. 4-7.

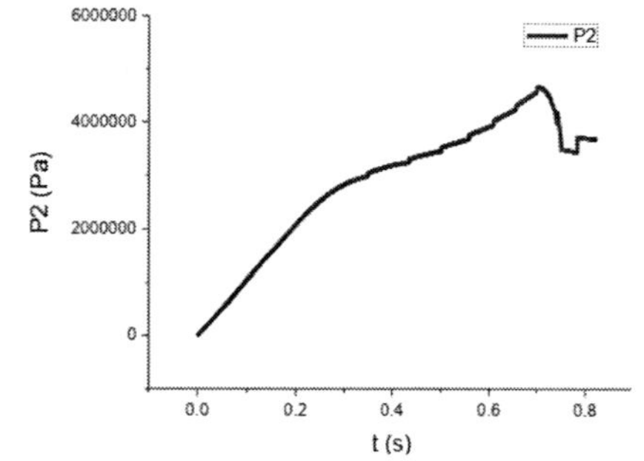

Figure 4. Pressure curve of the LPC.

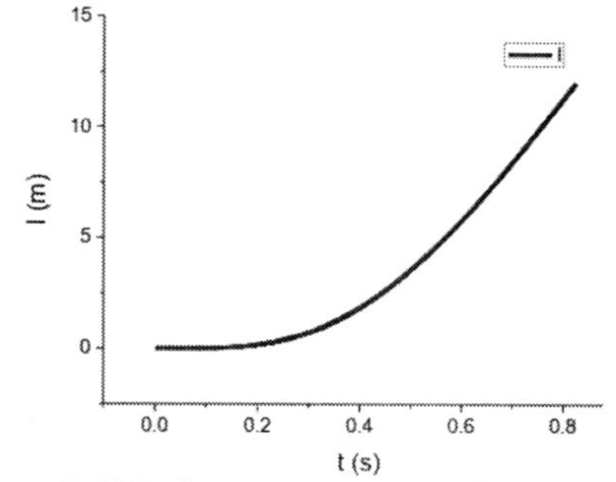

Figure 5. Displacement curve of the payload.

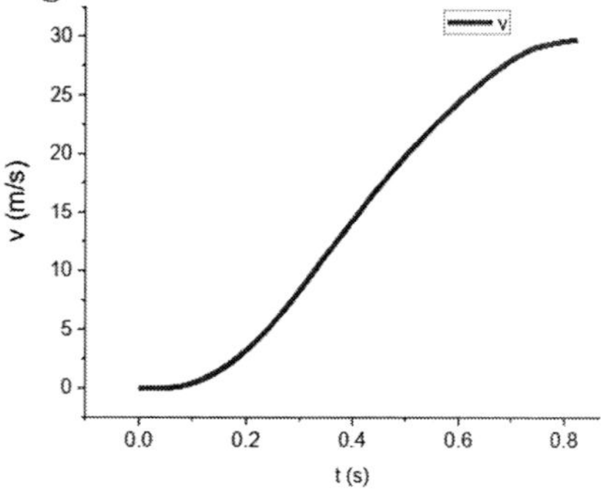

Figure 6. Velocity curve of the payload.

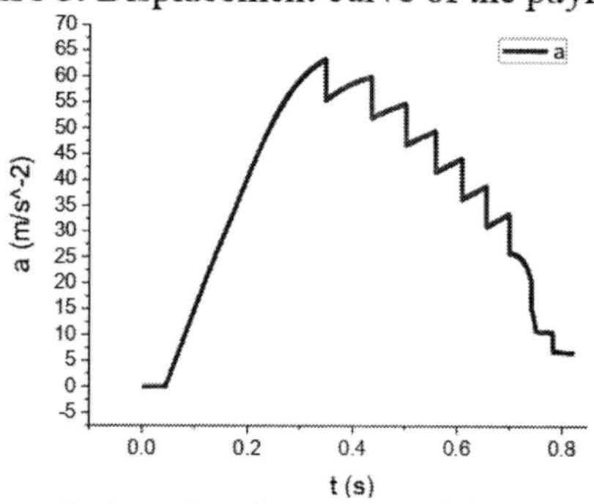

Figure 7. Acceleration curve of the payload.

In Fig. 4, due to the fact that the contact area between the multi-stage cylinder and the payload is gradually decreasing with the full extension of each stage, the pressure in the LPC is on the rise, and the maximum pressure appears at about 0.8s and the peak value is about 5.5MPa. As the decrease of the contact area is faster than the increase of the pressure in the LPC, the overall acceleration of the payload tends to decrease. So from Fig. 7, it can be seen that the acceleration of the payload declines after a certain stage cylinder is fully extended.

From the above figures, the total time of payload ejection is about 0.83s. In Fig. 6, the leaving velocity of the payload is about 30m/s. It can be seen from Fig. 7 that the maximum acceleration of the payload occurs at about 0.3s and it is less than 6g. In brief, the three meet the design requirements.

4.3 Dynamics analysis

As is shown in Fig. 8, the dynamics analysis simulation model is built in ADAMS. In the dynamics simulation, it is supposed that the motion between every stage cylinder is linear relatively to the axis of the cylinder. In addition, the payload is only subjected to the contact force at the bottom while the radial direction is unconstrained. And friction is small and can be neglected.Among them, the cylinder collision coefficient is set as steel to steel shown in Table. 2.

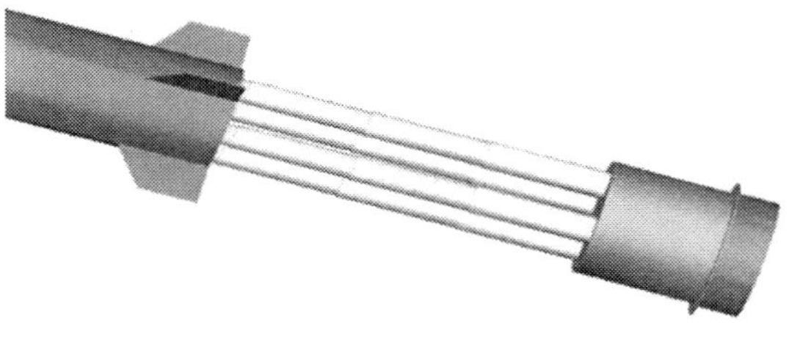

Figure 8. Dynamic model.

Table 2. The cylinder collision coefficient.

Parameter	Value
impact stiffness	1.0e5N/mm
damping	50Ns/mm
index	1.5
depth	0.1mm

Combined with zero dimensional interior ballistic curve obtained before, the results of dynamics simulation are shown in Fig. 9-12.

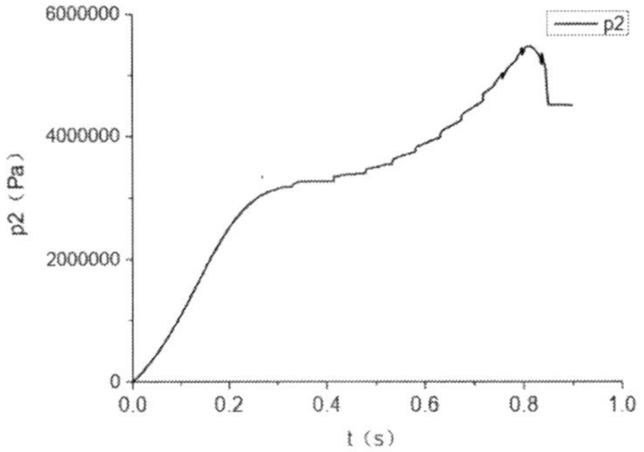

Figure 9. Pressure curve of the LPC .

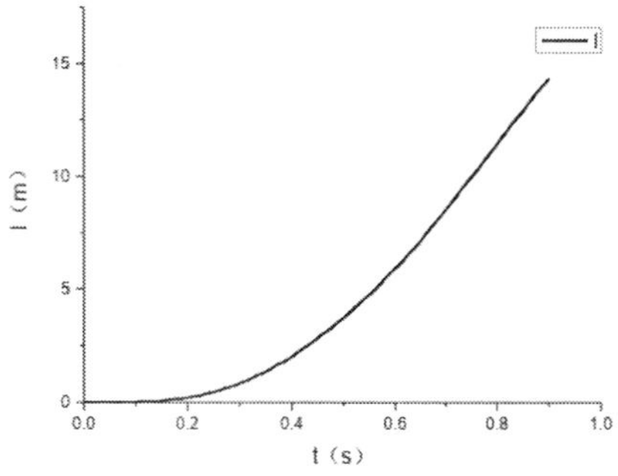

Figure 10. Displacement curve of the payload.

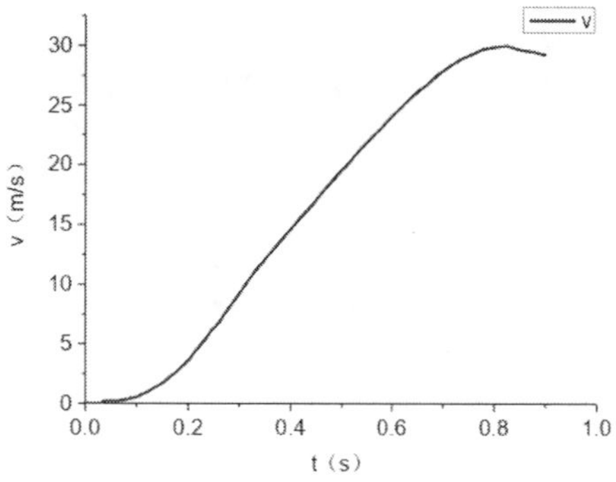

Figure 11. Velocity curve of the payload.

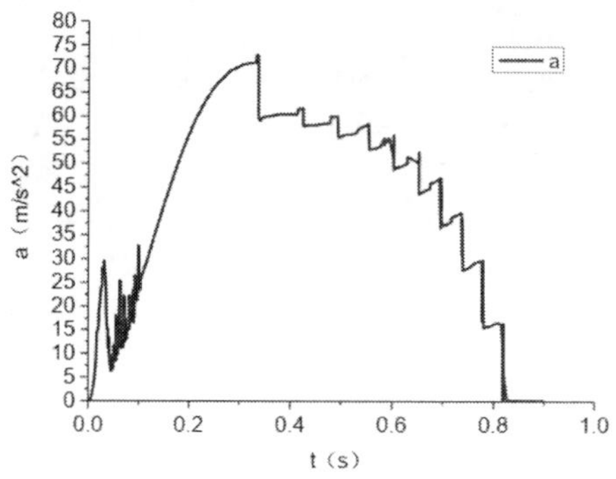

Figure 12. Acceleration curve of the payload.

As shown in Fig. 10, it takes 0.81s to leave the launching tube for the payload and its displacement is 11.8m. It can be seen from Fig. 11 that the velocity of the payload increases smoothly and reaches the maximum at the end of the ejection phase with a speed of 29.7m/s. The maximum overload of the payload is about 7.3g, which appears at the end of the first stage and then decreases progressively shown in Fig. 12.

From the above analysis, the gas multi-cylinder ejection proposed in this paper, without changing the length and diameter of the launching tube, which increases the leaving speed while maintaining a low overload, achieves the design requirements.

5 Conclusion

In the ejection process for payload, high leaving velocity and low overload are contradicted. In view of this problem, a new gas multi-stage cylinder ejection is proposed in this paper. Through preliminary design, interior ballistic calculation and dynamic analysis, it is shown that the ejection can achieve the expected goals to solve the problem in a certain extent. Furthermore, it provides the valuable reference and inspiration to further optimization for new ejections.

References

1. S.Z. Rui and Y.M. Xing. Development Rearch on Aircraft Launching Power system. *T. M. T.* **5**, 4-9 (2009)
2. R. Ren, D.W. Ma, L. Yao, Z. Liu and Q. He. Theoretical Modeling and Performance Research on Multi-stage Pneumatic and Hydraulic Catapult Device. A. A **37**, 1365-72 (2016)
3. F. Yan, S.Y. Shi and Y. Jiang. Rearch on Interior Ballistic of Gas-steam Ejection. S. S. A. T **39**, 118-22 (2017)
4. S.Z. Rui and Y.M. Xing. Comparative Studies of Interial Ballistic Performance among Several aircraft Eject Power System. J. B. U. A. A **35**, 766-770 (2009)
5. P.Y. Bai, J. Qiao. Analysis about the Interior Trajectory of Two-Step Cylinder Ejection Device. M. D. T **35**,44-49 (2007)
6. D.C. Tan 2015 Interior Ballistics for Ejection (Beijing: Beijing Institute of Technology Press)
7. F.B. Yang, D.W. Ma, F. Yang, et al. Interior ballistics modeling and Calculation of high-pressure ejection device. A. A **34**, 527-534 (2013)
8. R.G. Jiang. Aircraft Ejection and Design of Gas Generator. J. M. D. T **26**, 9-16 (1998)

Unmanned Aircraft System Pilot/Operator Qualification Requirements and Training Study

Shengjun Qi[a], Feng Wang and Li Jing

Chinese Flight Test Establishment, 710089 Xi'an, China

Abstract. With the rapid development of Unmanned Aircraft System (UAS) technology, UAS has become indispensable in military and civilian field. Growth in number of UAS and activity has required soaring demand for UAS pilot/operator who has become a hot profession in aviation. In this paper, the UAS classification and unique characteristics of operation are stated firstly. Then, considering current requirements about UAS Pilot/Operator qualification issued by aviation authorities in U.S, UK and China, the general and special qualification requirements are analyzed, which includes professional quality, medical requirements, psychological evaluation, training requirements, operating experience and cooperation. Furthermore, based on differences between UAS and manned aircraft pilot training, the training contents and methods, especially human factor and physiological health, are discussed, which is meaningful and helpful to selection and training of UAS pilot/operator.

1 Introduction

With rapid development of technology, Unmanned Aircraft System (UAS) has been developed from auxiliary force to main air power which could execute surveillance, reconnaissance, electronic intelligence, attack and strike and so on. Meanwhile, UAS has been widely used in civilian application, such as environmental monitoring, aerial mapping, meteorology, firefighting, border patrol, fishery protection and law enforcement. Nowadays, UAS has become indispensable in military field as well as civilian.

Data from FAA showed that over 19000 unmanned aircrafts flew in U.S. skies to conduct aviation activities in 2012. FAA predicts that 30000 unmanned aircrafts will be flying in U.S. skies in 2030[1]. Almost one out of every three aircraft used by the U.S. military is a drone [2]. The U.S. Air Force trains more drone pilots than it trains pilots of manned bombers and fighters combined. It also trains more drone pilots than any other branch of the military. But 1 out of 5 PRV pilot training slots went unfilled [3]. The UAS industry is expected to generate an estimated 100000 to 200000 jobs in next decade. Meanwhile, universities like Kansas State University, University of North Dakota, Embrey-Riddle Aeronautical University are offering bachelor degree in UAS. Some aviation enterprises such as CAE and Northrop provide UAS operator training program as well. In China, Civil Aviation Administration of China (CAAC) estimated that the number of UAS pilot is going to soar over 30000 in near future, but currently there only 700 operators having UAS pilot certificate, which could not meet the demand of market. Now only 27 certification bodies authorized by CAAC have been

[a] Corresponding author : saintknight372@hotmail.com

qualified to entitle UAS pilot certificate, which is mainly for small UAS. As demand increasing, UAS pilot/operator will be a hot professional career.

This study is mainly for UAS pilot/operator qualification and training. Firstly, classification and operation characteristics of UAS are stated. Then, according to requirements in current regulations, the requirements of UAS pilot/operator qualification are discussed. Finally, the last section proposes the training method and content, especially in human factor and psychological health. All of these will be meaningful for UAS pilot/operator selection and training.

2 UAS Classification and Operational characteristics

2.1 UAS Classification

Currently, due to variety of weight, size, performance and operating characteristics, there is no widely accepted classification of UAS. Generally, UAS are classified in terms of maximum take-off weight, altitude, endurance, typical use and operational radius, as shown in table 1[4].

Table 1. UAS Classification

Category	Weight	Altitude	Radius	Endurance	Typical Use
Micro	<2kg	Up to 200ft AGL	<5km	<1h	Reconnaissance, Inspection
Mini	2-20kg	Up to 3000ft AGL	<25km	1-2h	Reconnaissance, Surveillance
Small	20-150kg	Up to 5000ft AGL	<50km	1-5h	Surveillance, Intelligence,
Tactical	150-600kg	Up to 10000ft AGL	100-300km	4-15h	Reconnaissance, Surveillance
MALE	>600kg	Up to 45000ft AGL	>500km	>24h	Surveillance, attack , Reconnaissance
HALE	>600kg	Up to 60000ft AGL	Global	>24h	Reconnaissance, signal relay, Surveillance

2.2 Operational characteristics

Now, as most of UAS are controlled through datalink, the UAS operation has its own unique characteristics. Firstly, since the UAS Pilot/Operator manipulated vehicle by datalink on the remote control station which is located in hundreds or thousands miles base, it is obvious that there is high demand for bandwidth and latency, which has significant effect on mission operation. The more autonomous the control is, the more demanding operator requirement is. Secondly, the endurance varies from several minutes to over 24 hours as well as the operating environment is also complex due to different circumstances. For example, for HALE, pilot/operator should monitor the screens showing vital parameters of the whole system which is boring and dull. Thirdly, any mission for UAS could not be done successfully by individual and must be completed cooperatively by a team of operators, which is shown in previous research [5]. It is well known that there are six operators who monitor 13 screens and work with 6 workstations in Global Hawk Control Station in order to finish intended mission.

Furthermore, compared with manned aircraft pilot, the UAS pilot/Operator, who is not on board and just controls vehicle in Ground Control Station (GCS), could not get the same quality and amount of information. In another word, for UAS pilot/Operator, who loses most sensory information like vision, sound and kinetic feeling, there is no direct feedback from flight environment such as weather, smoke, vibration of aircraft and pull of gravity when he moves the controls, while he supervises the aircraft through on-board sensors and cameras down linking information displayed on the screen of GCS. All the information and status of Unmanned Aerial Vehicle (UAV) can only be felt by UAS pilot via radio telemetry or overhead satellite communication and all maneuvers that pilot does are through up link to transmit orders to the UAV. In conclusion, UAS operator feels isolated from vehicle and lacks traditional feedback. Due to various types and unique characteristics, there is more demanding on UAS pilot/operator.

3 UAS Pilot/Operator Qualification Requirements

3.1 Current requirements review

Nowadays, aviation authorities around the world are committed to legislate on rules and regulations for UAS, not only in operation, but also in approval process. European Aviation Safety Agency (EASA) published A-NPA 2015-10 introduction of a regulatory framework for the operation of Drones in 2015, stating that UAS pilot must gain operation certificate. In order to ensure flight safety, Federal Aviation Administration (FAA) formulated interim Operational Approval Guidance 08-01: Unmanned aircraft systems operations in the U.S. National Airspace system in 2008, proposing UAS pilot/operator requirement for the first time, which says that pilot must hold a license and other operator could not have a certificate. In 2011, FAA published Order 8130.34B Airworthiness certification of unmanned aircraft systems and optionally piloted aircraft, which requests that UAS pilot must hold and be in possession of , at a minimum, an FAA private pilot and a valid second-class(or higher) airman medical certificate. Section 333 of the FAA Modernization and Reform Act of 2012 also addresses that UAS pilot must hold pilot license, medical certificate or valid driving license. Small UAS Notice of Proposed Rulemaking (NPRM) issued by FAA in 2015 claims clearly qualification and obligation of pilot controlling lighter UAS under 25kg. For military, the US Army Regulation 95-23: Unmanned aircraft system flight regulations, requiring that UAS operator should be trained and evaluated.DEF STAN00-970 Design and Airworthiness requirements for service aircraft Part 9 UAV system issued by UK military defense in 2013 points out that UAS operator does not have the experience of manned aircraft . In 2002, the Civil Aviation Safety Authority (CASA) in Australia had announced Civil Aviation Safety Regulations (CASR) AC101-1 Unmanned Aircraft and Rockets: Unmanned Aerial Vehicle (UAV) Operations, Design Specification, Maintenance and Training of Human Resources, which requests that an operator intending to conduct commercial operations utilizing UAV must be the holder of an operating certificate. In China, considering UAS development, CAAC issued Civil UAV Pilot Management Interim Regulation in 2013 that is the guidance for Civil UAV pilot and gives specific requirements about pilot certificate, knowledge and operation. In 2015, CAAC issued Interim Administration Procedure for Civil UAS Operation for Commercial General Aviation Activities, which outlines that Operator, who controls UAS greater than 7kg, climbing over 500 meters, radio radius more than 500 meters, must hold training certificate or license. In 2016, Civil Aviation Flight Standard Management Requirements requests that UAS operator should possess corresponding technical qualification and attend professional training and the UAS should be incorporated into daily management of general aviation.

It can be seen that current UAS regulations only request general requirements of UAS pilot/operator. There are no agreed and detailed requirements about UAS pilot/operator qualification. Thus, it is vital to explore qualification for UAS pilot/operator career development.

3.2 Qualification Requirements

According to current airworthiness regulations and interaction between operator and system during operation, several factors should be considered for UAS pilot/operator qualification, including professional quality, medical requirement, psychological evaluation, training requirements, operating experience and coordination.

3.2.1 Professional Quality

A professional and qualified UAS pilot/operator should possess five aspects of quality, including professional dedication, sense of duty, self-control, enthusiasm for work, logical and rational. Besides, UAS pilot/operator must be at least 18 years old, physical fitness, know well about aeronautical theory and related aviation regulations.

3.2.2 Medical Requirements

Compared with manned aircraft pilot, UAS pilot/operator should have medical certificate, but it is not strict. Of course, it is necessary to have good eyesight, which is not good as manned aircraft's. Medical requirement could be different due to variety of UAS. For small UAS operated in line of sight, a driving medical license could be sufficient. Nevertheless, for larger UAS, operator must hold a valid class two or three medical certificate outlined in 14 CFR part 67 Medical Standards and Certification. For operators, who are responsible for mission design, planning, datalink and payload, the medical requirement could be relaxed due to particular position.

3.2.3 Psychological Evaluation

In recent years, the numbers of accidents caused by pilot psychological problems is climbing, but psychological health is not concerned seriously by airlines. Although International Civil Aviation Organization has paid much attention on this issue, there is still less regulation or research on it. In 2013, FAA published an investigation report about accidents, which shows that there are 8 accidents caused by pilot suicide in 2758 accidents between 2002 and 2012 [6]. According to final investigation report by French Bureau of Enquiry and Analysis for Civil Aviation Safety (BEA), the crash of Germanwings Flight 9525 was deliberately caused by the co-pilot, who showed symptoms of psychotic depression, had been treated for suicidal tendencies [7]. These accidents make airlines and authorities pay more attentions to pilot psychological health.

There is limit about research literature regarding the psychology of UAS pilot/operator, and released UAS regulations do not refer to psychological requirements in qualification requirements. In 2014, U.S Air Force conducted research of psychological evaluation for Remote Piloted Vehicle (RPV) pilot training candidates, and proposed pre-training standardized personality testing for the first time. Personality testing employed the NEO-PI-R, which measures candidates from five major personality domains: neuroticism, extraversion, openness, agreeableness and conscientiousness. The study evaluated 7682 participants who entered manned or unmanned pilot training between 2009 and 2013 [2]. The study revealed that RPV pilot training candidates are more likely to remain calm under pressure, are slower to anger, less prone to feelings of sadness and hopelessness, have higher self-esteem, and see themselves as more capable of handling difficult situations. Now, psychological evaluation has already been used in USAF RPV training candidates, rated pilots and pilot selection, which would help pilot/operator work well and efficiently in long endurance and complex mission environment.

As the level of UAS informatization and automation advancing, the physical workload for UAS pilot/operator is declining, but the mental workload is increasing, especially in emergency, special and crucial mission, Furthermore, psychological problems like stress or anxiety will cause health problem mentally and physically. If these problems are not solved promptly, it will have significant influence on mental health, flight safety and mission accomplishment to a certain degree. Hence, during UAS pilot/operator selection, it is necessary to evaluate psychological state of candidates which contains several factors, such as high levels of conscientiousness, responsibility, achievement orientation, sense of cooperation, emotional stability, resilience, self-confidence, self-esteem, risk tolerance, self-discipline and excitement seeking, and conduct psychological consultant at regular intervals, which is vital for safety of vehicle, people and asset on the ground.

3.2.4 Training Requirements

For small UAS pilot, it is necessary to pass aviation theory test, participate in simple adaptable training and acquire certificate or license as FAA regulation requests UAS pilot must hold a driving license. For UAS above tactical level, pilot/operator must have four-year degree of aviation or engineering, complete professional pilot training program, learn aeronautical knowledge comprehensively and systematically, and earn enough flight hours to qualify for an entry-level, because complexity and risk of UAS mission increase as technology and application expanding. For UAS operator serving in military, police and government, it is indispensable to hold a pilot license with a certain flight hours. Moreover, confronting with battlefield and crime scene directly, military or police operator should attend psychological health consulting to ensure the mental state achieving required level to perform assigned task better. For larger UAS pilot, manned aircraft pilot should be given top priority to candidate selection, and aircrew training should also be arranged in training program.

3.2.5 Operating Experience

Since UAS operating crew are not onboard, operating characteristics and control mode of UAS are to some extent different from traditional piloted aircraft. UAS crews locate on GCS hundreds or thousands of miles away from the vehicle, while UAS relies on more pre-programmed software instead of control device. To some degree, UAS pilot must finish special training to ensure safety of vehicle under circumstance that crew control system in an isolated environment. For small UAS, operator could lack experience due to operating in visual line of sight. But for larger UAS, especially for application for military or governmental purpose, it is essential to having sophisticated experience to ensure mission success and flight safety.

3.2.6 Coordination

As development and application of UAS has been rapidly expanding in recent years, UAS have more chance to cooperate with UAS, manned aircraft, unmanned ground vehicle, and unmanned underwater vehicle in special mission. It is obvious that UAS pilot/operator should not only be familiar with operation and knowledge of UAS and operating characteristics of other vehicles, but also fulfill systemic cooperation training prior to mission execution.

To conclude, a career as a UAS pilot/operator will be very different in the future and there will be more detail about UAS pilot/operator requirements to be released.

4 UAS Pilot/Operator Training

To some extent, there is difference between UAS and manned aircraft pilot training. Firstly, without human onboard, the UAS operator locates GCS to control vehicle. Secondly, comparing with manned aircraft pilot, there is no physical limitation for UAS pilot and the operating skills are different, which require totally different requirements. Thirdly, UAS is controlled by one crew team, including pilot, payload operator, datalink operator, mission planning operator, and other staff. That needs more demanding on responsibility and training content. Fourthly, different types of UAS highlight different aspects of training. All of these determine that UAS pilot/operator training content and method should be considered carefully case by case.

Like manned aircraft pilot training, UAS pilot/operator also needs more training. Relying on real UAS training has two disadvantages. The first one is the risk. Unlike manned aircraft, when manned aircraft has malfunction, pilot could take effective actions to mitigate risk. However, the most severe consequence will be crash if failure or fault happens during UAS flight. The second is growing cost when complexity increasing, especially for large UAS. Using real UAS for every training subject will cost more and need more crew assistance. At the same time, training will also result in worn or damage due to improper operation, which could increase maintenance expenditure.

Considering all these factors, the UAS pilot/operator training should include theoretical training, simulator training, small UAV operation, and specialized training, which could cover almost all the training subjects, reduce training risks and costs, improve efficiency and decrease unnecessary damage due to human errors. The basic content and method of UAS pilot/operator is shown in Table 2.

Table 2 the basic content and method of UAS pilot/operator training

No.	Course	subject	Content	Method and equipment
1	Theoretical training	Basic knowledge	Aerodynamic, the principle of Flight, Flight Control, Airspace Management, Flight Rules, Critical meteorological condition, etc.	Theoretical lecture and visit, every topic is taught by experts. Psychological health is taught by professional aviation psychologist, and targeted psychological training was integrated in simulation and flight. According to investigation reports of accidents caused by human factor, introducing human factor issues in UAS operation, giving relevant guidance in operation training.
		Introduction about UAS	UAS development, category and application, challenges and future technology	
		Components of UAS	Unmanned Vehicle, GCS, Datalink, Payload, Support Equipment, Human Crew	
		UAS missions	Typical task and intelligence analysis, Mission Planning, Mission Termination Procedures	
		Safety	Risk Analysis, Safety Management, Emergency Scenarios	
		Regulations	FAA, NATO, EASA, UK, Australia, etc.	
		Case study	Typical investigation reports of UAS accidents, like Global Hawk, Predator,Hunter and Pioneer	
		Psychological Health	Basic Knowledge of aviation psychology, Psychological training, Stress management, Case study	
		Human Factor	Basic knowledge, Human error and safety, Ergonomics, Case study	
2	simulator training	Introduction	Components, Functionality, Operation Procedures	Training simulator, stimulating from mission analysis to path planning, from control to de-brief, guided by professional instructor.
		Operating skill	Basic Operation, mission planning, take-off and landing, emergency disposal, etc.	
		Typical task simulation	Brief and De-brief, simulating, data analysis, paper report	
		Cooperative training	UAS operator team cooperative training on typical mission	
3	Small UAS operation	Flight skill	Technical skill, take-off, climb, cruise, landing and recovery, emergency disposal	Rotor-wing or small fixed-wing UAS, investigation reports from U.S military
		Cooperation	Simulating cooperation with commander, air traffic controller, and other staff	
4	Specialized training	Special subject training	Aiming at special purpose UAS, for example, for UAS designed for forest fire-fighting, extinguishant launch, reloading, payload exchange, fire detecting, etc.	Using special site, special UAS Using trainer aircraft,

		Operation difference	For larger UAS operator, piloting manned aircraft, comparing manipulative difference between UAV and manned aircraft	such as Cesena 172 or TB-20

According to statistics and investigation reports, psychological health and human error are the two main factors affecting safety. But for UAS pilot/operator training, there is limited research of these. So psychological health and human factor in UAS pilot/operator training will be discussed thoroughly.

A. Psychological Health

Better psychological health could help aircrew relieve fatigue, enhance working efficiency and confront the burdens preferably. Experts in Europe and U.S. point that psychological problem during flight could not be detected in recruiting psychological evaluation. Manned aircraft pilot receive psychological training and counseling at regular intervals. In the same way, the psychological health training should be added in UAS pilot/operator training program. All these training should be instructed by professional aviation psychologist, and the curriculum provision should be set on the basis of mission characteristics, age and current mental health condition, which could help pilot master relevant method and eliminate negative induced emotion. The contents of psychological health training include three parts. Firstly, aviation psychologists introduce fundamental content of aviation psychology. Then on the basis of investigation reports of UAS accidents led by human crew, how important psychological factors in UAS operating are should be discussed. Secondly, aviation psychologists explain current mental training methods, such as Risk Perception and Decision-Making Training, Willpower Mastery, and Self-Psychological Regulating. It is important to train UAS pilot/operator from aspects of feeling, attention, memory and thinking, using flight simulation and real flight. Finally, UAS pilot/operator would make adequate preparation to manage emergency through these training program, which could finally raise working efficiency as well as mitigate mental workload. Lastly, the methods of stress management should be interpreted, which will teach UAS pilot/operator how to adjust his emotion, state, sleep, exercise, self-relaxation regularly. Consequently, UAS pilot/operator could execute designed mission remarkably.

B. Human Factors Issues

Statistics show that nearly 40% of UAV accidents could be attributed to human errors or operating mishap [8]. The study by Barnes and Matz indicated that poor situation awareness was related to operator errors in a series of simulated Unmanned Combat Aerial Vehicle (UCAV) missions [5]. For instance, in which the UCAV crashed, a narrowed attentional focus, especially during targeting activities was partly responsible for the crashes. Simultaneously, pilot/operator's heart rate and subjective assessment of mental workload also increased. The contrast of single versus two operators pointed that only one person could not control UAS to complete assigned task effectively. Thus, it is necessary to offer human factor program in UAS pilot/operator training project, which is helpful for improving flight safety and performance efficiency, enhancing operator control level and reducing crew mishap. The syllabus of human factor includes four parts. Firstly, from the perspective of UAS pilot/operator, the basic content about human factor should introduce the limitation of human crew in cognitive process, such as vision, auditory, situation awareness, decision-making, attention, workload and attitude. Secondly, from the aspects of design and operating, instructor should present ergonomic influences on UAS operation, such as the human-machine interface, control system, display system, and internal environment in GCS. Thirdly, combining the concept of human error with investigation report of typical UAS accident such as Global Hawk and Predator B accidents, instructor will analyse how human error affecting operation and propose suggestion and measurement for preventing human error. Last part should mainly concerns analysing human factor issues that would affect flight safety and operation efficiency, such as workload, situation awareness, human-machine function distribution, teaming issue and operator control level, and giving relevant advice and training.Through human factor syllabus, UAS pilot/operator will realize the importance of human factor in operation, which will mitigate mishap or accident in UAS operating and enhance the safety, reliability and efficiency of UAS.

From above discussion, it can be seen that the basic training methods and contents are proposed and human factor and psychological health are also introduced thoroughly, based on the current development and application of UAS. As UAS technology advancing and automation increasing, the definition and responsibility of UAS pilot/operator will also be changed, and all of these will propose new requirements on UAS pilot/operator training.

5 Conclusions

As UAS technologies advance, military and civil aviation would rely more on UAS, the future for UAS pilot/operator looks to be more busy and promising than ever before. Before the levels of safety on UAS achieving equivalent to manned aircraft, UAS pilots are still irreplaceable. Now engineers and researchers are paying more attentions to study UAS pilot/operator qualification and training, which could improve level of flight safety and promote development and application of UAS. The Tendency of UAS control mode is that UAS has ability of fully autonomy in the future that reduces workload of UAS pilot/operator and raises more demanding on qualification. It is a long way to discuss UAS pilot/operator qualification and training with more exploration and continuous improvement. As rapid development and wide application of UAS, UAS pilot/operator will play a key role in future aviation.

References

1. Department of Defense, FY2011-2036, 25(2011)
2. W. Chappelle, J. Swearengen, T. Goodman,W. Thompson, **AFRL-SA-WP-TR-2014-0001,** 6(2014)
3. J.J. Best, *UAV Pilot*, 11(U.S, 2013)
4. Department of Defense, FY2013-2038,18(2013)
5. J.Barnes,F.Matz, Proceedings of the Human Factors and Ergonomics Society 42nd Annual Meeting,143(1998)
6. R.J.Lewis, E.M.Forster, J.E.Whinnery, N.L.Webster. **DOT/FAA/AM-14/2**,3(2014)
7. K.Williams, **DOT/FAA/AM-04/24**,5(2004)
8. French Bureau of Enquiry and Analysis for Civil Aviation Safety, **BAE2015-0125**,95(2016)

Impact Anlysis of Bow Flap on LHA-1 Ship Airwake

Jinling Wang[1,2,a], Guangwen Jiang[1], Jun Shen[1] and Chujun Hu[1]

[1] Systems Engineering Research Institute, CSSC, 100094 Beijing, People's Republic of China
[2] Beihang University, 100191 Beijing, People's Republic of China

Abstract: The sharp 90°corner of bow found on American Tarawa class general amphibious assault ship LHA-1 will produce large separated region in the airwake over the ship, and the turbulence in the separation region seriously affects the operation safety of the helicopter. In order to reduce the separation in bow region of the LHA-1, and optimize the helicopter operating environment, the numerical simulation method is used to study the influence of the bow flap on the airwake of LHA-1. The study results show that: the ANSYS k-ε two equation model based on the hybrid grid can be used to predict the steady-state characteristics of the ship airwake; the bow flap can improve the flow field downwind of the flap installation position, and the impact range can cover two spots; in headwind, the optimal installation angle of the bow flap is 15°~25°. The research results of this paper can provide modification and improvement reference for the amphibious assault ship in-service and under construction, so as to achieve the purpose of optimizing the ship airwake, thereby improving the safety of helicopter landing process and extending the service life.

0 Introduction

The amphibious assault ship that is the helicopter carrier, is available for launch and recovery operations of the helicopter formation, while the launch and recovery of helicopters on ships, especially the recovery, has become a worldwide technical issue. Not only does the pilot have to track a target moving with multiple degrees of freedom, but one must also land the helicopter in the highly unsteady airwake of the various pieces comprising the ship superstructure. The airwake contains relatively small length scales which will pass through the rotor, being felt as vibrations, and larger length scales that will affect the performance of the rotor as well as the handling characteristics of the helicopter[1]. Therefore, it is of great significance to study the ship airwake characteristics and take measures to optimize and control the airwake environment.

At present, many scholars have made airwake optimization and control research on the rear flight deck ship with the hangar. In 2002, Kumar[2] conducted subscale wind tunnel experiments to assess the effectiveness of passive control involving a porous surface and cavity underneath for reducing the level of surface pressure fluctuations in reattaching flows, and it demonstrated the design can significantly reduce the peak surface pressure fluctuations in the reattachment zone, a maximum reduction of as much as 35% has been observed; in 2003, Tai[3] modified the SFS ship configuration by adding a rounded bow section at the front end to avoid complete separation along the side walls,

[a] Corresponding author : bellincssc16@163.com

the concluding remarked that the complete separation along the hull surface is avoided; in 2005, Shafer[4] explored both active and passive flow control techniques on flow field improvement in the helicopter's final decent onto the flight deck, results show that it would cause a reduction in unsteadiness in the landing region of 6.6% at 0 degrees wind-over-deck (WOD) and 8.3% at 20 degrees WOD by injecting air through the porous surfaces; in 2010, Forrest[5] found that it can decrease the adverse pressure gradient and the turbulence at the hangar edge by installing a 30°flap at the hangar edge of starboard side or applying a chamfer with an angle of 30°; in 2013, the midshipman LaSalle[6] of the United States Naval Academy examined the effects of passive flow control techniques on the ship airwake of YP, and the investigation showed that the particular passive flow control fence produce a less favorable ship airwake for helicopter launch and recovery operations due to an increase in shear, turbulent kinetic energy density, and mixing within the helicopter landing region.

Different from the rear flight deck ship with the hangar, the amphibious assault ship has several landing spots on the straight flight deck, and the related research show that the flow separation of the bow edge will seriously affect the safety of the helicopter operating at the front spots of LHA-1. In order to reduce the separation zone of the sharp 90-degree corner found on LHA-1, Czerwiec and Polsky[7] investigated the effect of adding a 22.5° downward-deflected flap to the bow. Results from the wind tunnel and CFD showed that the flap reduced the turbulence levels related to pilot workload and aircraft fatigue. While Czerwiec and Polsky only discussed the airwake characteristics with 22.5° downward-deflected flap in headwind. This paper will continue research the relationship between flap mounting angle and the airwake characteristics in different wind angles.

1 Simulation Method

1.1 Geometric Model and Grid Generation

Figures 1 show the original LHA-1 (left) numerical model and the one with bow flap (right), respectively. For both original and the modified models, the cargo door and elevators are closed and the antennas are neglected. The LHA-1 flight deck has 10 landing spots shown in figure 2.

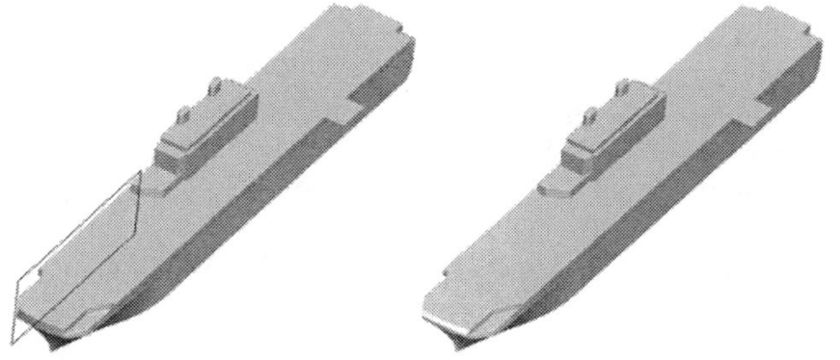

Figure 1. CFD model of LHA-1

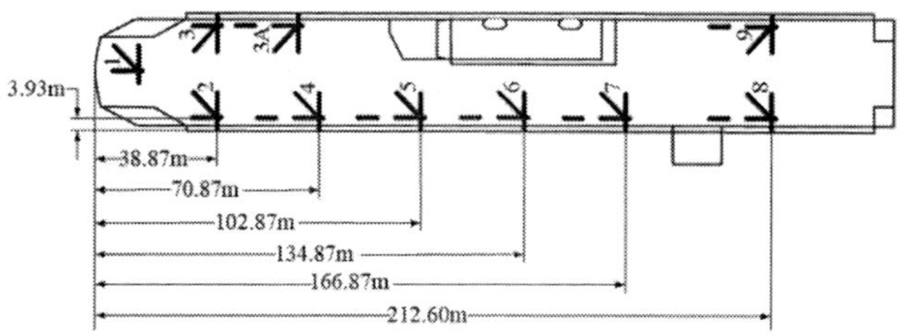

Figure 2. Location of LHA-1 landing spots

Hybrid grids are generated using ANSYS ICEM, the height of the first boundary layer grid on the hull surface is set to 0.003m to ensure that the dimensionless distance $y+$ is between 30 and 60; the maximum size of tetrahedral grid in the key area around the hull is 0.1m; the peripheral prismatic grid size is increasing exponentially from the periphery of the key area to the outer boundary. After verifying the grid independence, the number of grids used for numerical computation is 1.93×10^7, and the hull surface grids and the grids on the vertical section are shown in figure 3.

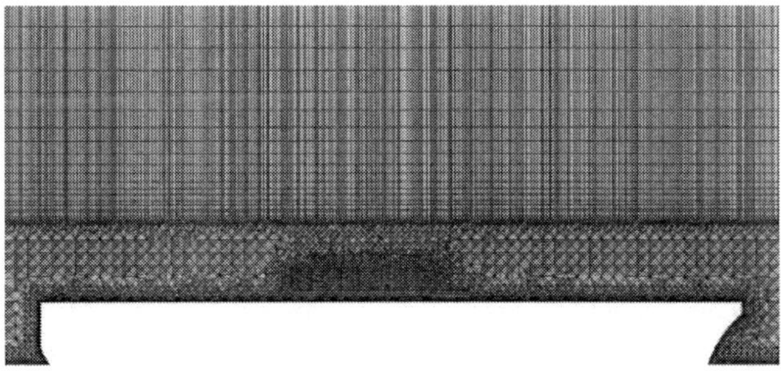

Figure 3. Distribution for different grids

1.2 Numerical Method and Boundary Conditions

The steady-state airwake of LHA-1 is simulated using commercial CFD solver ANSYS FLUENT, and the traditional Reynolds-average Navier-Stokes k-ε two equation turbulence model is applied. At the inlet, free-stream condition of 20.6m/s is imposed; at the exit boundary, a pressure far field outlet is defined and viscous, no-slip conditions are applied to all surfaces of LHA-1; the flat stationary water surface is designated as frictionless surface.

1.3 Simulation Conditions

The CFD analysis is performed for a relative wind of 20.6m/s with a head wind and wind from the port bow (or relative wind angle θ) of 15° . The simulation objects include the original LHA-1model and the model with bow flap. 12 different bow flap installation angles(φ, the angle between the plane of the flap and the flight deck plane) are designed ranging from 5° to 90° , and the original LHA-1 can been seen as the flap installation angle of 90° .

2 CFD Analysis and Results

2.1 Validation of CFD with Wind Tunnel Results

The experimental data to verify the numerical method of the CFD results come from NASA Ames Research Center's Fluid Mechanics Laboratory[8] using 1/48th-scale LHA-1model, and the tunnel's test section extends 2.1m high by 3m wide. Particle image velocimetry, 3 component velocity measurements were acquired at several landing spots along the ship deck for various wind speeds and ship yaw angles.

Figure 4 presents a comparison between CFD calculations and wind tunnel measured upwash distributions across the deck at the rotor tip-path-plane height of V-22 on the spot 2 and 4. Both the variation trend and the upwash velocity value calculated by CFD are in agreement with the wind tunnel measured results for headwind and the green 15°wind angle, which demonstrates that the eddy range and strength predicted by CFD are the same as the wind tunnel experimental results. The verification results proves that the ANSYS k-ε two equation model based on the hybrid grid can be used to predict the steady-state characteristics of the ship airwake.

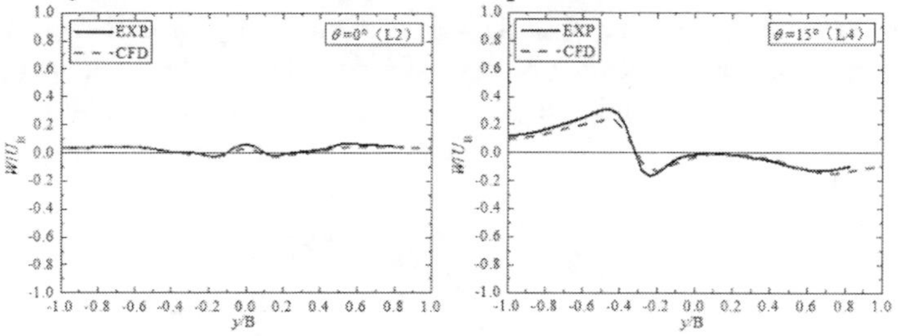

Figure 4. Comparison of CFD and experimental z-component velocity at spots 2 and 4

2.2 The Influence of the Flap Installation Angle on LHA-1 Airwake

Figure 5 provides the normalized vertical velocity contours predicted by CFD for the rotor disk of a UH-60 helicopter sitting on the flight deck in each of the three forward landing spots for 12 different flap installation angles φ in headwind.

The contrast results show that: when $5°\leqslant\varphi\leqslant15°$, as the φ increasing, the vertical velocity gradient gradually decreases, and the flow field is even more uniform, the improvement effect of the helicopter landing area is more and more obvious, and the improvement effect of the airwake in the spot 1 is most obvious; when φ=15° , the downwash of spot 1 disappears, and that in spot 2 and 3 are very weak; when $15°<\varphi\leqslant25°$, the flow field is kept constant with the increase of φ value; when φ=30°, compared with which when$15°<\varphi\leqslant25°$, the flow field in the spot 2 and 3 keep invariant, while the upwash region is smaller, and the uniformity of flow field becomes worse; when $30°<\varphi\leqslant90°$, the improvement effect of the helicopter landing area is getting worse, and the velocity gradient is getting larger as the value of φ increases.

The φ interval in this paper is 5° , which is not enough to accurately capture the critical φ that causing a change of the flow field. While it can be judged that the improvement of the flow is the best when the flap installation angle in the range of 15° to 25° according to the airwake variation trend of the LHA-1. It is observed from figure 5 that the flow field when φ=22.5°is the same with that when φ=15°, φ=20° as well as φ=25°. As it happens, Czerwiec and Polsky chose a flap installation angle which right locates in the range of the best installation angle for the improvement of the airwake.

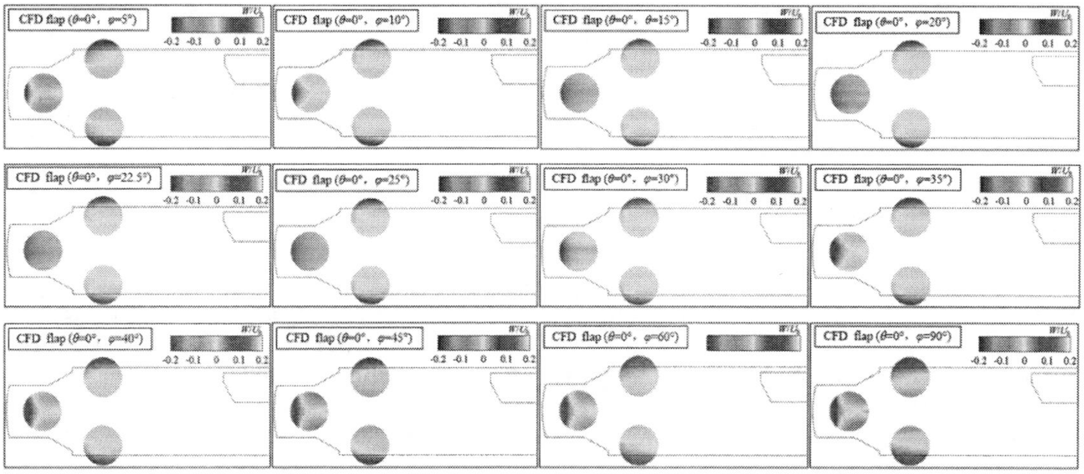

Figure 5. z-component velocity contours at the UH-60 rotor plane for LHA-1 with different bow flap angles in headwind

2.3 The 22.5° Bow Flap Effects on ship airwake of LHA-1

In this section, it continues to study the ship airwake characteristics of LHA-1 with and without 22.5° bow flap. Figure 6 displays the vorticity isosurfaces on the flight deck of two models in headwind and red 15° wind angle. For both WOD angles, the original LHA-1 model has larger coherent structures from the bow corner and the deck edges, the bow flap produces a significant reduction in the amount of vertical flow down the center of the flight deck. As a result, much of the bow and the center of the flight deck experience more uniform, steady flow for improved aircraft operation characteristics[7].

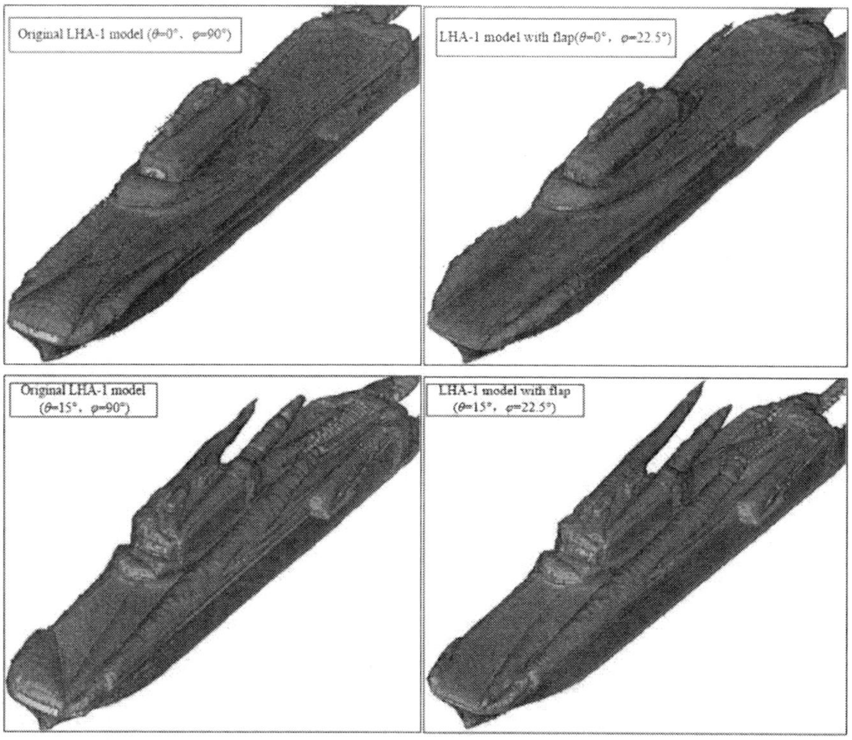

Figure 6. Iso-surface of vorticity of LHA-1 with and without bow flap in headwind and red 15°WOD angle

Figure 7 displays the pressure contour distribution over the bow for both CFD and wind tunnel results in headwind and red 15°, respectively. For original LHA-1 model, the large separated regions causing by the bow corner produce a low pressure, and the pressure in spot 1 is almost constant in headwind. For the modified LHA-1 model, the pressure increases gradually, and the pressure gradient decreases; especially for red 15° wind angle, the low pressure zone in the port bow deck disappeared.

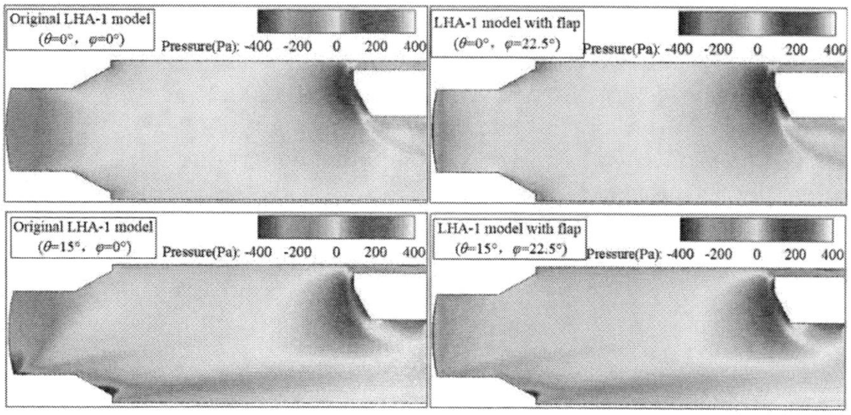

Figure 7. Pressure of flight deck surface for LHA-1 with and without bow flap in headwind and red 15°

Figure 8 supplies normalized vertical velocity contours for the rotor disk of UH-60 on three forward landing spots. The bow flap clearly produces much more uniform flow for spot 1, and eliminates the downwash flow regions for three spots of headwind, and the downwash velocity in spot 2 reduced obviously for red 15° wind angle. Combined with the calculation results in figure 7, it can be found that the bow flap has obvious improvement effect on the downstream region of the flap installation position and its influence on the flow field can cover two spots.

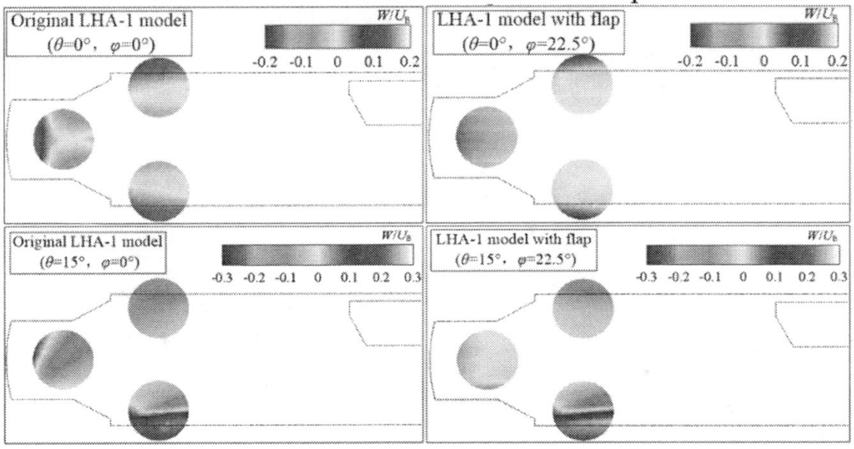

Figure 8. z-component velocity contours at the V-22 rotor plane for LHA-1 with/without bow flap in headwind and red 15° WOD angle

3 Conclusion

The numerical simulation results of LHA-1 ship airwake with flap in different installation angles draw the following conclusions:

(1) The ANSYS k-ε two equation model based on the hybrid grid can be used to predict the steady-state characteristics of the ship airwake;

(2) The bow flap can improve the flow field downwind of the flap installation position, and the impact range can cover two spots;

(3) In headwind, the optimal installation angle of the bow flap is located in range of 15° to 25°.

This study shows that the bow flap with a specific installation angle will improve the uniformity of the ship bow airwake and reduce the flow separation. Future work will investigate more wind angles conditions, and the flap will be extended to the whole ship deck edge to further verify or extend the conclusion of this paper.

References

1. G. F. Syms. J. Wind, Eng. Ind. Aerodyn, **96**, 1198(2008).
2. R. Kumar, R. Viswanath. Aer. J, **106**, 673(2002).
3. T. C. Tai. NATO RTO Symp. Vort. Fl. High Angl. Att, 10(2003).
4. D. Shafer. Virginia Polytechnic Institute and State University, 95(2005).
5. J. Forrest, C. Kääriä, I. Owen. Am. Hel. Soc. 66th Ann. For. Techn. Displ, 15(2010).
6. LaSalle, N. Robert. Annapolis, 72(2013).
7. R. Czerwiec, S. Polsky. 22nd Appl. Aer. Conf. Exh, 2(2004).
8. G. Rajagopalan, D. Schaller, A. Wadcock, Yamauchi G, Heineck J, Silva M. 43th AIAA Aer. Sci. Meet. Exh, 2(2005).

Research on Key Techniques of the Verification and Testing System for Spacecraft Propulsion

Wang Junwei[1,a], Zhang Lei[1], Xiao Qingsheng[1], Li Gao[1], Han Xiao[1] and Bai Bing[1]

[1]*Beijing Institute of Spacecraft Environment Engineering, Beijing, China*

Abstract. Electric propulsion is an important technology in the future space missions. The verification and testing system is mainly used for the manufacturers to fully understand the propulsion behavior in the most diverse conditions, and to indentify and eliminate potential failure modes to advance its validity and reliability, explore the applied potential for space missions. The paper presents a new verification and testing system for spacecraft propulsion and the key technologies needed to be developed are studied. The debugging data and research indicate that ultimate pressure of system is better than the 5.0×10^{-6}Pa. Both the cold plate used for xenon pumping and sputtering protective system get an excellent performance in system. The DSMC numerical simulation method is verified that is beneficial to the research of optimization design and plume effect of electric propulsion in future.

1 Introduction

In order to meet the space missions of high performance, long life spacecraft propulsion in future, electric propulsion technology is getting increasingly attention in many countries. The United States, Japan, Russia, the European space agency are strengthening in electric propulsion technology research and has been successfully applied in many space missions, electric propulsion technology for the task of deep space exploration and satellites in orbit has great application prospect.

Electric propulsion system needs to pass a comprehensive testing to verify their performance level, service life and compatibility with the environment before practical application. Through lots of environmental testing, researchers can completely understand the performance under different environmental conditions to promote the performance of the system parameters and usage, identify and eliminate potential risks, improve the effectiveness and reliability. At present, advanced research institutions and companies abroad have established the ground experiment test system used in all kinds of electric propulsion system [1-3]. Verification and testing technology of electric propulsion will become the important supplement to electric propulsion technology research in the future.

2 Typical compositions of verification and testing system

1) Vacuum vessel system

Typical design of vacuum vessel in verification and testing system for electric propulsion is horizontal structure, which can be designed as the main vessel and auxiliary vessel while a vacuum

[a] Corresponding author : tobenol@126.com

valve connecting between two segments. The main vessel is the main testing environment, so the design needs to consider the sputtering deposition and pollution effect of working medium. The length design of the main vessel generally should be 4 to 5 times of the radius. Furthermore, taking into account the sputtering effect of main vessel material and magnetic shielding, non-magnetic stainless steel is selected for the vessel.

2) Vacuum pumping system.

At present, the electric propulsion verification and testing system generally adopts the four-stage vacuum pumping configuration to exhaust gas. The typical vacuum system configuration includes the four-stage oil free pumps, the first stage is the rough vacuum system, and the main equipment is the Roots units. The second stage vacuum pumping system is the molecular pump pumping system. The third stage is the high vacuum pumping system, the main equipment includes the cryogenic pump and the fourth stage is the xenon pumping system, which is generally used to remove the working medium of xenon.

3) Temperature control system.

The temperature control system of the electric propulsion verification and testing system consists of the shroud and temperature control unit, wherein the shroud material is made of stainless steel. As shown in figure 1, the structure of shroud is in the diamond honeycomb shape, in which the bulge surface is oriented toward the test piece and the back plate faces the wall of the vessel. Using this kind of shroud can achieve better temperature uniformity.

Figure 1. Diamond honeycomb shape shroud using in testing system of electric propulsion

Heat transfer oil refrigeration and heating system can quickly achieve heating and cooling of shroud in the temperature range of -75℃ to +100℃. Temperature control system adopts water-cooling two-stage refrigeration technology, which provides the system refrigeration through the efficient plate heat exchanger and heat source through the electric heating device for the system.

4) Motion platform system

In order to realize the movement of electric propulsion inside the testing vessel, the motion platform system shown in figure 2 is equipped in general. The system design considers the usage under the high and low temperature environment, so the selection of components and structural materials of the platform should meet the harsh conditions of high vacuum and temperature change. In order to meet the requirements of high vacuum environment, the oil parts such as bearings is cleaned and treated with MoS_2 solid lubrication.

Figure 2. Motion platform system using in testing system of electric propulsion

5) Sputtering protection system

The high-energy particles produced by electric propulsion can produce sputtering corrosion on the structure and materials of testing system when it's working, especially for the high-power and long-life propulsion systems. The sputtering effect will have an impact on the test results, and on the other hand, it also pollutes the thruster itself. Typical sputtering protection system can implement sputtering protection of the whole testing system, which is generally realized through the style of beam target. The beam target adopts the materials with low sputtering rate and achieves full system protection through special incident angle design.

3 Key technologies of the system

3.1 The pumping technology of working medium

The verification and testing of spacecraft propulsion has its particularity. On the one hand, electric propulsion must be injected the working medium continuously, and the gas medium is transformed into high-energy beam in the vacuum vessel after ionization. Especially for the high power propulsion system, the higher power means that it will consume more gas medium. So the testing system needs to have greater extraction capacity than traditional thermal vacuum testing equipment. On the other hand, the testing should be specially design in vacuum system because of the particularities of xenon. The traditional vacuum pump can't be fully qualified for the extraction of xenon: the molecular pump is relatively small and expensive to the pumping system. The working principle of the diffusion pump may cause oil pollution to the propulsion system. The Cryopump is main pump of the traditional thermal vacuum test, but its pumping capacity for xenon is weak. Due to the higher ion energy of electric propulsion in working, the Cryopump using in electrical propulsion experiment as main pump has many limitations.

In our testing system, the vacuum system adopts cold plate driven by cold head to pumping working gas, and the testing data shows that the method can well meet the needs of ground verification. The cold plates installed in the vessel are shown in figure 3. In theory, the maximum pumping speed on the surface of cryogenic cold plate can be expressed by the following formula:

$$S_{max} = \frac{1}{4}\bar{V}_1 A = (\frac{RT_g}{ZwM})^{1/2} A = 3.638A(\frac{T_g}{M})^{1/2} \tag{1}$$

Where A is the cold plate area, cm^2; T_g is the temperature of the pumping gas, K; M is the molar mass of the gas molecule; R is the gas constant. In fact, due to the influence of gas pressure and condensation coefficient, the actual pumping speed of the cold plate is:

$$\begin{aligned} S_\alpha &= \alpha S_{max}[1 - \frac{P_s}{P_g}(\frac{T_g}{T_s})^{1/2}] \\ &= \alpha 3.638A(\frac{T_g}{M})^{1/2}[1 - \frac{P_s}{P_g}(\frac{T_g}{T_s})^{1/2}] \end{aligned} \tag{2}$$

In which P_g is the gas pressure, Pa; P_s is the vapor pressure, Pa; T_s is cold plate temperature, K; α is the condensation coefficient. For the calculation of gas characteristics of xenon, set $P_g >> P_s$, $\alpha = 1$, and the pumping speed of the cold plate is as follows:

$$S_{\alpha,Xe} = 3.638\sqrt{Tg / M} = 5.44L / s \cdot cm^2 \tag{3}$$

The pumping ability of cold plate is proportional to its area. Therefore, in the process of vacuum system configuration, we can calculate the need pumping speed according to the requirement of electric propulsion system such as working flow and working pressure, calculate and equip with enough area of cold plate.

In the design process of the cold plate, it is also necessary to consider the influence of the system thermal load to the final cooling power required in the system. The thermal load can be calculated by the following formula:

$$Q_{th} \approx \sigma\varepsilon_r (T_2^4 - T_1^4) A_1 \tag{4}$$

In which, $\sigma=5.67\times10^{-12}W/cm^2K^4$, ε_r is the emissivity of cold plate, T_2, T_1 is the ambient temperature and the cold plate temperature, and A_1 is the cold plate area.

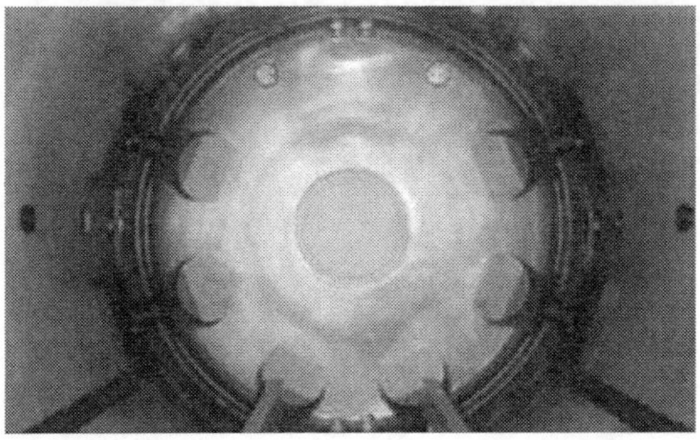

Figure 3. Cold plate driven by cold head in pumping system of the testing vessel

Therefore, in order to guarantee the normal working of the cold plate, the refrigeration power and the cold plate area required by the system has a direct relationship with the environment temperature. The pumping speed is proportional to the area of cold plate, but at the same time the cold plate area is limited by the heat load and the refrigeration power. Thus to improve the pumping ability, the effort should be on the minimization of heat load or selection of high-power refrigeration system of cooling-plate.

To reduce the heat load, it is recommended that the shroud and insulation materials are installed in the vessel while the shroud can be cooled by liquid nitrogen. On the one hand, liquid nitrogen shroud can pump part of gas, on the other hand, the shroud greatly reduces the system radiation to cold plate, which improving the abilities of pumping speed of cold plate and running time[4-5].

3.2 Sputtering protection research

In order to meet the requirement of long life testing to electric propulsion, the sputtering protection system should be added in system to realize the protection of xenon particles in the ignition process. The design of sputtering protection system is various and pivotal [6-7]. The figure 4 is kinds of protection system using in the testing system abroad.

Figure 4. The protection system using in the testing system abroad

The sputtering protection system in our system adopts the similar structure with shroud, which is titanium structure or graphite coated material. In general, sputtering protection system is divided into xenon pump protection unit, the main vessel protection unit and the beam dump installed in front of thruster. The beam dump adopts the structure of the louver while the vessel protection unit is fish skeleton structure. For the convenience of subsequent maintenance, the inside of vessel protection unit is detachable type titanium plate protection layer and the outer layer is the welded titanium plate. The structure of xenon pump protection unit is the same with beam dump. The overall structure is shown in figure 5 and the structure of beam dump is shown in figure 6.

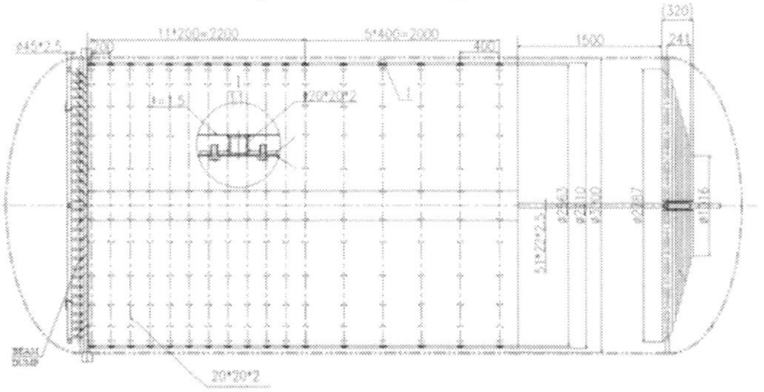

Figure 5. The overall structure of sputtering protection system

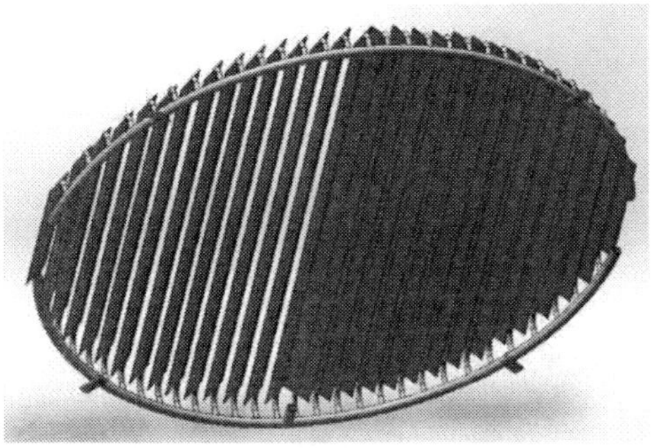

Figure 6. The beam dump installed in front of the thruster

3.3 System optimization based on DSMC method

Due to the complex pressure and temperature flow field in the system, the layout of the cold plate and other equipment is essential. The distribution of internal pressure flow field is related to the area, position, structure form of cold plate and sputtering protection system. Due to the difficult by traditional calculation method to carry on the design, it is recommended that the direct Monte Carlo (DSMC) method is used to simulate the vacuum flow field to achieve the optimal design of complete testing system.

The core of DSMC method is using a small amount of simulating molecules instead of the number of gas molecules to simulate the actual physical process. In the simulation, the calculating system tracks simulated molecular, records the change of their position and velocity, calculates the collision between the molecules and the molecules with surface. Through statistical average, various flow parameters can be obtained finally. It can better guide the configuration optimization of electric propulsion testing vacuum system through the numerical simulation of DSMC, which has been applied in advanced research institutions abroad [8-9].

All particles collide with the wall transfer into the neutral particles, then reflect them according to Maxwell's model.In addition to the xenon cold plate and the thruster wall are the exit of particle, the particle can't escape from the other boundary surface. The basic input parameter of simulation is shown in table 1.

Table 1. Basic input parameter of thrusters

Outside diameter of EP	100mm
Inner diameter of EP	60mm

Gas flow	160sccm
Divergence angle	45°
Ion velocity of Xe^{+}	18000m/s
Ion velocity of Xe^{++}	25456m/s
Atomic temperature of Xe	273K
Ion temperature of Xe	4eV
Wall temperature	300K
Reference temperature of isothermal model, T_{ref}	3.5eV
Reference electron density	$1.0e^{16}/m^3$

The background pressure in the vessel is the main factor which affect the measurement data of electric propulsion.The internal pressure of the vessel is made up of Xe atoms, Xe^{+} ion and Xe^{++} ion density. Due to the slower velocity of Xe atomic, the accumulation effect is obvious. At the same time, due to collision with wall, Xe atomic number is higher than that of Xe ions, which also cause the Xe atomic is the main part of the pressure distribution.

The simulation is shown in the figure 7, (a) is the original schemes without any sputtering protection devices. the position of the cold plate in (b) and (c) is the same, and the different is that a circular protection baffle is arranged in front of the cold plate in the latter section of the vessel in (c). By contrast, it can be seen that the blocking effect of the circular protection baffle structure on the reflux particles is obvious, and lots of particles are adsorbed by the pieces of xenon cold plates set in the latter section of the vessel, which reduces the overall pressure of the latter section. However, the circular protection baffle has no obvious effect on the particles in front part of vessel, so the pressure distribution in the front of the vessel is similar in the two optimization schemes.

The beam target is setted up in the end head of the testing system in Figure (d), The particle enteres the back head of vessel after collision with beam dump, which leads the particle's accumulation and increasement of pressure in end head, which illustrates that the Xe^{+} ions with higher speed and CEX particle collide with beam dump and lose its charge. The motion speed dropped to level of thermal motion and particles accumulates in this aera. The pressure in front of baffle is lower than other schemes, so the scheme can provide a better ignition environment in the place of thrusters. It also can be seen from the diagram that the xenon cold plate should be added to reduce the pressure in the end head.

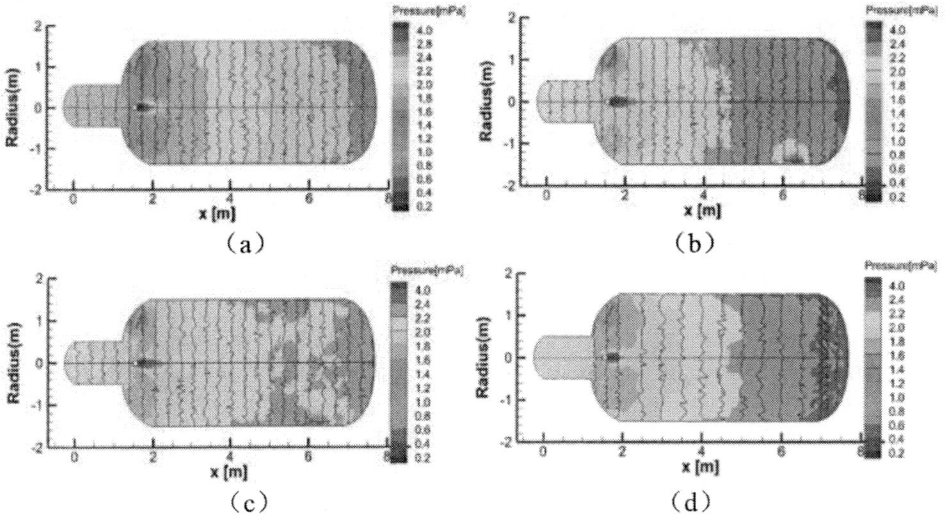

Figure 7. The simulation and optimization design of testing system

4 Conclusions

Based on the research of the key technology, the final validation and testing system for spacecraft electric propulsion get an excellent performance, and the ultimate pressure of vacuum system is better than the 5.0×10^{-6}Pa. The pumping capacity of cold plate can satisfy the demand of high-power electric propulsion using xenon gas while the sputtering protective system has played a certain protective effect to all system. Through the DSMC numerical simulation, it indicates that the distribution of pressure field is related to area, position, structure of cold plate and sputtering protective system. The numerical simulation method can better guide the configuration optimization of electric propulsion testing system. The method has been applied in advanced research institutions abroad and proved that it's beneficial to the research to the optimization design of vacuum system and plume effect research of electric propulsion.

References

1. Neumann A, Holz A, Dettleff G, et al. The new DLR high vacuum test facility STG-ET[C]. The 32nd International Electric Propulsion Conference, Wiesbaden, Germany. 2011:1-10
2. M. Saverdi, M. Signori, L. Milaneschi, et al. The IV10 Space Simulator for High Power Electric Propulsion Testing: Performance Improvements and Operation Status[C]. The 30th International Electric Propulsion Conference, 2007:1-13
3. M. Andrenucci, M. Berti, L. Biagioni, et al. The New EP Test Facilities at Centrospazio and Alta[C]. The 28th International Electric Propulsion Conference, 2003:1-9
4. Andrew D. Ketsdever. Design Considerations for Cryogenic Pumping Arrays in Spacecraft–Thruster Interaction Facilities [J]. Journal Of Spacecraft And Rockets Vol. 38, No. 3, May–June 2001
5. Leonardo Biagioni, Andrea Passaro, Mariano Andrenucci. Particle Simulation of Tailored Vacuum Pumping Configurations for Electric Propulsion Testing[C]. Fourth International Symposium Environmental Testing for Space Programmes, ESA SP-467, 2001
6. Passaro A, Vicini A, Biagioni L. Plasma Thruster Plume Simulation: Effect of Vacuum Chamber Environment[C]. The 35th AIAA Plasmadynamics and Lasers Conference 28 June-1 July 2004, Portland, Oregon
7. Andrew D. Ketsdever. An Overview of Ground Based Spacecraft-Thruster Interactions Studies: Facility Design Issues[C]. The 38th Aerospace Sciences Meeting & Exhibit. AIAA 2000-0463
8. Biagioni L, Kim V, Nicolini D, et al. Basic issues in electric propulsion testing and the need for international standards[C]. The 28th International Electric Propulsion Conference. Toulouse, France. IEPC Paper. 2003, 230: 17-21
9. L. Biagioni, L. Boccaletto, S. Marcuccio, D. A Large Space Simulator for Electric Propulsion Testing: Design Requirements and Engineering Analysis[C]. The 36th AIAA/ASME/SAE/ASEE Joint Propulsion Conference. AIAA 2000-3750

Is Engineering Ethics Important for Aerospace Engineers?

Sakineh Haghighattalab[1,a], An Chen[1,b] and Mohammadreza Saghamanesh[2]

[1]*Institutes of Science and Development, Chinese Academy of Sciences, Beijing 100190, China*
[2]*School of Aerospace Engineering, Tsinghua University, Beijing, 100084, China*

Abstract. Engineering as a profession has a direct effect on society and the environment. Engineering ethics is a part of the essence of engineering. One of the important branches of engineering profession is aerospace engineering. Furthermore, aerospace industry achievements play an undeniable role in our lives. Research and development in the aerospace domain have contributed to the progress of some new technologies in the last decades. The purpose of this study is to emphasize the importance of engineering ethics as an essential part of aerospace engineering. Engineering ethics examines professional responsibilities and ethical decision making of engineers. Moreover, codes of ethics help the engineers to apply ethical principles in critical conditions. The poor ethical decision-making of engineers leads to engineering failures which jeopardized human life and the environment. This paper by examining two case studies related to the field of aerospace engineering (Challenger and Columbia disasters) describes the role of the negligence of engineering ethics on the occurrence of engineering disasters.

1 Introduction

Science and ethics are inseparably connected. Ethics is a necessary basis for science. Some ethical issues have a fundamental impact on all scientific activities [1]. Carl Mitcham notes "when science takes the world into its laboratory, engineering takes the world for a laboratory" [2]. In addition, the role of the engineering profession in creating advanced technologies has caused that the engineers pay more attention to the consequences of their designs and decisions. [3].

Engineering plays a key role in society and the environment. Moreover, engineering ethics is considered as a component of the essence of engineering. Engineers often face conditions in which they do not have enough data and cannot be completely certain that their design will never damage anyone or cause hurtful changes to society [4]. Hence, the design of engineers should be based on the highest standards of integrity and honesty. According to NSPE code of ethics for engineers "the services provided by engineers require honesty, impartiality, fairness, and equity, and must be dedicated to the protection of the public health, safety, and welfare. Engineers must perform under a standard of professional behavior that requires adherence to the highest principles of ethical conduct"[5]. The evidence within the engineering ethics literature indicates that disregard for ethics by engineers have been lead to many engineering disasters [4, 6, 7]. Some well-known disaster cases

[a] haghighattalab61@gmail.com
[b] Corresponding author: anchen@casipm.ac.cn

which are resulting immorality in engineering are Challenger Disaster (1986), Bhopal Disaster (1984), Chernobyl nuclear disaster (1986), Columbia Disaster (2003), Hydrolevel Corporation (1971) [6, 8]. One of the important branches of engineering is aerospace engineering. Aerospace achievements have an undeniable role in our lives. Furthermore, the progress in many industries such as the transportation field, health domain, safety, and autonomy has seen significant advances in the last decades due to the aerospace research and development [9].

The goal of this study is to emphasize the importance of engineering ethics as an essential part of the engineering activities especially in the field of aerospace engineering. In this paper, by exploring two space shuttle case studies (Challenger and Columbia), we will illustrate the role of disregarding to ethics on the occurrence of disasters. The studies in Challenger accident have revealed a violation of engineering ethics by managers and engineers as a significant cause of the disaster [10, 11]. Seventeen years later, the Columbia disaster repeated with ethical reasons similar to the Challenger. For more clarification, the descriptions of ethics and engineering ethics are presented in next sections.

2 Ethics

Ethics is defined as "the study of the characteristics of morals" [4]. Ethics is a communication behavior based on respecting the rights of others. Ethics is being accountable for individuals' rights [12]. Ethics are divided into three categories in an initial division [6] (see Fig. 1).

Common ethics: the series of common moral ideals. In fact, lie, break promises, cheat, steal, harming others physically are some concepts that we think about them when we imagine ethics.

Personal ethics: the series of ethical beliefs that are usually obtained from religious, education or family.

Professional ethics: "the set of standards adopted by professionals insofar as they view themselves acting as professionals." Every profession such as architecture, law, pharmacy and, engineering has its professional ethics [6].

Figure 1. Categorization of Ethics [6]

3 Engineering ethics

Engineering faces many ethical issues and challenges in its development process and needs to resolve its dilemmas with the help of ethical guidance related to the engineering profession [13, 14]. Engineering ethics "is the rules and standards governing the conduct of engineers in their role as Professionals"[4]. Engineering ethics has developed in the last three decades with two goals: the prevention of engineers' unethical behavior, and the prevention of technology threats to the public health and safety [15, 16]. Professional engineering ethics can be considered in two parts: 1) preventive ethics that focuses on preventing professional misbehavior and engineering disaster 2) aspirational ethics that inclined to make a better life via technology [6].

Many engineers are salaried employees of corporations or organizations. Since companies usually try to gain more profit, some managers may pay little attention to the safety in their decisions. Engineering ethics helps to solve the conflicts between management cost-benefit calculations and the engineer obligations to public safety [11]. Often, the correct responses to ethical dilemmas in engineering profession are difficult and cannot be solved by engineers and managers. Engineering ethical codes help engineers to solve such ethical dilemmas.

Professional codes of ethics are one signs of the modern societies. Lynch and Kline believe that the perception of the significance of ethical codes is main elements of engineering ethics instruction [11]. For example, one of the most famous ethical codes of the engineering is the IEEE (Institute of Electrical and Electronics Engineers) Code of Ethics. Some of the IEEE codes are as follows: "1) to hold paramount the safety, health, and welfare of the public, to strive to comply with ethical design and sustainable development practices, and to disclose promptly factors that might endanger the public or the environment; 2) to avoid real or perceived conflicts of interest whenever possible, and to disclose them to affected parties when they do exist; 3) to be honest and realistic in stating claims or estimates based on available data; 4) to reject bribery in all its forms" [17]. Engineers have to pursue their profession's code of ethics and also know that their professional responsibilities are beyond these codes [18].

A significant point of engineering ethics literature is the study of disaster cases [4, 15, 19]. Case studies have some advantages such as permitting engineers to know the essence of a problem and learning to choose the best decision in the prevention of disasters [20]. The Challenger and Columbia disasters as two engineering ethics cases are presented in the next sections.

4 The Challenger disaster

The Challenger disaster because of its complexity is analyzed by many ethicists as well as managers and engineers to understand professional misconduct (see Fig. 2). To know about the details of the Challenger disaster, please refer to [21-22].

Figure 2. The Challenger disaster

"On January 28, 1986, the space shuttle Challenger exploded 73 seconds into its flight, killing the seven astronauts aboard" [21]. The technical cause of the incident was due to "a failure of the o-rings, the seals in the connecting joint between two segments of the rocket booster, to seal one of the boosters thereby creating the environment for the fuel explosion that resulted" [22].

Three groups of human factors played principal roles in the occurrence of the incident: Thiokol engineers and managers who were directly responsible for the launch and also NASA officials as signatories the launch [22-23].

Before the Challenger launch, a team of NASA held a meeting with engineers of Morton Thiokol to talk about the safety of launching. Morton Thiokol engineers pointed to problems with O-rings in 7 of the past 24 shuttle launches and warned a connection between low temperatures and O-ring problems. They stated that the launch should be stopped, but the warning of engineers was not accepted by NASA officials. The four senior managers ignored the advice of engineers and voted for the launch. NASA quickly accepted the managers' opinion that led to a human and technical failure [21].

NASA accepted more risks due to pressure to keep the congressional budget. NASA management contravened rules because of the shuttle program's continued economic viability. The unfortunate issue was that the NASA managers who recommended the launch did not inform the astronauts of the risk. They easily violated the human and ethical rights of astronauts.

In fact, the Challenger accident was the result of the difference in priorities between the engineers and management at Thiokol and NASA, some conflicts in organizational culture and, the failure of managers and engineers in the application of individual ethical responsibility [22]. In addition, self-deception had been penetrated in the NASA space program. The case demonstrates how the self-deception can affect the engineers and limit them to predict risk correctly [6]. Boisjoly et al. revealed that the governing organization of the Challenger undermined the individual decision maker's responsibilities and became a tool for avoiding real and efficient accountability in the whole management system [21]. In summary, disregarding ethical responsibility is revealed with reflecting on some cases: decision-making processes were not in accordance with the standards of NASA or communication specialists; the design and practice process was flawed; the final decision to launch was a huge risk regardless of the importance and respect for the lives of astronauts; the decision makers sacrificed the moral concepts in their actions to attain their purposes [24].

5 The Columbia disaster

Seventeen years later, the Columbia disaster as another significant aerospace accident occurred. Like Challenger, the non-technical causes were insufficient attention to ethical principles, especially at managerial levels (see Fig. 3). On February 1, 2003, NASA's space shuttle Columbia during returned to the earth, broke and all seven astronauts were killed. The Columbia Accident Investigation Board stated that technical and organizational causes had a role in the Columbia disaster [25].

The technical cause was "damage to Columbia's left wing by a 1.7 pound piece of insulating foam that detached from the left "bipod ramp" that connects the External Tank1 to the orbiter, and struck the orbiter's left wing 81.9 seconds after launch. The foam strike created a hole in a Reinforced Carbon-Carbon (RCC) panel on the leading edge of the wing, allowing superheated air (perhaps exceeding 5,000o F) to enter the wing during reentry. The extreme heat caused the wing to fail structurally, creating aerodynamic forces that led to the disintegration of the orbiter"[25-26].

Figure 3. The Columbia disaster

The organizational causes were "...rooted in the Space Shuttle Program's history and culture, including the original compromises that were required to gain approval for the Shuttle, subsequent years of resource constraints, fluctuating priorities, schedule pressures, mischaracterization of the Shuttle as operational rather than developmental, and lack of an agreed national vision for human space flight. Cultural traits and organizational practices detrimental to safety were allowed to develop, including: reliance on past success as a substitute for sound engineering practices..., organizational barriers that prevented effective communication of critical safety information and stifled professional differences of opinion; lack of integrated management across program elements; and the evolution of an informal chain of command and decision-making processes that operated outside the organization's rules" [25-26]. Unfortunately, in the Columbia program, without enough budget and schedule, quality was seriously compromised, so that safety also faced with baleful consequences.

To summarize, the Challenger and Columbia have some resemblances in the unethical decision making of senior-level managers that lead to the similar tragedies. In both accidents, the managers did not pay attention to the experience of engineers and experts' consultations due to organizational considerations [27]. In both the Challenger and Columbia disasters, budget and schedule were limited. The organizational and bureaucratic structure was also a common issue between both incidents. It caused a barrier to the receiving of important data. Furthermore, self-censorship as a significant factor in the organization resulted in having incorrect information for decision-making [28]. Finally, it must be said that with regard to ethical issues that led to these two cases, applying engineering ethics to decision making within crucial conditions can prevent the similar engineering disasters in the future.

Since the 1950s, aerospace engineering and technology have been considered as a national issue in many countries because it has crucial consequences for defense and economics. For this reason, many governments and private companies are working more than ever in aerospace investigations. Hence, engineering ethics should be more considered in the aerospace field due to the serious role of the aerospace technology on many current technologies. As mentioned above, ethical codes play an important role in engineers' decision making. Therefore, this paper suggests that further investigation should be done to design codes of ethics for the various specialties within aerospace engineering.

6 Conclusion

The aerospace as a significant and effective branch of the engineering profession has contributed to the development of many new technologies. In addition, the aerospace engineering has a serious role in the defense and economics for many countries in the last decades. Moreover, the case studies in this paper demonstrated that ethical weaknesses such as mismanagement and negligence of safety issues play a crucial role in the occurrence of engineering disasters. Therefore, the attention to engineering ethics should be bolstered in the aerospace engineering. At last, this study suggested that future research is needed to design ethical codes as a framework for ethical decision making of engineers and managers in the field of aerospace engineering.

Acknowledgments

This work has been supported by CAS-TWAS President's Fellowship for International Ph.D. Student.

References

1. H. Zwart, Frontis, 67-80 (2005)
2. C. Mitcham, *Thinking ethics in technology: Hennebach lectures and papers,* 138 (1997)
3. J. M. Basart, M. Serra, Sci. Eng. Ethics. **19** 179-187 (2013)
4. C. B. Fleddermann, *Engineering ethics*, **4** (1999)
5. NSPE Code of Ethics (1993)

6. C. E. Harris Jr, M. S. Pritchard, M. J. Rabins, R. James, E. Englehardt, *Engineering ethics: Concepts and cases* (2013)
7. D. P. Michelfelder, N. McCarthy, D. E. Goldberg, *Philosophy and engineering: Reflections on practice, principles and process,* **15** (2014)
8. J. Ladd, in *Ethical issues in the use of computers*, 8-13 (1985)
9. W. Staszewski, C. Boller, G. R. Tomlinson, *Health monitoring of aerospace structures: smart sensor technologies and signal processing* (2004)
10. M. W. Martin, R. Schinzinger, *Ethics in engineering,* 23 (2005)
11. W. T. Lynch, R. Kline, Sci. Technol. Human. Values. **25** 195-225 (2000)
12. A. Gharamaleki, *Introduction to professional Ethics* (2009)
13. D. Nieusma, D. Riley, Eng. Stud. **2** 29-59 (2010)
14. J. Beever, A. O. Brightman, Sci. Eng. Ethics. **22** 275-291 (2016)
15. C. E. Harris Jr, Sci. Eng. Ethics. **14** 153-164 (2008)
16. R. D. Hollander, In *Engineering ethics for a globalized world,* 55-67 (2015)
17. IEEE Code of Ethics (2006)
18. M. Davis, Philos. Public. Aff. 150-167 (1991)
19. M. Davis, *Thinking like an engineer: Studies in the ethics of a profession* (1998)
20. C. J. Abaté, Sci. Eng. Ethics. **17** 583-596 (2011)
21. R. P. Boisjoly, E. F. Curtis, E. Mellican, J. Bus. Ethics. **8** 217-230 (1989)
22. P. H. Werhane, J. Bus. Ethics. **10** 605-616 (1991)
23. B. S. Romzek, M. J. Dubnick, Public. Adm. Rev. 227-238 (1987)
24. R. C. Kramer, J. A. Jaksa, Eric. (1987)
25. C. A. I. Board, **1** (2003)
26. M. S. Smith, *NASA's Space Shuttle Columbia: Synopsis of the Report of the Columbia Accident Investigation Board* (2003)
27. T. M. Garrett, Am. Rev. Public Adm. **34** 389-402 (2004)
28. R. Dimitroff, L. Schmidt, T. Bond, PMJ. **36** 28-38 (2005)

Configurations and Key Issues of Low Thrust Measurement for Microspacecrafts

Wei-jing Zhou[1], Ji-fei Ye[1], Nan-lei Li[1] and Hao Chang[1,a]

[1]*State Key Laboratory of Laser Propulsion & Application, Space Engineering University, Beijing 101416, China*

Abstract. Microthrusters on microspacecrafts are of interest and its output thrust must be characterized for a particular applicaton. The technology of micro thrust measurement can evaluate the performance of microthrusters. The measurement principles of steady-state thrust, pulse impulse, average thrust and instantaneous thrust are introduced. The classical thrust stands at home and broad are summarized and they are classified into torsional arm type, balance type, hanging pendulum type and flexure vibration type according to their dynamics configurations. The characters, purpose, advantages and disadvantages are analyzed. The key problems, such as error influence factors and its countermeasure, calibration, and measurement automatization, are pointed out.

1 Introduction

Microspacecrafts can potentially reduce launch cost and can complete a variety of new space-based missions, so it has attracted much attention in recent years. And then, miniaturization of every component on the microspacecraft is required. Among all the components, microthruster, which can produce very small (on the order of μN or nN) and highly accurate thrust for stabilization, pointing and position keeping of microspacecraft is of current interest. Solid Propellant Microthruster (μSPT), Pulsed Plasma Microthruster (μPPT), Field Emission Electric Propulsion (FEEP), Colloid Thruster, Laser Ablation Microthruster (μLAT) and so on is generally recognized as potential microthrusters. But, the accurate measurement of their performance is the basic problem to be solved at first. And the most critical performance measurement is that of thrust. For the purpose of thrust measurement, many thrust measurement stands already exist. In this paper, the principles of thrust measurement are concluded. Several typical configurations of microthrust measurement system are summarized. And the key issues of microthrust measurement are discussed. The goal of this paper is to help users choose best configuration for a particular microthruster, and provide issues which are must pay attention to for particular micro thrust measurements.

2 Principles of Microthrust Measurement

Microthrusters can be classified into continuous thruster and pulse thruster according to the thrust output mode. For continuous thrust, the steady-state thrust is generally measured. For pulse thruster,

[a] Corresponding author: changhao5976911@163.com

pulse impulse, average thrust and instantaneous thrust is generally measured. The measurement principles are summarized in Table 1.

Table 1. Measurement principles of micro thrust

Thrust	Measurement principle	Measurand	Measurement ristrictions
steady-state thrust	Thrust is linear with the steady-state deflection angle of the executive unit.	steady-state deflection angle	——
	Thrust is compensated with feedback force in time.	deflection angle	The responsive bandwidth is low because of the inertia.
pulse impulse	Impulse is linear with the maximum deflection angle of the executive unit.	maximum deflection angle	Pulse width is several orders of magnitude smaller than the natural period of the thrust stand.
average thrust	Average thrust is linear with the average value of the quasi steady-state deflection angle of the executive unit.	quasi steady-state deflection angle	Pulse repetition frequency is several orders of magnitude higher than the natural frequency of the thrust stand.
instantaneous thrust	Thrust is linear with the deformation of the elastic element.	deformation	A compromise between the sensitivity and the responsive bandwidth are needed to take into count.
	Thrust is linear with the angular acceleration of the executive unit.	deflection angle	The precision is restricted by the sampling rate, the numerical differentiation algorithm, etc.
	Thrust is compensated with feedback force in time.	deflection angle	The responsive bandwidth is low because of the inertia.

3 Conclusion and Analysis of Classical Microthrust Measuring System

For the purpose of thrust measurement, many thrust measurement stands already exist. And they can be classified into torsional arm type, balance type, hanging pendulum type and flexure vibration type according to their dynamics configurations. Typical applications in various configurations are shown in Table 2 to Table 5. And test characteristics, advantages and disadvantages are analyzed in Table 6.

Table 2. Thrust stands based on torsional arm

Time	Research team	Application	Displacement/Angle measurement	Calibration method	Performance
1971[1]	NASA Goddard Space Flight Center	Pulse Plasma Thruster	differential capacitance displacement transducer	Permanent magnet and multiturn coil	Range: 4.4μN~50mN Max error: 4.5%
1997[2]	NASA Lewis Research Center	Pulse Plasma Thruster	LVDT	Steady-state force calibration: weight Impulse calibration: impact pendulum	Average thrust: 573μN Impulse bit: 266μNs
1999[3]	C. R. Phipps team	Micro Laser Plasma Thruster	mirror and laser beam	—	Range: 20μNs Precision:: 1nNs
2000[4]	NASA Golddard Space	Colloid and Field Emission	autocollimator targeting a mirror	Measuring Natural frequency and	Range: 1~100μN Precision: better than 0.1μN

	Flight Center	Electric Propulsion thrusters		moment of inertia	
2004[5]	University of Tokyo	Liquid Propellant Pulsed Plasma Thruster and Diode Laser Ablation Microthruster	LVDT	Impact pendulum	For LP-PPT, range: 20~80μNs, Resolution: 2.1μNs For μDLAT, range: 1~10μNs, Resolution: 0.7μNs
2006[6]	C. R. Phipps team	msμLPT	Interferometer	Helmholtz coil	Force response: 1.25mN/rad Range: 100nN~500μN, Precision: 25nN
2012[7]	Key Lab of Renewable Energy and Natural Gas Hydrate	MEMS micro nozzle chip	LVDT	Steady calibration: weights Impulse calibration: ball	Resolution: 0.1mN
2012[8]	Institute of Mechanics, Chinese Academy of Sciences	—	light	Calibrating spring constant with period method	Range: 54~323μN Precision: 2μN

Table 3. Thrust stands based on balance

Time	Research team	Application	Displacement/Angle measurement	Calibration method	Performance
2005[9]	University of Colorado	Pulsed Plasma Thruster	LVDT	Impact hammer	—
2011[10]	University of Southern California	Micro Cathode Arc Thruster	LVDT	—	Impulse bit: 0.72±0.02μNs
2012[11]	Beijing Aerospace Institute for Metrology and Measurement Technology	Micro Thruster	Displacement transducer	—	Range: 0.1~1mN Expend uncertainty: 0.05mN（k=2）

Table 4. Thrust stands based on hanging pendulum

Time	Research team	Application	Displacement/Angle measurement	Calibration method	Performance
1996[12]	Princeton University	—	Laser interferometry	Force transducer	Range: 100μNs~10Ns Steady-state thrust: as low as 20μN Error: <2%

2004[13]	NASA Marshall Space Flight Center	Electric Propulsion System	LGDT	Weights	Range: 100μN~1N

Table 5. Thrust stands based on flexure vibration

Time	Research team	Application	Displacement/Angle measurement	Calibration method	Performance
2006[14,15]	Johns Hopkins University Applied Physics Laboratory	Micro-scale Pulsed Plasma thruster	Interferometer	Ball	Min impulse bit: 0.3μNs

Table 6. Characters of all types of thrust stands

Type	Characters	Measuring	Advantages	Disadvantages
Torsional arm type	The thruster is mounted on the arm. The arm response in horizontal plane by rotating about its axis. The restoring force is provided by elastic element or feedback force.	steady-state thrust pulse impulse average thrust	The restoring force is not influenced by the weight of the thruster or the arm. And the relationship between the restoring force and the deflection angle is not influenced by any nonlinear factors. And the structure is not sensitive to vertical vibration.	The spring constant can vary or the thermal deformation can happen as the temperature varies.
Balance type	The thruster is mounted on the arm. The arm response in vertical plane by rotating about its axis. The restoring force is provided by elastic element or feedback force.	steady-state thrust pulse impulse average thrust	Calibration is easy to carry out. And the ablation mass can be measured.	The thrust is not separated from the gravity.
Hanging pendulum type	The thruster is mounted on the pendulum. The pendulum response in vertical plane by rotating about its axis. The restoring force is provided by gravity.	steady-state thrust pulse impulse average thrust	The restoring force is not influenced by the temperature.	The restoring force is provided by the gravity so that the nonlinear effect exists. And the variation of the thruster can influence the restoring force.
Flexure vibration type	The thruster is mounted on the elastic element. And the elastic element vibrates when the thruster fires.	pulse impulse instantaneous thrust	The configuration is simple.	The noise interference is serious so that it is difficult to filter.

4 Key issues of Microthrust Measurement

(a) Error Analysis. Environmental factors that may have an impact on measurement are: (i) Vibration. Footsteps, passing vehicles, air flow and so on can cause measurement signal noise. So, firstly, the good testing environment, such as at night, must be taken into account. And then corresponding vibration reduction measures, such as vibration isolation foundations and vibration isolation platform, can be considered. At last, data processing model for noise reduction, such as the use of Fourier transform spectral analysis, adaptive filtering and so on, can be researched. (ii) Temperature. Temperature changing can cause changes in the spring coefficient of the elastic element. And this can affect measurement accuracy. So the characteristics of the impact must be studied. And the higher heat capacity and thermal conductivity material can be taken into account. Also, the spring coefficient can be calibrated before each experiment. (iii) Electromagnetic environment. Electromagnetic environment has a certain effect on some component (such as the electromagnetic dampers, coil permanent magnet) or thrusters (Pulsed Plasma Microthruster). Electromagnetic intensity can be monitored real-time so that the interference source can be checked out and removed. A certain amount of shielding measures can be carried out. Because of different configurations of thrust stand, the environmental interference factors must be analyzed specifically and then corresponding denoising measures is to be carried out.

(b) Automation Control. Thrust measuring system involves level control, zero control, calibration and so on. Because of particular measurement requirements in vacuum and the flexibility, the above operations had better automate as much as possible.

(c) Calibration. Calibration is the key issue in the validation of the measurement system. For thrust measurement, calibration commonly uses weights, electrostatic force and electromagnetic force. The friction caused by pulley can arouse error when the weight method is carried out. Electrostatic force and electromagnetic force has no mechanical contact, and can be controlled easily by program. But they are sensitive to electromagnetic interference. For pulse impulse measurement, calibration commonly uses ball falling method, the impact pendulum method and the force hammer method.

5 Summary

With the development of the microspacecraft, micro propulsion becomes a research field of current interest. The micro thrust measurement is very important for the performance validation of the microthrusters. But it is a complex and challenging technology. The working mode and the configuration of all kinds of microthrusters are different from each other so that the dynamical configuration of the measuring system must be chosen carefully according to the specific case. Interference is easy to be introduced into the measuring data. So denoising is the key technology to be solved all the time. And lastly, the implementation of the micro thrust measuring system is a complex and repeating work which needs patience and confidence.

Acknowledgements

This research was supported by the National Natural Science Foundation of China (Grant Nos. 11602304 and 11502301).

References

1. Stark K W, Dennis T, McHugb D, and Williams T, NASA TN D-7029 (1971)
2. T W Haag, Rev Sci Instrum, **68(5)**, 2060-2067 (1997)
3. C. R. Phipps. Micro Laser Plasma Thrusters For Small Satellites. STTR 1999-F49620-98-C-0038 (1999)
4. S. M. Merkowitz, P. G. Maghami, A. Sharma, et al, Classical and Quantum Gravity, **19(7)**,1745-1750 (2002)
5. H. Koizumi, K.Komurasaki, and Y.Arakawa, Rev Sci Instrum, **75(10)**, 3185-3190 (2004)

6. Claude R. Phipps, James R. Luke, Wesley Helgeson, et al. *Fourth International Symposium on Beamed Energy Propulsion* (2006)
7. Cen Jiwen, Fu Jian, Jiang Fang, etc, *Proceeding of the 8th Chinese Electric Propulsion Conference* (2012)
8. Shen Bo, Zhao Fengming, Gao Hui, Kang Qi, *Proceeding of the 8th Chinese Electric Propulsion Conference* (2012)
9. Anthony P. Pancotti, Taylor Lilly, Andrew D. Ketsdever, et al, AIAA 2005-4415 (2005)
10. Patrick Vail , Anthony Pancotti , Taisen Zhuang, Alexey Shashurin, *47th AIAA/ASME/SAE/ASEE Joint Propulsion Conference & Exhibit* (2011)
11. Li Jing, Mei Hongwei, Wang Geng, *Proceeding of the 8th Chinese Electric Propulsion Conference* (2012)
12. E.A.Cubbin, J. K. Ziemer, E. Y. Choueiri, et al, Rev. Sci. Instru., **68(6)**: 2339-2346 (1997)
13. T. E. Markusic, J. E. Johns, M. D. Cox, AIAA 2004-3441 (2004)
14. Daniel H. Simon, and H. Bruce Land, AIAA 2005-4264 (2005)
15. Jerold W. Emhoff, Daniel H. Simon, H. Bruce Land, AIAA 2007-5299 (2007)

Numerical Simulation of Rotor in Hover and Forward Flight

Qinghe Zhao[a]

School of Energy and Power Engineering, Beihang University, Beijing 100191, China

Abstract. The flow around rotor is numerical simulated in hover and forward flight based on multi-structured grid. In hover the flow field can be transformed into a steady-state flow field in the rotating coordinate system. The experimental data of Caradonna and Tung rotor is used to verify the numerical simulation result. The numerical results compare well with the experimental data for both non-lifting and lifting cases. Non-lifting forward flight is simulated and the prediction capabilities have been validated through the ONERA two-blade rotor. The pressure distributions of different positions under different azimuth angles are compared, which is in good agreement with the experimental data. There is unsteady shock wave when forward flight. Dual-time method is used to obtain unsteady flow field with rigid moving grid in the inertial system.

1 INTRODUCTION

The flow around rotor blade has many different characteristics from the fixed wing. The local velocity of rotor varies linearly with the radius of rotation. Across the tip of the transonic speed and near the hub of the incompressible flow area often occur at the same time. Furthermore, the induced velocity produced by the rotor wake vortex system has an important influence on the flow field. The flow field is unsteady and characterized by strong non-linear and three-dimensional effects. Transonic regions near the advancing blade tip leading to shock-induced separation, as well as extensive regions of separated flow due to dynamic stall occurring on retreating blades.

For the hover state, the coordinate system is established on the blade and rotates together with the rotor. The stationary flow field with rotational symmetry is observed. In the rotational coordinate system to solve the equation can greatly improve the hover flow field calculation efficiency. Unlike in hover, the forward flight involves asymmetric flow fields and cannot be solved in a rotating coordinate system. It should be simulated unsteady flow field, which require more computational resource.

A large number work had been carried out to solve flow field in hover with structural meshes [1,2]. High order scheme such as ENO is used to distinguish the flow characteristics [3]. The Chimera method was used to solve the flow field. The blade-fixed grids are embedded in the background grid. Along the chimera boundaries, a tri-linear interpolation method is used to couple the solution of the blade-fixed and background grids. Unstructured grid is used to refine the flow field [4]. Adaptive mesh refinement involves automatically refining the mesh to resolve important flow characteristics. A detailed overview was introduced on the hovering rotor [5]. in hover simulation [6]. Multi-block

[a] Corresponding author : zhaoqinghe2001@163.com

structured grid combined deforming mesh method is used to simulate the forward flight flow field [7]. Blade-vortex interactions acoustic is introduced [8,9]. Forward flight filed is simulated [10]. Simulation progress in Europe is introduced [11]. Despite these methods, multi-block structured grids are still the most commonly used form and are widely used in solving rotor flow field.

In the present work, multi-block structured grids are used. The structure of the paper is as follows. Section 3 presents the governing equations in an rotational coordinate system, as used in the hovering simulations. The spatial discretization and temporal integration method are described, followed by the forward flight in a inertial frame of reference. Validation results are presented in Section 3 for both hovering and forward flight cases. In this paper, the work of Caradonna and Tung, the experiments of ONERA were used. Conclusions are finally drawn in Section 4.

2 NUMERICAL SOLUTION PROCEDURES

For hover cases, the flow around a hovering rotor can be treated as a steady problem. For forward flight simulations the flow field is asymmetry and unsteady. So the simulation process is introduced separately.

2.1 Methodologies for hover case

For hover cases, the flow around a hovering rotor can be treated as a steady problem. For forward flight simulations the flow field is asymmetry and unsteady. So the simulation process is introduced separately.

Due to the symmetry of the flow in rotational coordinate system, The computational domain can be replaced only by a domain around one blade. The periodicity boundary condition is used to taken into the influence of other blades. Therefore, the computational expense can be reduced greatly. The government equation is given below

$$\frac{d}{dt}\int_{V(t)} \vec{w}\,dV + \int_{\partial V(t)} (\vec{F}(\vec{w}) - \vec{F}_v(\vec{w}))\vec{n}\,dS = \vec{S} \tag{1}$$

Where $\vec{w}$ is the conserved variables. $\vec{F}$ and $\vec{F}_v$ are the inviscid and viscous fluxes, respectively. Two-equation $k-\omega$ turbulence model is used to simulate the viscous flow field. There is a source terms in the non-inertial frame of reference. It means the influence of the centripetal and the Coriolis force. For an hovering rotor rotating around the y-axis with the angular velocity $\vec{\omega}=[0,\Omega,0]$, the source term $\vec{S}$ can be expressed by

$$\vec{w}=\begin{bmatrix}\rho\\ \rho u\\ \rho v\\ \rho w\\ \rho E\end{bmatrix},\quad \vec{F}=\begin{bmatrix}\rho(q-q_\omega)\\ \rho u(q-q_\omega)+pi_x\\ \rho v(q-q_\omega)+pi_y\\ \rho w(q-q_\omega)+pi_z\\ \rho h(q-q_\omega)+pq_\omega\end{bmatrix},\quad \vec{S}=\begin{bmatrix}0\\ -\rho\Omega w\\ 0\\ \rho\Omega u\\ 0\end{bmatrix} \tag{2}$$

Where q_ω is the grid velocity, it produce a grid flux. The finite volume discretization with central differencing for the flux approximation leads to 2 order scheme. The implicit Lower-Upper Symmetric Gauss Seidel operator(LUSGS) is applied to obtain the update solution.

2.2 Methodologies for unsteady flows

For forward flight, the flow field should be simulated in unsteady condition. The dual time stepping approach is introduced, and the system can be solved by an iteration through pseudo time to the steady state. The implicit dual-time approach introduced by Jameson has become a very attractive method for solving such problems as it allows an implicit discretisation to be used in real time. The steady state is reached at a pseudo time in each time step.

$$\frac{3V_{i,j,k}^{n+1}w_{i,j,k}^{n+1}-4V_{i,j,k}^{n}w_{i,j,k}^{n}+V_{i,j,k}^{n-1}w_{i,j,k}^{n-1}}{2\Delta t}+R_{i,j,k}(w_{i,j,k}^{n+1})=0 \tag{3}$$

$$V_{i,j,k}^{n+1}\frac{w_{i,j,k}^{n+1,m+1}-w_{i,j,k}^{n+1,m}}{\Delta\tau}+\frac{3V_{i,j,k}^{n+1}w_{i,j,k}^{n+1,m}-4V_{i,j,k}^{n}w_{i,j,k}^{n}+V_{i,j,k}^{n-1}w_{i,j,k}^{n-1}}{2\Delta t}+R_{i,j,k}(w_{i,j,k}^{n+1,m})=0 \tag{4}$$

$$V_{i,j,k}^{n+1}\left(\frac{1}{\Delta\tau}+\frac{3}{2\Delta t}+A+B+C\right)\Delta w_{i,j,k}=-R_{i,j,k}(w_{i,j,k}^{n+1,m})-\frac{3V_{i,j,k}^{n+1}w_{i,j,k}^{n+1,m}-4V_{i,j,k}^{n}w_{i,j,k}^{n}+V_{i,j,k}^{n-1}w_{i,j,k}^{n-1}}{2\Delta t} \tag{5}$$

Where $\Delta w_{i,j,k}=w_{i,j,k}^{n+1,m+1}-w_{i,j,k}^{n+1,m}$, Δt is the physical time step, $\Delta\tau$ is the pseudo time step. A,B,C is the Jacobi metrics of the inviscid flux. The accelerate convergence methods such as local time step can be used in the pseudo time cycles.

2.3 Mesh movement

There is no change of the relative position of points in the same grids if the rotor moves rigidly. Assuming it rotate around the y-axis, the transformation matrices are introduced for rotation. And the grid coordinate and grid velocity can be calculated directly from formula 6. Where 0 denotes the original grid position and n denotes the updated grid position. The mesh velocity introduced is essentially the mesh rotation velocity.

$$\begin{bmatrix}x\\y\\z\end{bmatrix}^n=\begin{bmatrix}\cos\omega t & 0 & \sin\omega t\\0 & 1 & 0\\-\sin\omega t & 0 & \cos\omega t\end{bmatrix}\begin{bmatrix}x\\y\\z\end{bmatrix}^0 \qquad \begin{bmatrix}u\\v\\w\end{bmatrix}^n=-\omega\begin{bmatrix}\sin\omega t & 0 & -\cos\omega t\\0 & 0 & 0\\\cos\omega t & 0 & \sin\omega t\end{bmatrix}\begin{bmatrix}x\\y\\z\end{bmatrix}^0 \tag{6}$$

There is no grid volume changing, so the geometric conservation law(GCL) is also kept.

3 THE HOVER CASE

An experimental study of a helicopter model in hover had been carried out by Caradonna and Tung in 1981 [12]. This experimental data is usually used to verify the CFD program. The blade had a NACA0012 profile, which had a diameter of 2.286 meter, and a chord length of 0.191m, according to the aspect ratio was 6. In addition, since the flow field is symmetric, only one blade can be modelled by periodic boundary conditions. Figure 1 shows the grid topology. The rotor rotation centre is on the z-axis, the rotor blades are on the y-axis, a quarter of the point is at x = 0 and the leading edge of the blade points to the positive y-axis. The rotor hub is a simple straight cylinder. In the calculation, flow field is initialled by static flow conditions.

In hover the pressure coefficient is calculated by formula 7, where ω is the rotate velocity and r is the span distance.

$$Cp=\frac{p-p_0}{0.5\rho(\omega r)^2} \tag{7}$$

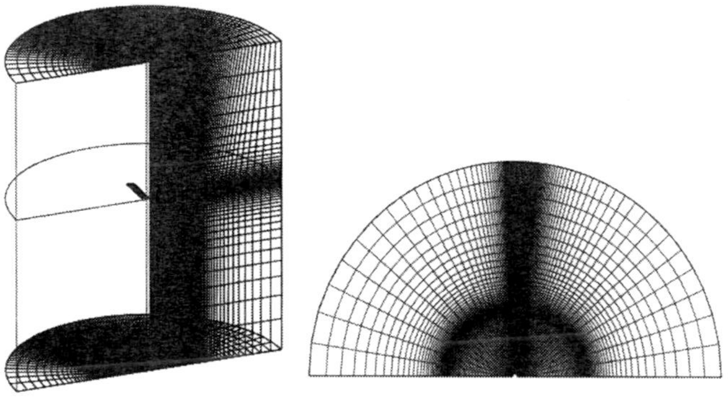

Figure 1. Grid topology of Caradonna and Tung

3.1 Non-lifting in hover

The non-lifting in hover case is simulated firstly. For this case the tip Mach number is 0.52 and the corresponding Reynolds number is 2.32 million. Figure 2 shows the comparison of CFD results to the experimental data for the pressure coefficient distribution at five different sections(r/R=0.50,0.68,0.80,0.89,0.96). The comparison shows a good agreement with the experimental data. At the same time the pressure contour is given, it can be seen that there is a low pressure zone at the rotor tip.

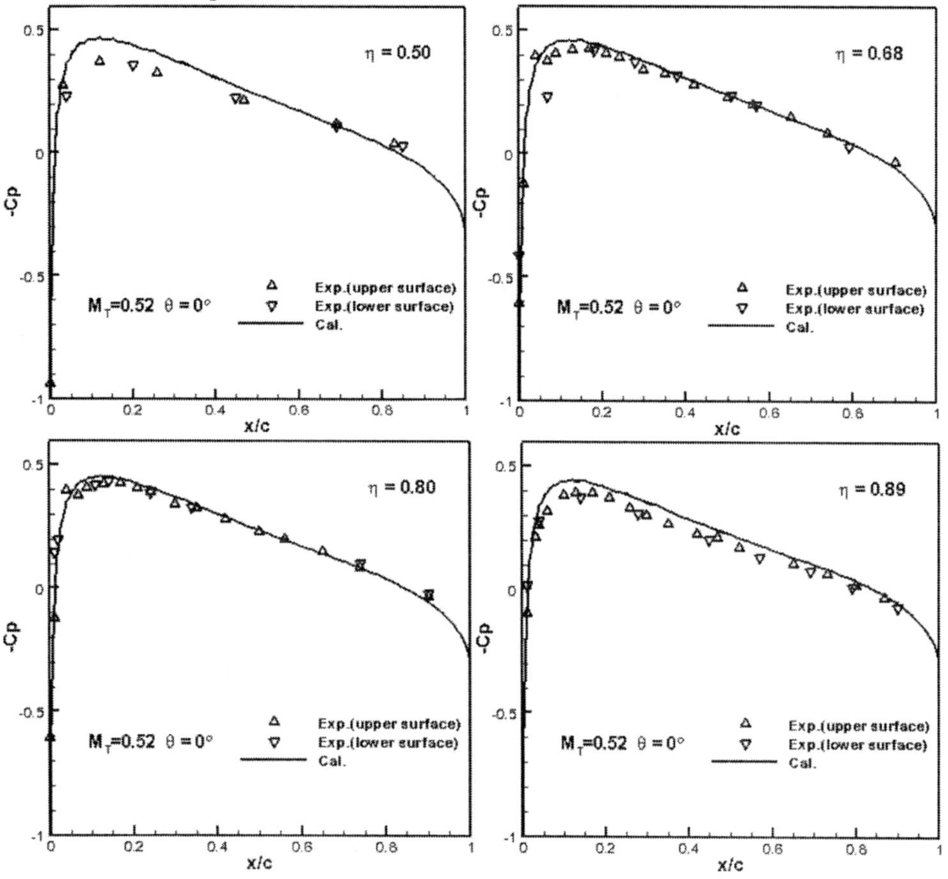

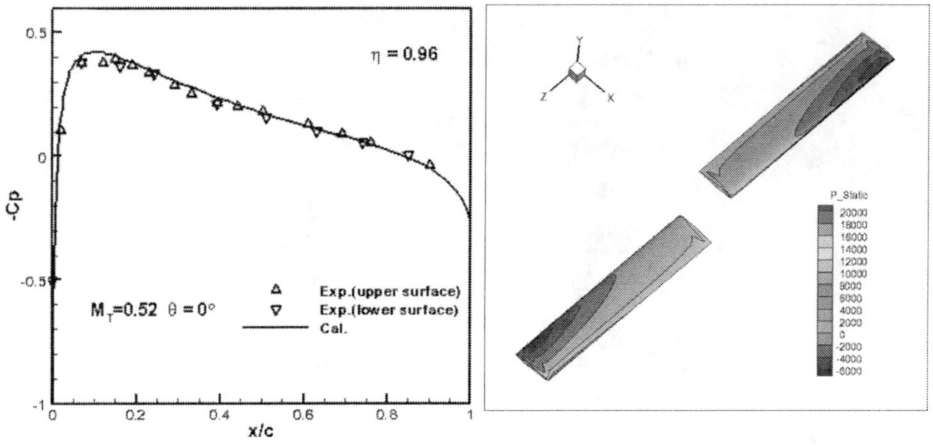

Figure 2. Pressure distribution comparison of flow field with non-lifting in hover

3.2 Lifting in hover

For transonic lifting in hover, the tip Mach number is 0.877, the total angle is 8 ° and the corresponding Reynolds number is 3.93 million. Figure 3 shows the comparison of pressure coefficient calculated with the experimental data at different sections, except that the calculated pressure coefficient on the first section is higher than the experimental values, and the calculated values of the other sections are generally in good agreement with the experimental values. From the pressure contour it can be seen that the supercritical flow conditions was dominated by stronger transonic shocks in the rotor tip region.

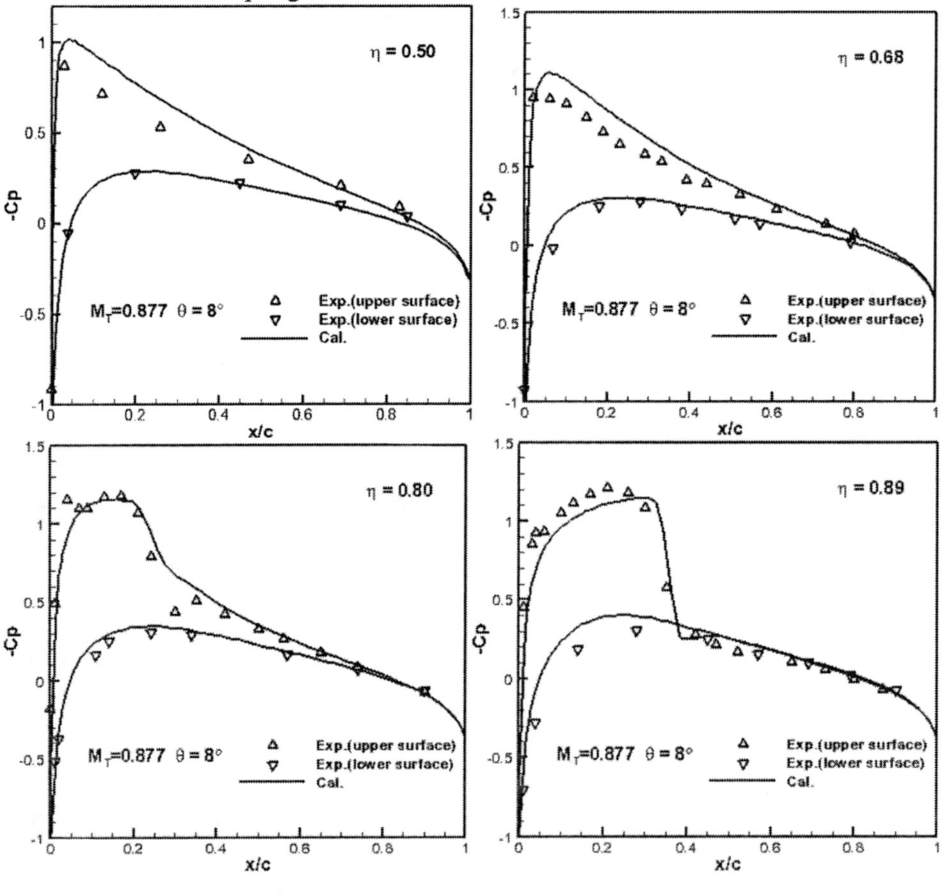

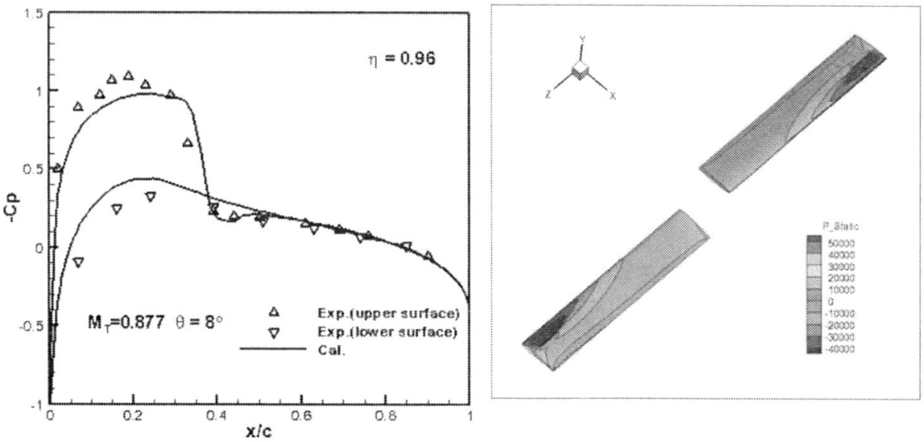

Figure 3. Pressure distribution comparison of flow field with lifting in hover

4. FORWARD FLIGHT CASE

This section presents the results for ONERA two-blade rotor in non-lifting forward flight [13]. The flight condition is advance ratio 0.5 and a tip Mach number 0.625. The profile of the rotor changes from NACA0017 at the root to NACA0009 at the tip. Since the forward flow field is unsteady, the entire grid is used to solve the field. The Figure 4 shows the topology of the grid.

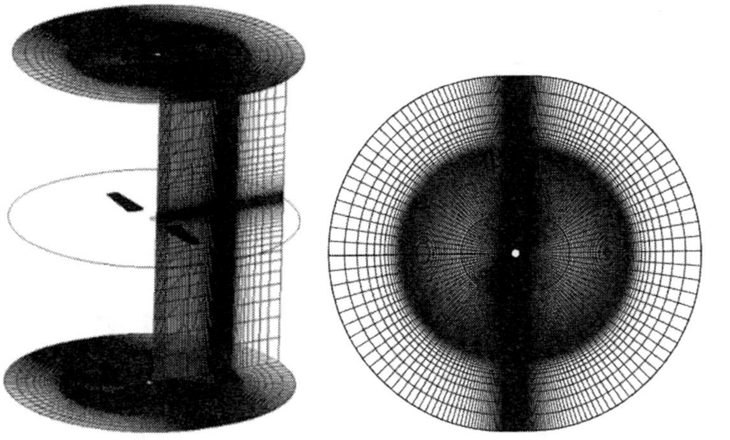

Figure 4. Grid topology of ONERA rotor

The pressure coefficient calculation of forward flight depend on not only the local rotation speed, but also forward flight speed, the formula is given by formula 8

$$Cp = \frac{p - p_0}{0.5\rho(\omega r + V_\infty \cdot \sin\varphi)^2} \tag{8}$$

The simulation results have been compared with the experimental data. The azimuthal step used for forward flight is $\Delta\psi = 0.25°$, which is equivalent to a rotational period includes 1440 time steps, 50 inner iterations is used in each time step. The figure 5 shows the pressure contour of several azimuths. Respectively 0,30,60,90,120,150,180, 210,240,270,300,330 degree. It can be seen that there is a clear shock wave when forward flight. The shock wave diminishes or even disappears when the blade is moving backward.

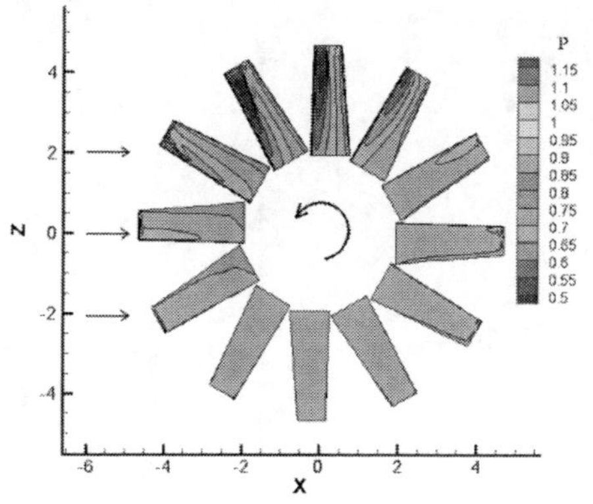

Figure 5. Pressure contour of the ONERA rotor forward flight

Figure 6 shows the comparison of the pressure distribution at 0.85 and 0.90 section with experimental values at different azimuths. The experimental data corresponds to the forward flight condition. There is no experiment data for the retreating side. The present method compares quite well with the experimental data for the azimuthal positions of $\psi = 60,90,120$. The biggest differences is encountered at $\psi = 150°$. At this azimuch the flow is highly unsteady and the shock is moving rapidly downstream.

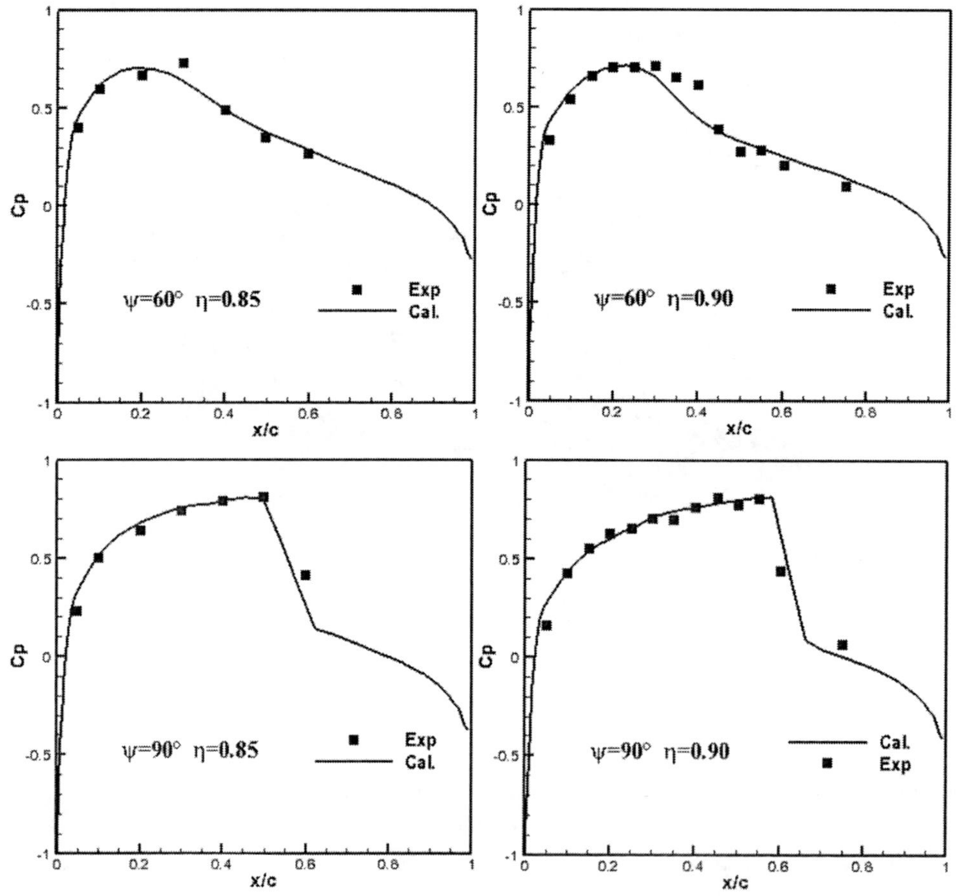

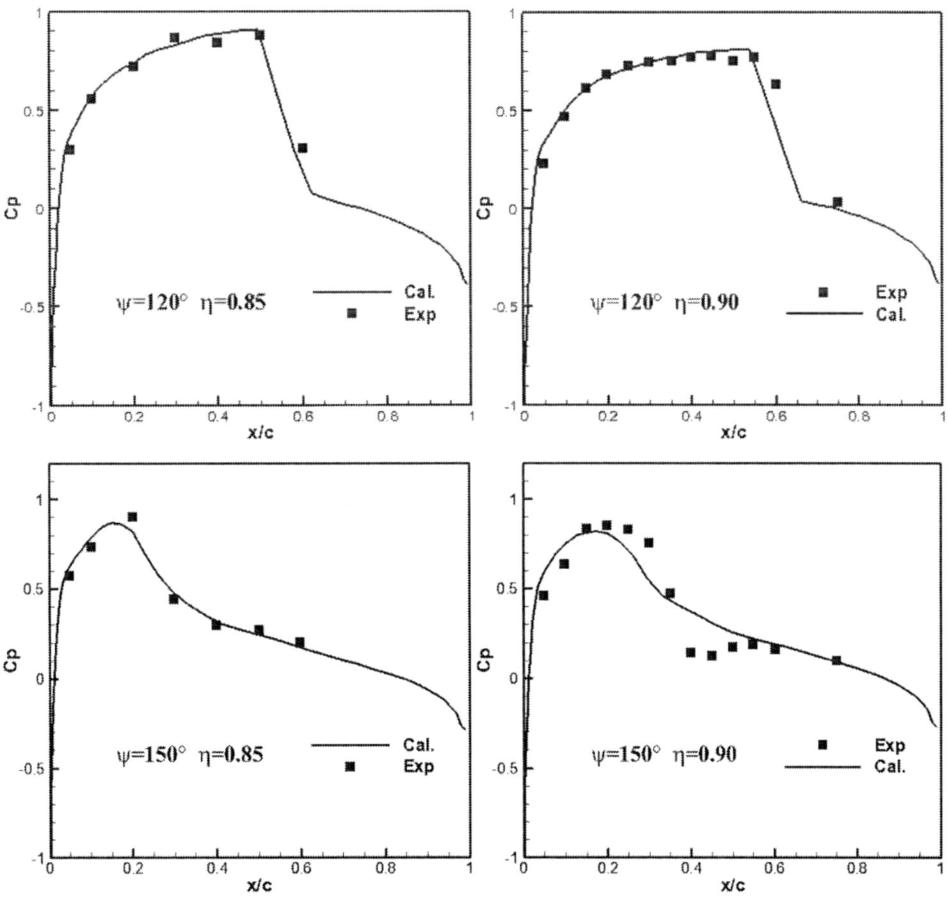

Figure 6. Chord-wise Cp distribution for the ONERA rotor at two radial stations

5. CONCLUSION

A numerical approach is presented for the simulation of hovering and forward flight flow field based on multi-block grid. In hover the flow field can be transformed into a steady-state in the rotating coordinate system. The numerical results compare very well with the experimental data for both non-lifting and lifting cases. Forward flight prediction capabilities have been validated through the ONERA two-blade rotor. The pressure distributions at different positions under different azimuth angles are compared, which is in good agreement with the experimental data.

References

1. Srinivasan, G. R., and McCroskey, W. J. (1988) Navier-Stokes Calculations of Hovering Rotor Flowfields *AIAA Journal* Vol 25 No 10
2. Srinivasan, G. R., and Baeder, J. D. (1992) Flowfield of lifting rotor in hover: A Navier-Stokes simulation AIAA Jour*nal* Vol 30 No 10
3. Hariharan, N., and Sankar, L. (2004) High-Order Essentially Non-oscillatory Schemes for Rotary-Wing Wake Computations, *Journal of Aircraft* Vol 41 No 2, 2004
4. Strawn,R.C., and Barth,T.J. (1993) A finite-volume Euler Solver for computing Rotary-Wing Aerodynamics on Unstructured Meshes *Journal of American Helicopter Society* Vol 38,(2) pp.61-67
5. Hariharan, N., Egolf, A., and Sankar, L. (2014) Simulation of Rotor in Hover: Current state and challenges *AIAA paper* 2014-0041

6. Pomin, H., and Wagner, S. (2002) Navier-Stokes analysis of isolated rotor flow in helicopter hover flight *Journal of Aircraft* 39(5) 813-821
7. Steijl, R., Barakos, G., and Badcock, K. (2006) A framework for CFD analysis of helicopter rotors in hover and forward flight International *Journal for numerical methods in fluids* 51:819-847
8. Yung, H.Yu., Chee, Tung., Judith, Gallman., Klaus J, Schultz., Berend, der. Wall., Pierre, Spiegel., and Bertrand, Michea. (1995) Aerodynamics and Acoustic of Rotor Blade-Vortex Interactions, *Journal of Aircraft* Vol 32 No 5
9. Lyaysan, Ildusovna. Garipova., Andrei, Sergeevich. Batrakov., Alexander, Nikolaevich. Kusyumov., and Sergey, Anatolievich. Mikhaylov. (2016) Aerodynamic and acoustic analysis of helicopter main rotor blade tips in hover, *International Journal of Numerical Methods for Heat & Fluid Flow* Vol.26 No.7
10. Yang,Z., Sankar,L.N., Smith,M.J., and Bauchau,O. (2002) Recent Improvements to a Hybrid Method for Rotors in Forward Flight, *Journal of Aircraft* Vol 39 No 5
11. Renzoli, P. D'Alascio., A. Kroll., N. Peshkin. D., and H,L.Hounjet. M. (2000) EROS-a common European Euler code for the analysis of the helicopter rotor flowfield *Progress in Aerospace Sciences* 36:437-485
12. Caradonna, F. X., and Tung, C. (1981). Experimental and analytical studies of a model helicopter rotor in hover *Vertica* 5(1):149-161
13. Philippe J-J, Chattot J-J. (1980) Experimental and theoretical studies on helicopter blade tips at ONERA. Sixth European Rotorcraft and Powered Lift Aircraft Forum, Bristol, September

Rotordynamic Characteristics Analysis for an aero-engine Low Pressure Rotor Rig Test Model

Huiying Song[1,a], Shaohui Wang[1] and Kai Sun[1]

[1]AECC Commercial Aircraft Engine Co., Ltd., South Lianhua Road 3998, 201108, Shanghai, China

Abstract. 3D whole engine finite element model of a low pressure rotor rig test model is established in this paper. Rotorynamic characteristics are mainly analysed with three models, which include rotor model with static stiffness, rotor model with dynamic stiffness, and whole engine model. The calculation results were compared and discussed deeply. Rotordynamic characteristics of rotor model with static stiffness are similar with the rotor model with dynamic stiffness, but the latter may have additional resonance peaks caused by dynamic stiffness. Whole engine model, which can capture the modes of casing and coupling vibration between stator and rotor, may have more critical speeds than rotor model only. The unbalance response amplitude and phase angle of the whole engine model are different with the only rotor model, in the values and distributions of the peaks. The result of rotor model with dynamic stiffness is closed to the whole engine model than the rotor model with static stiffness. The peak values of the whole engine model are smaller than the only rotor model. Rotordynamic characteristics with whole engine model are more accurate than rotor model only, so it necessary to analyse rotordynamic characteristics with whole engine model in detail design stage.

1 Introduction

Finite element Aero-engine rotordynamic model usually does not include detailed casing model. Instead, it has been a common practice to represent casing using spring model. There are two types of spring casing models in traditional rotor dynamics analysis. One only considers static stiffness K_s and the other considers dynamic stiffness K_d. In the K_s method, no casing modes nor coupling modes between rotor and casing are considered. As for the K_d method, casing modes are considered, and the spring constant calculation is similar to that for the static stiffness. The only difference is the force is applied with a frequency, in other words, the rotor support stiffness changes with rotor speed, especially near the stator resonant frequencies. Consequently, the frequencies in Campbell diagram change severely with rotor speed.

Thin shell structure casing is used in most modern aero-engines. Therefore, there are coupling effects between the dynamic characteristics of the rotor and the casing, which has to be properly characterized in rotordynamic analysis. The coupling effect cannot be properly reflected by the two methods mentioned above, but can be addressed by a 3D whole engine finite element model[1-5],

[a] Corresponding author : songhy02@163.com

including major engine structure characteristics and capturing the casing modes and coupled vibrations between the casing and the rotor.

In this paper, 3D finite element model of an aero-engine low pressure rotor rig test system is established. Rotordynamic characteristics are simulated using three models mentioned above. The results are compared and analysed, recommendations for the future studies are made in the conclusion.

2 3D whole engine finite element modelling and Modal analysis

The 3D whole engine finite element model of a low pressure rotor rig test model is shown in figure 1, which includes the casing and the rotor. The rotor is supported by 3 bearings. Both the rotor and casing are modelled using SOLID185 elements. The keyoption for the SOLID185, KEYOPT(2)=2 is chosen to overcome shear locking[6]. MASS21 elements are used to modelling blades, which is connected with disk by RBE3. The COMBI214 elements are used to modelling bearings and supports, which connect the rotor to the casing.

MEMF(Modal Effective Mass Fraction)[7], obtained by dividing the modal effective mass by the model total mass, is a common indicator for measuring the significance level of a mode. Therefore, a mode with a large effective mass, which is usually global mode, will contributes significantly to the system's response. Modal analyses are performed for both the casing and the engine. The global vibration mode shapes of casing and engine in frequency domain are demonstrated in Figures 3 and 4 respectively, where frequency ratio replaces the actual frequency. The global vibration mode shapes of the whole engine are shown in figure 4, including coupling modes between the rotor and the casing.

Figure 1. The 3D whole engine finite element model.

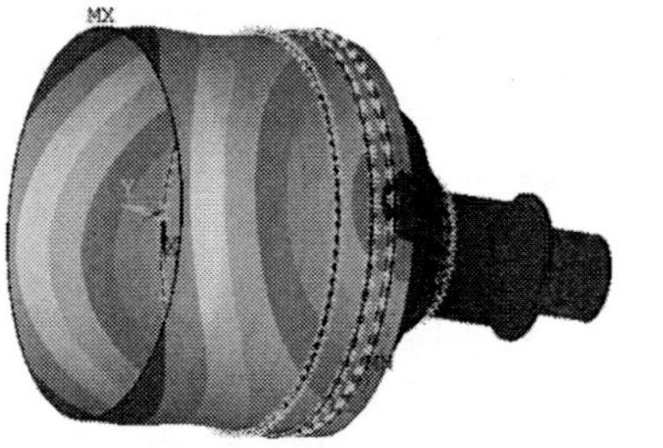

Figure 2. Mode of casing (frequency ratio 0.40).

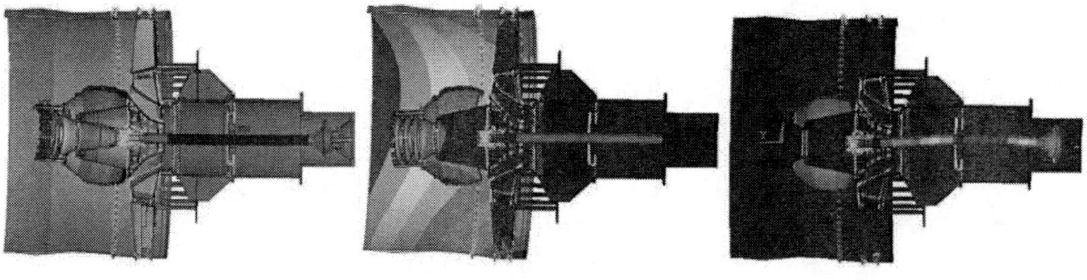

(a) mode 1 (frequency ratio 0.372) (b) mode 2 (frequency ratio 0.424) (c) mode 3 (frequency ratio 0.731)

Figure 3. Modes of engine

3 Static stiffness and dynamic stiffness

If we look at the rotor model only, instead of the whole engine model, the static stiffness and dynamic stiffness can be determined by experiments or calculations[8,9]. The static and dynamic stiffness of the casing can be obtained by the linear static analysis and the harmonic response analysis respectively. Since the casing is axisymmetric, the stiffness in the two orthogonal directions is assumed to be same. The static stiffness and dynamic stiffness of the casing support as functions of frequency ratio are shown in figure 4, with peak around 0.4. The casing mode around frequency ratio at 0.4 is excited and the mode shape is shown in Figure 2. The mode is excited by the bearing excitations. Dynamic stiffness as a function of frequency, severely impacting the casing resonant frequencies.

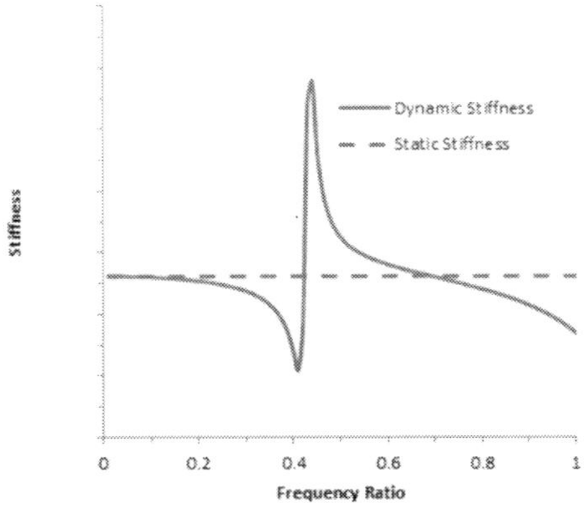

Figure 4. The static stiffness and dynamic stiffness of a support in the casing.

4 Rotorynamic Characteristics

4.1 Rotorynamic Characteristics based on rotor model only with static stiffness

In the K_s method, only the stiffness but no modes of the casing is considered, which means only rotor modes are represented and no casing modes nor coupling between rotor and casing is considered. The Campbell diagram based on the static stiffness is shown in figure 5. Here the speed ratio is used to replace the actual speed. FW represents forward whirling, BW represents backward whirling. As demonstrated in Figure 6, there are two critical speeds in the rotor speed range, the first one (CS1) is 0.527 and the second one (CS2) is 0.760. Figure 6 shows the mode shapes of the rotor (mode 1 and mode 2). The unbalance load is applied at the fan disk, and a harmonic force response analysis is performed in the frequency range of interest. The displacement response amplitude and phase angle are presented in the following figures 7 and 8 respectively. There are two distinct peaks in the exciting frequency range, the first distinct peak(peak1) and a sharp phase shift about 180 degree occur around 0.52, corresponding to CS1, the second distinct peak(peak2) with a sharp phase shift about 180 degree occur around 0.77, corresponding to CS2.

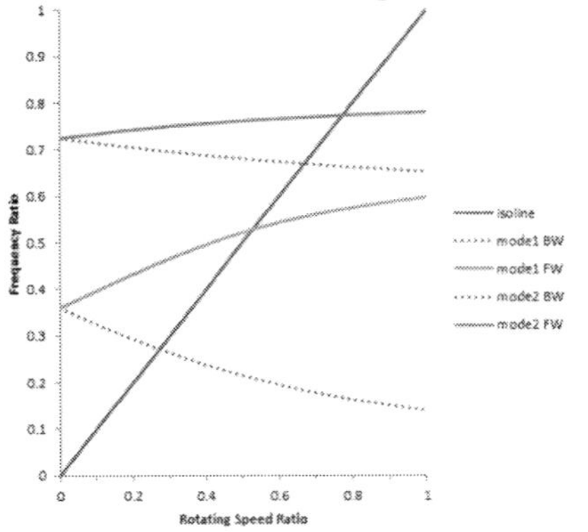

Figure 5. The Campbell diagram based on rotor model with static stiffness.

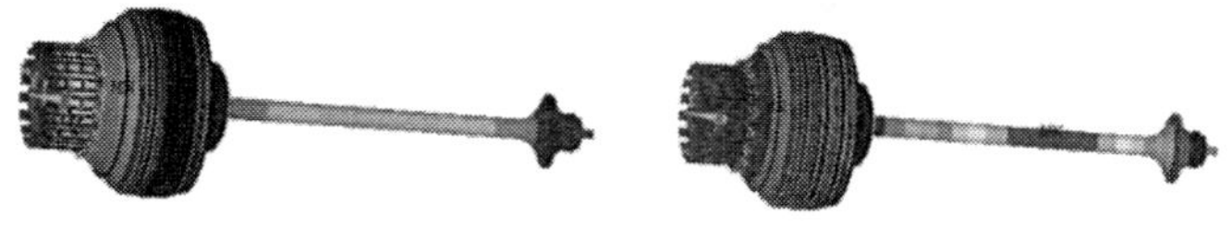

(a) Mode 1 (b) Mode 2

Figure 6. Modes of rotor.

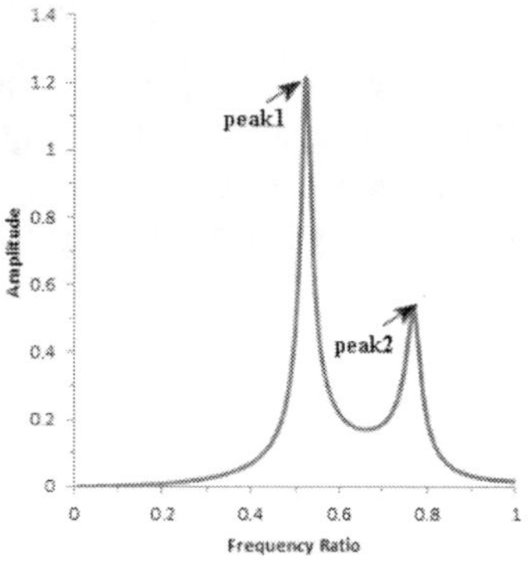

Figure 7. The displacement response amplitude based on rotor model with static stiffness.

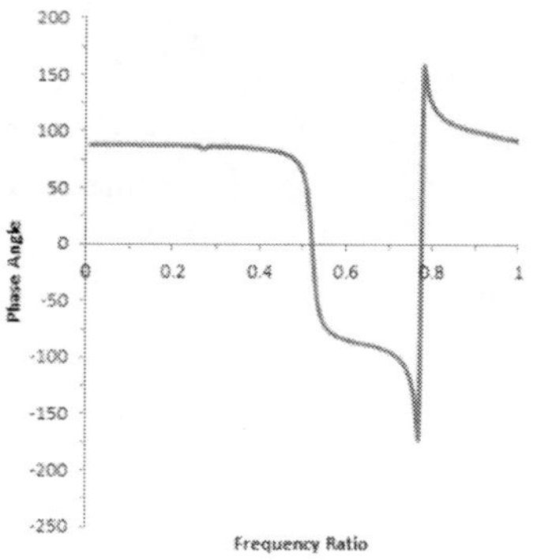

Figure 8. The displacement response phase angle based on rotor model with static stiffness.

4.2 Rotorynamic Characteristics based on rotor model only with dynamic stiffness

In the K_d method, both casing stiffness and modes are considered, but coupling vibration between rotor and casing is not included. Therefore, the rotor support stiffness changes with the rotor speed, especially at the resonant mode of the casing. As a result, the frequencies in Campbell diagram change severely with the rotor speed. The Campbell diagram based on the dynamic stiffness is shown in figure 9, which shows natural frequencies of many modes change severely around 0.4, the resonant frequency of casing. It can be seen from the Campbell diagram, the first critical speed (CS1) is 0.536 and the second critical speed (CS2) is 0.770. The unbalance load is applied at the fan disk, and a harmonic force response analysis is performed in the frequency range of interest. The displacement response amplitude and phase angle are presented in the following figures 10 and 11 respectively. There are three distinct peaks in the exciting frequency range, the second distinct peak(peak2) and a sharp phase shift about 180 degree occurred around 0.53, which is correspond to CS1, the third distinct peak(peak3) and a sharp phase shift about 180 degree occurred around 0.77, which is correspond to CS2. However the first peak distinct peak and a smaller phase angle change occurred around 0.41, which is not critical speed, and induced by dynamic stiffness only.

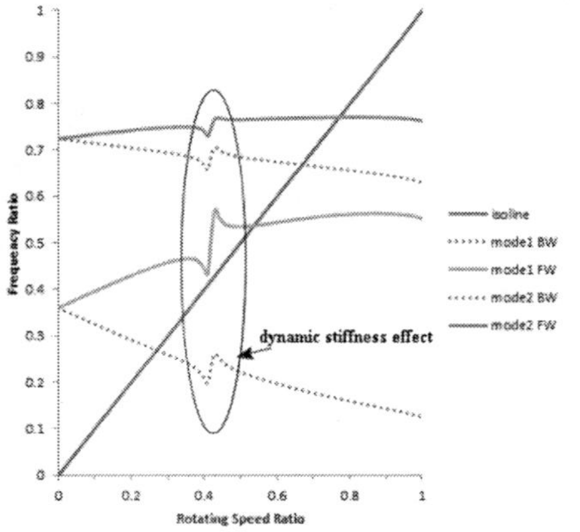

Figure 9. The Campbell diagram based on rotor model with dynamic stiffness.

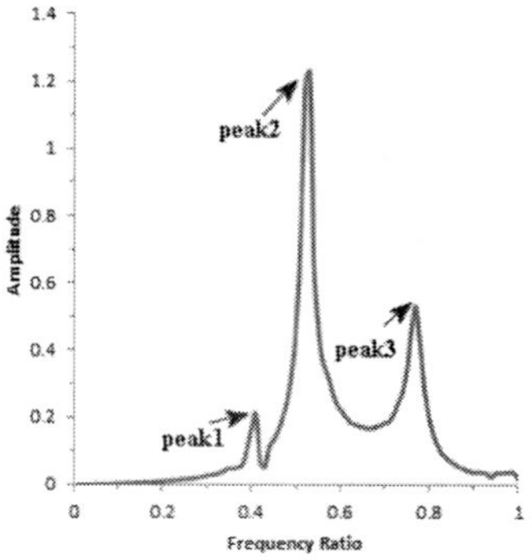

Figure 10. The displacement response amplitude based on rotor model with dynamic stiffness.

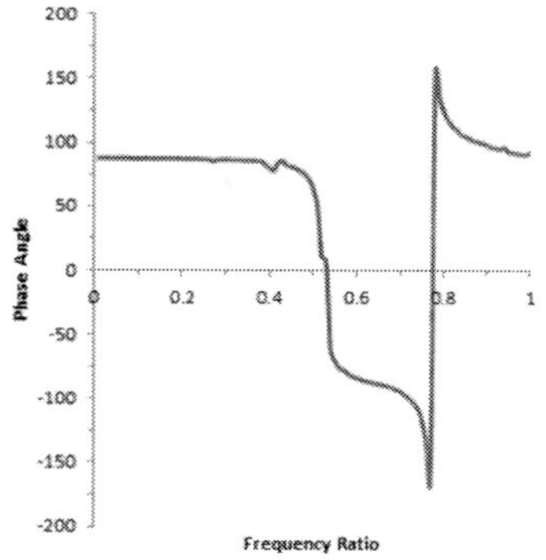

Figure 11. The displacement response phase angle based on rotor model with dynamic stiffness.

4.3 Rotorynamic Characteristics using 3D whole engine model

The Campbell diagram using 3D whole engine model is shown in figure.12. The whole engine model has a lot more modes than the standalone rotor model. The modes of the whole engine model include rotor modes, coupling modes between casing and rotor, as well as casing modes (mode 1-mode 5 and mode 10-mode 14), which do not change with the rotating speed. For clarity, casing modes are excluded and only rotor modes are retained, the modified Campbell diagram is plotted and shown in figure 13. As Figure 13 shows, there are two different mode frequencies, representing two different vibration mode shapes shown in figure 3(a) and (b). Both mode shapes are coupled vibration between the casing and rotor, meaning when the rotor is vibrating as mode 1 shown in figure 6(a), the casing may be vibrating as either of the two mode shapes. As a result, the first critical speed has two values, one is (CS1(1)) 0.405, the other (CS1(2)) is 0.540, the second critical speed (CS2) is 0.764. The unbalance load is applied at the fan disk, and a harmonic force response analysis is performed in the frequency range of interest. The displacement response amplitude and phase angle are presented in the following figures 14 and 15 respectively. There are three distinct peaks in the exciting frequency range, the first distinct peak (peak1) and a sharp phase shift about 180 degree occurred around 0.42, which is correspond to CS1(1), the second distinct peak (peak2) and a sharp phase shift about 180 degree occurred around 0.54, which is correspond to CS1(2), the third distinct peak (peak3) and a sharp phase shift about 180 degree occurred around 0.77, which is correspond to CS3. One can also find that one critical speed based on the whole engine model is near with the casing resonant frequency, so the engine casing needs to be improved.

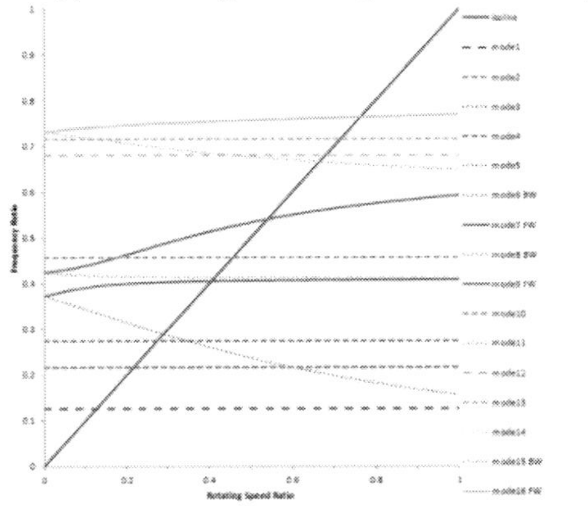

Figure 12. The preliminary Campbell diagram based on whole engine model.

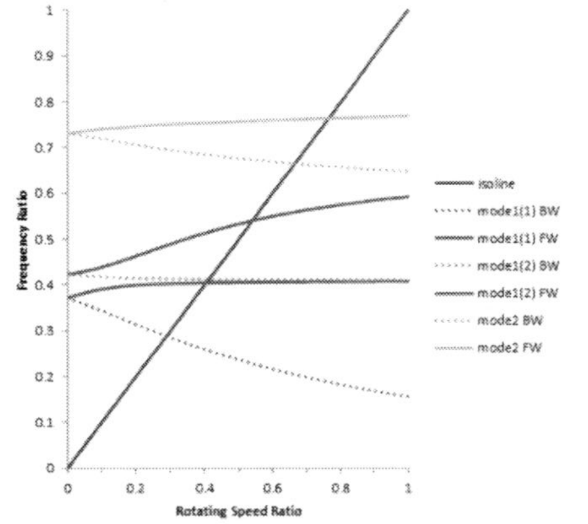

Figure 13. The modified Campbell diagram based on whole engine model.

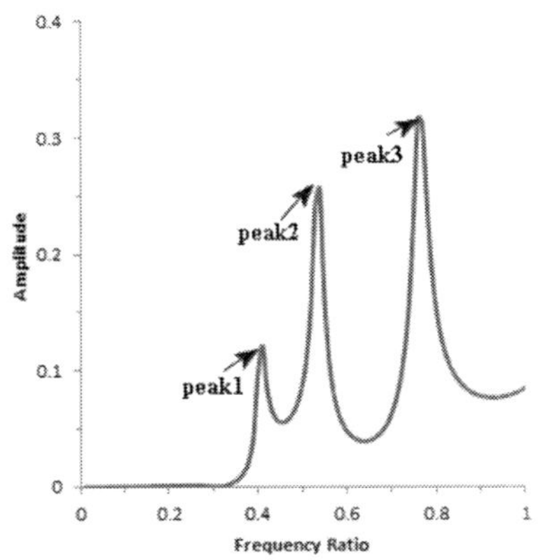

Figure 14. The displacement response amplitude based on whole engine model.

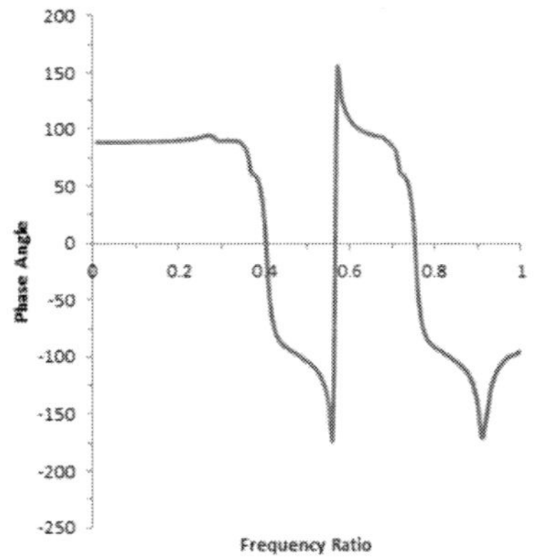

Figure 15. The displacement response phase angle based on whole engine model.

5 Results analysis

Table 1 shows the natural frequencies of different models at 0 rpm. It can be seen that the natural frequencies of the three models are consistent, except that the whole engine model has an additional mode frequency. The critical speeds predicted by the three models are shown in table 2. While the critical speeds predicted by the three models are comparable, the critical speeds based on the rotor model with dynamic stiffness are closer to the whole engine model than the rotor model with static stiffness, since the dynamic stiffness is dynamically more representative of the whole engine model. Because of the coupling effects between the casing and rotor, there is an additional critical speed based on whole engine model. Figure 16 shows the displacement response amplitudes based on three different models. We can see that the results based on the standalone rotor models are similar, while the rotor model with dynamic stiffness has an additional peak induced by the stator casing model. The results based on whole engine model and the standalone rotor models are different in the peak values and distributions. In addition, the peak values of the whole engine model are smaller than the standalone rotor models.

Table 1. Summary of natural frequencies at 0 rpm.

Mode	Rotor model(K_s)	Rotor model(K_d)	Whole engine model
1	0.360	0.360	0.372 0.424
2	0.724	0.724	0.731

Table 2. Summary of critical speeds.

	Rotor model(K_s)	Rotor model(K_d)	Whole engine model
CS1	0.523	0.536	0.405 0.540
CS2	0.760	0.770	0.767

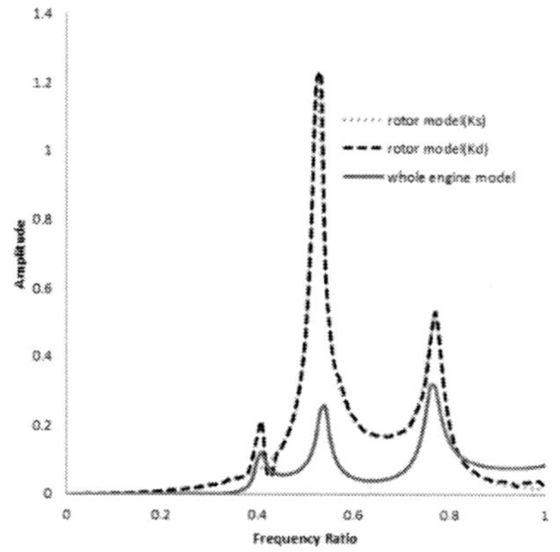

Figure 16. Summary of displacement response amplitudes.

6 Summary and conclusions

In this paper, 3D finite element model of an aero-engine low pressure rotor rig test system is established. The Campbell diagrams and unbalance harmonic responses are calculated respectively with static stiffness, dynamic stiffness and whole engine finite element model. The various calculation results were compared and analysed. From the simulation results, the following conclusions can be gained:

(1) The natural frequencies and critical speeds of the three models are consistent, except that the whole engine model can reveal additional mode frequencies and critical speeds induced by coupling effect between the rotor and stator casing.

(2) The unbalance response amplitude and phase angle of the whole engine model are different from that of the standalone rotor model, in that the peak values and distributions are different, and the peak values of the whole engine model are smaller than the standalone rotor model.
(3) The critical speed calculations based on the rotor model with dynamic stiffness are closer to the whole engine model than the rotor model with static stiffness only, because the dynamic stiffness represents the whole engine model better than the rotor model only considering static stiffness.
(4) With the whole engine model, the casing modes and the coupling between the rotor and casing should be considered, so to improve the accuracy of the rotordynamic characteristics.
(5) Generally speaking, critical speed should avoid the stator casing resonant frequency. Because whole engine model can reveal more critical speeds than the standalone rotor model, whole engine model should be used in the detail design stage, and it is necessary to verify that critical speed avoids the stator casing resonant frequency.

Acknowledgments

The authors would like to thank Dr. Hu Shoufeng, Dr. Ming Cao for their guidance throughout the entire research process. The research is funded by the "Independent innovation special fund of AECC(ZZCX-2017-010)".

References

1. Chen Meng, Ma Yanhong, Liu Shuguo, Journal of Beijing University of Aeronautics and Astronautics, **33**(9), 1013(2007)
2. Jeffrey J, Vannini G, Camatti M, Proceedings of ASME Turbo Expo 2006, 90481 (2006)
3. Seong Min Jeon, Hyun Duck Kwak, Suk Hwan Yoon, Journal of Propulsion and Power, **24**(3), 433 (2008)
4. Hong Jie, Chen Meng and Liu Shuguo, Proceedings of ASME Turbo Expo 2007, 27162(2007)
5. Jeffrey J, Vannini G, Camatti M, Journal of Engineering for Gas Turbines and Power, **132**, 082401(2010)
6. Eric Qiuli Sun, MSC 2006 Conference(2006)
7. ANSYS Documentation 2013 ANSYS Inc
8. Masaharu Shinozaki, Osamu Funatogawa and Masao Kobayashi, Proceedings of ASME Turbo Expo 2001, 0254(2001)
9. Gu Jialiu, Rotor dynamics (National Denfence Industry Press, Beijing, 1985)

Increasing the Efficiency of Storage of Cryoproduct through the Introduction of Technical Solutions

S V Zaika[1,a], Aksenova D.K.[1], E A Shatokhin[1] and S S Zaika[1]

[1]*Samara University, Department Of Heat Engineering And Heat Engines, 443086 Moskovskoye shosse, 34, Samara, Russia*

Abstract. In various systems of aerospace engineering, transport, as well as in energy complexes for various purposes, cryogenic liquids are now widely used. Liquefied natural gas (LNG) is an environmentally friendly energy source that has several advantages over petroleum products. LNG using reduces the weight and size characteristics of tanks for storage and transportation. Reducing the loss of cryoproducts using low-potential power machines to ensure efficient storage of cryoproducts. Significant problems associated with the loss of its low-potential energy (caused by heat from the environment). There is a need to improve solutions to compensate for losses and expenses. This study examined the possibility of reducing the loss of cryoproducts when stored in tanks and transportation systems. Details were studied: assessment of losses of the cryoproduct and its capabilities for low-temperature energy, calculation of the loss compensation loop, calculation of the contour for generating additional energy based on a heat power plant, analysis of low-potential power plants, compensating losses of cryoproducts during storage. The main results of the research are the proposal to organize the LNG Storage, calculate the loss compensation loop and the supplementary energy circuit.

1 Introduction

The use of LNG in many branches of the agro-industrial complex as an energy carrier has obvious advantages due to a number of features, and first of all, its wide application possibilities, relatively low cost, environmental cleanliness, the possibility and convenience of transportation [1].

LNG as an energy carrier and natural gas derived from it can be used for:

- heat supply (production of steam and hot water) in boiler installations of residential and public buildings, as well as consumer services. The range of these units is quite wide and depends on the required heat output;

- heating of premises and supply of the population with hot water.

- power supply to remote deep areas by creating mobile gas turbine power stations capable of operating on gaseous fuels. When operating a gas turbine power plant, the heat of the exhaust gases can be used for the heating of nearby residential areas. The subject of the study is the thermophysical processes occurring during the storage of cryoproduct (LNG), and the object of investigation is a low-potential power plant [2].

[a] Corresponding author : zaika@ssau.ru

2 Estimation of storage losses

When storing a cryoproduct, its losses are due to the low efficiency of thermal insulation and the heat input from the environment. It is possible to calculate the loss of the low-temperature heat of the cryoproduct lost during its storage by the following equation:

$$\Delta m = m_{\Sigma} \cdot \frac{\theta}{100} \quad (1)$$

$$Q_{evap} = \Delta m \cdot r + c_p \cdot \Delta m \cdot (T_c - T_0) \quad (2)$$

where Δm - mass of evaporated cryoproduct per day, m_{Σ} - mass of stored cryoproduct; Θ - loss of product due evaporation, % Q_{evap} - low-potential heat of the evaporated cryoproduct; r - heat of evaporation of cryoproducts; T_c – cryoproduct temperature; T_o – ambient temperature.

Cryoproduct has a high specific low-temperature energy potential, which can be used as a "refrigerator" in a thermal machine operating along the Curzon-Alborne cycle. The loss of disposable energy L per day, provided that the low-potential heat of the cryoproduct is used according to the Curzon-Alborne cycle, is calculated as follows:

$$L = \frac{Q \cdot \eta}{1-\eta} \quad (3)$$

$$\eta = 1 - \sqrt{\frac{T_2}{T_1}} \quad (4)$$

Losses from evaporation in tanks, depending on its dimensions and designs. When choosing the insulation, the required liquid used and the fact that the thickness of the insulation increases with an increase in the volume of the vessels more slowly than the diameter of the vessel. When choosing the isolation of tanks, economic factors should be considered. The use of vacuum-multi-layer insulation allows a 10-fold reduction in oxygen loss compared with vacuum-powder insulation. However, the difference in their cost is so great that the payback is calculated in tens of years.

Specific volumetric daily losses of low-grade heat of the cryoproduct depending on the volume of the reservoir and specific volumetric daily losses of available energy under the condition of using low-potential heat along the Curzon-Alborn cycle are determined by the equation:

$$q_V = Q/V \quad (5)$$

$$l_V = L/V \quad (6)$$

where q_V - specific daily volumetric losses of low-grade heat of the cryoproduct; l_V - the specific daily volume losses of the available energy, provided low-potential heat is used in the Carnot cycle; V – tank volume.

Even though the relative losses of the cryoprotectant decrease with the increase in the volume of the reservoir, the absolute values of losses of low-grade heat increase, which is quite logical.

The calculation results show a significant amount of the low-potential energy of the cryoproduct lost during storage. LNG has the greatest potential of cryogenic energy that lost during storage. Based on these results, the problem arises of determining the ways of solving the reduction in mass losses of the cryoproduct due to evaporation in tanks.

Calculation of specific inflows per running meter of pipelines of various diameters. The calculated values of specific heat inflows are presented in Table 1 and in Figure 1.

Table 1. Values of specific inflows per running meter of pipelines

Diameter of the pipeline	Size of inner tube, DxS	Casing diameter, DxS	Specific heat input, [W / m]
15	18x1,4	100x2	0,9
25	28x1,4	100x2	1,1
50	56x2	150x2,5	1,2
65	70x2	150x2,5	1,3
100	100x2	200x2,5	1,5
150	150x2,5	250x3	2,2
200	200x2,5	300x3	3,0
250	250x3	350x3,5	4,0
300	300x3	408x4	5,2
400	408x4	500x4	8,0

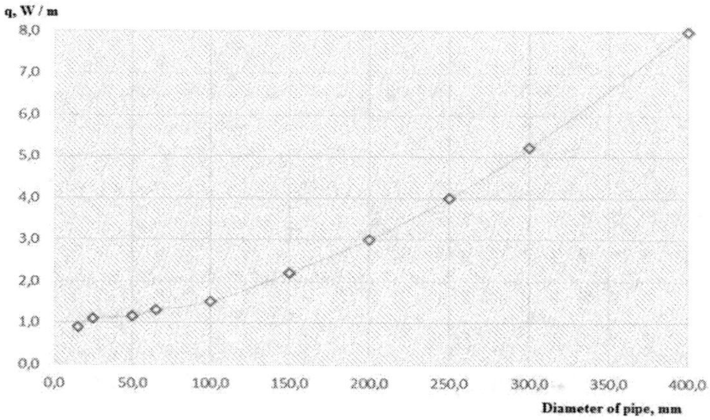

Figure 1. Dependence of specific heat input per 1 meter of pipeline from its internal diameter

Figure 2 shows the scheme for transporting a cryogenic liquid over distances. In such a system, the heat supplied to the cryoproduct in the pump is given to a low-pressure liquid in the heat exchanger, and the injected liquid is already supplied to the pipeline even at a pressure (greater than the saturation pressure). The problem was considered under the condition that the liquid reaches the gate in a saturated state. Numerous examples have been calculated for various flow rates, pump discharge pressures, pipe diameters and pipeline lengths.

3 Options for organizing the LNG storage facility

Three different options for organizing LNG storage were selected. The first combination consists of 4 tanks with a capacity of 112 m^3, the second - from 2 tanks with a capacity of 248 m^3, the third - from 8 tanks with a capacity of 49.50m^3. The logic of the organization consists of selecting a more advantageous option, based on the considerations of the least loss of evaporation while the cylinders are empty. The number of tanks of certain volumes was selected on the basis of the mass of fuel required to operate the gas piston plant (GPU) with a capacity of 1 MW for 28 days, which is 168 480 kg. LNG preparation and storage systems are shown in Figure 2.

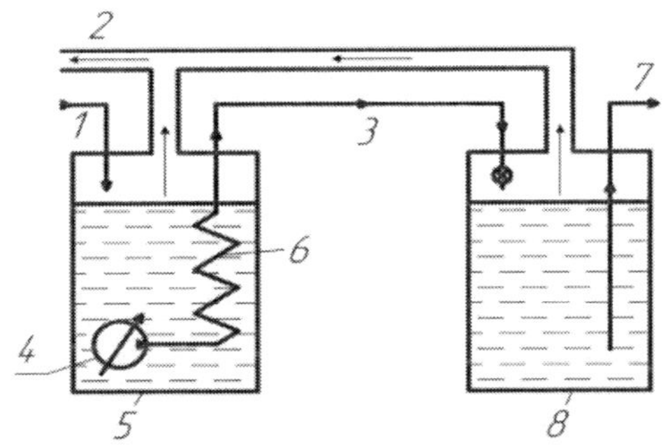

Figure 2. Liquefied gas transmission system
1- liquid intake; 2 - return or release of steam;3 - pipeline; 4 - the pump; 5 - pump reservoir; 6-heat exchanger;7-to the consumer of the liquid; 8 - remote storage

During operation, tanks are used alternately. Losses in the mass of the cryoproduct in the used tank are not considered, while in the remaining tanks losses occur while they are in the standby mode. The amount of loss depends on the selected storage scheme and the characteristics of the tanks involved (Table 2).

Considering the processes of heat transfer, the efficiency in the Curzon-Alborne cycle is calculated as follows:

$$\eta' = 1 - (T_2 / T_1)^{1/2}; \tag{7}$$

where T_2 - cold source temperature, K; T_1 - temperature of the heat reservoir, K.

With an increase in the maximum permissible temperature of fuel elements, the average temperature of the working fluid T 'and the efficiency increase. At the same time, the average temperature of the working fluid increases slowly, and the efficiency is fast enough. efficiency can be improved by regeneratively heating the working fluid to a high temperature. When you receive even a small amount of heat from the environment, the liquid (LNG) partially passes into the gaseous form, this leads to gas leaks. The amount of evaporated liquefied natural gas (evaporation loss) is rapidly becoming significant, their energy value can not be neglected. Therefore, in general, several solutions are proposed to reduce the losses of the liquid phase of the cryoproduct, based on the use of a heat-negative power plant having a loss compensation contour.

Table 2. Options for organizing the LNG storage facility

The scheme	**a**	**b**	**c**
Number of tanks	2	4	8
V, m^3	248	112	49,50
M, kg	84 240	42 212	21 060
Loss per day, %	0,09	0,12	0,12
Losses in mass in one tank per day, kg	75,8 16	50,6 5	25,272
Losses for all time in all tanks, kg	106 1,42	212 7,3	2476,66

4 Possible technical solutions for loss reduction cryoproduct at its storage

Technical solutions are proposed for reducing the loss of cryoproducts during storage by using a plant that consists of two circuits, one of which is a loss compensation contour and the other is a contour for generating additional energy. The circuit for generating additional energy can be realized based on the Stirling engine.

The loss compensation contour includes:

- cryogenic tank with liquefied natural gas;
- throttle valve. This is a special valve used in refrigeration units and heat pumps in which a process of throttling occurs, that is, an irreversible expansion of the liquid with a decrease in pressure and temperature.
- a gas receiver intended for the accumulation of an evaporating liquid phase;
- heat exchanger - evaporator, which is necessary to cool the gas heated after the compressor with liquid methane to the temperature at which it will be completely in the liquid state;
-centrifugal compressor [3].

The gas from the suction chamber located near the axis of rotation is sucked into the impeller, enters its interblade channels and is pushed through these channels by the centrifugal forces to exit the impeller. With this movement, a constantly increasing centrifugal force acts on the gas, that increase gas velocity. From the thermodynamics of the flow, we know that when the gas pressure flow is accelerated, it decreases. The blade channels of the impeller have an expanding shape, and this leads to an increase in pressure.

The total effect of these two influences leads to an increase in pressure, so that at the exit of the impeller the absolute gas velocity becomes maximum and its pressure rises. Then the gas enters the fixed vane diffuser, where its braking takes place with a corresponding increase in pressure. Then, the flow in the reverse guiding device passes through the interscapular channels with increasing cross section and enters the next stage.

The principle of operation of the loss compensation circuit during evaporation is as follows. Liquefied natural gas is charged into the tank with a capacity in accordance with the selected scheme of the storage park at a temperature of 138 K and a pressure of 0.6 MPa. The evaporated natural gas flows through the drain valve to the receiver. Accumulated losses per day are compressed by a centrifugal compressor up to 1 MPa.

The operating time of the compressor is 20 minutes. The drive of the centrifugal compressor is an electric motor that operates on the energy generated by the Stirling engine, depending on the chosen circuit. After the compressor, the heated gas is cooled by liquid methane in the heat exchanger-evaporator and then throttled to 0.65 MPa. Liquefied gas returns to the tank.

5 Scheme of plant based on the Stirling engine

The drive of the centrifugal compressor can be an electric motor running on the energy generated by the Stirling engine. After the compressor, the heated gas is cooled by liquid methane in the heat exchanger-evaporator and then throttled to 4.5 MPa. Liquefied gas returns to the tank.

The second circuit of the heat-negative installation based on the Stirling engine includes:

- Stirling engine, which includes three heat exchangers: a heat exchanger for the expander cavity, a compressor cavity, a regenerative heat exchanger. In Stirling engines, the heat exchanger of the expander cavity, perceiving heat from an external source at a high temperature, transfers it to the engine's working body located in the working cavity adjacent to the expansion cavity [4].

The principle of operation of the circuit based on the Stirling engine is as follows (Figure 3). The compressor cavity of the Stirling engine is cooled by liquefied natural gas to a temperature of 158 K, and the expander cavity is heated from the exhaust gases of the gas piston unit to a temperature of 400 K. The actual engine power is 4.8 kW. Liquefied gas, passing through the heat exchanger of the compressor cavity of the Stirling engine, enters the heat exchanger-utilizer, where it is heated by exhaust gases from the GPU. After full gasification of gas, it is disposed of in the GPU.

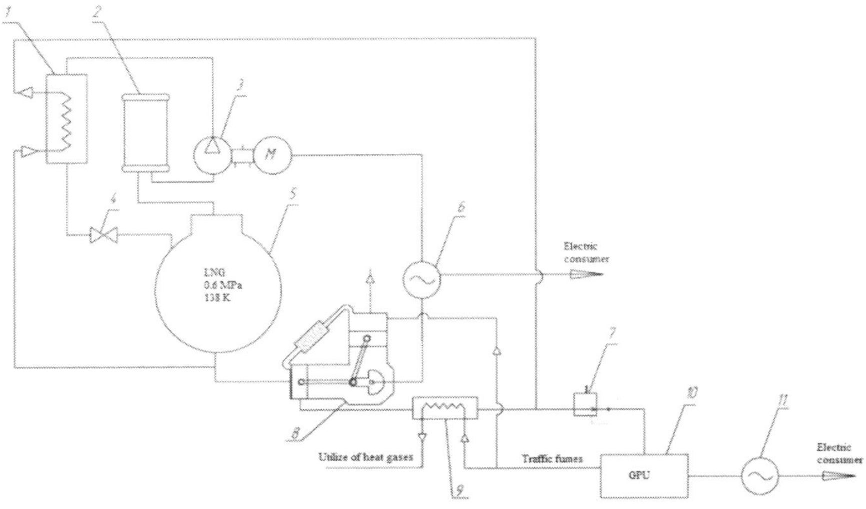

Figure 3. Scheme of installation based on the Stirling engine

1 - heat exchanger-evaporator; 2 - Gas receiver; 3-compressor; 4 - throttle valve; 5 - LNG tank; 6 - electric generator; 7 - reducer; 8 - Stirling engine; 9 - heat exchanger-utilizer; 10 - GPU; 11 - power generator.

The advantage of the proposed low-temperature power plant based on the Stirling engine:
- high energy efficiency;
- proven technology of small-scale production of engines.

Disadvantages of the proposed low-temperature power plant based on the Stirling engine:
- the difficulty of ensuring the tightness of the Stirling engine;
- The complexity of the design of the Stirling engine [5].

6 Low-potential energy converters

It is proposed to use thermoelectric generator modules (TGM) to realize the low-temperature potential of LNG, which are now widely used. The power generated by such generators ranges from a few milliwatts to several kilowatts and is ultimately determined by the economic feasibility of choosing this method of energy conversion [6].

At present, the line of modern thermoelectric generators provides the possibility of obtaining electric power with an efficiency of up to 15% (depending on the temperature levels of hot and cold junctions). In the proposed complex it is proposed to install Peltier elements on the surface of the evaporator in such a way that the cold junction is in contact with the surface of the evaporator heat exchanger, and the hot one is in contact with the environment (Figure 4).

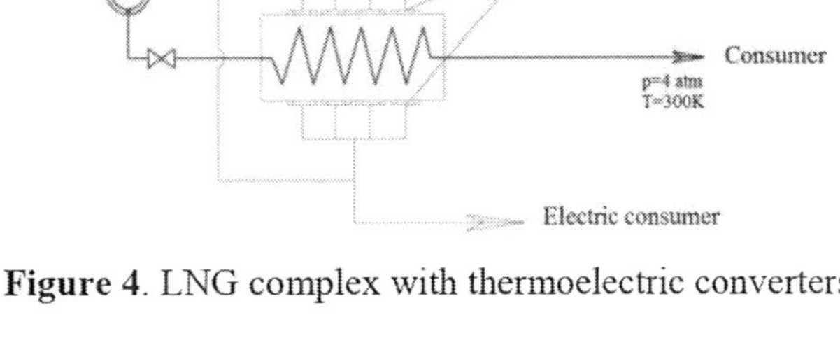

Figure 4. LNG complex with thermoelectric converters

With a LNG flow rate of G = 1000 kg / h and a given efficiency, the heat flux at the cold junction Q_h is about 32 kW, and the net electrical power P = 4.2 kW obtained, with the number of 40 W modules amounting to about 104 units.

7 Conclusion

As a result of this work, an analysis of energy losses in storage systems of cryoproducts through horizontal and vertical cylindrical tanks was carried out, energy potential losses were determined, as a result of which it was revealed that the greatest potential for cryogenic energy lost during storage is LNG. The scheme of a low-potential power plant compensating losses of a cryoproduct at its storage is analyzed.

References

1. Kakumoto, Y., Koyamatsu, Y., Shiota, A., Qudaih, Y., Mitani, Y. Application of Geographic Information System to Power Distribution System Analysis (2016) Energy Procedia, 100, pp. 360-365.
2. Burtsev, S.A., Karpenko, A.P., Leontiev, A.I. A method for distributed production of liquefied natural gas at gas-distribution stations. (2016) High Temperature, 54 (4), pp. 573-576.
3. A.M. Arkharov, I.D. Kunis. *Cryogenic refueling systems of launch rocket and space complexes*, publishing house MSTU. NE Bauman, 2006.
4. G. Zhang, W. Xu, Y. Yang, D. Zhang, Energy Procedia, *Utilization of LNG cryogenic energy in a proposed method for inlet air cooling to improve the performance of a combined cycle*, **Volume 61**, p. 899-903, (2014).
5. Y. Xiong, P. Luo, B. Hua, Energy Procedia, *A novel CO2-capturing natural gas combined cycle with LNG cold energy utilization*.**Volume 61**, pages 899-903, (2014).
6. E.V. Blagin, A.I. Dovgyallo, S.O. Nekrasova, D.V. Sarmin, D.A. Uglanov, Applied Thermal Engineering *Estimation of the energy efficiency of cryogenic filled tank using in different system and devices*. 2016. **Volume 101**, 25 May 2016, Pages 537-544

Numerical Analysis of Bypass Mass Injection on Thrust Vectoring of Supersonic Nozzle

Md Shafiqul Islam[1], Md. Arafat Hasan[1] and A.B.M. Toufique Hasan[1, a]

[1]*Department of Mechanical Engineering, Bangladesh University of Engineering and Technology (BUET). Dhaka-1000, Bangladesh*

Abstract. High speed aerospace applications require rapid control of thrust (i.e. thrust vectoring) in order to achieve better manoeuvrability. Among the existing technologies, shock vector control is one of the efficient ways to achieve thrust vectoring. In the present study, bypass mass injection (passive control) was used to generate shock vectoring in a planar supersonic Converging-Diverging (CD) nozzle. Two diffenrent bypass lines were used to inject mass in the diverging section varying their dimension in the span wise direction (10 mm ×10 mm^2 square channel and 2.68 mm×38 mm^2 rectangular channel) in such a way that, the mass flow ratio in both the case remain the same (4.9%) in order to compare the effect of bypass channel dimension in the resulting thrust vector angle and thrust performance. Reynolds-averaged Navier-Stokes (RANS) equations with k-omega SST turbulence model have been implemented through numerical computations to capture the three-dimensional steady characterstics of the flow field. Results showed a significant change in the shock structure with the fromation of recirculation zone near the bypass injection port in both the case with a variation of shock structure and thrust performance for different geometry bypass lines. It was found that, thrust vector angle increases as injection length increases in the span wise direction.

1 Introduction

Shock vector control (SVC) is one of the fluidic thrust vectoring (FTV) methods and utilizes the shock wave to control the thrust direction. This method using a bypass line was researched in this study. The geometry of nozzle and the experimental data for validation was taken from the Hunter [1] experiment carried out at NASA Langley research center. Two distinctive flow separation phenomena, the free shock and restricted-shock separation, which were observed in experiments with nozzles by Manuel Fray [2] was found on different walls simultaneously and separately in different spans of the studied nozzle. Investigation of Hadjadj [3] gives some key insight on Mach disk, triple point, slip line found on shock structures of separated nozzle flow. In the numerical simulation carried out by Nasuti and Onofri [4] provide an analysis of Mach reflection of the separation shock in over expanded nozzle. Studies of Mittal [5] found interesting phenomenon that the thrust vectoring automatically occurs at low NPR in modified Hunter [1] nozzle. To achieve thrust vectoring Waithe and Deere [6] adapted mass injection by multiple and single port at different NPR to compare their effectiveness. Lately Deng et al.[7] conducted Large Eddy Simulation to achieve thrust vectoring in an asymmetric nozzle

[a] Corresponding author: toufiquehasan@me.buet.ac.bd

using bypass mass injection and showed that effective thrust vectoring can be achieved using bypass mass injection keeping the mass flow ratio (ratio of mass flowing through the bypass line to the total mass flow) below 7%. In the present study thrust vectoring was achieved using bypass mass injection on a planar nozzle.

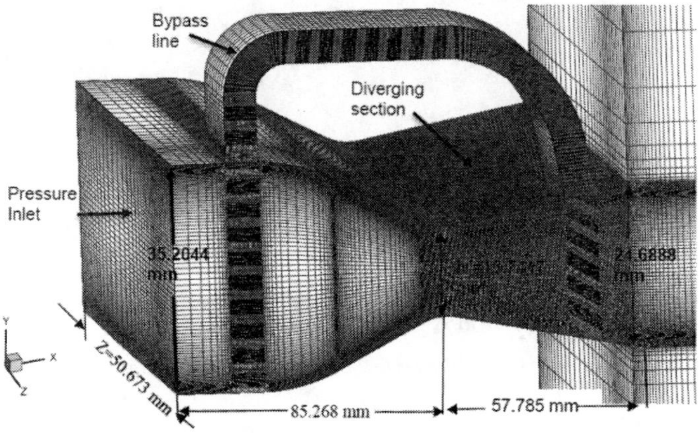

Figure 1. Computational domain with10×10 sq mm bypass line.

2 Computational Tool

Reynolds-averaged Navier-Stokes (RANS) equations with k-omega SST turbulence model have been implemented through numerical computations to capture the three-dimensional steady characterstics of the flow field. Steady state computaion was performed as suggested by Hunter [1] that the flow structure in the nozzle assumes steday state after NPR 2. The commercial computational fluid dynamics (CFD) code ANSYS Fluent (version 14.5) was used to simulate the flow within the nozzle and the exhaust plume near the nozzle exit using density based solver.

3 Computational Domain

Half of the nozzle was modelled to reduce number of nodes thus computational power and the data obtained was mirrored along the symmetry to get the full flow field inside the nozzle. Computational domain was modeled keeping a 27.483 mm horizontal extra domain prior to inlet and 1258.63 mm horizontal extra domain after the exit to capture the flow structure beyond the exit region of the main nozzle. Grid independency test was done using two dimensional model. Figure 2(b) shows the $y+$ value of different mesh resolution, in this study medium mesh with near wall treatment of figure 2(b) having 20000 node in two dimensional model was taken as the base and in three dimension 35 grids was given in the 50.673 mm extrusion in Z direction. The final three dimensional structured mesh contained 1.6 million hexahedral cells and the first length normal to the wall was about y= 0.4 μm near both for the nozzle wall and nozzle side wall to catch the boundary layer properties accurately. Figure. 1 shows the computational domain having 10×10 sq mm bypass line with the boundary conditions.

4 Boundary Condition

The bypass line is so designed that the mass flow ratio remains below 7% to avoid reduction in the resultant thrust which was found in a study by Deng et. al [7]. In the present study with bypass lines of 10×10 mm^2 and 2.68×38 sq mm^2, mass flow ratio is 4.9% was found at NPR 2.4 which is below the maximum percentage of mass flow ratio as suggested by Deng et al.[7]. The operating condition of this study is 0 pa with inlet outlet pressure equal to 2.4e5 and 1e5 pa respectively. Upper and lower nozzle wall and side nozzle was as well as the bypass line walls were considered as wall while running simulations. The temperature in all the conditions was fixed to 298.15 K so there was no heat flux through any wall. The pressure far field was assigned with pressure equal to 1e5 at Mach number 0.1. Turbulent intensity and viscosity ratio at inlet is .1% and 1 respectively whereas at outlet is 5% and 10 respectively. Air was used as the working fluid whose density is assumed vary according to the ideal gas law and its viscosity are assumed vary according to Sutherlan's law. Specific heat capacity and thermal conductivity of the fluid are kept constant for simplicity.

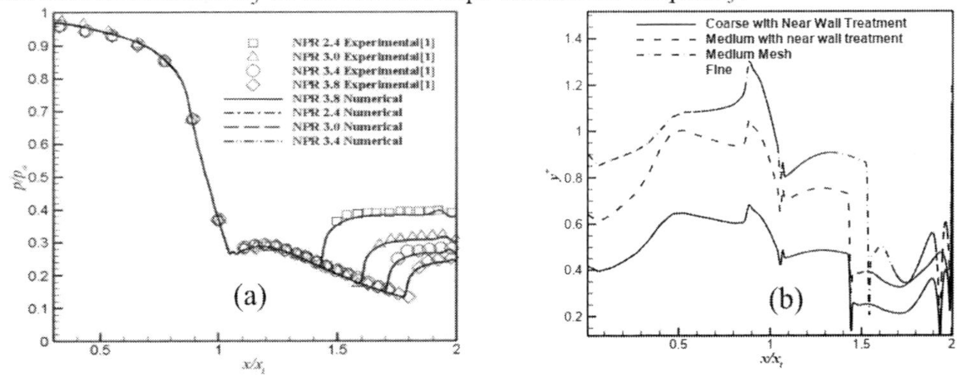

Figure 2. Pressure distribution at nozzle wall (a) and wall $y+$ at different mesh resolution (b).

5 Validation

Numerical results found form the simulations in the original nozzle at different NPR was plotted in the same graph with the experimental results conducted by Hunter [1] and a complete match in the pressure distribution along the nozzle wall is shown in figure 2(a). After validating the main nozzle, nozzle with bypass line was drawn and meshed following the earlier procedure giving special attention to the bypass channel meshing to ensure better result. Pressure distribution and shock structure is also similar to the study carried out by Hasan [8] using RANS equations with k-omega SST turbulence model in the same nozzle geometry.

6 Analysis of 10×10 mm^2 bypass line (20% span wise)

6.1 Shock Structure.

All the results presented here are computed at NPR 2.4. Shock structure drastically changes due to bypass mass flow injection. The arrow in all the figure shows the mass flow direction in the flow field where x_t is the abscissa of the throat. Lambda shock structure is still present but location of Mach stem has been shifted towards the throat to adjust the incoming flow. So, the flow becomes asymmetric and thrust vectoring is achieved at NPR 2.4. Within 0 to 30mm there is restricted shock separation (RSS) on lower wall and free shock separation (FSS) on upper wall evident from the shock structure presented in figure 3(a). Symmetric shock structure is observed at 31 mm from the symmetry plane shown in figure 3(b). At this injection any slice far from 31 mm from symmetry plane shock structure is reversed. That is FSS on lower wall and RSS on upper wall shown in figure 3(c). As flow approaches nozzle side wall shock gradually diffuses and eventually fades away.

6.2 Streamlines

At the symmetry plane it is clearly seen that, after the bypass mass injection in the diverging sections, streamlines are tilted towards the lower nozzle wall hence the flow is separated from the upper wall. There are 2 pairs of recirculation zone, each pair consisting a clockwise and an anticlockwise vortex, which are located on the opposite side of the injection port shown in figure 3 (d). Another anticlockwise recirculation zone is also evident at the lower nozzle wall at the end of diverging section. But after the edge of the bypass line, the flow direction gradually changes and reattaches to the upper wall, still reverse flow is observed at the nozzle lower wall at the end of the diverging section. The flow becomes almost symmetric at 31 mm from symmetry plane apart from a small recirculation at the lower wall shown in figure 3 (e). Up to this the main bulk flow is slightly downward but after that main bulk flow is deflected slightly in the upward direction. At 45 mm from symmetry the bulk flow is heavily tilted upward shown in figure 3(f). As moved near the nozzle side wall, streamline direction becomes unpredictable due to the boundary layer interaction of the fluid particle with the nozzle side wall.

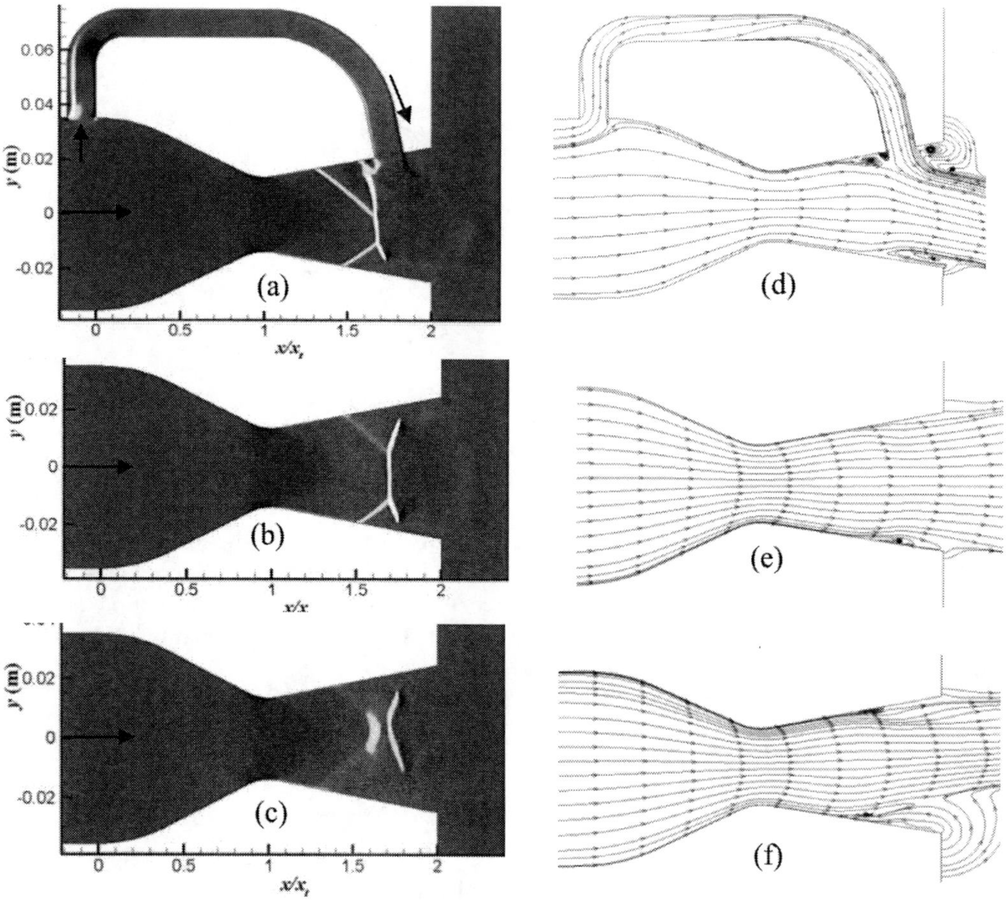

Figure 3. Density gradient contour at intermediate slices starting from the symmetry (a), at Z=31 mm (b), Z=45 mm (c), and streamlines at intermediate slices starting from the symmetry (d), at 31 mm (e), and at 45 mm (f).

7 Analysis of 2.68×38 mm^2 bypass line (75% span wise)

7.1 Shock Structure.

Lambda shock structure is present but location of Mach stem has been shifted towards the throat to adjust the incoming flow. Within 0 to 38 mm (throughout the extension of bypass channel) there is restricted shock separation (RSS) on lower wall and free shock separation (FSS) on upper wall evident from this shock structure presented in figure 4(a) and figure 4(b). Near the nozzle wall complex shock diffusion occurs where shock gradually diffuses and eventually extinguishes at the side wall of the nozzle as shown in figure 4(c). Throughout the extension of the bypass channel flow pattern is uniform and thrust vectoring is confined within 38 mm from symmetry.

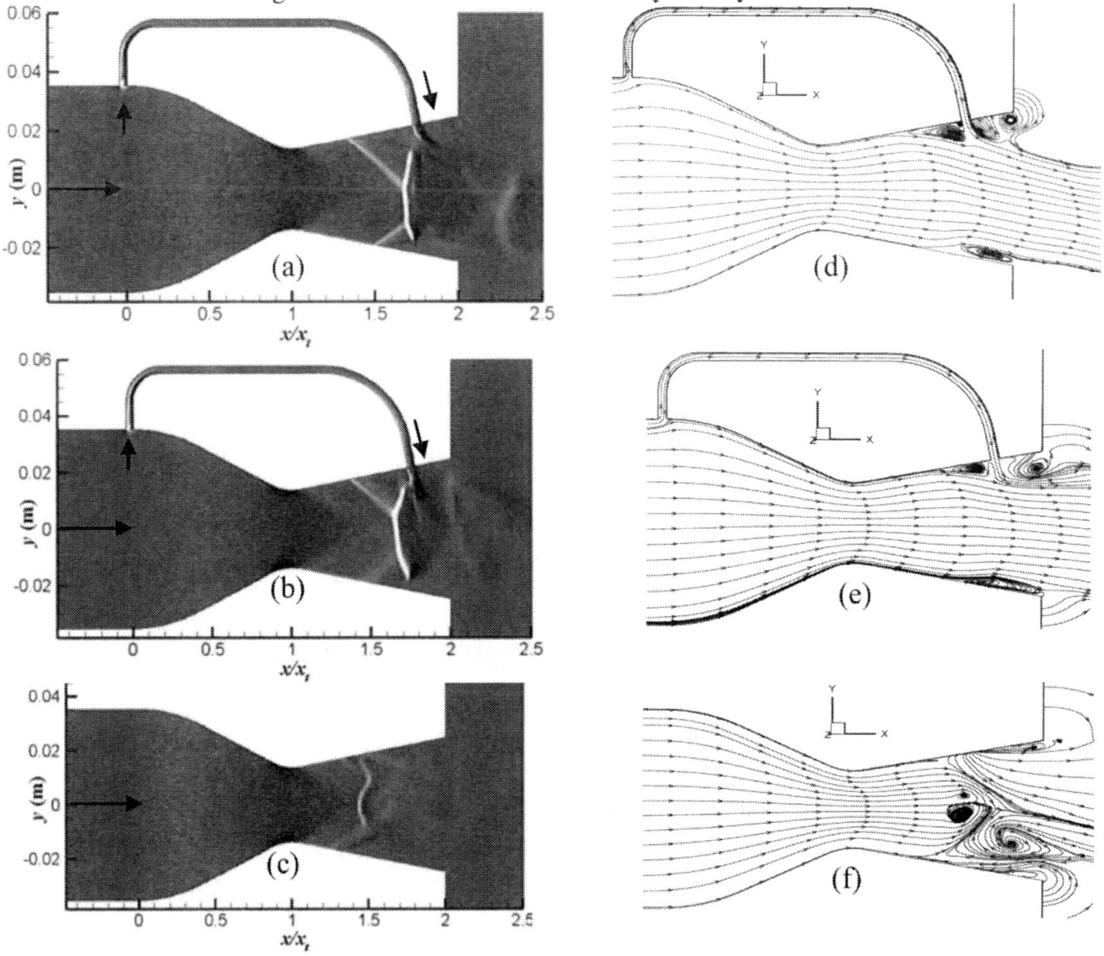

Figure 4. Density gradient contour at intermediate slices starting from the symmetry (a), at Z=35 mm (b), at side wall (c) and streamlines at intermediate slices starting from the symmetry (d), at 35 mm (e), at the side wall (f).

7.2 Streamlines

At the symmetry plane it is clearly seen that, after the bypass mass injection in the diverging sections, streamlines are titled towards the lower nozzle wall. There are 2 pairs of recirculation zone, each pair consisting a clockwise and an anticlockwise vortex, which are located on the opposite side of the injection port shown in figure 4(d). Another anticlockwise recirculation zone is also evident at the lower nozzle wall at the end of diverging section. At 35 mm from the symmetry in figure 4(e), the streamline is almost the same as observed in the symmetry. But after the edge of the bypass line, the flow direction gradually changes and as moved near the nozzle side wall, streamline direction becomes unpredictable due to the boundary layer interaction of the fluid particle with the nozzle side wall which is shown in figure 4(f).

8 Thrust Vector angle and Thrust Performance

The area averaged horizontal and vertical component of velocity at the nozzle exit, mass flow rate and velocity magnitude was calculated from fluent at the nozzle exit. The thrust vector angle is determined by $\delta = \tan^{-1}(F_y/F_x)$ [2] and thrust is the product of mass flow rate and velocity magnitude at the nozzle exit. A comparison of thrust vector angle and thrust performance is shown is Table 1.

Table 1. Comparison of thrust vector angle and thrust performance

Configuration	Mass Flow Rate (kg/s)	Bypass mass flow ratio (%)	Velocity Magnitude (m/s)	Thrust (N)	X Velocity (m/s)	Y Velocity (m/s)	Thrust Vector Angle
No Bypass	0.812	----------	239.3	194.3	239.3	0.00	-------
10×10 Sq mm Bypass	0.813	4.92	258.27	210	235.59	-14.1413	3.14°
2.63×38 Sq mm Bypass	0.813	4.92	259.34	210.8	237.57	-38.62	9.23°

9 Concluding Remarks

Thrust vectoring using bypass mass injection is a passive mode of thrust control as the injected mass controlling the thrust direction comes from the same combustion chamber producing the thrust. Here in this present study injecting mass by a bypass line at NPR 2.4 on the Mach disc at two different bypass configuration keeping mass flow ratio constant around 4.9% came up with some interesting findings those noted below

- Shock structure changes from symmetric to asymmetric due to bypass mass injection just beneath the injection port for 20 %(10×10 Sq mm bypass) but there is no sign of reverse shock at 75%(2.63×38 Sq mm bypass)
- Due to the presence of reverse shock, integrated effect of thrust vectoring is minimized in 20%
- Recirculation region is observed within the vicinity of injection port and on the nozzle lower wall on both case.
- Flow is span wise more uniform in nature in case of 75% than 20%
- Resultant thrust is vectored by an angle 3.14° for 20% but for 75% it is 9.23°.
- Thrust performance slightly increases from 210 N in case of 20% to 210.8 N in case of 75%.
- It appears that for thrust vectoring 75% bypass will be a better choice than 20%

Acknowledgement

The present work has been carried out with computational resource support from Higher Education Quality Enhancement Project (HEQEP), AIF (3rd Round)-Sub-Project CP 3111, UGC, MoE, GoB.

References

1. C.A.Hunter, *Journal of Propulsion and Power* **20**, 527-532 (2004)
2. M.Frey, *Journal of Propulsion and Power* **16**, 478-484 (2000)
3. A. Hadjadj and M. Onofri, Shock Waves **19**, 163–169 (2009)
4. F.Nasuti and M. Onofri, Shock Waves **19**,229-237(2009)
5. V.M.K. Kotteda and S. Mittal, Shock Waves **27**, 441–455(2017)
6. K. A. Waithe and K. A. Deere, "Experimental and Computational Investigation of Multiple Injection Ports in a Convergent-Divergent Nozzle for Fluidic Thrust Vectoring. AIAA paper. 2003 Jun 23;**3802**:2003.
7. R. Deng, T. Setoguchi, H.D. Kim, *International Journal of Heat and Fluid Flow* **62**, 474-481 (2016)
8. Hasan, ABM Toufique, *International Journal of Heat and Fluid Flow* 46 70-83(2014)

Research on Performance Seeking Control of Turbofan Engine at Cruise Status Based on Sequential Quadratic Programming Algorithm

Zhao Shufan[1,a], Wu Heng[1], Shi Liming[2], Huang Cheng[3] and Yang xinyi[1]

[1] *Aviation Foundation College, Naval Aviation University, Yantai Shandong 264001,China*
[2] *Unit 92198 PLA Force, Huludao Liaoning 125109,China*
[3] *Unit 78525 PLA Force, Chengdu Sichuan 611630,China*

Abstract. The sequential quadratic programming algorithm was applied to the performance seeking control of multi-variable turbofan engine. According to the actual condition of aircraft cruise status, the constant thrust condition is added in the constraint functions of nonlinear programming. Based on the mathematical modes of turbofan engine, the SQP algorithm is adopted to the minimum fuel consumption mode of cruise condition. The simulation results show that after the optimization process, the engine fuel consumption rate decreased by nearly 2%, and the turbine temperature dropped by about 1.2%, the dual optimization benefits have been achieved. There are great potentials in engineering applications.

1 Introduction

Since the nineties of last century, the aero-engine performance seeking control technology is the latest development of integrated control of aircraft and aero-engine. In traditional design, each component of the aero-engine is reserved for enough margin space, to ensure that can work under the worst working condition. The components designed for the worst working conditions are too conservative to exploit full performance in some normal working conditions. The performance seeking control real-time online estimates the current safety situation, gives full play to the performance of components, and avoids wasting available margin space, to improve one performance or comprehensive performance indicators of aero-engine and to make engine works safely and efficiently.

Aero-engine is a strong nonlinear, multivariable time-varying system. When the nonlinear aero-engine was linearized, it will inevitably result in the conversion error, and the final global optimal solution will deviate from the actual global optimal point. With the development of optimization methods, some non-linear optimization methods are maturing [1]. SQP algorithm is a non-linear optimization method, it can maintain the overall convergence and a local super-convergence at the same time [2]. In this paper, the SQP algorithm was introduced into the aero-engine performance seeking control, which solves the problem of constraint condition and optimization algorithm under the minimum fuel consumption mode. The simulation results were given. The results showed that the SQP method can achieve the global optimal point, make a good optimization effect.

[a] Corresponding author : lightoffreedom@163.com

2 Sequential Quadratic Programming Algorithm

For a nonlinear programming problem, it can be expressed as follows.

Objective function: $\min f(x)$

Restrictions: $g(x) \geq 0, h(x) = 0$

The main idea of algorithm of SQP is that to solve the problem by changing it to a series of quadratic programming (QP) [3][4][5].

$$L(x,\lambda) = f(x) + \sum_{i=1}^{m} \lambda_i g_i(x) \tag{1}$$

For the general constraint optimization problem, a QP sub-problem can be constructed at the current point (x_k, λ_k):

Objective function: $q_k(d) = 0.5d^T B(x_k, \lambda_k)d + \nabla f(x_k)^T d$

Restrictions: $\nabla g_I(x_k)^T d + g_I(x_k) \geq 0$, $\nabla h_I(x_k)^T d + h_I(x_k) = 0$

The SQP consists of three parts.

First, the Hessian matrix of the Lagrange function is updated in each iteration.

B is the positive definite quasi-newtonian approximation of the Hessian matrix of Lagrange function and is calculated by BFGS method. The iteration equation is:

$$B_{k+1} = B_k + \nabla B_k \tag{2}$$

Where ∇B_k is correction matrix.

Second, the solution of the quadratic program.

A new QP problem will occur in each main iteration of the SQP method.

The target function:

$$\min \frac{1}{2} x^T Bx + c^T x \tag{3}$$

Restrictions:

$$A_i x = 0, i = 1, \ldots, m_e \tag{4}$$

$$A_i x \leq b, i = m_e + 1, \ldots, m \tag{5}$$

Third, one-dimensional search and calculation of target function.

The target function:

$$\min_{\gamma \in R, c \in R^n} \gamma \tag{6}$$

$$A_i x = 0, i = 1, \ldots, m_e \tag{7}$$

$$A_i x - \gamma \leq b, i = m_e + 1, \ldots, m \tag{8}$$

The calculation flow was shown in figure 1.

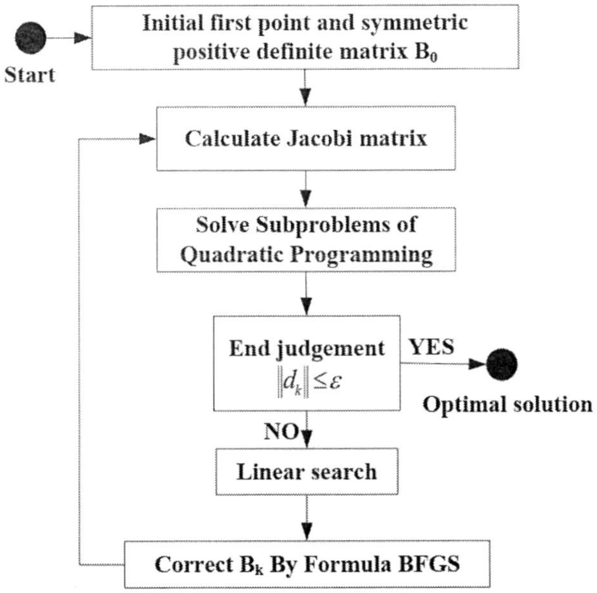

Figure 1. Calculation process of SQP algorithm

3 Engine Performance Seeking Control Mode

Integrated control system based on the aero-engine performance seeking control technology is shown in figure 2, the system is a typical upper and lower computer structure. The upper computer is a closed loop real-time simulation system which mainly has the flight controller, the rudder surface mechanism, the aircraft and aero-engine, which mainly gives the flight control instruction, the aero-engine control instruction and the aero-engine optimization control mode. The lower computer is mainly the calculation module, which is composed of the aero-engine airborne real-time mode, the mode selector and the optimization calculation module. The upper computer gives the flight status information and the optimization control mode command to the lower computer. The lower computer optimizes the control law by the airborne mode and the optimization calculation module, and then gives it to digital electronic controller to obtain a more favorable thrust response.

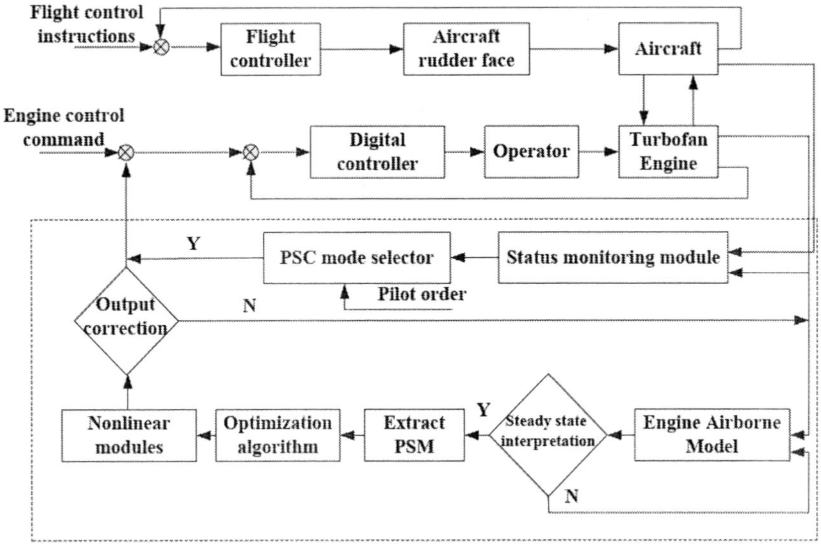

Figure 2. Comprehensive control of aircraft and engine

The relationship between the control variable and the state variable is $X = Pu$. The performance seeking control technique optimizes the control variable u by the onboard engine mode to optimize one parameter or the combination of several parameters of state parameter X. Therefore, the performance seeking control technology is expressed as a non-linear programming form.

$$\text{Performance: } \max(\min) f(u)$$

$$\text{Restrictions: } g_{i,\min} \le G_i(u) \le g_{i,\max}, i = 0,1,2\cdots$$

In which $G_i(u)$ are restriction, usually include: the range of control variable, the limit position of the control structure, the safety limit to ensure the normal operation of the engine, such as no overheating, no over-rev and so on.

The actual work of the engine is subject to the following constraints: high and low pressure rotor speed limit, fan and compressor surge margin limit, maximum fuel supply limit and its rate of change, turbine temperature limit, compressor guide blade angle restrictions and its rate of change restrictions. The constraint function is normalized to avoid the calculation error due to the different dimension of each constraint function.

$$g_i = \left| u_i / u_{i,\max} \right| - 1 \le 0 \tag{9}$$

Performance seeking control mode is closely related to the mission, the demand of different tasks on engine is different. There are three main modes, the maximum thrust mode, the minimum fuel consumption mode and the lowest turbine temperature mode.

The minimum fuel consumption mode is mainly used for aircraft steady-state flight process. Relevant data shows that for a normal routine flight mission, there are nearly 85% of the time and 45% of the fuel spent on the cruise process. By optimizing control law of the engine in the minimum fuel consumption mode, the fuel consumption of the engine will reduce while the engine thrust meets the needs of steady-state flight. It will help improving the flight performance of aircraft. Of course, the mode must ensure that the engine work safely, otherwise there is no practical significance. The corresponding non-linear programming is:

$$\begin{array}{l} \min sfc \\ wf_{\min} \le WF \le wf_{\max} \\ A8_{\min} \le A8 \le A8_{\max} \\ pnf_{\min} \le PNF \le pnf_{\max} \\ pnc_{\min} \le PNC \le pnc_{\max} \\ t_{5\min} \le T_5 \le t_{5\max} \\ F = const \end{array} \tag{10}$$

Where sfc is specific fuel consumption, WF is fuel flow of combustor, $A8$ is outlet area of nozzle, PNF is relative speed of fan, PNC is relative speed of compressor, T_5 is outlet temperature of low pressure turbine.

4 Simulation Analysis of Turbofan Engine Performance Seeking Control

4.1 Analysis of Thrust Need

In the engine performance seeking control technology, the minimum fuel consumption mode is much more difficult than the maximum thrust mode. There is a strong equality constraint in the constraint condition. It becomes more difficult [6] from the algorithm design or running time. On the other hand, the relevant research is optimized for the control law [7], either the maximum thrust mode [8] at the fixed height, the speed, or the minimum fuel consumption mode at the fixed speed [9].

Before the optimization of the control law in minimum fuel consumption, the thrust demand F_R of the current flight condition should be obtained according to the actual task demand of the aircraft, so

$F = F_R$. This ensures that the turbofan engine produces the corresponding thrust under the current control law and the thrust can match the actual mission requirements.

4.2 Minimum fuel consumption mode of aircraft cruise

Flight conditions were flying height $11000m$, flight speed 0.5Ma, engine state parameters at cruise flat flight are: $X = \{F, sfc, PNC, PNF, T \cdots\}$, engine control variables are: $u = \{WF, A_8, dGvc, dGvf\}$, $dGvc$ is angle of compressor guide vane, $dGvf$ is angle of fan guide vane.

Objective function: engine performance parameters $\min sfc$

Equality Constraints: the engine power of the state is satisfied $F_R = 2005daN$

Inequality constraints:

$$\begin{aligned} & wf_{\min} \le WF \le wf_{\max} \\ & A8_{\min} \le A8 \le A8_{\max} \\ & dGvf_{\min} \le dGvf \le dGvf_{\max} \\ & dGvc_{\min} \le dGvc \le dGvc_{\max} \\ & pnf_{\min} \le PNF \le pnf_{\max} \\ & pnc_{\min} \le PNC \le pnc_{\max} \\ & t_{5\min} \le T_5 \le t_{5\max} \\ & SML \ge SML_{\min} \end{aligned} \tag{11}$$

Where SML is surge margin of low pressure.The status and control variable changes of turbofan engine performance seeking control were shown in figure 3.

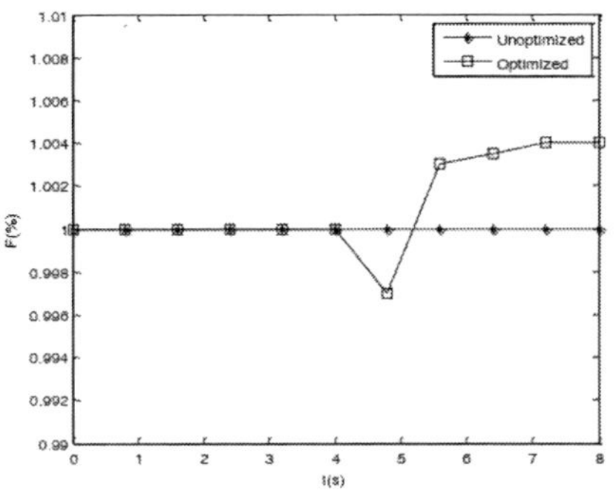

(a) F in the PSC process

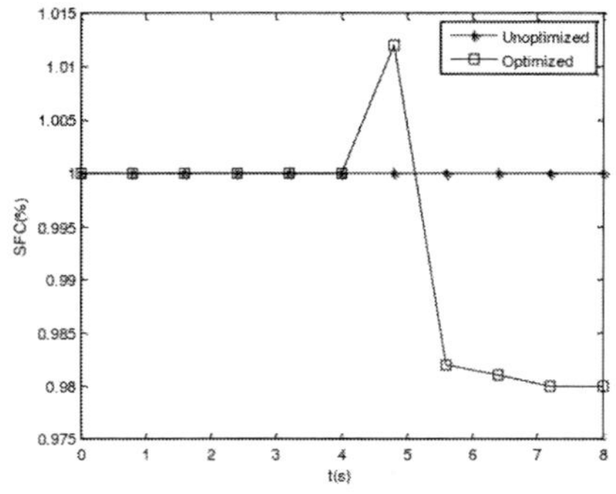

(b) SFC in the PSC process

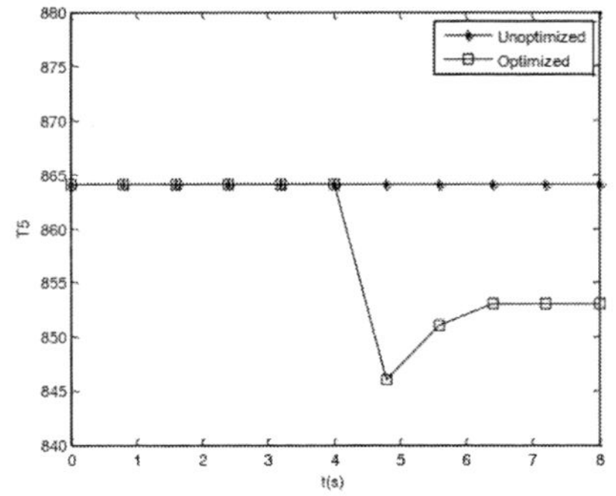

(c) T5 in the PSC process

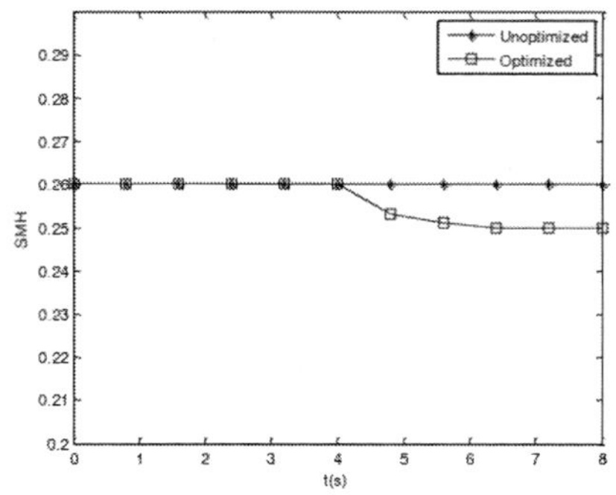

(d) SMH in the PSC process

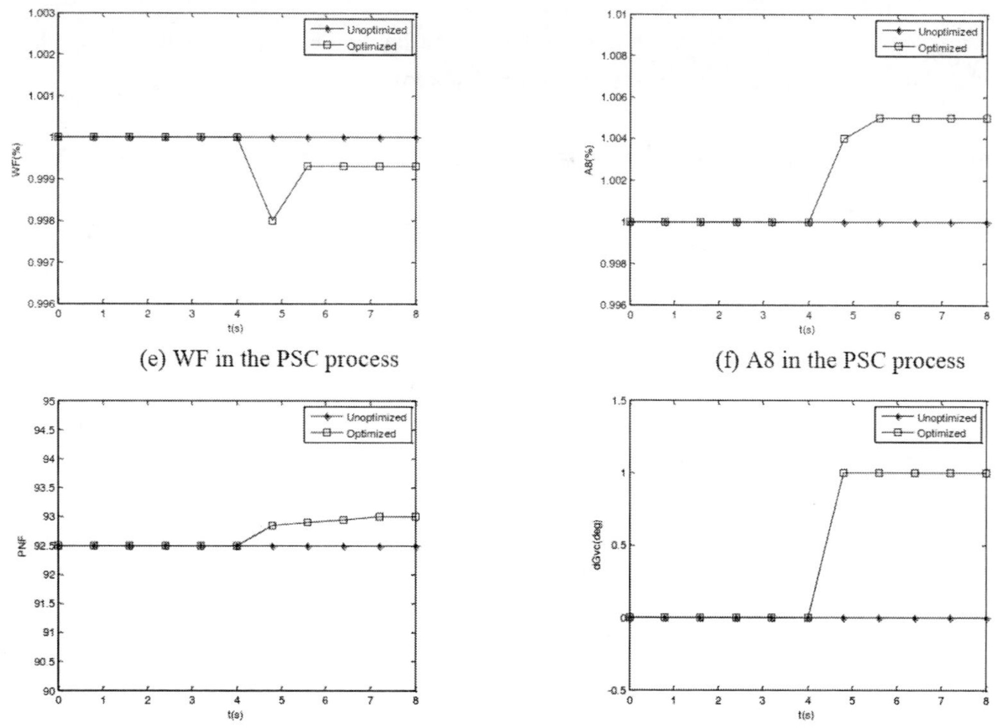

(e) WF in the PSC process (f) A8 in the PSC process

(g) PNF in the PSC process (h) dGvc in the PSC process

Figure 3. Curve of engine PSC process of aircraft cruise

Due to existence of the equation constraints of thrust constant, the simulation time is longer. The control variables start changing at fourth second, the minimum fuel consumption mode reduces the main fuel flow, changes the nozzle area and angle of guide vane to achieve control goal. The trends of control variables change are in line with the principle of the mode, and it tends to be stable after 6 seconds. After the optimization process, the engine thrust increased slightly, the fuel consumption decreased by nearly 2%, the rotor system speed increased slightly, and the turbine temperature dropped by about 1.2%, the dual optimization benefits have been achieved.

5 Conclusion

The results showed that the addition of constant thrust condition conforms to the actual situation of aircraft cruising condition, and the application of SQP algorithm in performance seeking control of multi-variable turbofan engine is feasible. It can reduce the fuel consumption rate and the turbine temperature while keeping the thrust of the engine unchanged. The dual optimization benefits have been achieved. There are great potentials in engineering applications.

Acknowledgments

This work is supported by special funds of Taishan Scholar Project, National Natural Science Fund (51505492) and Natural Science Foundation of Shandong Province(ZR2016FQ19).

References

1. Matthew C T, Declan G B and Ian P. 2001 Input/Output Conditioning of Robust Integrated Flight/Propulsion Control Journal of Guidance Control and Dynamics, p 6

2. James W D 2002 Integrated Flight and Propulsion Control: Current Successes and Remaining Challenges AIAA p 6021
3. Wang R X 2011 Research on Modeling and Control of Flight / Push Integrated System (Nanjing: Nanjing University of Aeronautics and Astronautics) p 20
4. Yuan Y X and Sun W Y 2001 Optimization theory and method (Beijing: Beijing Science and Technology Press) p 33
5. Liu Y, Tang L S and Chen Y P 1996 Nonlinear parallel algorithm genetic algorithm (Beijing: Beijing Science and Technology Press) p 22
6. Sun F C 2009 Research on Optimal Control Technology of Aeroengine Performance (Nanjing: Nanjing University of Aeronautics and Astronautics) p 15
7. Wu D 2004 Research on Nonlinear Performance Optimization of Aviation Propulsion System (Xian: Northwestern Polytechnical University) p 50
8. Li P Z 2007 Application of Genetic Algorithm in Cruise Optimization Control of Turbofan Engine (Xian: Northwestern Polytechnical University) p 29
9. Liu X D 2007 Research on Performance Optimization of Turbofan Engine Based on Genetic Algorithm (Xian: Northwestern Polytechnical University) p 37

A Review for Aircraft Landing Problem

Li Zipeng[1,a] and Wang Yanyang[1,b]

[1] *School of Aeronautic Science and Engineering, Beihang University, Beijing, China*

Abstract. Substantial increase in the amount of civil aviation traffic and other factors have led to frequent congestion and adverse effects in terminal area. Therefore, the landing sequence of aircrafts need to be managed so as to reduce adverse effects. This type of problem is called Aircraft Landing Problem (ALP), and is an optimization problem. In this paper, the current situation of ALP research is analysed. The specific process of landing is described, and the existing research problems are classified. At the same time, on the aspect of optimization, this paper discusses the optimization processing of ALP, including the establishment of optimization model and the design of optimization algorithm. In the last, the future research trend of ALP is proposed as conclusions.

1 Introduction

With the growth of civil aviation transportation, numerous aircrafts need to land within a short time. Due to shortage of aviation resources, flights delay is caused, which resulting in reduced efficiency, economic losses and other adverse effects.

However, expanding the airport runway facility simply is limited by various factors, so the need for reasonable management of aircrafts to reduce the adverse effects increases. Such problem is called Aircraft Landing Problem (ALP), which is an important issue in air traffic control.

2 Description of ALP

The landing process is described as follows. The approaching aircrafts flew to the terminal area from different directions. The airspace of the terminal area is divided by the initial and the final scheduling boundaries, between which is the scheduling window. If an aircraft enters this area, the algorithm accepts the following inputs: the specific time aircraft enters, the estimated time of arrival (ETA) which is predicted by the traffic control tool of ATC, the weight level of the aircraft and so on. After calculating, the scheduling algorithm allocates the runway and arrange scheduled time of arrival (STA) for aircrafts. After scheduling, aircrafts reach the converging point and enter the frozen area in which landing sequence is not changed unless exceptional cases.

ALP is an optimization problem, belonging to the NP-hard problem [1]. The model of ALP can be divided into basic or detailed models [2]. The basic model only considers runways as constraint of the airport capacity in the terminal area, and it can be abstracted as single or parallel machine scheduling

[a] Li Zipeng : lzplzp37@buaa.edu.cn
[b] Wang Yanyang : Wangyanyang@buaa.edu.cn

problem in the machine scheduling problem [3]. The detailed model takes runways and other related terminal resources into account, which can be abstracted as a job shop scheduling problem.

3 Classification of ALP

Using the basic model, ALP can be divided into single and multiple runway problems. Using detailed models, considering the airport resources, ALP can be transferred into corresponding problems in multi-airport environment.

3.1 Problems for single or multiple runways

The airport can be divided into single and multiple runways airport according to the number of runways. Single runway problem (SRP) means that all aircrafts are scheduled to land on the same runway. Multi-runway problem (MRP) need to do runway assignment as well as sequencing. In essence, SRP can be abstracted as a single server queue in queuing theory, and MRP can be abstracted as multiple servers queue, including sequencing and distribution.

In China, most airports have one runway, so the study of SRP still makes sense. Its trend is to improve the accuracy and efficiency of the solution constantly, as well as to be the basis of research in multi-runway problem.

The current research mainly focuses on MRP. As a result of the emergence and continuous improvement of heuristic algorithms, the scale of problem solved grows, and precision and efficiency of calculation are gradually improved.

Multi-runway airports have different runway configurations, which will affect the airport capacity, and the current study is mainly based on parallel runways configuration. In China, most multi-runway airports use parallel runways configuration, but there are still some airports using other configurations like new airport in Chengdu, using two parallel runways and a lateral runway, and it will have an impact in wake interval.

The parallel runways also have different modes of operation, including: independent or relative parallel instrument approach and so on. In independent parallel instrument approach mode, the various runways are independent. In the relevant parallel instrument approach mode, aircrafts landing on the adjacent runways need to meet a certain radar interval. The current research is mainly based on independent parallel instrument approach mode.

3.2 Problems for multi-airport

Multi-airport terminal area refers to the terminal area including two or more airports, and arrivals and departures of these airports interrelate.

In addition to the multi-runway sequencing at the airports, the multi-airport problem also includes sequencing in the entry point of terminal area. According to the basic principle of queuing theory, if entry points, airports, runways and other key nodes can be considered as a multi-server system, the multi-airport problem can be abstracted as multi-server parallel queuing events.

The study of multi-airport problem starts late, and many researches are incomplete. For example, someone considering the space resources of entry point and runways simply, combined the queuing theory and designed algorithm for ALP without considering the intersections in the airspace.

However, the current study of multi-airport, in addition to the proposed shortages like single optimization goal, ignoring the safety interval of cross-route, also lose sight of the number of runways of different airports, complex runway configuration and the impact of takeoff flow.

4 Method

In ALP, according to the model, allocation and sequence of landing should be provided ensuring constraints and specific need of different stakeholders. Based on above requirement, the key elements of ALP are extracted and the conceptual model is constructed as Fig.1.

The ALP covers the establishment of optimization model and the design of algorithm. The optimization model is composed of decision variables, optimization goals and constraints.

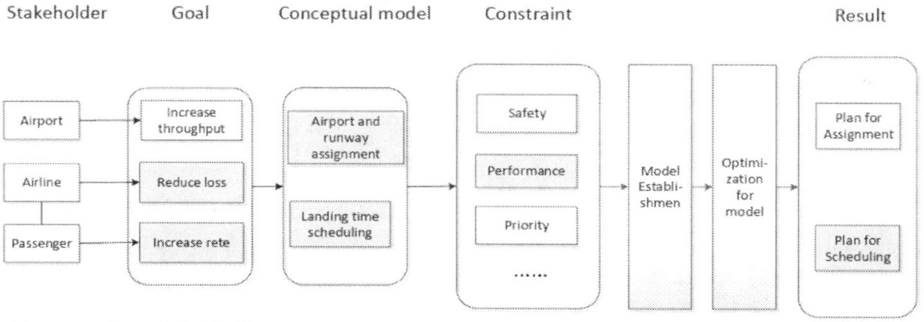

Figure 1. Conceptual model of ALP

4.1 Optimization goals

ALP involves aspects of benefits. ATC concerns about the safety; airline concerns about its economic losses; airports concerned about the delay rate; passengers concerned about the flight punctuality rate. Different angles result in the different optimization goals.

The current optimization goal can be divided into single and multi-objective goal. Common expression of single goal are time minimization and cost minimization.

Time minimization focuses on the ability of solving terminal congestion, mainly with forms of minimal delay or landing completion time. Minimal delay is the most common goal for ALP, which represents the shortest waiting time in the air and achieves maximal airport throughput. Minimal landing completion time is suitable for short-term terminal congestion.

Cost minimization focuses on airport operating costs and airline benefits. The cost is sometimes expressed as a linear or square relationship with time, which cannot adequately reflect the actual complex situation, and it can be divided into various expenses according to considerations. The study at Delft University divides expenses into passenger spending subsidies, passenger inconvenience, fuel consumption, etc. Similarly, someone divides expenses into operating expenses of resource path, flight delays expenses and flight cancellation expenses.

Other angles can also be chosen as goals. For example, Gini coefficient is used as optimization goal to ensure equitable distribution of resources between airlines.

As the number of concerns increases, the optimization goal in recent years has evolved into multiple to balance requirements of various aspects, and ALP is transformed into a multi-objective optimization problem (MOO).

The common solving of MOO problem is to change it into a mature single-objective optimization problem, such as setting the weight coefficient for different optimization goals. However, it is difficult to evaluate the solution objectively.

In the multi-airport environment, multiple optimization goals are proposed and the goal of fairness of public airspace resources use is added to reduce large delays of certain airport. Non-dominated sorting genetic algorithm with elite strategy is used to solve MOO and obtain the Pareto optimal solution.

4.2 Constraints

ALP belongs to multi-constraint problems. Constraints can be divided into multiple types, including safety constraints which maintain flight safety, performance limits which characterize the performance of aircrafts to avoid the results beyond the actual range, priority constraints which permit certain flights to be prioritized, fairness constraints which ensure the fairness of sequencing and cost constraints which can reduce loss to some extent.

Common typical constraints are as follow.

First one is the minimum interval constraints which belongs to safety constraints. It mainly refers to the horizontal interval in space, and can be divided into wake or radar interval constraint according to different causes.

Two successive aircrafts can reduce the threat of wake maintaining certain distance, which introduced the wake interval constraints. The minimal distance is real-time calculated based on speed and weight. In order to simplify the problem, ICAO specifies the minimum wake space interval between different weight types of aircrafts, and the distance interval is converted into the time interval in order to meet the time-based needs of ATC. The time-base process may bring some defects, such as lack of accuracy, but nowadays, the standard is widely used when dealing with ALP.

Radar interval constraints are unique to MRP. When the mode of relative parallel instrument approach is used, two successive aircraft landing on different runways need to meet the lateral radar interval.

Second one is the time window constraint belonging to performance constraints. Each aircraft has earliest and latest landing time, which are related to its flight performance and its position. Aircrafts will land during the time window.

Third one is runway capacity constraints belonging to safety constraints. It means that each runway can only serve an aircraft at one time.

Fourth one is the maximal position shift constraint (MPS) and the relative position shift constraint (RPS). They refer to the range of aircrafts position shift relative to the original sequence and the last sequence. These constraints limit the range of movements to a certain extent, which can ensure that aircrafts will not be led to fuel exhaustion or affecting landing order, which relates to safety and performance. This constraint can also avoid large delay, ensuring the fairness of the sequencing.

Among them, the minimum wake interval constraint is vital, which will affect runway capacity at peak hours, and runway capacity constraints must also be taken into account. Other constraints are used to improve the landing requirements better.

4.3 Algorithms

Since ALP is an NP-hard problem, listing all possibilities will introduce large calculations, and is not suitable for practical use. Currently, common algorithms solving ALP are divided into three categories: FCFS algorithm, exact algorithm and heuristic or meta-heuristic algorithm.

Due to the lack of effective accessorial decision-making tools, FCFS is a method currently used in the manual scheduling environment, especially for single runway problems. FCFS sequences the aircrafts according to their ETAs. It is simple and quick, and can obtain good solution in the uncrowded terminal area. However, it may bring greater delay in rush hours and has been unable to meet the current needs of air traffic management.

Subsequently, FCFS is improved to time advance algorithm (TA) and delay exchange algorithm (DE). TA accelerates the first flight to reduce the sum of delay time without changing order, and DE only advances urgent flights. These methods do not contain idea of optimization, and not practical when the number of flights increased.

Exact algorithms are used to deal with linear programming problems, including branch and bound algorithm, column-and-row generation [4], branch-and-price algorithm [5], dynamic programming algorithm and so on. It always builds 0-1 mixed model to solve ALP [6]. However, such algorithms cannot even solve the problem of medium size, which makes people begin to consider algorithms of new type [7].

As a result, heuristic algorithms are gradually applied to ALP. They are usually designed by imitating some phenomena or processes in nature, having the advantage of high parallelism, and can deal with nonlinear complex model. Although the optimal solution cannot be guaranteed, the approximate solution can be obtained within reasonable time, and under reasonable computing resource conditions, the solution will be of high quality. Common heuristic algorithms include ant colony algorithm, genetic algorithm [8], particle swarm algorithm [9], cellular automata algorithm, bat algorithm and so on. They can handle many optimization problems. However, they all have their own shortcomings, and need to be improved or used in combination to make the problems better solved.

For example, classical genetic algorithm (GA) cannot search the global optimal solution unless the optimal individuals of each generation are preserved. With the complexity of problems, the defects of classical GA are emerging. Therefore, it is improved to maintain the diversity of species, or be combined with other algorithms to improve the performance, such as adaptive annealing GA. GA used by ALP is also modified with the development of GA. A non-dominated sorting genetic algorithm is designed with elite strategy to ensure the optimality and abundance of population so that efficiency of calculations can be improved. Someone used the adaptive evaluation vector to guide the search of the results for the short-term ALP and adopt the multi-objective genetic algorithm based on inequality. These improvements make the algorithms more advantageous.

Other heuristic algorithms can also have better performance through a certain improvement. In the literature [10], the improved strategy is proposed for the traditional particle swarm algorithm which is easy to fall into local optimum and has shortcomings in solving combinatorial optimization problem.

The improved algorithm generates the initial particle swarm through tabu search strategy to meet the multiple constraints of ALP, introduces the operations of cross and mutation in GA to enhance the information exchange ability between particles, redefines the inertia weight and acceleration factor to enhance the diversity and convergence of particles, adding the evolutionary crossover operator to enhance the local search ability. Through comparison, it is proved that the abilities of local and global search are enhanced.

Combining heuristic algorithms can improve the accuracy or efficiency to a certain extent. For example, immune concentration adjustment mechanism is added and tabu search algorithm is combined with particle swarm algorithm, which avoids being easy to fall into the local optimum and improves the efficiency. Artificial fish swarm algorithm and particle swarm algorithm are also combined together. Artificial fish swarm algorithm is used for global search, and particle swarm optimization algorithm is used to find the local optimum, which overcomes the shortcomings of slow convergence in artificial fish swarm algorithm and of being easy to fall into the local optimum in the particle swarm algorithm, thus improving the search speed and convergence accuracy.

4.4 Static and dynamic situation

Dynamic and static situations involve whether there are new aircrafts flying into terminal and whether results are to be modified in real time in the sequencing process.

From static to dynamic, there are three common ways, one is to change the optimization model, such as adding reordering cost into the goals. When aircrafts need to be rescheduled, Beasley used shift penalty function to evaluate the expense of changing, and control shifts within a certain range. Second, according to the definition of the dynamic situation, the new flight joining the queue can be regarded as a trigger event model, and flights should be reordered after trigger happens. Third, receding horizon control (RHC) strategy [11] can be added to achieve dynamic. For example, B.S.Girish applied the hybrid particle swarm algorithm in the framework of a receding horizon, and Marcella Samà combined the receding horizon algorithm with the branch and bound algorithm to achieve dynamic and improve the efficiency of computing. The frequency of new flights joining the queue is not regular, so the sequencing frequency is also irregular to follow, and use of RHC in a fixed frequency to implement the algorithm can meet its dynamic needs to a certain extent.

5 Tendency

Many runways use a mixed mode of operation, so it is necessary to consider takeoff aircrafts on the same runway to avoid accidents [12]. However, most studies ignored departing flights in the terminal area. Some studies take departing flights into account without considering the trigonometric inequality of interval. Ghoniem A [13] solved this problem by introducing new simple interval constraints in two aircrafts of identical weight level and operation to ensure the inequality. However, the above method can get better results for arrivals, and departures do not get a good arrangement. Therefore, the need to consider arrivals and departures together is increased so that the appropriate solution for both sides can be obtained.

ALP is usually more complicated. Runways have different modes of operation, and they may be mixed operated or can only be used for take-off or landing [14], which is seldom taken into account in research. In future studies, similar complex situations should be considered.

In addition, optimization goals need to be improved. Minimization of time or economic losses does not adequately meet actual needs. In practice, the delay or economic loss of different aircrafts cannot be treated equally, and the weight factor need to be considered.

At the same time, the aircraft may have the problem of insufficient fuel due to conversion problems. However, the existing research does not take these factors into account.

6 Conclusions

ALP is gradually converted into a dynamic, multi-airport, multi-runway problem. Multi-runway, multi-airport is the future trend of construction. Meanwhile, hub airports have growing traffic and busy terminal area, which need to be considered as a dynamic problem.

Optimization goals are changed gradually from single to multiple [15]. They are no more simply time delay or economic loss. Fuel coefficient and other aspects of comprehensive consideration should also be added. Constraints will become more comprehensive, and multi-runway problem make some new constraints need to be introduced. With the frequent use of heuristic algorithms, the improvement and combination of algorithms has been ongoing, and the direction of which is always to enhance its accuracy and efficiency.

7 Acknowledgement

This research was supported by the National Science and Technology Major Project (Grant No. 2016ZX04004006).

References

1. B. S. Girish, *Applied Soft Computing* **44** 200-21 (2016)
2. M. Samà, A. D'Ariano, D. Pacciarelli, *Transportation Research Part E: Logistics and Transportation Review* **60** 140-55 (2013)
3. S. Vadlamani, S. Hosseini, *J. of Air Transport Management* **40** 144-8 (2014)
4. Maher S J, *Transportation Science* (2015)
5. A. Ghoniem, F. Farhadi, M. Reihaneh, *Eur. J. of Operational Research* **246(1)** 34-43 (2015)
6. A. Jacquillat, A. R. Odoni, *Operations Research* **63(6)** 1390-410 (2015)
7. A. Lieder, D. Briskorn, R. Stolletz, *Eur. J. of Operational Research* **243(1)** 61-69 (2015)
8. H. Zhou, X. Jiang, *Mathematical Problems in Engineering* (2015)
9. A. A. A. Esmin, R. A. Coelho, S. Matwin, *Artificial Intelligence Review* **44(1)** 23-45 (2015)
10. M. Yingjun, S. Xiaona, Z. Dongfang, *Application Research of Computers* **7** 030 (2015)
11. F. Furini, M. P. Kidd, C. A. Persiani, et al, **18(5)** 435-447 (2015)
12. G. Hancerliogullari, G. Rabadi, A. H. Al-Salem, *J. of Air Transport Management* **32** 39-48 (2013)
13. A. Ghoniem, H. D. Sherali, H. Baik, *INFORMS Journal on Computing* **26(3)** 514-30 (2014)
14. A. Lieder A, R. Stolletz, *Transportation Research Part E: Logistics and Transportation Review* **88** 167-88 (2016)
15. W. Malik, Y. C. Jung, *16th AIAA Aviation Technology, Integration, and Operations Conf.* 4072 (2016)

Research of Spacecraft Assembly Simulation with Ground Mechanical Equipments

HE Wen-xing[1,a] and CHEN Chang-yu[1]

[1]Beijing Institute of Spacecraft Environment Engineering, Beijing 100094, China

Abstract. In order to overcome the shortcomings of the physical test conditions, research and development of the spacecraft container simulation verification platform to enhance the level of digital development of the spacecraft container. Based on the theory of rigid-flexible coupling dynamics, the dynamic model of the box system is established. The dynamic external excitation is determined by obtaining the track spectrum and identifying the condition of the line. Based on the computational fluid dynamics, the thermal model of the spacecraft container is established, and the container is realized by simulating the natural convection and air conditioning control spacecraft container thermal simulation. Using modular design, C/S architecture and navigation process to achieve the simulation platform architecture versatility, ease of maintenance and scalability and other needs. Finally, the mechanics and thermal performance verification of the spacecraft container and physical test are carried out. The results show that the simulation results of the box dynamic system and the thermal system are in good agreement with the experimental results.

1 Introduction

The spacecraft container is the electro-mechanical equipment to realize the function such as spacecraft ferry transport, made up of multiple subsystems, in the process of designing and using process, involving many professional departments such as force, heat, electricity, which need to be used in structural design and perform a large number of analysis and simulation work when planning formulation. Limited to the limitations of the physical tests, the current sports car test can not meet the requirements of the increasing security requirement of spacecraft. To overcome the limited physical sports car test conditions and defection of measuring point are little, based on the demand of improving digital level of container, simulation verification platform for spacecraft container is urgently needed for research and development.

There are a number of research on aircraft container dynamics and thermal in the domestic. Bai Qiaofeng, et al[1] built the physics model of the container suspension system and built the model using the dynamic energy method . Tan Dekun, et al[2-3] developed spacecraft container vibration test system by using virtual instrument, and analyzed the spectral analysis, key point highlight and extremum retrieval of the collected data. Hu Yupeng,et al[4] put forward BWBZX-1 heat preservation container design scheme, and then through the numerical simulation research structure of thermal physical parameters on the container material and the influence of the temperature field and heat

[a] Corresponding author : hewenxing@163.com

transfer characteristics of the final pass environmental test for validation. Through the whole modeling of the spacecraft and the container, Su Xinming, et al[5] used the numerical simulation of the passive thermal insulation performance of the container in different working conditions. Xiao gang ,et al[6]introduced the design and development features of the container cases and the results of the airlift test results based on a certain type of spacecraft airlift container, and the results showed that the container design was reasonable.

Nowadays, in the process of the container design and simulation, the container mechanics, thermal analysis model lacked unified management in the simulation data and the simulation project is single and insufficient. A simulation analysis method for the optimization and use of the working condition environment of the container is lacked. In addition, the design of the spacecraft container should be evaluated through the commissioning to evaluate the performance of the container, and decide whether it meets the required quality requirements .It is necessary to develop the simulation and verification platform for spacecraft transport container based on the demand of digital level. According to the container and the spacecraft, container digital mechanics simulation model and the thermal simulation model is established respectively, which can realize ,in the process of railway transport ,the simulation analysis of spacecraft container by simulation of the stress and temperature conditions.

2 SIMULATION MODEL OF THE CONTAINER SYSTEM

The spacecraft transport container system consists of several subsystems, including box, bracket, shock absorber, satellite and railway transport vehicles. The dynamics and thermal model of the container system are described as follows:

2.1 Dynamic modeling of container system

Because of the complexity of the container system and the large difference of stiffness of different parts, such as the structure stiffness of the container and other structures is large, and the stiffness of the bracket and astral structure is low, therefore, the dynamics modeling for stiffness large parts adopt equivalent rigid body, the structure of low stiffness and elastic deformation is large equivalent for the flexible body, container system eventually formed a coupled multi-body system, the complete system equations [7] as follows:

$$\dot{p} = T(p)v \tag{1}$$

$$M(p)\dot{v} = F(p,v,c,s,t,u,\lambda) - G^{\mathrm{T}}(p,c,s,t,u)\lambda \tag{2}$$

$$\dot{c} = F_c(p,v,c,s,t,u,\lambda) \tag{3}$$

$$0 = g(p,c,s,t,u) \tag{4}$$

$$0 = b(p,v,c,s,t,u,\lambda) \tag{5}$$

Among them: p is the position vector of the hinge position; T is the angular transformation matrix; V is the state vector of hinge velocity; M is the mass matrix; F is the force and torque vector generated by the element; c is the dynamic state vector of the control unit and the control unit; s is the algebraic state of displacement and acceleration; t is for time; u is the external input of displacement and velocity level; λ is the binding force and torque; G is the jacobian matrix for the constraint equation ;Fc is dynamic state equation of the force and control unit; g is the constrained algebraic equation of constraint; b is the constraint algebraic equation related to the algebraic state quantity.

2.2 Thermal modeling of container system

Due to differences in container inside and outside temperature and the container system internal air flow with heat transfer, the container internal flow field can be approximately treated as the three dimensional incompressible turbulent flow. The control equation of the transport equation as follows:

$$\frac{\partial(\rho\varphi)}{\partial t} + div(\rho u\varphi) = div(\Gamma grad\varphi) + S \tag{6}$$

Among them: t is the time; ρ is the air density; u is the velocity vector; φ is the flow field flux; S is the source term; Γ is the diffusion coefficient.

In addition, the state equation of association pressure, temperature and density should be added, and the expression is as follows:

$$p = \rho RT \tag{7}$$

Among them: p is time; R is the molar gas constant; T is the temperature.

3 DEVELOPMENT OF BOXED SIMULATION PLATFORM

3.1 Platform architecture design

Since the simulation platform application involved spacecraft container the whole life cycle, including the spacecraft container design, analysis, manufacture, application and so on. At present, it is necessary to fully consider the extension requirements of various functional modules in the framework design of the platform. Therefore, module division, call relation, data link need to be applied in the design, simulation, optimization, manufacturing, test and other aspects of the basic flow. Aiming at the general usability, maintainability and extensibility of the simulation platform architecture, the following solutions are used in architecture design:

(1) The system adopts the concept of modular design, and adopts the matching management of the user - role - permission - function module to establish the mapping relationship between the user and the function module.

(2) Because of the complexity and diversity, the whole platform uses C/S architecture in which the client application deployment aims at different user groups and the centralized management of data is conducted by the server. This architecture can solve the problem of localization of complex commercial software applications. What's more, software license management can adopt the floating management pattern to increase effective utilization of software;

(3) Considering the user's ease of use, the design fully adopts the navigational process, organization of the structure tree and the Tab page to make users easier to understand and master the operating process.

The design, optimization, manufacturing and other modules in the diagram provide the interface for the simulation of the whole process platform simulation.

3.1.1 The function module of the front end

The front function module includes four modules: project, working condition simulation, test data and user information. The working condition simulation is the most core business module and it is also the user's main application interface. Considering the late scalability requirements, the front business module can be configured in the background where the role is associated with each business module, and the user is associated with role again. And then they achieve good expansion function of the system.

The simulation module is divided into dynamic simulation and thermal simulation module. The simulation module includes three parts: simulation calculation, historical working condition query and working condition library management. "Simulation calculation" is a simulation analysis of a module according to the process. The "historical working condition query" is a query for all working conditions of the user's history simulation, and it can check the setting parameters of each working condition and corresponding simulation results. "Working condition library management" is to modify and delete all the conditions of management which were simulated by users.

The experimental data module is also divided into dynamic simulation and thermal simulation submodule. The test module mainly provides functions of viewing and downloading the test data. " viewing Test data " means to select a project to display the test data with the table and curve; "Downloading test data" refers to that test data are segmented and recombined according to the selected project and processing requirements of the data, and then the test data are output and downloaded to the required location according to requirements.

User information module is mainly to change passwords for users.

3.1.2 The function module of the background

The background function module includes four modules:the project database management, working condition bank management, test library management and user management. The project library management module is the creation, modification and deletion of the project. Working condition bank management module is divided into dynamic simulation bank, and each thermal simulation bank, each module contains the function of three parts: the simulation template bank management, simulation model bank management and simulation working condition library management. Test library management module is divided into dynamic simulation and thermal simulation, and the function of each module consists of three parts: test data entry, test data modification and test data deletion; The "user management" module includes the functions of adding, modifying, and deleting users.

3.2 Platform data management

The diversified design data of the container includes model data, working condition of dynamic simulation, dynamic simulation results data and dynamic test data, working condition of the thermal simulation data, the thermal simulation results and the thermal test data. How to realize the different stages of heterogeneous data management and quick retrieve, how to make a reasonable correlation of different types of application data, how to extend product instance data, and how to design the data structure and data relations are based on platform build experience to analysis and design for specific applications.

4 THE SIMULATION PLATFORM OF THE CONTAINER

To verify the accuracy of the dynamic model of the container system and considering the total transportation, dynamic modeling of the container system was carried out, and the simulation results were compared with the test data. The whole container model included container body, L bracket, satellite and the model of vibration damping.

Figure1and Figure 2 are the comparison between the acceleration curves and the spectral curves of the container pedestal and the satellite certain location. The dynamic simulation outputs the acceleration of the selected measurement points and compares the results of 1700s to 2500s. The comparison results were showed that the simulation and test results have good consistency, and the error of container and satellitic acceleration were small. Container station within 15 Hz consistency is very good. But the satellitic points could be seen that the frequency of the test data between 12-15Hz is abundant. From the comparison of the frequency domain, the low frequency part consistency is better, and the high - frequency part experiment of 20Hz was slightly larger than the simulation.

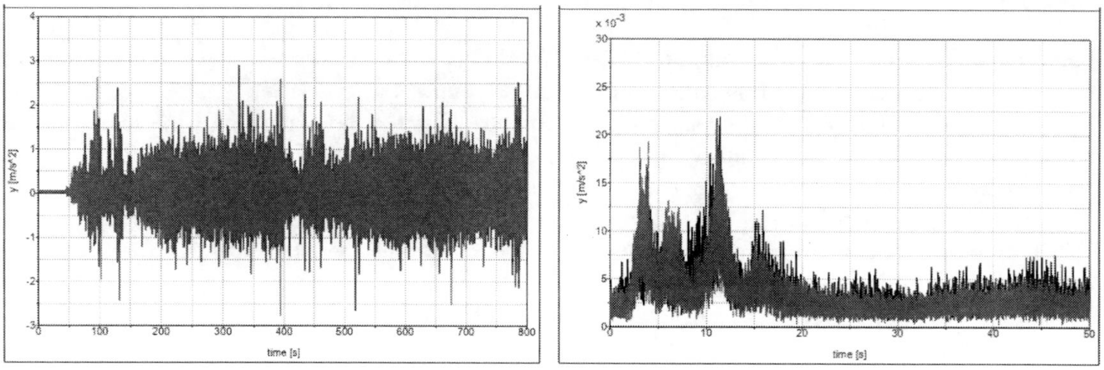

Figure 1. transverse acceleration and spectrum comparison of the container pedestal

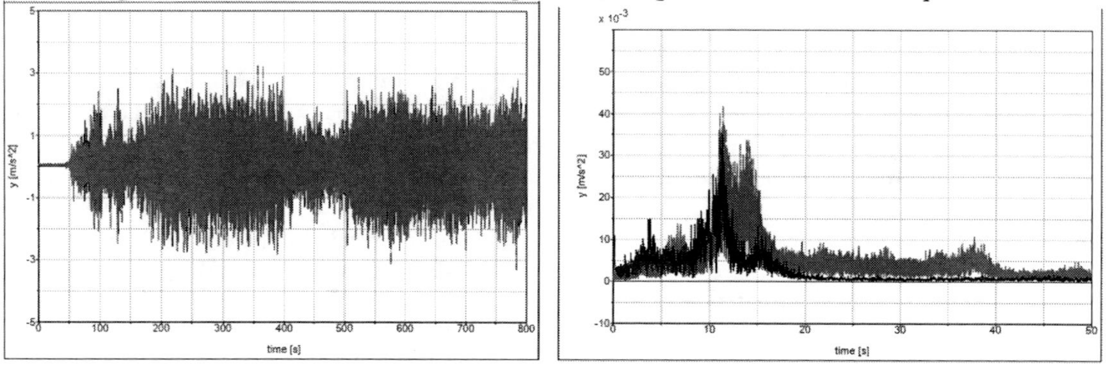

Figure 2. transverse acceleration and spectrum comparison of a satellite certain location

5 CONCLUSION

Spacecraft transport container multi-functional platform aims to the business process of container design, modifying the existing design methods, introducing advanced digital simulation analysis and management, comprehensive covering container structure design, manufacturing process planning, and working condition analysis of simulation requirements, significantly improving the level of digital development of the container. As a multi-functional platform, the container simulation platform is the most important part of the container, in order to realize the virtual commissioning based on the container of the fixed condition of railway transportation experiment as the core, through the platform design and research, the function of the following are realized:

(1) Establish the digital mechanical simulation model and thermal simulation model of the container, according to the container and spacecraft product structure, materials and transportation methods;

(2) The dynamic and thermal-environment excitation of the actual railway physics commissioning test is applied to these simulation models, using the results of simulation analysis to verify the design of the container quality, carrying out the simulation analysis of the force and temperature of the railway transportation in the container cases;

(3) Manage the data of mechanics and temperature conditions in the actual transportation of railway transport vehicles, and conduct data analysis.

Acknowledgements

This research is supported by Natural Science Foundation of China (No. 51405025), and Beijing Spacecraft Smart Assembly and Equipment Engineering Research Center.

Referrence

1. Qiaofeng BAI, Hailong LIANG, Xiaolong MEI. Simulation Analysis of Suspension System of Spacecraft Transport Packing Container[J]. Mechanical Engineering & Automation, 2013, 42(2): 19-21.
2. Dekun TAN, Lifan MENG, Lefu WANG.Spacecraft packing chest vibration measurement system[J]. China Measurement & Testing Technology, 2008, 34(1): 28-30.
3. Dekun TAN, Spacecraft Packing Chest Vibration Measurement System Based on Virtual Instrument[D]. North University of China.2008
4. Pengyu HU, Qunsheng LUO, Xia YIN. Design and Verification of BWBZX-1 Insulated Packaging Container[J].Packing Engineering, 2016, 37(1): 33-38.
5. Xinming SUN.Shiming FU,Yifei PEI. Passive Heat Preservation Analysis on Spacecraft Transportation Package[J]. Journal of Astronautics 2012, 33(9): 1334-1340.
6. Gang XIAO,Wenyu HAO,Guofen ZHANG. The development of containers for air transport of spacecraft and an evaluation[J]. Spacecraft Environment Engineering 2010, 27(6): 795-800.
7. Hao GAO, Study 0n Rigid-Flexible Coupling Dynamic Simulation Method And Platform For Railway Vehicle[D] Southwest Jiaotong University.2013.

Calculation Method of Stability Derivatives and Verification for Power-lift Aircraft

Shi Yazhou[1] and Wang Weijun[2,a]

[1,2]*School of Aeronautic Science and Engineering, Beihang University, Beijing, China*

Abstract. It is difficult to realize the STOL and high-speed cruise at one time when designing an aircraft, whereas the application of power-lift technology has successfully solved this technical problem. The research about the dynamic characteristics of power-lift aircraft in hover and low speed flight regime is key important for the selection of control strategy in transition process. However, if adopting the traditional aerodynamic calculation method based on thrust coefficient concept, the hover and low speed forward flight of power-lift aircraft cannot be studied as a unified process and the stability derivatives cannot be accurately calculated in low speed. In this paper, a method for calculating the stability derivatives and model the aircraft by using full dimensional aerodynamic data is proposed, which can not only solve the problems mentioned above, and it is very convenient for the design of the control law in the transient process.

1 Introduction

The traditional calculation of the stability derivative is first calculating the dimensionless derivatives, and then translating the dimensionless derivatives into the dimensional derivatives by their relations [1]. This method is perfectly applicable to conventional aircraft, but it is not applicable for VTOL aircraft used power-lift technology under the conditions of hovering and low speed.

The traditional aerodynamic estimation method of power-lift aircraft is based on the concept of thrust coefficient C_T (C_T=T $(q*S)^{-1}$, T is the engine thrust, dynamic pressure $q=\rho V^2/2$, S is the reference area of the wing). The method can calculate aerodynamic forces accurately at the range of the specified thrust coefficient, however, due to there exists the problem of aerodynamic divergence under the condition of small velocity and large thrust coefficient [2], the dimensional stability derivatives calculated by the traditional method is inaccurate. Especially, when the aircraft is in hover, the traditional method can't be used. Therefore, when making the simulation of the power-lift aircraft with stability derivatives calculated by the traditional method, the dimensional stability derivatives of hovering, low-speed state and high-speed state will be discontinuous; this will influence the accuracy of the simulation.

In this paper, the dimensional aerodynamic data is directly used to solve the dimensional stability derivatives of the aircraft. The longitudinal linear time-varying simulation system for a STOL aircraft and a new type of VTOL aircraft is established, which can verify the effectiveness of the proposed method.

[a]Corresponding author : wangwj@buaa.edu.cn

2 Modeling method and the calculation of stability derivatives

2.1 Modeling method

There are many aircraft dynamics modeling ways, such as nonlinear all-variables flight dynamic modeling, small disturbance linear modeling, linear time-varying simulation modeling [3-5]. In this paper, the dynamic model of the target aircraft is built by using a linear time-varying system. This method can convert the nonlinear dynamic equation into the linear time-varying dynamic equation, which can describe the motion process of aircraft accurately and simplify the design process of the aircraft control system when using the designing method of the linear control system.

The linear time-varying dynamics simulation of the aircraft starts from a baseline trim point, and a new benchmark is set at regular intervals. The time interval between any two-benchmark points is small enough to ensure that the changes of the state variables are within the range of small perturbations. The aerodynamic forces under arbitrary conditions can be calculated by using the following equations.

$$\begin{aligned} X &= X_0 + \Delta X \\ Z &= Z_0 + \Delta Z \\ M &= M_0 + \Delta M \end{aligned} \tag{1}$$

X_0 is the aerodynamic force at the benchmark, ΔX is the aerodynamic force increment compared with the reference point. The calculation formula of aerodynamic increment is shown in equation (2).

$$\begin{aligned} \Delta X &= X_u \Delta u + X_w \Delta w + X_q \Delta q + X_{\delta_c} \Delta \delta_c + X_T \Delta T \\ \Delta Z &= Z_u \Delta u + Z_w \Delta w + Z_q \Delta q + Z_{\delta_c} \Delta \delta_c + Z_T \Delta T \\ \Delta M &= M_u \Delta u + M_w \Delta w + M_q \Delta q + M_{\delta_c} \Delta \delta_c + M_T \Delta T \end{aligned} \tag{2}$$

Where δ_c is the deflection angle of control surface.

Note that, X_i, Z_i and M_i ($i=u, w, q, \delta c, T$) are the stability and control derivatives at the benchmark points, and their accuracy and continuity will affect the accuracy of the simulation results.

Finally, a longitudinal linear time-varying simulation model is shown in figure 1.

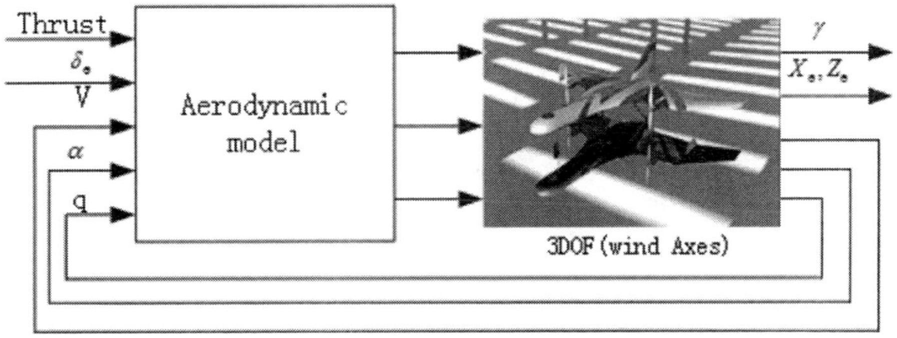

Figure 1. Linear time-varying simulation model of power-lift aircraft.

2.2 Calculation of stability derivatives

2.2.1 Traditional calculation method

Taking X_α as an example, firstly the force coefficient C_x about the X-axis of the Body Axis system can be calculated by the following formula.

$$C_x = C_L \alpha_x - C_D \tag{3}$$

Where C_L is the lift coefficient, C_D is the drag coefficient.

Secondly, the dimensionless derivatives $C_{X\alpha}$ can be calculated by formula (4).

$$C_{x\alpha} = \frac{\partial C_x}{\partial \alpha} = C_L + \alpha_x \frac{\partial C_L}{\partial \alpha} - \frac{\partial C_D}{\partial \alpha} \quad (4)$$

When the stability axes are chosen as the Body Axis system, α_{x0}=0, the above formula can be transformed into:

$$C_{x\alpha} = \left(\frac{\partial C_x}{\partial \alpha}\right)_0 = C_{L0} - \left(\frac{\partial C_D}{\partial \alpha}\right)_0 \quad (5)$$

Where the subscript zero indicates the reference flight condition.

Finally, X_α can be obtained by formula (6).

$$X_\alpha = C_{x\alpha} \times \frac{1}{2}\rho V^2 S \quad (6)$$

The stability derivatives X_α calculated by the traditional method at high speed is quite accurate.

However, when the aircraft is in a low speed flight, especially, when the speed tends to zero, the corresponding thrust coefficient tends to infinity, which has exceeded its applicable range ($0.48 \le C_T \le 5.5$).Therefore, a small velocity error will cause a huge deviation of the thrust coefficient C_T, which will cause the inaccurateness of C_x [6-7]. Therefore, the stability derivative calculated by the formula (5) and the formula (6) is also inaccurate.

In addition, when the aircraft is in hover, the inflow speed Va is 0, the aerodynamic data of aircraft cannot be expressed in the form of dimensionless coefficients, so the traditional methods cannot be used.

2.2.2. New calculation method.

In order to unify the calculation method of the stability derivatives of aircraft under high-speed flight, low speed flight and hovering condition, the calculation method by using dimensional data is presented.

Taking X_α as an example. The dimensional aerodynamic data under trim conditions of the aircraft is given in table 1. Note that the engine thrust and its effects on aerodynamic forces have been included in the aerodynamic data.

Table 1. The axial force X of the aircraft vary as the angle of attack α and thrust T.

		α, (deg)						
	T(N)	-10.0	-2.0	6.0	12.0	18.0	24.0	28.0
	0	-11146	-14690	-21669	-26862	-27924	-46898	-49583
	42214	-2.1934	-13058	-27013	-40816	-53042	-58868	-62629
	83993	12770	-8625.4	-29531	-50022	-66440	-73630	-77779
X	126207	26988	13627	-26924	-51545	-76015	-88716	-94341
	167551	41690	13475	-19343	-46637	-80537	-106400	-117760

From the above table, supposing the trim thrust is 82191.38, the relation between α and X is shown as figure 2.

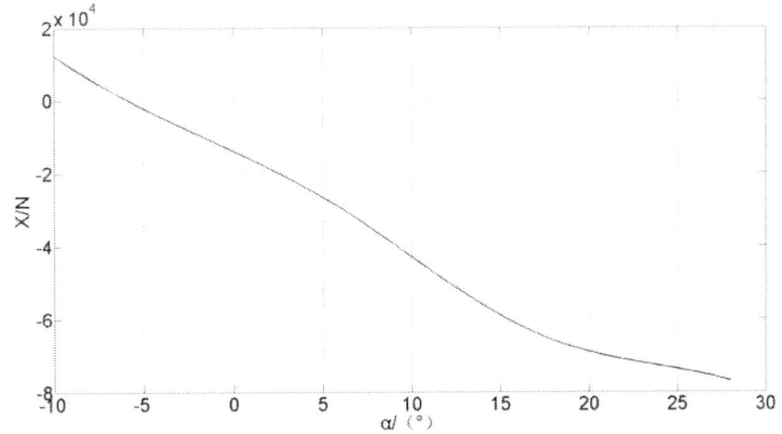

Figure 2. The curve about X varying as α.

Differentiating the curve in figure 2, figure 3 can be got as below.

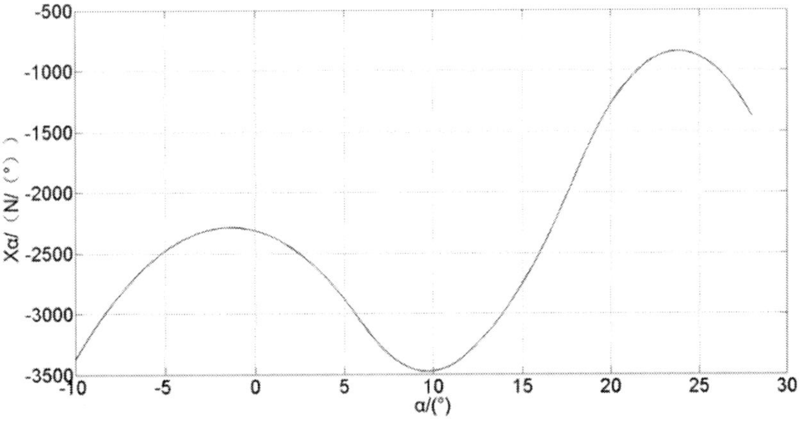

Figure 3. The curve about X_α varying as α.

The value of the X_α of the aircraft at the trim angle of attack α can be obtained from figure 3.

Other dimensional stability derivatives of aircraft, such as u derivatives, q derivatives, $\dot{\alpha}$ derivatives and so on, can be conveniently obtained by the above methods.

3 Simulation verification

3.1 Simulation for a STOL aircraft

In this part, a linear time-varying simulation of the EBF aircraft in reference [8] is made in Simulink, and a step response analysis compare with the step response curve(as shown in figure 4) in the reference is done.

The initial status parameters of the aircraft are selected the same as it in reference 8, as shown in equation (7).

$$\begin{aligned} &V_0 = 36.01111\text{m}/s, \alpha_0 = 4.1550°, \gamma_0 = \text{-}6.5°, T = 82191.38N, \delta_{\text{ddc}0} = 5° \\ &\delta_{\text{sp}0} = 10°, \delta_{\text{h}0} = \text{-}3.2084° \end{aligned} \tag{7}$$

Where δ_h is the deflection angle of horizontal tail, δ_{ddc} is the deflection angle of direct drag control surface, δ_{sp} is the deflection angle of spoiler.

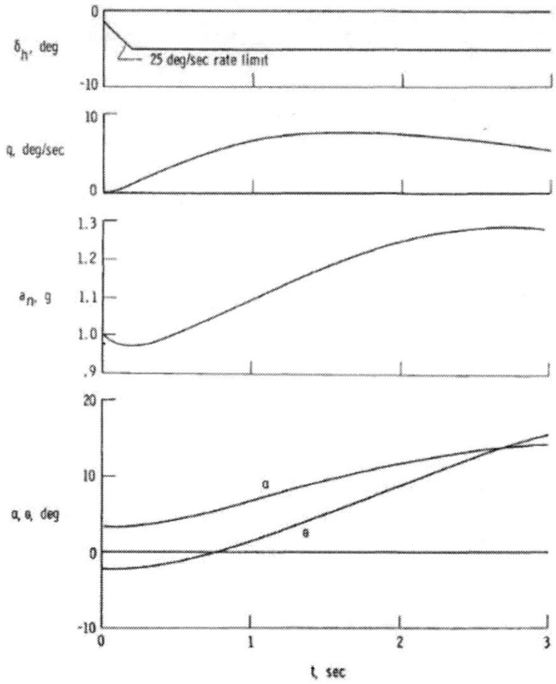

Figure 4. Time history of EBF airplane response to a longitudinal control column step input.

Deflecting the horizontal tail from reference point to -7°, the step response curves generated from the simulation is shown in figure 5.

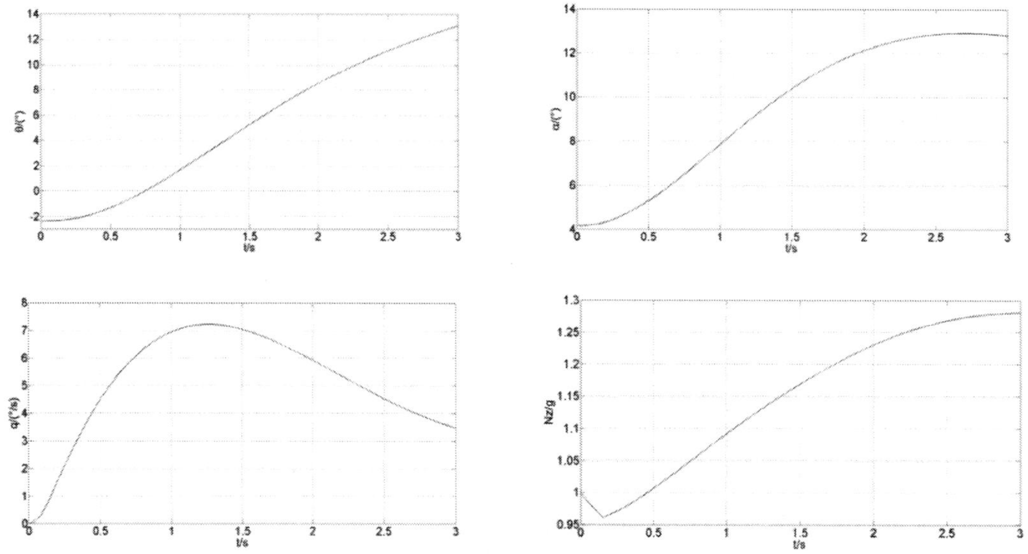

Figure 5. Step response curves of linear time-varying simulation.

Contrasting figure 4 and figure 5, it is found that either the magnitudes or the trends of the two curves are basically consistent. Considering the influence of other factors, the result is ideal, which proves the correctness of the calculation method of stability derivatives proposed in this paper.

3.2 Simulation for a VTOL aircraft

Next, a simulation is made for a new type of VTOL power-lift aircraft. The configuration of the aircraft is shown in figure 6.

Figure 6. The configuration of the VTOL power-lift aircraft.

The aircraft can obtain the Lift-Enhancement by using the flap to deflect the propeller slipstream, but this will cause the serious coupling of aerodynamics and engine forces. Therefore, the acquisition and processing of aerodynamic data is essential. Nevertheless, the wind tunnel test with power system is very complex and costly, and the CFD calculation method [9-11] with power system is also complex. Here we use the below method to integrate the aerodynamic forces and engine forces.

Take the calculation of lift L as an example. The individual aerodynamic data can be obtained by CFD calculation, and the direct forces generated by propeller slipstream can be obtained by ground test [12].The integrated lift L can be represented as the following form.

$$L=\eta * L_a + L_T \quad (8)$$

Where L_a is the aerodynamic force, L_T is the direct forces generated by propeller slipstream; η is the engine efficiency factor, which indicates the influence of propeller slipstream on aerodynamic force. η≥1 and increases with the increase of the throttle. The other aerodynamic data of the aircraft can also be obtained by this method.

The modeling simulation method introduced in the 2.1 section is used. The initial states of the aircraft as shown in formula (9).

$$V\text{=0m/s},\ \theta=45.6242°,\ Throttle\text{=0.9406},\ \delta_f=59.4847,\ T_f\text{=24.9N} \quad (9)$$

Where δ_f is the deflection angle of the flaps, T_f is the pulling force of the lift fan.

The step response of the aircraft as the throttle increases to 1 is shown in figure 7.The response process of each parameter is similar to the actual flight-test. It is proved that the stability derivative calculation method and simulation modeling method proposed in this paper can achieve the simulation of VTOL aircraft from hovering to low speed flight.

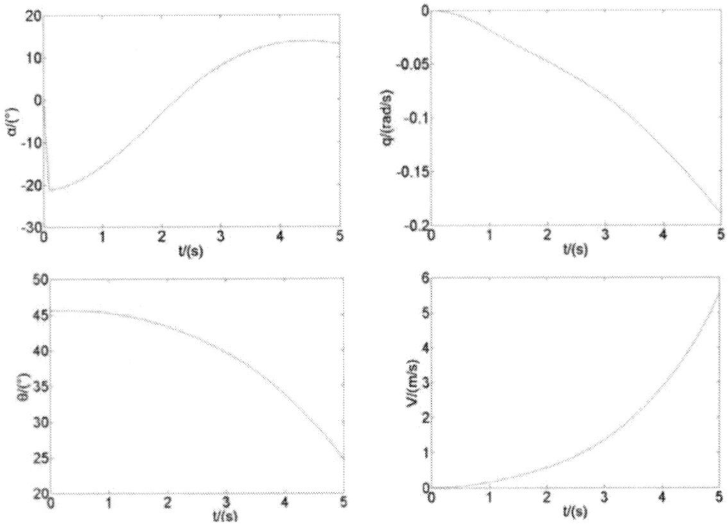

Figure 7. The step response of the VTOL power-lift aircraft.

4 Conclusion

A method for calculating stability derivatives by using dimensional aerodynamic data is proposed and the linear time-varying simulation models are established to evaluate the accuracy of the derivatives calculation method and verify the effectiveness of the modeling method to unify the modeling process of VTOL aircraft from hovering to low speed flight.

The simulation results show that the method can avoid the inaccuracy of the stability derivatives due to the limitation of C_T in the low speed flight. In addition, the method can unify the calculation method of stability derivatives at high speed, low speed and hovering condition, which maintains the continuity of stability derivatives, ensures the accuracy of the linear time-varying modeling and simulation method, and lays a solid foundation for the design of the subsequent longitudinal control system.

Reference

1. Etkin B and Teichmann T 1982 *Dynamics of Flight: Stability and Control 3rd edition* ed Cliff Robichaud and Catherine Beckham et al (New York: John Wiley &Sons, Inc) chapter 5 pp 129-148
2. Shengwei Z 2014 Research On Method of Powered Lift Aircraft Aerodynamics Calculation [J]. *Aeronautical Science Technology,* 2014(**3**):14-17.
3. Yue T, Wang L and Ai J 2013 Longitudinal Linear Parameter Varying Modeling and Simulation of Morphing Aircraft [J]. *Journal of Aicraft,* 2013, **50**(**6**):1673-1681
4. Wang H, Dong X, Xue J, Liu J and Wang J 2016 Modeling and simulation of a time-varying inertia aircraft in aerial refueling [J].*Chinese Journal of Aeronautics,* 2016, **29**(**2**):335-345
5. Johnson W, Samuel J, Craig IL and Ashkenas 1973 *Analysis and moving base simulation of transition configuration management aspects of a powered lift aircraft* NASA Technical Report by Ames Research Center, NASA-CR-114698, STI-TR-1015-2
6. Shengwei Z and Wei W 2017 Method for evaluating powered high-lift effects of externally blown flap [J]. *Acta Aeronautica et Astronautica Sinica*, 2017, **38**(**6**): 220689-220689.
7. Bobbitt P and Margason R 2007 *45th AIAA Aerospace Sciences Meeting and Exhibit, issue January(2007) pp1-43*
8. Bruce G. Powers and David A. Kier 1973 *Simulator Evaluation of the low-speed flying qualities of an experimental STOL configuration with an externally blown flap wing or an augmentor wing* NASA Flight Research Center Technical Note,NASA TN D-7454. pp 21-40
9. Abbas J and Narducci R 2010 *Analysis of CFD modeling techniques over the mv-22 tiltrotor* [C]*//Phoenix: American Helicopter Society 66th Annual Forum, May 11-13*
10. HY Xu, ZY Ye and AM Shi 2012 Numerical Study of Propeller Slipstream Based on Unstructured Dynamic Overset Grids [J]. *Journal of Aircraft,* 2012,**49** (**2**) :384-389.
11. Sweeten BC 2010 *CFD Analysis of UAVs using VORSTAB, FLUENT, and Advanced Aircraft Analysis Software*, (University of Kansas)
12. Yuxing Z 2017 *Lateral Decoupling Control Design of VSTOL Joined-Wing* (Beihang University) pp 24-32

Analysis of Operation of Throttle Micro-cooler at Nitrogen Use with Near-critical Parameters

Dmitry Uglanov[a], Dmitry Sarmin, Alina Akulova, Daria Aksenova, Sophia Dostovalova

Department of Heat Engineering and Heat Engines, Samara University, 443086, Moskovskoye shosse, 34, Samara, Russia

Abstract. The work of a throttle cooling system with a standard balloon and a throttle cooling system from a cryogenic filling balloon is considered. A comparison between the geometric and hydraulic characteristics of the micro heat exchangers of the two systems is made. The prospects of using the latest system for cooling of photodetector devices (PD) of the infrared range of navigation and research systems for aerospace applications are also shown.

1 Introduction

In the modern aircraft for aviation and space applications various opto-electronic system (OES) and navigation, research and military equipment (laser rangefinders, rifle scopes, guidance systems, lidar, gyroscopes) are increasingly used [1,2]. One of the devices included in these systems, regardless of their purpose, is a photodetector (PD) of the infrared range, which works effectively only when it is cooled to a temperature level of 80K and below. For cooling photo-detectors (PD) IR range throttle systems are widely used as a closed or open cycle. The operation time of the throttle balloon cooling systems depends on the weight of the working fluid and the required cooling capacity.

There is a constructive solution that allows to significantly increase the time of operation of the throttle cooling system and not to overload the balloon with high pressure. So, it is proposed to use the so-called cryogenic filling balloon (patent of Samara state aerospace University № 2163699, Russia) as a part of the throttle cooling system [3,4]. Its feature is cryogenic filling, which allows for the same volume of the balloon and the mass of the refilled working fluid to increase the time of operation of the throttle system by 1,5 times.

2 Method of calculation

In this work, the main characteristics of the heat exchangers of the throttling cooling systems with a standard balloon and a balloon with cryogenic filling (universal gas tank) with nitrogen are calculated using a working medium having a temperature of 130 ... 135K at a pressure of 4MPa [5]. Figure 1 shows the working cycles and schemes of these systems. In the first case, a standard balloon is used in the throttle system, filled with nitrogen with an initial pressure $p_0 = 30\,\text{MPa}$ and a temperature $T_0 = 293\text{K}$. In the second case, the above universal gas balloon is used as part of the throttle system.

[a] Corresponding author : dmitry.uglanov@mail.ru

Previously performed thermodynamic calculations showed that in case of filling of the balloon with cryogenic products (nitrogen) 30…40hours the equilibrium state of the balloon-working system sets in. The temperature amounts to $T_0 = 130 \dots 135$K at pressure $p_0 = 4$MPa. In both cases, the initial mass of the working fluid is the same.

The work of the throttling micro-cryogenic installation in the real conditions of use is characterized by the presence of a number of non-stationary modes, "deforming" workflow micro-coolers. The main initial parameters of the process - the pressure of there frigerant p_{input}, its temperature T_{input} and the flow M (or one of these parameters) - are changed not only during the start of the installation, but also in the cooling mode of the object.

The cooling mode with variable flow M is used in micro-coolers with cooling capacity regulators and with "start-up" sections of the heat exchanger, the mode with variable p_{input} is typical for open cooling systems, and the $T_{input} = var$condition is typical for micro-cryogenic systems, since they usually operate at variable ambient temperature and temperature change in the balloon.

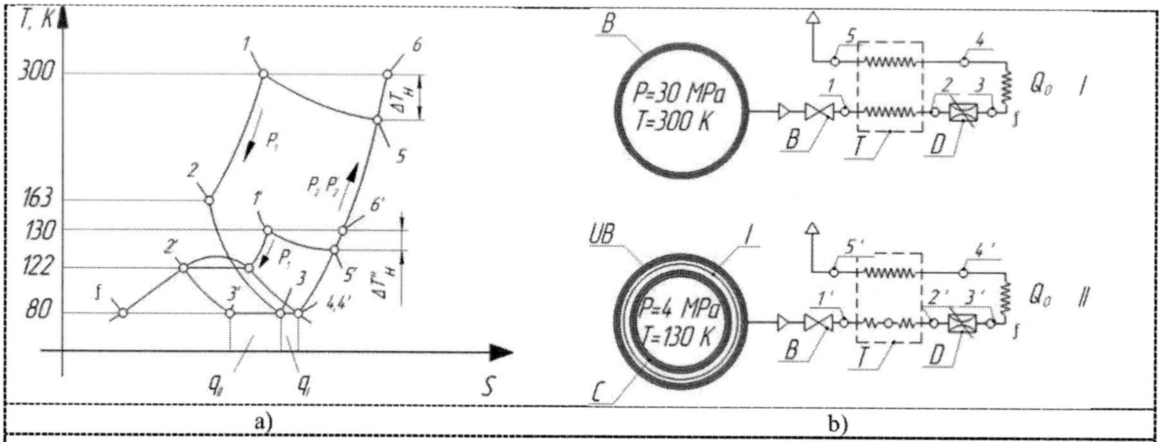

Figure 1. The work of throttle cooling systems: a) the workflow in the T–S diagram; b) – circuit throttle cooling systems: I–conventional balloon system; II – a system with a universal balloon; B –balloon; B –shut-off valve; D –throttle valve; I – thermal insulation; C – vessel(thermos); UB –universal balloon, T –heat exchanger.

Experience in the development of micro-cryogenic systems has shown [6] that the throttle heat exchanger is characterized by exceeding the available cooling capacity in relation to the thermal load on the system. Micro-cryogenic installation is usually expected to work in extreme conditions, that is, either for the minimum possible in the cooling mode of the working refrigerant pressures and maximum ambient temperatures, or for the start-up period conditions, when it is necessary to provide a relatively short time for the system to go into operation.

To calculate the characteristics of the heat exchanger, a method was chosen for calculating the "excess" cooling capacity Q at a variable flow rate of the refrigerant [7]. In this case, $M = mM_o$ (where the exponent$m > 1$); Q_0^*- thermal load on the microcooler; Mo is the minimum refrigerant flow (at the beginning of the cooling system with $m = 1$), M is the actual flow rate.

In the initial mode, i.e. at $m = 1$, the temperature pressure at the warm end of the heat exchanger is equal to ΔT_{H_0}, for the throttle cooling system with a universal gas balloon $\Delta T_{H_0}{}'$ (all letters denoting the thermodynamic parameters for the second system in the future go with a stroke) and is determined by the total heat exchange surface F_o and the average integral heat transfer coefficient k_o from the high-pressure direct flow to the reverse flow. The temperature of the gas before the throttle is equal to $T_{2_0}$and can be found for the enthalpy of point 4:

$$i_{20} = i_4 + \frac{Q_0^*}{M_0} \tag{1}$$

If the refrigerant flow rate increases by m times, the difference $i_4 - i_{3j}$ (in calculation 3_{ij} - the enthalpy of the working fluid after the throttle D) shall decrease by the same factor.This difference increases over time as a result of the removal of the working fluid mass from the balloon at the calculation step j, $i_4 - i_{3j}$, necessary to compensate the load Q_0^*, which corresponds to an increase in the gas temperature before the throttle from T_{2o} to T_{2j}. However, the actual temperature value of T_2 is determined by the joint action of a number of factors related to change M, i.e.an increase of about m times the thermal load on the heat exchanger, an increase in the heat transfer coefficient between the forward and backward gas flows is proportional to $M^{0,7-0,9}$, an increase in the temperature head between the flows due to the growth of $\Delta T_{\text{н}}$ and, to a lesser extent, the change in the heat inflow of the environment, from the gas microcooler and additional cooling of the parts in this zone of the object cryostat.

Thus, at $T_2 < T_{2j}$, the surface of the heat exchanger F_o can be divided into three zones: F_p - provides cooling of the direct flow to a temperature corresponding to the constant heat curve, which passes through point 4 (Fig. 1a); F_x - necessary for further lowering the temperature of the gas to the maximum permissible value T_2; ΔF_x - "excess" surface, which in principle in this regime micro-cooler cannot be used. Therefore

$$F_0 = F_p + F_X + \Delta F_X \tag{2}$$

The inequality $\Delta F_x \geq 0$ is a condition for the normal operation of the microcooler. The values $m = 1$ and $m = m_{max}$, at which $\Delta F_x = 0$, correspond to the minimum and maximum refrigerant flow rate, providing the required operating mode of the cooling device; in the range m from 1 to mmax ΔF_x reaches a maximum, and some other parameters - extreme values. In the heat exchange zone corresponding to the ΔF_x, the return flow of the refrigerant contains the liquid phase.

Since the thermal load on the microcooler $Q_0^*$is constant, the microcooler leaves the operating mode when the flow value is $M > m_{max}\,Mo$

$$M(\Delta i_T - \Delta i_{\text{H}}) < Q_0^*, \tag{3}$$

Δi_{H}–loss of cold from under exercise.

For the studied mode, the dependence is determined by the $\Delta F_x = f(m)$. Heat transfer in different zones of the heat exchanger can be described by equations:

$$\begin{cases} k_X \cdot \Delta T_X \cdot F_X = Q_0^* \\ k_P \cdot \Delta T_P \cdot F_P = Q_P = Q_{PX} - Q_0^* \end{cases} \tag{4}$$

Here $\Delta T_X, \Delta T_P$– mean temperature differences between streams, and k_X, k_P – medium-integral value of the heat transfer coefficients respectively for the areas of F_X and F_P; $Q_P = M(i_1 - i_4)$ – heat load to the area F_P of the heat exchanger; Q_{PX} is the same in the area of $F_P + F_X$.

In accordance with [5], and taking that $\frac{k_X}{k_P} = z$ and$\frac{k_P}{k_{P_0}} = m^\beta$, the following formulas are used for calculation:

$$F_X = F_P \cdot \frac{k_P}{k_X} \cdot \frac{\Delta T_P}{\Delta T_X} \cdot \frac{Q_0^*}{Q_P} \tag{5}$$

$$\Delta T_X = F_0 - F_P \cdot \left(1 + \frac{\Delta T_P \cdot Q_0^*}{z \cdot \Delta T_X \cdot Q_P}\right) \tag{6}$$

$$F_{P_0} = \frac{F_0}{1 + \frac{\Delta T_P \cdot Q_0^*}{z \cdot \Delta T_X \cdot Q_P}} \tag{7}$$

$$\frac{Q_P}{Q_{P_0}} = \frac{M}{M_0} = m = \frac{k_P \cdot \Delta T_P \cdot F_P}{k_{P_0} \cdot \Delta T_{P_0} \cdot F_{P_0}} \tag{8}$$

$$\frac{\Delta F_X}{F_0} = 1 - \frac{1}{m^\beta} \cdot \frac{\Delta T_{P_0}}{\Delta T_P} \cdot \frac{\Delta T_{X_0}}{\Delta T_X} \cdot \frac{\Delta T_X \cdot z \cdot m \cdot (Q_{PX_0} - Q_0^*) + \Delta T_P \cdot Q_0^*}{\Delta T_{X_0} \cdot z \cdot m \cdot (Q_{PX_0} - Q_0^*) + \Delta T_{P_0} \cdot Q_0^*} \tag{9}$$

$$i_2 = i_4 \cdot \left(1 + \frac{1}{m}\right) + \frac{i_{20}}{m} - \frac{\bar{z} \cdot \Delta \bar{T}}{m^{1-\beta} \cdot M_0} \cdot \frac{i_1 - i_{2_0}}{i_{5_0} - i_1} \cdot \frac{Q_0^*}{\Delta T_0} \cdot \frac{\Delta F_X}{F_0} \tag{10}$$

The heat exchanger of the microcooler is single-row, transverse-counter flow (β is a quantity that depends on the design of the heat exchanger in accordance with the calculated data of [6] β= 0,87); the throttle is made in the form of a micro-hole of constant cross-section.

3 Results of calculation

In the calculations it is assumed that the temperature along the length of the heat exchanger varies linearly. The hydraulic losses through the heat exchanger channels were calculated in accordance with the procedure [7,8]. Figures 2-6 show the results of calculating the characteristics of the microcoolers from the parameter m, which shows the change in the required flow rate to ensure a constant cooling capacity (Table 1).

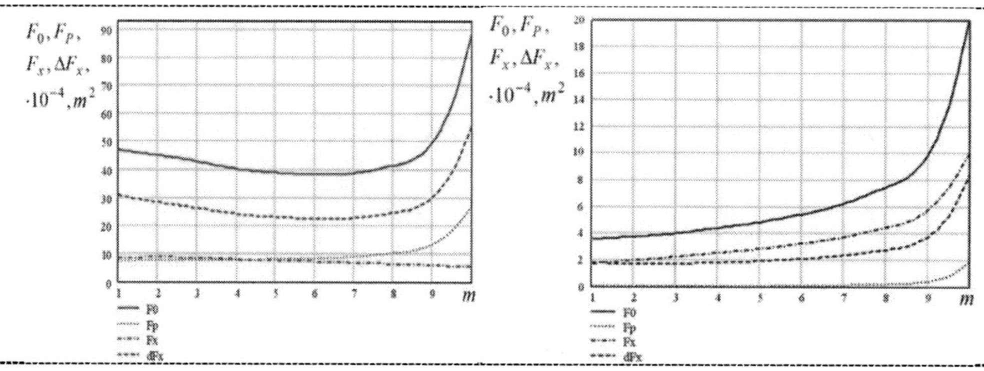

Figure 2. Dependence of the ratio $\Delta F_x/F_o$ of the heat exchanger on the parameter m in the regime of excessive cooling capacity with variable refrigerant flow rate.

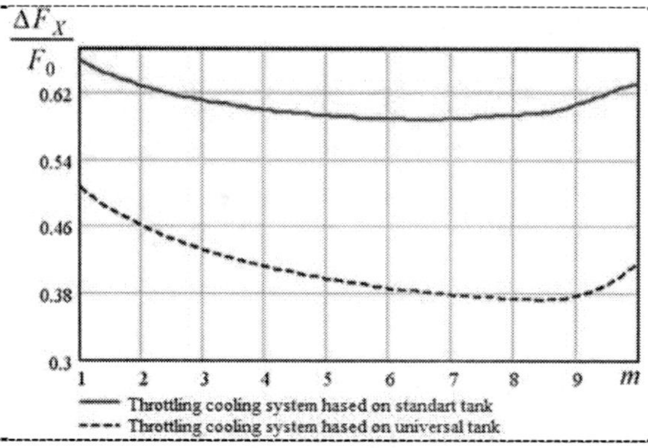

Figure 3. The graph of the area dependence of the heat exchanger on the exponent m in the regime of excessive cooling capacity with a variable flow rate of the refrigerant: $p_o = 30\ MPa, T_o = 293\ K$ (a); $p_o = 4\ MP$, $T_o = 130\ K$(b).

Fig. 2a and 2b show the dependencies of the required surface of the microcooler as a function of the exponent m, which varies during the operation of the balloon throttle system. Comparison of the obtained data allows to make a conclusion that the required heat transfer surface of the microchiller of the throttle system with a universal gas balloon is 5 ... 10 times smaller than that of a similar system with a standard balloon. The maximum difference is observed at low flows of the working fluid, i.e. at the beginning of the cooling system operation.

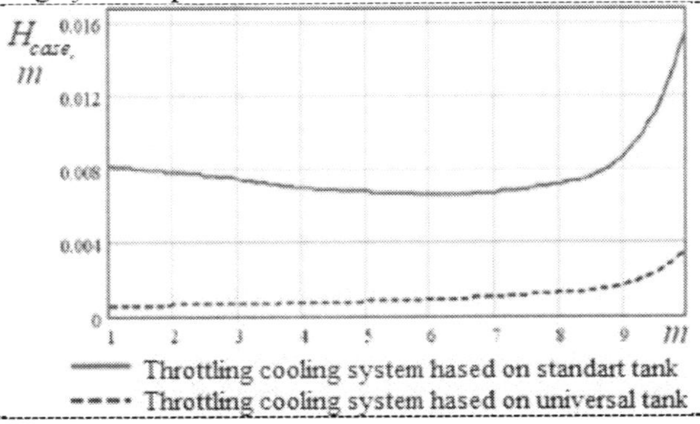

Figure 4. The dependence of the height of the casing of the heat exchanger H_{case} from indicator m mode of excess cooling capacity with variable refrigerant flow.

Table 1. Characteristics of micro coolers depending on parameters m.

m	The throttle system with a standard balloon		The throttle system with the universal gas bottle	
	τ, c	$G \cdot 10^4$, kg·c	τ, c	$G \cdot 10^4$, kg·c
0	0	0	0	0
1	410	3,597	630	1,140
2	820	3,709	1260	1,276
3	1230	3,863	1890	1,474
4	1640	4,072	2520	1,790
5	2050	4,361	3150	2,136
6	2460	4,776	3780	2,620
7	2870	5,416	4410	3,296
8	3280	6,557	5040	4,371
9	3690	9,155	5670	6,580
10	4100	21,012	6300	16,73

The dependence of the ratio of the excess surface area of the microcooler to its full surface on the indicator m is shown in Fig. 3. Comparison of this indicator of both systems shows that for a throttle cooling device with a universal gas balloon requires 60 ... 70% less than the excess surface of theΔF_X. For each curve there is a minimum point (0,37 – for a throttle system with a universal balloon; 0,58 – for a throttle system with a standard balloon), which is due to a change in the heat transfer coefficient during the operation of the throttle cooling system. The change in the heat transfer coefficient is due to the change in the basic thermodynamic and thermophysical parameters of the working fluid at different stages of the cooling system.

The size of the casing along the length is determined by the flow rate of the working fluid and similarly to the surface area of the microcooler varies with an increase in m (Fig. 4).

Fig. 5 shows the dependence of the temperature of the refrigerant before the throttle on the indicator m for two throttle systems is presented.

The temperature in the throttle system with a standard balloon is higher than for the second system at 20 ... $40K$. This is objectively due to the difference in the initial temperature of the working fluid in the balloon.

Fig. 6 shows the dependence of the hydraulic resistance p in the channels of high and low pressure of the microcooler on the indicator m is an order of magnitude lower for the throttle cooling system with a universal gas balloon. This result is due to the fact that in the throttle system with cryogenic filling to ensure the required cooling capacity there are significantly lower costs and, accordingly, the speed of the working fluid in the channels of the microcooler.

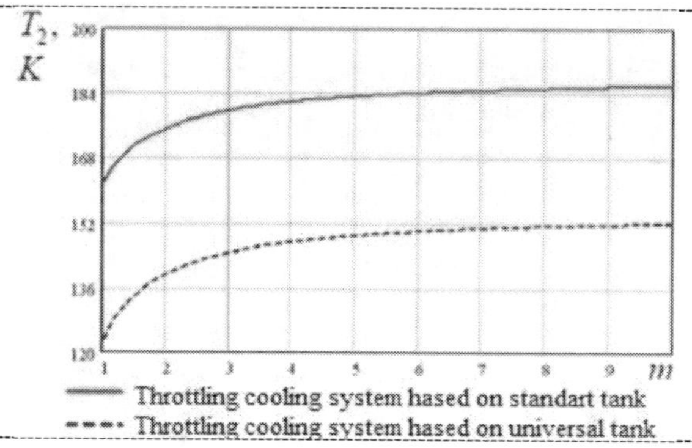

Figure 5. Dependence of the coolant temperature before the throttle on the indicator m in excess cooling capacity at variable flow.

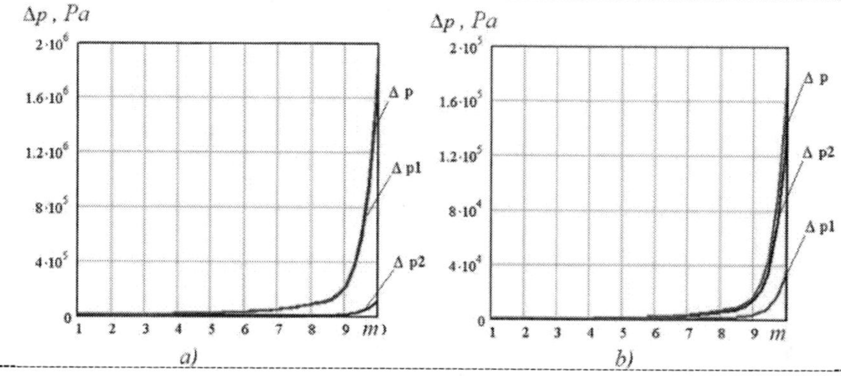

Figure 6. The dependence of the pressure loss Δp from the value of m in excess of the cooling capacity with variable refrigerant flow (a – $p_o = 30\ MPa$, $T_o = 293\ K$; b –$p_o = 4\ MP$, $T_o = 130\ K$).

Conclusions

Thus, the throttle cooling system with a balloon with cryogenic filling is obtained in 2 times more compact and has a working time of 1,5 times more than the throttle system with a standard balloon at the same initial weight of the working fluid.

References

1. B.I. Vasiliev, U.M. Manum, *Infrared LIDAR for ecology monitoring*, 71 (2005)
2. V.V. Protopopov, N.D. Ustinov, *Infrared laser location systems*, 175 (1987)
3. A.I.Dovgyallo,S.V.Lukachev and others,*Fuel tank*,Patent №2163699, Russian,MPK 7F17C9/02, 99114577//06
4. E.V. Blagin, A.I. Dovgyallo, S.O. Nekrasova, D.V. Sarmin, D.A. Uglanov, Applied Thermal Engineering Estimation of the energy efficiency of cryogenic filled tank using in different system and devices. 2016. Volume 101, 25 May 2016, Pages 537-544
5. A.I. Dovgyallo, A.P. Logashkin, D.V. Sarmin, D.A. Uglanov, Applied physics: Moscow physical society,*Throttle board coolingsystem based on cryogenic filling tank*,**6**, 75-78 (2008)
6. A.K. Grezin, V.S.Zinoviev, *Micro cryogenic technique*, 232 (2008)
7. V.M. Keis, A.L. London,*Compact heatexchangers*, 157 (1962)
8. V.V. Sichov, A.A. Vassermann, A.D. Kozlov, G.A. Spiridonov, V.A. Tsimarnii, *Thermodynamics nitrogen parameters*,(1977)

Investigation of Flow over the Airfoil NACA 0010-35 with Various Angle of Attack

Shiming Xiao[1,a] and Zutai Chen[1]

[1]Department of Mechanical Engineering, Engineering Faculty, University of Portsmouth, The United Kingdom

Abstract. This report is using the finite element analysis method to investigate the performance of an airfoil (NACA 0100-35), which uses 2D computational fluid dynamics simulations based on ANSYS to find the lift and drag coefficients under different conditions. The boundary conditions are defined as follows: The density is $1kg/m^3$, gauge pressure is 0, and airflow velocity is 50m/s. Also the Reynolds number is considered during the simulation. As a result, the simulation shows the stalling points and efficiencies of the airfoil under different angles of attack. The simulation also plots the contours and stream lines in these conditions.

1 Introduction

Considering the important role the aircraft manufacturing has been played, selection of airfoils which is the section side of the airplane wings should be determined with certain conditions. One of the most popular softwares in this field is ANSYS, and it is used to obtain accurate results for simulating airfoils. The objective of this report is to investigate an airfoil (NACA 0010-35) with different angels of airflows based on ANSYS with 2D CFD (Computational Fluid Dynamics) simulation. The simulation is required to find the values of drag and lift coefficients in order to get stalling points and efficiencies.

2 Parameters of the airfoil

2.1 Lift and Drag

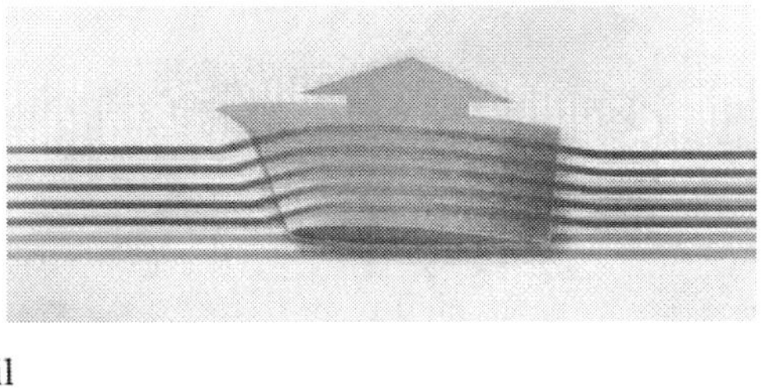

Figure 1. Airflow through airfoil

[a] Corresponding author: flora901120@gmail.com

According to Tom Benson's article[1], when an airfoil moves through airflow (as it is showed in figure 1), an aerodynamics force will be generated on it (as it is showed in figure 2 and figure 3). This aerodynamic force contains of the following:

$$L = \frac{1}{2}\rho v^2 C_l A \tag{1}$$

$$D = \frac{1}{2}\rho v^2 C_d A \tag{2}$$

Where D is the drag force, L is the lift force, ρ is the density of the fluid, v is the speed of the object relative to the fluid, A is the cross-sectional area, and C_d is drag coefficient and C_L is lift coefficient.

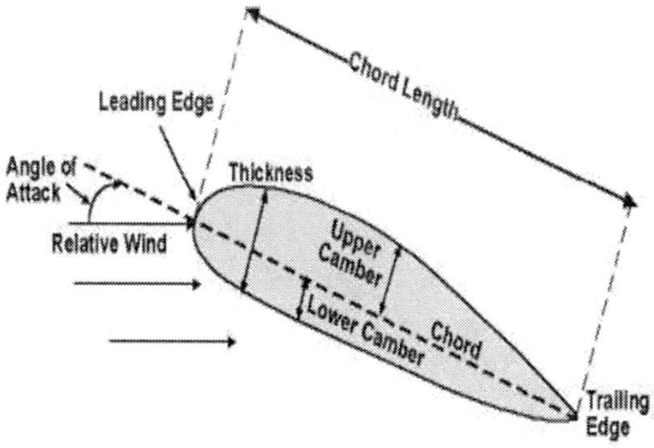

Figure 2. wing side view (airfoil)

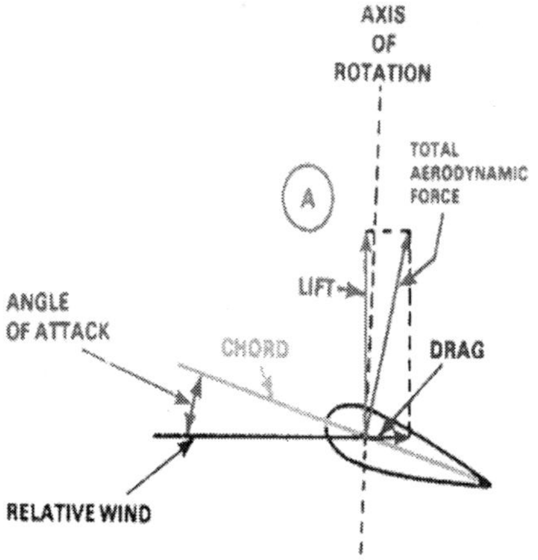

Figure 3. Lift and Drag forces Vs Angle of Attack

From the equation above, it can be illustrated that the lift force and drag force on any object are depend on the density of the fluid and the relative speed between the object and fluid proportionally.

2.2 Lift and Drag

The Lift Coefficient equation which could be rearranged from the Lift equation is as follow:

$$C_l = \frac{2L}{\rho v^2 A} \tag{3}$$

Where L is the lift, ρ is the density of the fluid, v is the velocity of the fluid and A is the relevant surface area [2]. For the airfoil, the relevant surface area, which is related to the drag coefficient types, is a plane form area [3]. T. Benson [4] mentioned from NASA, the lift coefficient is usually determined experimentally, meanwhile it is a number which could describe all the complex dependencies of shape, inclination and some flow conditions on aircraft lift. In this report, C_l is calculated from ANYSIS.

Also, the Drag Coefficient equation could be rearranged from the Drag equation as follow:

$$C_d = \frac{2D}{\rho v^2 A} \tag{4}$$

Where D is the drag force, which is by definition the force component in the direction of the flow velocity, ρ is the mass density of the fluid, v is the speed of the object relative to the fluid and A is the reference area.

In this report, Cl and Cd could be calculated from the ANSYS system.

2.3 Reynolds number

The equation for Reynolds number is showed in following equation:

$$R_e = \frac{\rho v D}{\mu} \tag{5}$$

Where ρ is the density for airflow (1kg/m^3), v is the velocity for airflow (50m/s), D is 1m, μ is the viscosity (1.460e-5) for the air.

Patil et al.[5] investigated that the Lift and Drag forces at various angle of attack for Reynolds number from 10,000 to 800,000 by CFD analysis and concluded that the lift and drag forces increased with the Reynolds number increased. In this simulation, calculated by the above equation, R_e is 3.42 x 10^6. So the flow is turbulent. This case only investigates the relationship between the angle of attack and its effects. The angle of attack, as it is showed in Figure 3, is the angle between the relative wind and the chord. As the leading edge which is front point of the airfoil turns up, angle of attack increases and connected to the increase of the lift force and drag force [3]. Investigations from former work from Sahin et al. [6] analysed the impact of stalling angle on lift and drag coefficient. Meanwhile, Bhat et al.[7] had also worked out the stalling angle of NACA0012 at a certain Reynolds number.

3 Simulation

3.1 Choose an Airfoil

Airfoil NACA 0010-35, as a symmetrical section, was chosen from the UIUC Airfoil Database(n.d.) to simulate. (showed in Figure 4)

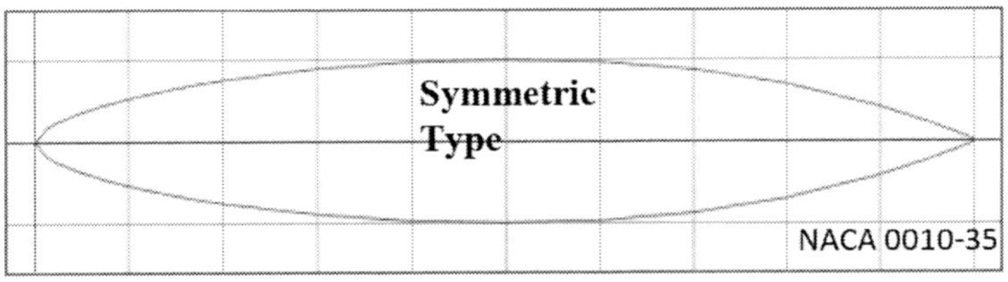

Figure 4. Geometry for NACA 0010-35

The data in Table 1 below were downloaded from the UIUC Airfoil Database(n.d.) and imported to the ANSYS workbench using Notepad.

Table 1. Data for NACA 0010-35 Airfoil

Point	x-cord	y-cord	z-cord
1	0.00000	0.00000	0.00000
2	0.0125	0.00878	0.00000
3	0.0250	0.01267	0.00000
4	0.0500	0.01844	0.00000
5	0.0750	0.02289	0.00000
6	0.1000	0.02667	0.00000
7	0.1500	0.03289	0.00000
8	0.2000	0.03789	0.00000
9	0.3000	0.04478	0.00000
10	0.4000	0.04878	0.00000
11	0.5000	0.05000	0.00000
12	0.6000	0.04867	0.00000
13	0.7000	0.04389	0.00000
14	0.8000	0.03500	0.00000

15	0.9000	0.02100	0.00000
16	0.9500	0.01178	0.00000
17	1.0000	0.00100	0.00000
18	1.0000	-0.00100	0.00000
19	0.9500	-0.01178	0.00000
20	0.9000	-0.02100	0.00000
21	0.8000	-0.03500	0.00000
22	0.7000	-0.04389	0.00000
23	0.6000	-0.04867	0.00000
24	0.5000	-0.05000	0.00000
25	0.4000	-0.04878	0.00000
26	0.3000	-0.04478	0.00000
27	0.2000	-0.03789	0.00000
28	0.1500	-0.03289	0.00000
29	0.1000	-0.02667	0.00000
30	0.0750	-0.02289	0.00000
31	0.0500	-0.01844	0.00000
32	0.0250	-0.01267	0.00000
33	0.0125	-0.00878	0.00000
34	0.0000	0.00000	0.00000

3.2 Meshing

A C-shape domain has been sketched around the airfoil, and the size was 12.5 m for both the semicircle and the rectangle. The geometry setup is illustrated in figure 5, the airfoil is in the center of the figure.

The Bias factor was set to 150.

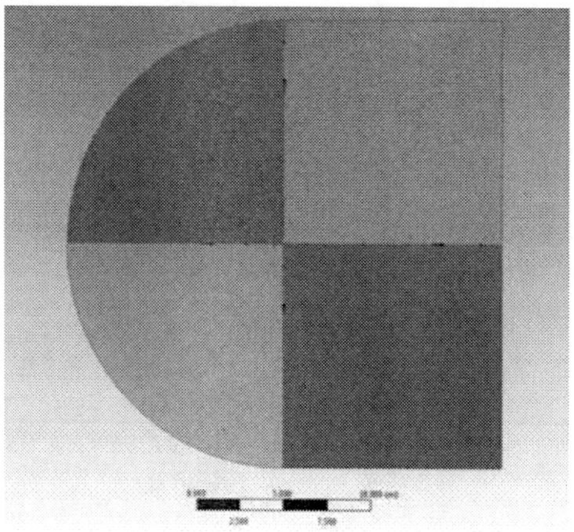

Figure 5. Geometry setup

$$\%Error = \frac{|Measured\ Value\text{-}Initial\ Value|}{Initial\ Value} \times 100 \qquad (6)$$

The Number of divisions was calculated using the above %Error equation. The angle of attack was set to 3 and run the calculations with three different numbers of divisions which in result showed different numbers of elements and drag forces. Then, the number of division with the lowest % error was chosen which is 70 as it is showed in Table 2, and the side edge sizing was set to 140, then applied this details to the system for the airfoil. The numbers of setting up were showed in Table 3.

Table 2. No. of divisions and %Error

Angle of attack	No. Divisions	No. Elements	Drag	% Error
3	50	15000	50.322226	
3	60	21600	51.549922	2.4396
3	70	29400	52.446637	1.7395
3	80	38400	54.092778	3.138

Table 3. Numbers of setting up

NACA 0010-35	
Grid Shape	C-Shape
Size	12.5
No. of Divisions	70
Side Edge Sizing	140
Bias Factor	150

3.3 Meshing

C-Mesh was applied to both airfoils in order to guarantee the accuracy of the model by getting more refined mesh over the edge of the trail and the surface of the airfoils. The below figure 6 shows the C-mesh of the simulation:

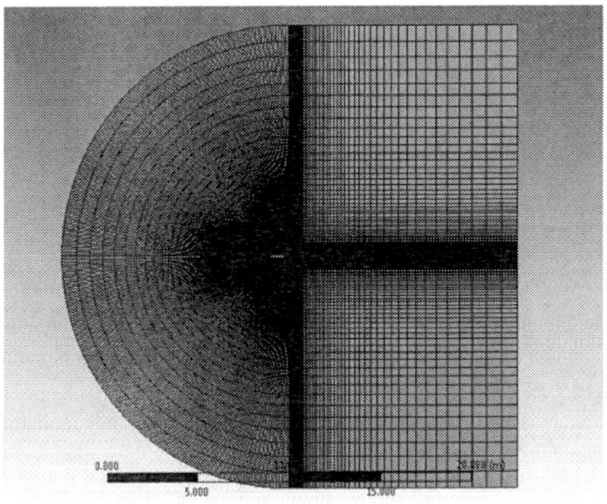

Figure 6. C-mesh

3.4 Set up the test

The below settings were used to set up the test:

Model - Solver: Different turbulence models like Spalart- Allmaras, Realizable k-ε and k-ω shear street transport had been compared by Eleni et al. [8]. The realizable k-ε was used because it is common and popular turbulence model which is mostly used in the industrial applications.

Materials: The Fluid air was used in the category

Boundary conditions zone: The zone condition was selected to be from Inlet.

Monitors: The Residual Convergence which was chosen for the monitors was 1e-6.

Solution method: Second order upwind was used.

Solution Initialization: The initial values have been computed from the inlet.

Boundary conditions for calculations: The velocity component for X and Y directions are calculated as below:

$$x = (\text{velocity}) * [\text{Cos (angle of attack)}] \tag{7}$$

$$y = (\text{velocity}) * [\text{Sin (angle of attack)}] \tag{8}$$

Other settings:

- Velocity: 50
- Density: 1
- Pressure: 0

3.5 Run the calculation

Number of Iterations : The number of iterations used was 10, 000 to make sure that all the residuals will converge with the increase of the angle of attacks.

4 Results

To make the simulations for the airfoil, there were more than 40 running tests have been done with different angle of attacks as the table 4 showed. The results will show the lift and drag coefficients, velocity and pressure behaviors towards the airfoil, and also the effect of airflow over the airfoil.

Table 4. Results

Angle	x-velocity	y-velocity	Cd	Cl	Efficiency (Cl/Cd)
-5	49.8097349	-4.5377871	0.0355915	-0.3825206	-10.7475169
0	50	0	0.0165975	-2.31E-04	-0.01391463
2	49.9695414	1.7449748	0.0192779	1.57E-01	8.133606551
3	49.9314767	2.6167978	0.0229464	0.23148598	10.08809686
4	49.878202	3.48782368	0.0282730	0.3023891	10.69532443
5	49.8097349	4.35778713	0.0352934	0.36860041	10.44388029
7	49.627307	6.09346	0.0547759	0.49069	8.958122961
9	49.384417	7.8217232	0.0810158	0.59171372	7.303681813
11	49.081359	9.5404497	0.1118332	0.65734985	5.877945356
12	48.90738	10.395584	0.1284684	0.67660827	5.266726303
13	48.718503	11.2475527	0.1456626	0.68685465	4.715379985
14	48.514786	12.0960947	0.1632566	0.68923882	4.221812901
15	48.2962913	12.9409522	0.1812625	0.68638258	3.786676387
17	47.8152377	14.6185852	0.2178388	0.67328093	3.090729788
18	47.5528258	15.4508497	0.236155	0.66704097	2.824589655
19	47.275928	16.2784077	0.2544698	0.66275165	2.604440659

4.1 Maximum Lift Coefficient CL (Stalling point) of the Airfoil

The above table shows that the angle of attack was increased to reach the maximum Cl (Stalling point). Referring to the table and the figure 7, it can be seen that as the angle of attack increases, the Cl increases as well until the angle of attack is set to 14 then the Cl starts to decrease, so the angle of attack 14 is called the stalling point where it has the maximum Cl of the airfoil.

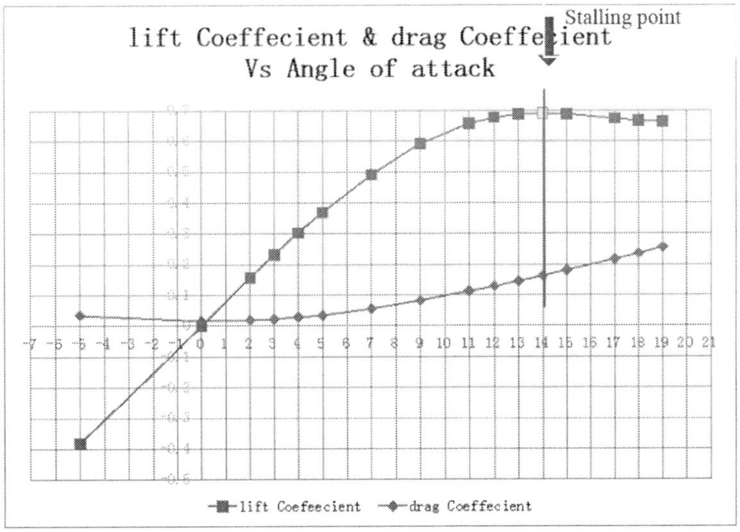

Figure 7. Lift and Drag Coefficients vs Angle of attacks

4.2 Maximum Efficiency of the Airfoil

The efficiency is defined as the ratio between Cl and Cd at each angle of attack, so referring to the figure 8, it can be seen that the maximum efficiency is reached at angle of attack (4) for airfoil1.

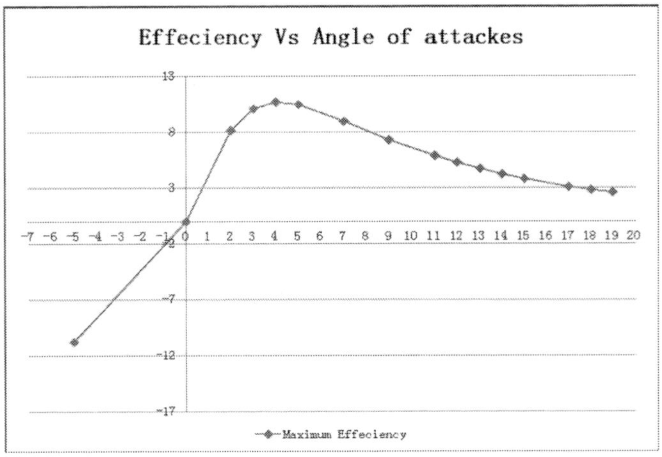

Figure 8. The efficiency vs Angle of attack

4.3 Contours of static pressure of the airfoil

It can be seen from the below pictures (figure 9,10,11,12) that when airfoil1 moves through an airflow and the angle of attack increases, then the pressure will increase gradually and will appear on the bottom surface of airfoil. According to Bernoulli's principle, the upper surface of the airfoil has low pressure while the lower surface has higher pressure, which leads to the flow accelerates on the upper surface and the decrease of the flow velocity at the lower surface.

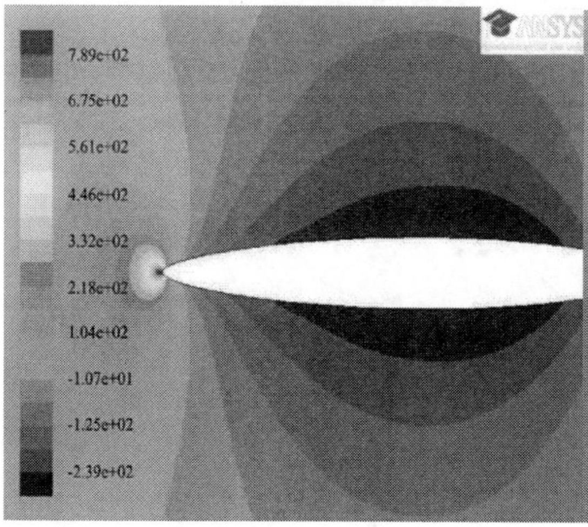

Figure 9. Contours of static pressure (Angle=0)

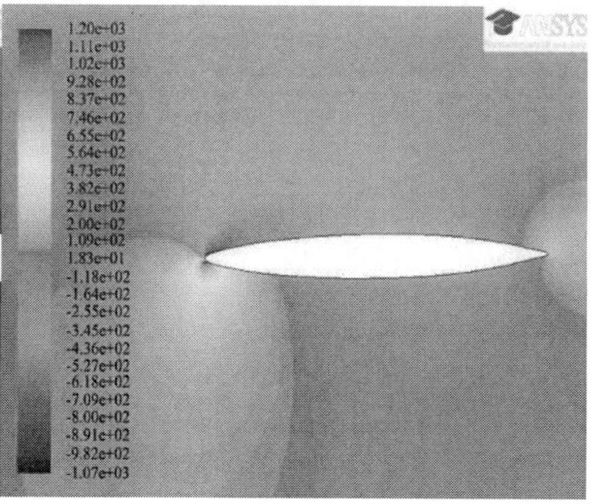

Figure 10. Contours of static pressure (Angle=4)

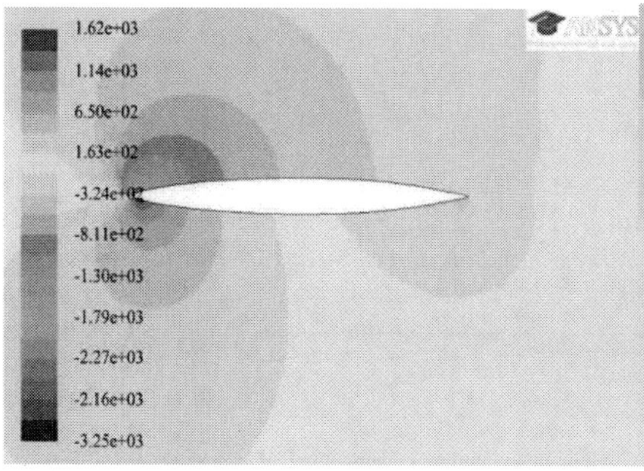

Figure 11. Contours of static pressure (Angle=14)

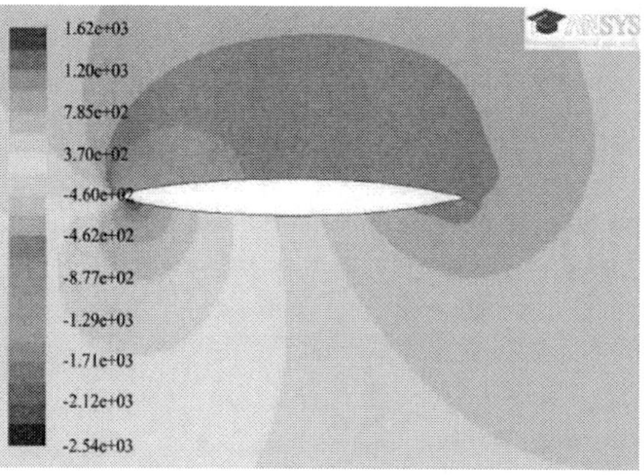

Figure 12. Contours of static pressure (Angle=19)

4.4 Contours of velocity magnitude of the airfoil

It can be seen from the below pictures (figure 13,14,15,16) that when airfoil1 moves through an airflow and the angle of attack increases, then the velocity will increase gradually and will appear on the top surface of the airfoil. Also, the air separation is going to be shown clearly when angle of attack increases after the stalling point.

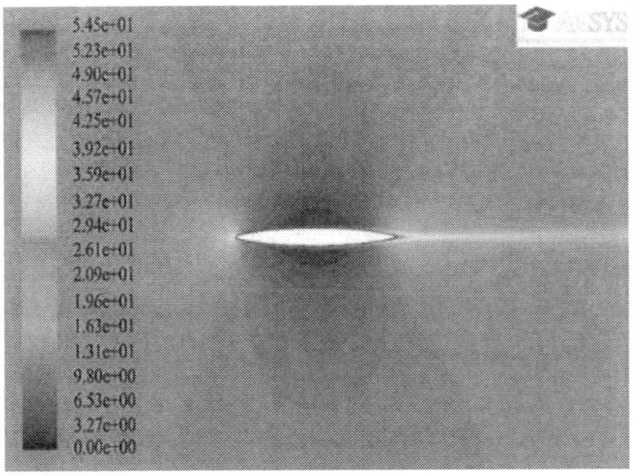

Figure 13. Contours of velocity magnitude (Angle=0)

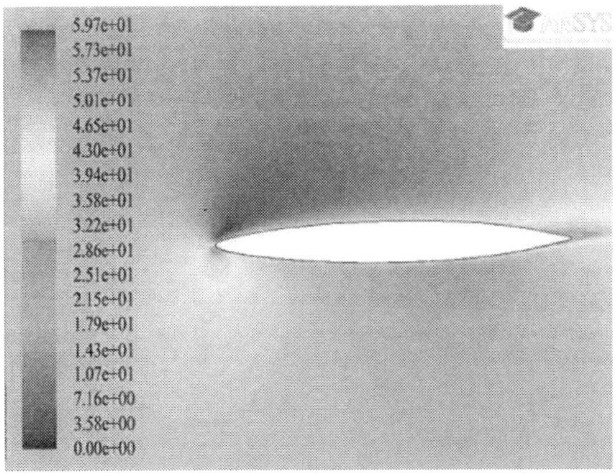

Figure 14. Contours of velocity magnitude (Angle=4)

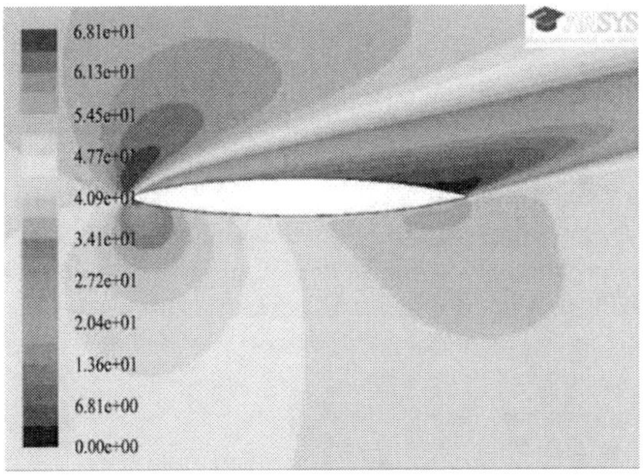

Figure 15. Contours of velocity magnitude (Angle=14)

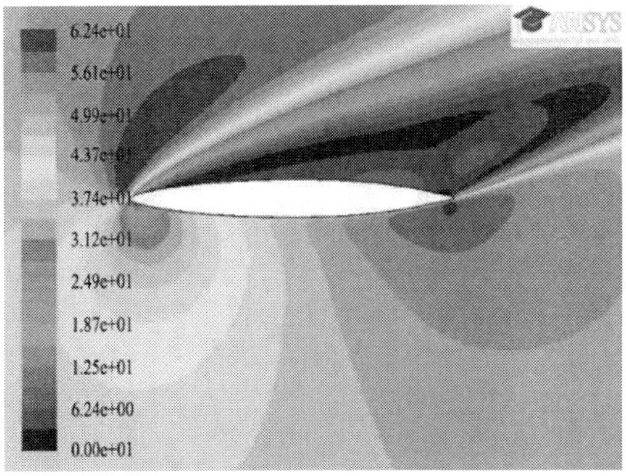

Figure 16. Contours of velocity magnitude (Angle=19)

4.5 Velocity Stream Function of the Airfoil

The figure 17,18,19,20 shows that when airfoil1 moves through an airflow and reaches the Maximum CL, then the air separation will start and a vortex will occur while the angle of attack increases.

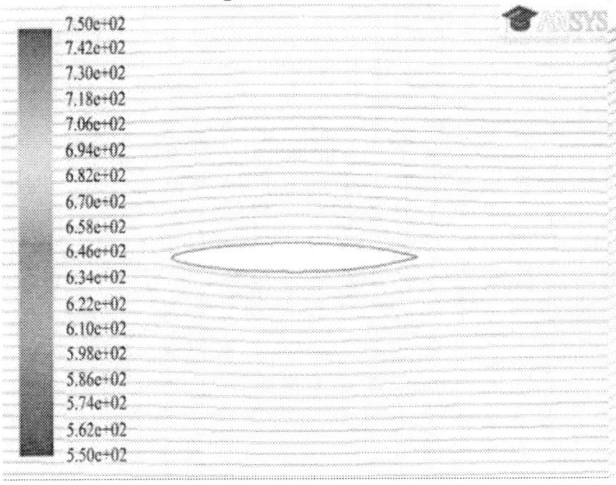

Figure 17. Velocity stream function (Angle=0)

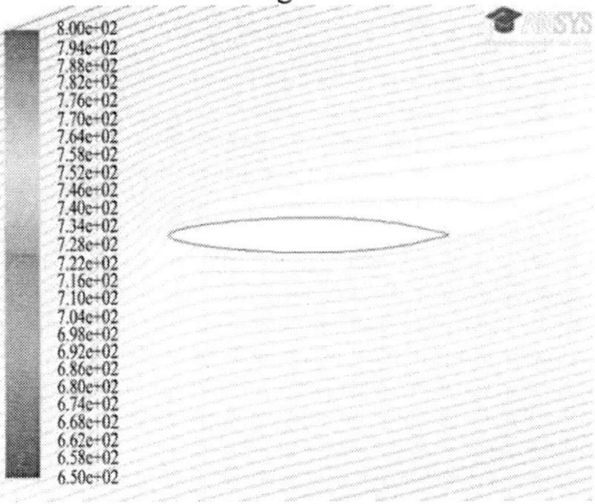

Figure 18. Velocity stream function (Angle=4)

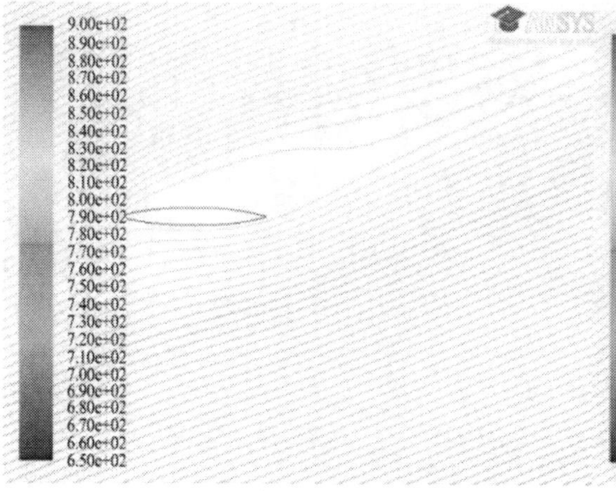

Figure 19. Velocity stream function (Angle=14)

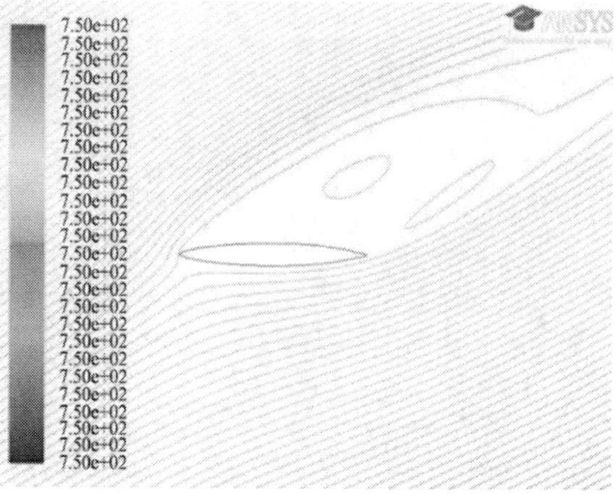

Figure 20. Velocity stream function (Angle=19)

5 Discussions and Conclusion

This Symmetrical airfoil was selected and investigated using ANSYS Workbench fluent program Simulation. The aim of this investigation is to reach the maximum Lift Coefficient (stalling point) and the maximum Efficiency for the airfoil through many running calculations with different angle of attacks. By comparing all the results for different angle of attacks, it can be noticed that NACA0010-35 has the stalling point at the angle of 14. The less stalling point means that the airfoil will have earlier air separation which will help in lifting the airplane. By contrast, when an airfoil reaches the maximum efficiency with angle of attack earlier means that its Maximum Efficiency is more.

Appendix

The data and geometry of the airfoil NACA0010-35 are downloaded from the UIUC Airfoil Database as follow:

http://m-selig.ae.illinois.edu/ads/coord_database.html#N

References

1. T Benson. National Aeronatics and Space Administration. Aerodynamic Forces, retrieved from http://www.grc.nasa.gov/WWW/k-12/airplane/presar.html (n.d.)
2. Cited from Wikipedia: http://en.wikipedia.org/wiki/Lift_coefficient
3. K S Patel, S B Patel, U B Patel, et al. IJER, **3** 154-158 (2014)
4. T Benson. National Aeronatics and Space Administration. Aerodynamic Forces, retrieved from http://www.grc.nasa.gov/WWW/k-12/airplane/liftco.html (n.d.)
5. B S Patil, H R Thakare, Procedia Engineering **127** 1363 – 1369 (2015)
6. I Sahin, A Acir. International Journal of Materials, Mechanics and Manufacturing **3** 1 (2015)
7. S S Bhat, R N Govardhan. Journal of Fluids and Structures **41** 166-174 (2013)
8. D C Eleni, T I Athanasios, M P Dionissios.JMER **4 (3)** 100-111(2012)

Preliminary Validation of the Wind Tunnel Based Flight Control System Evaluation Method

Min Huang[1,a], Zhong-wei Wang[2], Xing-Bao Yang[1], Zhen-yun Guo[2] and Yao-bin Niu[2]

[1]*Dalian Naval Academy, Dalian, Liaoning, 116018, China*
[2]*National University of Defense Technology, Changsha, Hunan, 410073, China*

Abstract. As mathematical models of aircraft aerodynamics and rudder loadings always have to be built in the hardware-in-the-loop simulation, wind tunnel based flight control system (FCS) evaluation methods were proposed in order to test and evaluate the flight control systems under real aerodynamic and rudder loading environment. To validate the evaluation method, a wind tunnel based flight control system test was performed in a hypersonic wind tunnel facility. As the aircraft support rig in the wind tunnel is static, the aircraft angle of attack cannot be changed in this test. During the test, the elevator response, the lift force and the pitching moment were measured. By analysing the measured data, the elevator control performance of the pitch control system was determined, and the pitch angle was successfully predicted, but the open-loop pitch control performance was not determined. These results validate the feasibility of evaluating the elevator control performance and predicting the pitch angle of a FCS by the wind tunnel based FCS evaluation method.

1 Introduction

Traditionally, after a flight control system is evaluated and confirmed by the hardware-in-the-loop simulation (HILS), it should then be tested and evaluated by the flight tests. However, the aircraft aerodynamics and rudder loadings in the HILS have to be mathematical models [1] which will make the HILS always have a distance away from the real flight tests. In order to replace these mathematical models with real factors, the aircraft installed with FCS hardware is directly put inside the wind tunnel which can make the wind tunnel airflow act on the aircraft body as in real flights. Then, the aircraft can rotate in the wind tunnel under the action of the aircraft flight control system [2]. Obviously, this kind of wind tunnel based FCS test is more realistic and closer to the flight tests than the HILS, especially in the case of modelling unsteady or nonlinear motions. If such a test is used to evaluate the FCS after HILS and before flight tests, more realistic FCS evaluation results will be obtained, and some FCS errors that can hardly be exposed in HILS, such as the inefficiency of the FCS under unsteady motions (in 2009, the hypersonic vehicle HTV-2 experienced a flight failure caused by the inefficiency of the FCS under the coupled pitch-roll motion [3]), can be detected by the wind tunnel based FCS tests. Therefore, the wind tunnel based FCS tests can make up the deficiency of HILS by putting the aircraft inside the wind tunnel, and reduce the risk of flight tests by exposing and solving some flight problems on the ground.

[a] Corresponding author : hmjtxy@163.com

Due to the advantages of wind tunnel tests with control, many departments were developing the wind tunnel test techniques with control [4] in an attempt to predict flight performances and evaluate the flight control laws in wind tunnels. The Arnold Engineering Development Center (AEDC) [5] built an 8-wire suspended 2-DOF (degrees-of-freedom) wind tunnel test system with control and the Physical Sciences Incorporation (PSI) [6] in USA built an 6-wire suspended 3-DOF (pitch, yaw and roll) wind tunnel test system with control to reappear and predict the coupling motions that may occur in the real flights. The University of Cambridge [7] and the University of Bristol [8] in UK, the SibNIA [9] in Russia, the Royal Institute of Technology [10] in Sweden, the French-German Research Institute of Saint-Louis [11] in France, the China Aerodynamics Research and Development Center (CARDC) [12-13] in China all had built wind tunnel test systems with control to evaluate the flight control laws. However, up to now, the wind tunnel tests for evaluating the flight control systems have not been developed yet. Therefore, the wind tunnel tests for evaluating the flight control systems should be researched.

According to the above demand for the wind tunnel based FCS evaluation methods, the wind tunnel based FCS evaluation method for FCS was proposed in references [14-15]. To preliminarily validate the evaluation method, a wind tunnel based flight control system test was performed in a hypersonic wind tunnel, and the test data was analyzed to judge whether the FCS performances can be determined by the evaluation method. If some or the whole performances of the pitch control system are determined, such as the elevator control performance and the open-loop or the closed-loop pitch control performances, the evaluation method can be partly or fully validated.

2 The wind tunnel based flight control system evaluation method

Before introducing the wind tunnel based flight control system test, the wind tunnel based flight control system evaluation method proposed in references [14-15] is generally introduced here.

A classic unmanned pitch control system (shown in Figure 1) was taken as the evaluation object in the wind tunnel based flight control system evaluation method. The followings are the details towards this method.

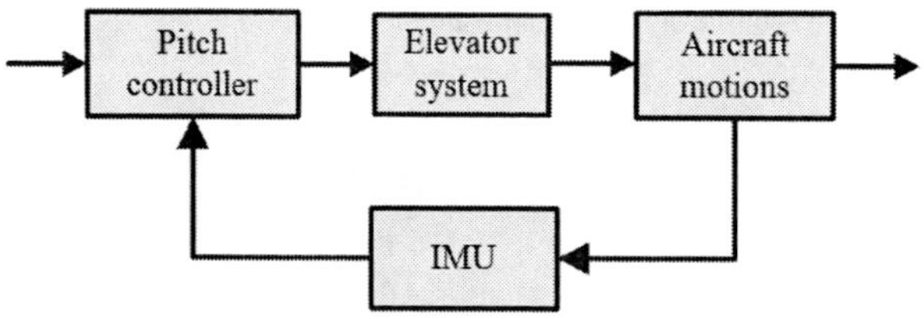

Figure 1. The structure of an unmanned pitch control system.

Firstly, the test system scheme of this evaluation method is shown in Figure 2. This system is mainly consisted of four parts: a wind tunnel facility, a full-scaled aircraft test article with FCS hardware installed inside, a test article support rig with an aerodynamic balance installed on it, a real-time data acquisition and control system (short as RDAS). The wind tunnel facility generates the same airflow as that of the real flight; i.e. the airflow velocity, density, viscosity, and static pressure in the wind tunnel are the same as that in real flight. The full-scaled aircraft test article which is the same as that in real flight is put inside the wind tunnel. In this way, the aerodynamic/motion/control in this wind tunnel test can be consistent with the real flight. The test article support rig is used for driving the test article to rotate according to the computed angle of attack command so as to provide an equivalent attitude response. The RDAS is used for collecting data measured in the wind tunnel, processing data and transmitting processed data to the support rig and the controller in the wind tunnel.

Secondly, the working principle of the above system for testing the FCS is introduced here. Before a wind tunnel test, a pitch angle command should be preprogramed into the controller. At the beginning of a wind tunnel test, the RDAS should send an airflow command to the wind tunnel facility so as to generate a desired airflow. During the test, the IMU measures the angle of attack α

and the aerodynamic balance measures the lift force L, which are then sent to the RDAS. With the measured α and L, the RDAS computes the pitch angle feedback with equation (1) which is then sent to the controller. The controller combines the pitch angle command and the pitch angle feedback, and generates an elevator angle command for the servo which will then drive the elevator to deflect. Once after the elevator deflects, the balance measures the pitching moment M_z which is then sent to the RDAS. The RDAS use the measured M_z to compute a new angle of attack command for the support rig with equation (2). Then the support rig will drive the aircraft test article to the new angle of attack. And the RDAS will compute a new pitch angle feedback for the controller until the feedback pitch angle reaches to the pitch angle command.

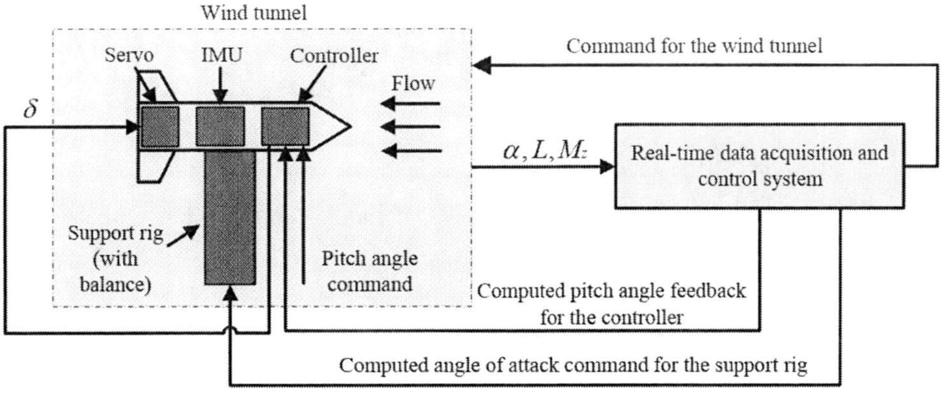

Figure 2. The wind tunnel based FCS test system scheme for the pitch control system.

$$\begin{cases} \dot{\theta} = \dfrac{T\sin\alpha + L}{mV} - \dfrac{g\cos\theta}{V} \\ \varphi = \theta + \alpha \end{cases} \tag{1}$$

$$\ddot{\alpha} = M_z / I_z \tag{2}$$

In the above equations (1~2), I_z is the pitching inertia moment, θ is the longitudinal velocity angle, T is the thrust (which is set as 0 in this paper), m is the aircraft mass, g is the local acceleration of gravity, V is the aircraft speed, φ is the pitch angle.

Then, according to the working principle of the test system introduced above, closed-loop or open-loop pitch control tests of the pitch control system can be performed. With the collected data from these tests, the pitch control performances can be determined.

To validate the above evaluation method, wind tunnel based FCS tests according to the evaluation method should be performed. In this paper, a hypersonic wind tunnel facility is used for the validation.

3 A hypersonic wind tunnel test system for validating the evaluation method

The hypersonic wind tunnel test system for validating the evaluation method includes a hypersonic wind tunnel facility, a full-scaled hypersonic aircraft test article with FCS hardware installed inside, a 5-degree-of-freedom abdominal support rig, a water cooling equipment installed on the support rig for cooling the rig, an aerodynamic balance installed on the rig, and a RDAS. The support rig can move on the angle of attack degree-of-freedom, the angle of sideslip degree-of-freedom, and three translational degree-of-freedoms as shown in Figure 3. As the rig is static during the tests, the rig should be adjusted to and fixed at an appropriate position before a wind tunnel test. The RDAS includes the aircraft data acquisition system (short as ADAS), the wind tunnel data acquisition and control system (short as WTDAS). The ADAS mainly collects the test data from the aircraft test article, such as the angle of attack and the elevator angle. The WTDAS mainly collects the test data from the wind tunnel test devices, such as the aerodynamic forces and moments, the airflow data.

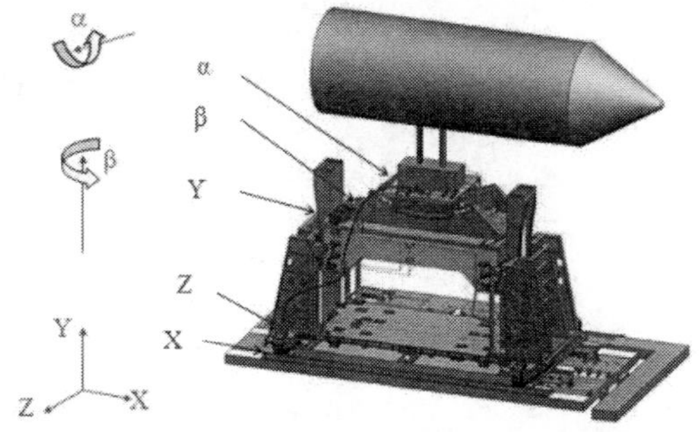

Figure 3. Sketch map of the 5-DOFs abdominal support rig in the hypersonic wind tunnel.

As the support rig cannot move during the tests, dynamic aircraft wind tunnel tests cannot be performed. Therefore, only the static open-loop pitch control test was performed to preliminarily validate the evaluation method. The open-loop pitch control test scheme is introduced as follows, including its test loop, pitch control law and test operation procedure.

4 The open-loop pitch control test based on the hypersonic wind tunnel system

4.1 The pitch control test loop

The open-loop pitch control test loop is shown in Figure 4. Before the test, the aircraft is fixed at a specific angle of attack α_i by the support rig which is set as 0 degree. And a control command α_c is preprogramed into the controller. The α_c is a sinusoidal input and equates to $2\sin(2\pi t / T_0)$. The time period T_0 is set as 3 seconds. As the support rig and the aircraft test article are static during the test, the RDAS just collects the elevator angle , L, and M_z, and doesn't feedback the computed data to the controller and the support rig.

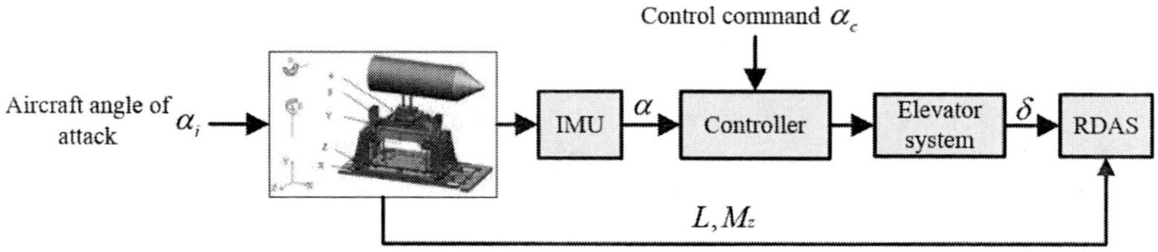

Figure 4. Sketch map of the open-loop pitch control test loop based on the hypersonic wind tunnel test system.

The collected elevator response is used to analyse and determine the elevator control performance of the pitch control system. And the collected L and M_z are used to predict the pitch angle with equations (1)~(2) and analyse the open-loop pitch control performance.

4.2 Pitch control law

4.2.1 The control law

The classic PID control method is used for pitch controller. The control law of the pitch controller [16] is:

$$\begin{cases} e(t) = \alpha_c(t) - \alpha_i \\ \delta(t) = K_p e(t) + K_i \int_0^t e(\tau) d\tau + K_d \dfrac{de(t)}{dt} \end{cases} \tag{3}$$

Thereinto, the PID control parameters are set as: $K_p = 2$, $K_i = 0.2$, $K_d = 1$.

4.2.2 Simulation validation of the control law

As the hypersonic wind tunnel test requires that the elevator angle should be less than 10 degree, the PID control law and the control parameters were validated by the mathematical simulation so as to ensure that the largest elevator angle is less than 10°. By inputting the values of α_i and α_c into the equation (3), the elevator response was computed. The computed largest elevator angle δ_{max} =5.63°<10°, which validate the PID control law and parameters.

4.3 Test operation procedure

The open-loop pitch control test was performed according to the following procedure:

(1) Adjust the position of the support rig and fix the aircraft test article at 0° angle of attack;
(2) Turn on the aircraft, and return the elevator angle to 0°;
(3) Turn on the wind tunnel, and let it generate airflow with the speed of 5Ma;
(4) After a while when the airflow gets steady, the elevator is controlled to deflect;
(5) After 6 seconds, the elevator stops deflecting;
(6) Turn off the wind tunnel, and finish the test.

5 Analysis of the test data and determination of the pitch control system performances

Based on the hypersonic wind tunnel test system in Section 3, and according to the test scheme in Section 4, the open-loop pitch control test was performed. The elevator response and the aerodynamic responses were obtained. The elevator response was used to determine the elevator control performance. If the elevator control performance is successfully determined, the feasibility of the wind tunnel based FCS evaluation method for evaluating the elevator control performance of a FCS can be validated. The aerodynamic responses were used to predict the pitch angle response and determine the open-loop pitch control performance. If the open-loop pitch control performance is successfully determined, the feasibility of the wind tunnel based FCS evaluation method for evaluating the open-loop pitch control performance of a FCS can be validated.

5.1 Analysis of the elevator response data and determination of the elevator control performance

δThe ADAS collected elevator response is shown in Figure 5. The collected elevator response curve is almost identical to the elevator command curve. But some error exists.

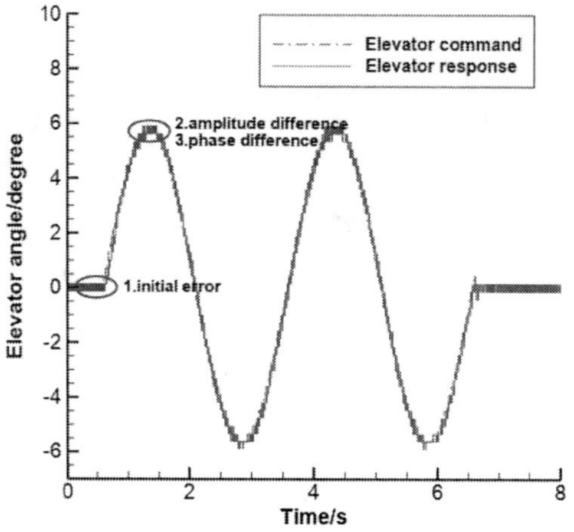

Figure 5. ADAS collected elevator response.

Firstly, before the elevator deflects, an initial error exists, which is mainly caused by the systemic errors, such as the elevator structure oscillation, the measuring error or the collecting error. The initial error is named as Δ_i. By subtracting the initial elevator command from the initial elevator response, the Δ_i can be obtained:

$$\Delta_i = 0.432° - 0.144° = 0.288°$$

Actually, such a noise error also always exists in the collected elevator response. Here, it is supposed that the collected elevator angles at all the time have the same noise error Δ_i. Therefore, when comparing the elevator command with the elevator response, the Δ_i should be subtracted from the elevator response.

Secondly, by comparing the elevator response and the elevator command, the amplitude difference A_δ and the phase difference T_δ can be obtained to determine the elevator control performance. By reading data from Figure 5, the maximum angle of the elevator response is 5.911°, the maximum angle of the elevator command is 5.63°, and then the amplitude difference is:

$$A_\delta = |5.911° - 5.63°| / 5.63° \times 100\% = 0.124\%$$

The time for the elevator response and the elevator command to reach to their maximum angles were chosen to compute their phase difference. The time for the elevator response to reach to its maximum angle 5.911° is 1.141s, the time for the elevator command to reach to its maximum angle 5.63° is 1.204s, and then the phase difference is:

$$T_\delta = |1.204\text{s} - 1.141\text{s}| = 0.063\text{s}$$

As the elevator control performance indexes A_δ and T_δ are not 0, the elevator response will have a difference with the elevator command. This elevator angle tracking error will lead to the attitude angle tracking error, which will then affect the attitude control performances. Specifically, the attitude control rapidity and accuracy will be reduced. But as the indexes A_δ and T_δ are very small and are quite allowable in the control system, it can be judged that the elevator response tracks the elevator command very well, which demonstrates that the pitch control system has a good elevator control performance. Through this successful determination of the elevator control performance, the feasibility of the wind tunnel based FCS evaluation method for evaluating the elevator control performance of a FCS was validated.

5.2 Analysis of the aerodynamic data and determination of the open-loop pitch control performance

5.2.1 Analysis of the aerodynamic data

The WTDAS collected aerodynamic response data including the lift force L and the pitching moment M_z is shown in Figure 6.

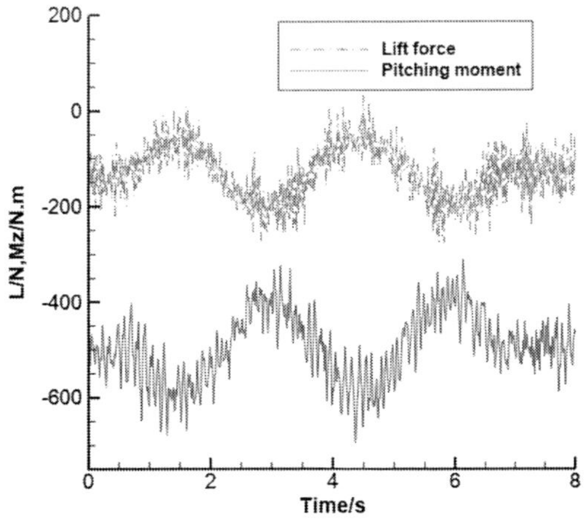

Figure 6. WTDAS collected aerodynamic responses.

The aerodynamic response data was analyzed as follows:

(1) The collected lift force response L curve in Figure 6 is just opposite to that of the elevator response in Figure 5, and the trend of M_z curve is correspondent to that of the elevator response. This shows that: within the deflection range of elevator angle, as the elevator angle changes, the lift force L will change negatively with the elevator angle and the pitching moment M_z will change positively with the elevator angle.

(2) As the collected lift force and pitching moment in Figure 6 contain obvious noise, data filtering is needed here. A low pass filter based on the FFT was used to filter these two data. The filtered aerodynamic response data without noise was shown in Figure 7.

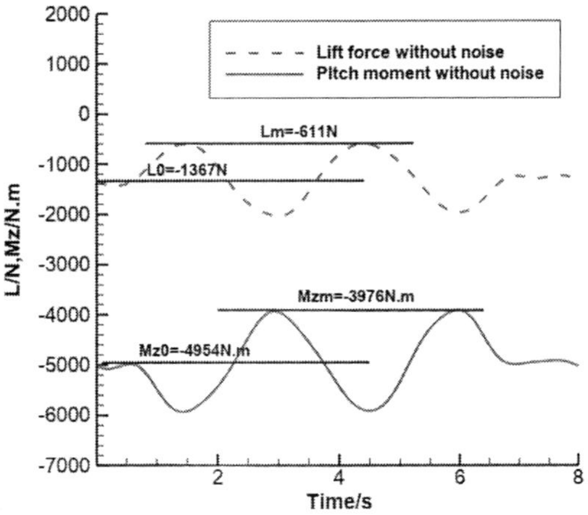

Figure 7. Aerodynamic responses without noise.

(3) For the convenience of computing the pitch angle response, the equations for computing the aerodynamic response should be constructed. For this purpose, firstly, the mean lines at the medium position and the maximum position of the aerodynamic responses were respectively drawn and their aerodynamic values were marked in Fig 8. Secondly, the lift force magnitude A_L and the pitching moment magnitude A_{Mz} were computed: $A_L = Lm - L_0$ =75.6N, $A_{Mz} = Mzm - Mz_0$ =97.8N.m. Thirdly, according to their initial values, magnitudes and time period T_0 (which is 3s), their standard aerodynamic response equations were constructed:

$$L = -136.7 + 75.6\sin(2\pi t / T_0) \tag{4}$$

$$M_z = -495.4 - 97.8\sin(2\pi t / T_0) \tag{5}$$

5.2.2 Prediction and analysis of the pitch angle response

By inputting the above aerodynamic equations (4)~(5) into the equations (1)~(2), the angle of attack and the pitch angle response was computed as shown in Figure 8.

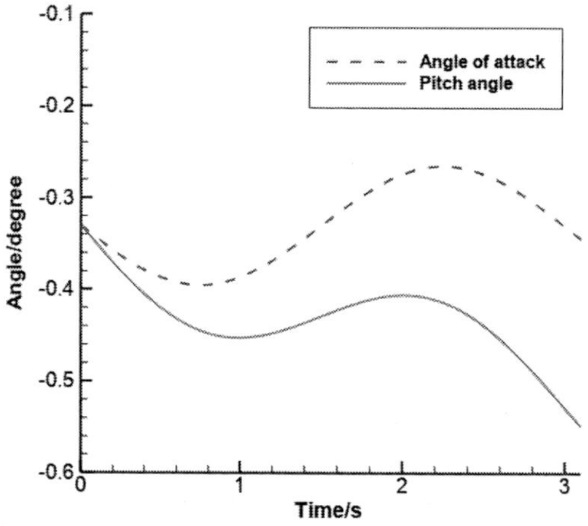

Figure 8. Predicted angle of attack and pitch angle responses.

From the curve, it can be clearly seen that the maximum pitch angle is far less than the maximum control command which is 2 degree, but the trend is consistent with the control command. As for why the pitch angle response is this small, the reason is that the aircraft angle of attack in this wind tunnel test is always 0 and the pitching moment is correspondingly very small which leads to small pitch angle response.

5.2.3 Determination of the open-loop pitch control performance

Although through the above test and data processing, the pitch angle was successfully predicted, it cannot be used to determine the open-loop pitch control performance. This is because the aircraft angle of attack in this test didn't change and the predicted pitch angle is not dynamically real and cannot be compared with the control command. Therefore, the current static support rig should be improved to a dynamic rig so as to perform dynamic wind tunnel based FCS tests, obtain the dynamic lift force and pitching moment and then the dynamically real pitch angle can be predicted. In this way, the open-loop pitch control performance and even the closed-loop pitch control performance can be determined.

To sum up the above analyses, the elevator control performance of the pitch control system was successfully determined, and the pitch angle response was successfully predicted. These validate the feasibility of part of the wind tunnel based FCS evaluation method, i.e. the feasibility of evaluating the elevator control performance and predicting the pitch angle of a FCS by this method. Although the open-loop pitch control performance was not determined, the test and data analysis work can lay foundations for future performing of more integrated tests and validating more of the evaluation method.

5 Conclusions

The wind tunnel based flight control system evaluation method can evaluate the FCS under real aerodynamics and rudder loadings, which make this method more realistic than the traditional hardware-in-the-loop simulation evaluation method. With a hypersonic wind tunnel test system, a static open-loop pitch control test of the pitch control system was performed and used to validate this method in this paper.

Through this test, the elevator response and the aerodynamic responses were collected. By analysing the test data, the elevator control performance of the pitch control system was successfully determined, and the pitch angle response was successfully predicted, which validate the feasibility of evaluating the elevator control performance by the wind tunnel based FCS evaluation method. As the support rig and the aircraft test article cannot move during the wind tunnel tests, the predicted pitch angle is not dynamically real and the open-loop pitch control performance was not determined. But the test and data analysis work in this paper can lay foundations for future performing of the dynamic open-loop or even closed-loop pitch control tests, successful determining the pitch control performances and fully validating the evaluation method.

References

1. Shan J Y, Meng X Y, Ding Y. Hardware in the loop simulation. Beijing: National Defence Industry Press, 2008. (in Chinese)
2. Li Z F. Special wind tunnel test techniques. Beijing: Aviation Industry Press, 2010. (in Chinese)
3. Li H F. Guidance and control techniques of hypersonic aircrafts. Beijing: Chinese aerospace press, 2012. (in Chinese)
4. Huang M, Wang Z W. A Review of Wind Tunnel Based Virtual Flight Testing Techniques for Evaluation of Flight Control Systems. International journal of aerospace engineering, 2015(1), 1-22.
5. Lawrence C, Mills B. Status update of the AEDC virtual flight testing development program. 40th AIAA Aerospace Sciences Meeting &Exhibit, AIAA 2002-0168, Reno, USA, 14-17 January, 2002.
6. Magill J C, Cataldi P, Morency J R, Hammer D X, Burgess R, Jeter E. Demonstration of a wire suspension for wind-tunnel virtual flight testing. Journal of Spacecraft and Rockets, 46(3): 624-633, 2009.
7. Papageorgiou G and Glover K. Two-degree-of-freedom control of an actively controlled wind-tunnel model. Journal of Guidance Control and Dynamics, 25(3): 510-516, 2002.
8. Richardson T S, Dubs A, Lowenberg M H, and Jones C D. Wind-tunnel testing of a dynamic state-feedback gain scheduled control system. Proceedings of the AIAA Guidance, Navigation and Control Conference and Exhibition, AIAA 2005-5976, San Francisco, California, USA, 15-18 August, 2005.
9. Sohi NP. Modeling of spin modes of supersonic aircraft in horizontal wind tunnel. Proceedings of the 24th International Congress of the Aeronautical Sciences, Yokohama, Japan, September, 2004.
10. Stenfelt G, Ringertz U. Yaw control of a tailless aircraft configuration. Journal of aircraft, 47(5): 1807-1810, 2010.
11. Strub G, Theodoulis S, Gassmann V, Dobre S. Pitch Axis Control for a Guided Projectile in a Wind Tunnel-based Hardware-In-the-Loop Setup. AIAA Modeling and Simulation Technologies Conference, AIAA 2015-0153, 5-9 January, 2015.

12. Zhao Z L, Wu J Q, Li H, Zhou W Q, Mao D Y, Yang H Y. Investigation of virtual flight testing technique based on 2.4m transonic wind tunnel. Acta Aeronautica et Astronautica Sinica, 37(2): 504-512, 2016. (in Chinese)
13. Nie B W, Zhu M H, Guo L L, Wen Y C, Jiang M. Key technique and design scheme of the wind tunnel virtual flight system. In: 2012 Chinese Guidance, Navigation and Control Conference, Beijing, China, 10-12 August, 2012.
14. Huang M, Wang Z W, Guo Z Y, Niu Y B. Design of the wind tunnel based virtual flight testing evaluation method for flight control systems. Proc IMechE Part G: Journal of Aerospace Engineering, 0(0): 1-13, 2016.
15. Huang M, Wang Z W, Guo Z Y, Zeng Q H, Niu Y B. Wind tunnel hardware-in-the-loop simulation techniques for flight control system evaluation. 2017 2nd International Conference on Control, Automation and Artificial Intelligence, 2017.
16. Huang Y R, Qu L G. Parameter adjusting and realizing of PID controllers. Beijing: Science press, 2010. (in Chinese)

Evaluation Abilities and Requirements of the Wind Tunnel Hardware-in-the-loop Simulation for Evaluation of Missiles' Flight Control Systems

Xing-bao Yang[a] and Min Huang
Dalian Naval Academy, Dalian, Liaoning, 116018, China

Abstract. In order to determine the flight control system (FCS) performances that wind tunnel hardware-in-the-loop simulation (WT-HILS) can evaluate, and lay foundations for designing the WT-HILS evaluation methods, the evaluation abilities and requirements of WT-HILS for FCS evaluation were analyzed. Firstly, the features of WT-HILS were pointed out. Then, according to the features of WT-HILS, what FCS performances WT-HILS can evaluate were analyzed, and the differences of evaluation abilities among the WT-HILS, the traditional hardware-in-the-loop simulation (HILS), and flight tests were compared. Lastly, from the aspects of testing, data processing, and FCS performance determination, the specific requirements of WT-HILS for evaluating FCS were analyzed.

NOMENCLATURE

φ, φ_i, φ_o, φ_m, φ_e : Pitch angle, pitch angle input command, pitch angle output, maximum value of the output pitch angle, steady-state value of the output pitch angle

e_{tr} : Steady-state error which is defined as the difference between φ_i and φ_e

t_s : Transient time, defined as the time required for the output pitch angle reaching to 98% or 102% of the φ_e

σ : Overshoot, defined as the difference between φ_m and φ_e

1 Introduction

Wind-tunnel hardware-in-the-loop simulation [1-2] was proposed aiming at improving the reality of the traditional hardware-in-the-loop simulation. By putting the aerodynamic objects in the wind tunnel airflow to perform rotational motions under the action of controllers, the aerodynamics and rotational motions can be reproduced realistically in WT-HILS. Compared to the traditional HILS using mathematical models to model the aerodynamics and motions, WT-HILS is closer to flight tests and can give more reliable FCS evaluation results. Based on this, many departments are developing such a

[a] Corresponding author : yxbjtxy@163.com

technique in an attempt to apply WT-HILS to evaluate FCS. Among these departments, some uses WT-HILS to evaluate the flight control laws with the method of rapid control prototyping (RCP) so as to early detect the FCS problems before the real FCS hardware is developed, such as the University of Cambridge [3] and the University of Bristol [4] in UK, the SibNIA [5] in Russia, the University of Texas [6] in USA, the China Academy of Aerospace Aerodynamics (CAAA) [7] and the China Aerodynamics Research and development center (CARDC) [8-10] in China. Some attempts to use WT-HILS to evaluate the hardware integrated FCS before flight tests and after HILS so as to detect those FCS problems that may occur in flight tests and are not detected in HILS and reduce the risks of flight tests, such as the Arnold Engineering Development Center (AEDC) [11] and the Physical Sciences Incorporation (PSI) [12] in USA.

The above researches mainly concentrated on the development of testing rigs and few researched the WT-HILS evaluation methods. However, the FCS evaluation methods are the key technology needed to be solved for applying WT-HILS to evaluate FCS [13-16]. Before designing the WT-HILS evaluation methods, what FCS performances can be evaluated by WT-HILS should be first known and specific requirements of the WT-HILS evaluation methods should be listed so as to indicate directions for the design of WT-HILS evaluation methods. Therefore, the evaluation abilities and the requirements of WT-HILS for the evaluation of missiles' FCSs were analysed in this paper.

2 WT-HILS features

Before researching the WT-HILS technique, features of WT-HILS should be first pointed out so as to propose practicable WT-HILS evaluation abilities and requirements of the WT-HILS evaluation methods. As the WT-HILS is based on wind tunnels, it has the following three obvious features:

(1) In WT-HILS, translational motions or the flight trajectory of the aircraft cannot be reproduced in wind tunnels during one test. In current wind tunnels, the support rigs can enable rapid change of the model positions. However, the space of the wind tunnel test section is limited, which makes the translational motions of the aircraft test articles are limited and the aircrafts cannot fly the whole flight trajectory as in real flights. In order to model the translational motions by WT-HILS, a common method is to calculate the translational motions in real-time with the measured data from wind tunnels.

(2) In WT-HILS, the amplitude of the aircraft speed cannot be changed in real time as in real flights during one test. During flight tests, the amplitude of aircraft speed may change rapidly. While in WT-HILS, the wind tunnel airflow speed cannot change as quickly as in real flights during one test at present. Besides, current wind tunnels are classified into wind tunnels with various range of speed, such as the low-speed wind tunnels, the transonic wind tunnels, the supersonic wind tunnels. There is not a wind tunnel that can generate an airflow that can vary from low speed to supersonic speed or even hypersonic speed. Therefore, it is quite hard for WT-HILS to reproduce the motions with speed change in real time at present. The FCS performance concerning speed changes will not be able to be evaluated by WT-HILS.

(3) In WT-HILS, the direction of the aircraft speed or the velocity directional angles of the aircraft cannot be changed in real time as in real flights during one test. In real flights, the attitude angles (the pitch angle and the yaw angle) consist of not only the aerodynamic angles but also the velocity directional angles. As the aerodynamic angles (the angle of attack and the angle of sideslip) are driven by the wind tunnel support rigs, aerodynamic angles can be reproduced in real time in the WT-HILS by moving the rig in real time or letting the aircraft freely rotate on the rig. While in current wind tunnels the wind tunnel airflow direction cannot change or cannot change as quickly as the change of the aircraft velocity direction in real flights, the aircraft velocity directions cannot be changed in real time in WT-HILS. To model the real attitude motions in real time, the velocity directional angles should be calculated in real time with the wind tunnel balance measured aerodynamic forces.

3 Evaluation abilities of WT-HILS

3.1 What FCS performances WT-HILS can evaluate?

Generally, a missile's FCS performances can be divided into the navigation/guidance performances, attitude stability and control performances. The navigation/guidance performances are used for representing the FCS performances concerning both translational motions and attitude motions. The indexes of navigation/guidance performances include velocity indexes, height indexes, taking off and landing indexes, endurance indexes, guidance precision and so on [17]. The attitude stability and control performances are used for representing FCS performances concerning attitude motions. The indexes of attitude stability and control performances include attitude stability indexes, attitude control indexes, and agility indexes and so on [17-18].

As introduced in the WT-HILS features, although the translational motions, the velocity directional angles and the attitude angles cannot be reproduced as in real flights in wind tunnels, velocity directional angles, attitude angles and translational motions can be calculated in real time with the measured data from wind tunnels. However, current wind tunnels generally cannot change airflow speed as in real flights and the influence of speed change can hardly be compensated by numerical methods, because once the airflow speed cannot be adjusted in real-time, the different airspeeds' aerodynamic loadings for making up the influence of airspeed change cannot be obtained. Therefore, before the wind tunnel airflow can be adjusted in real-time, the FCS performance involved with speed changes cannot be evaluated by WT-HILS. The following is a brief description of what FCS performance WT-HILS can evaluate.

(1) Navigation/guidance performances under steady flights. As in steady flights the translational accelerations are 0, navigation/guidance performances under such flight conditions can be evaluated by WT-HILS, such as the target tracking indexes under steady flights, the hovering radius and cycle indexes under steady hovers. While due to the limitation of speed change, the velocity indexes, taking off and landing indexes, guidance precision cannot be evaluated.

(2) Attitude stability performances. The WT-HILS can evaluate the stability performances under various kinds of attitudes, whether it is the normal motion or the unsteady motion. Besides, an important evaluation ability of WT-HILS is to evaluate the stability boundaries of FCS, which can also be called the FCS control limitations. When the aircraft's attitudes exceed the stability boundaries, the FCS will become invalid. Therefore, by obtaining the stability boundaries with WT-HILS, those manoeuvres out of the attitude stability boundaries can be avoided to perform in flight tests and the risks of flight tests will be reduced.

(3) Attitude control performances under normal motions. Under normal motion conditions (motions with small amplitude), the aircraft motions can be decoupled into longitudinal motions and lateral-directional motions. The longitudinal motions include the short-term mode and the long-term mode. The short-term mode is mainly relative to the angle of attack and the pitch angle, while the influence of speed change is small and can be neglected. The long-term mode is mainly relative to the velocity and flight path angle. The lateral-directional motions include the Dutch roll mode, roll mode and spiral mode. These modes are mainly relative to the roll angle, angle of sideslip, yaw angle, while the influence of speed change is usually small and can be neglected. Therefore, attitude control performances under longitudinal short-term motions and lateral-directional motions can be evaluated by WT-HILS, while those under longitudinal long-term motions cannot be evaluated by WT-HILS.

(4) Attitude control performances under unsteady manoeuvres. The unsteady manoeuvres are mainly relative to large angle of attack or large angle of sideslip, and have quite obvious nonlinear aerodynamic characteristics, which will result in strong couplings between aerodynamics and motions. To modelling these manoeuvres realistically, the WT-HILS rigs should permit free rotations or free manoeuvres so as to avoid the calculating of aerodynamic angles with measured moments and the aerodynamics and motions can counteract with each other as that of real flights. In this way, FCS performances under unsteady manoeuvres can be effectively evaluated by WT-HILS, while it is quite difficult for HILS to evaluate such performances. Some departments had developed rigs to evaluate the control algorithms of limit cycles [4], the free spin [5] and the pitching-rolling coupling motions [9].

(5) Heading agility indexes. Aircraft agility can be divided into the flight path agility and the heading agility. The flight path agility means "the ability to change the direction and amplitude of velocity with FCS". Apparently, it cannot be evaluated by WT-HILS. The heading agility means "the ability to change the head directions (such as pitch and yaw) with FCS". As the heading agility is mainly relative to the attitude motions and the manoeuvring time, it can be evaluated by WT-HILS.

Several sample indexes that WT-HILS can evaluate are listed in the Table 1. These indexes are used to represent the attitude tracking performance of the missile's pitching control system (shown in Figure 1), which belongs to the attitude control performances under normal motions.

Table 1. Sample indexes that WT-HILS can evaluate.

FCS performances	Indexes	Required range demonstration
Attitude tracking performance	e_{tr}	[0, 3%]
	σ	[0, 145%]
	t_s /s	(0, 1]

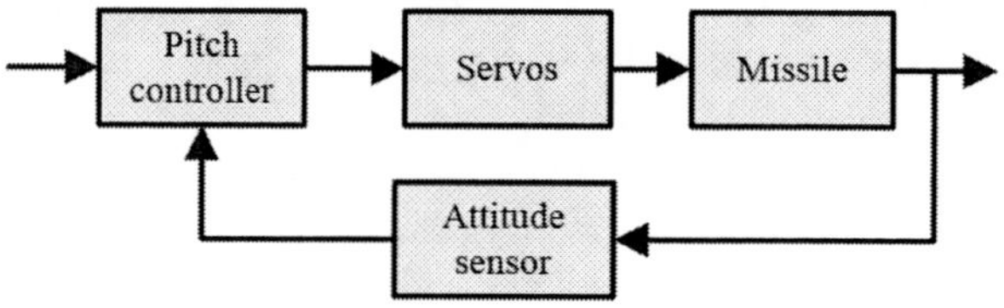

Figure 1. The missile's pitching control system.

3.2 Comparisons of the evaluation abilities among WT-HILS, HILS, flight tests

As analyzed above, WT-HILS cannot evaluate all indexes of FCS. The evaluation abilities among WT-HILS, HILS and flight tests are compared as follows.

3.2.1 Comparisons of the evaluation abilities between WT-HILS and HILS

By putting the hardware integrated aircraft in the wind tunnel airflow, WT-HILS can evaluate the FCS more realistically than HILS and can evaluate some FCS indexes that are difficult to be evaluated in HILS. Though less realistic than WT-HILS, some FCS performances can only be evaluated by HILS on ground.

(1) The WT-HILS can evaluate the navigation/guidance performances under steady flights, attitude stability and control performances under longitudinal short-term motions and lateral-directional motions, and heading agility indexes more realistically than HILS.

(2) The WT-HILS can evaluate attitude stability and control performances under unsteady maneuvers accurately which can hardly be evaluated by HILS because it is quite difficult for HILS to model unsteady aerodynamics and motions accurately.

(3) HILS can evaluate those FCS performances related to velocity change, such as the velocity indexes, taking off and landing indexes, guidance precision, and longitudinal long-term performances, while these cannot be evaluated by WT-HILS at present.

(4) HILS can evaluate the robust performances of FCS by performing a series of simulation tests in a short time with many kinds of combined disturbances added to various system parameters. While the robust performances of FCS can hardly be evaluated by WT-HILS because on the one hand, the disturbances, such as the center of gravity disturbance, the moment of inertia disturbance, and the wind disturbances are hard to add into the WT-HILS, on the other hand, it is difficult for WT-HILS to perform a lot of wind tunnel tests due to the high cost and long period of wind tunnel tests.

3.2.2 Comparisons of the evaluation abilities between WT-HILS and flight tests

(1) WT-HILS can evaluate FCS performances with almost zero risks only if the aircraft test article protection measures are taken. Therefore, the aircraft test article can be totally released so as to test the control limitations of FCS while it is of high risks and costs to do so in flight tests.

(2) Though WT-HILS is less risky, it cannot replace flight tests. On the one hand, the test environment in flight tests is completely real and the FCS evaluation results of flight tests are absolutely reliable. On the other hand, many FCS performances relating to the speed change cannot be evaluated by WT-HILS while flight tests can evaluate all the attitude and guidance control abilities during the whole flight trajectory.

4 Requirements of the WT-HILS evaluation methods

4.1 The technical content of the WT-HILS evaluation methods

The technical content of a WT-HILS evaluation method can be generally divided into three steps (shown in Figure 2): 1) aiming at the evaluated FCS indexes, design the WT-HILS test method, which should definite the test system scheme, and the test performing scheme (including the test types, test inputs and test procedures); 2) design the data processing method, which should transfer the measured data to the indexes; 3) design the performance determination method, which should give the index level criterions so as to determine the index levels and judge whether the indexes are satisfied or have enough performance levels.

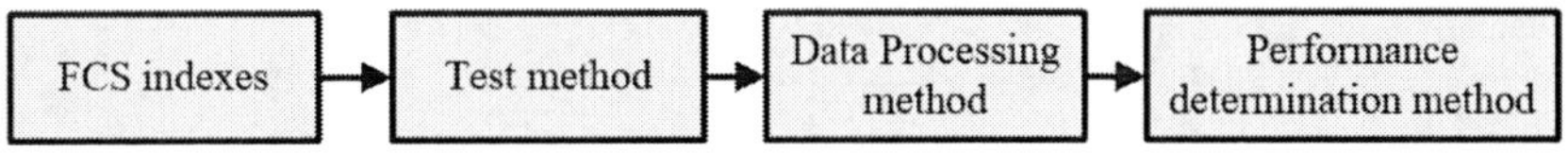

Figure 2. Procedures of a WT-HILS evaluation method.

4.2 Requirements of the WT-HILS evaluation methods

For the convenience of analysis, a missile's guidance control system (shown in Figure 3) is taken as the sample evaluation object here. The requirements of the WT-HILS evaluation methods for evaluating the guidance control system are introduced as follows.

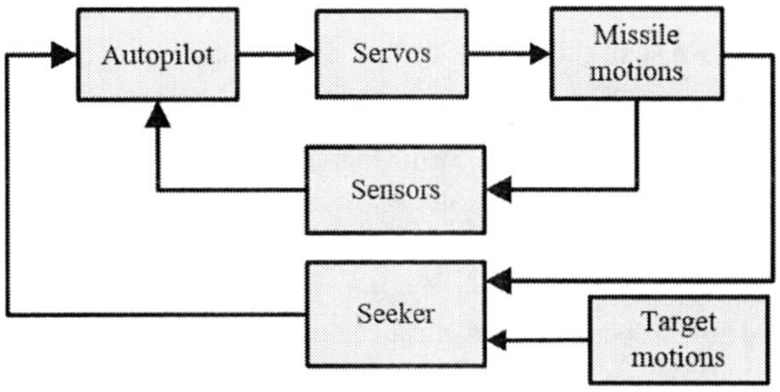

Figure 3. A missile's guidance control system.

4.2.1 Requirements of the test method

(1) Requirements of the WT-HILS test system. To let the wind tunnel airflow act on the real aircraft, the aircraft test article containing the FCS hardware should be put inside the wind tunnel test section. Due to the limitation of translational motions, attitude angles and translational motions should be calculated numerically with the measured aerodynamic forces and aerodynamic angles. The aerodynamic angles should be measured by the original gyro located inside the test article or other

angular sensors, and the aerodynamic forces should be measured by an aerodynamic balance. To permit rotations of the aircraft test articles, special model support rigs are needed. These rigs are different from the traditional ones. The traditional model support rigs are usually used to ensure 1-DOF free oscillations or predefined forced motions, while the WT-HILS rigs should allow 3-DOF (pitch, yaw, and roll) rotations by responding to the actions of servos. If some or all of the 3-DOF aircraft rotations need to be driven by the support rigs, aerodynamic moments should be measured by the balance in real-time to calculate the aerodynamic angle commands for the support rigs.

To provide the measuring environment for the linear accelerometer, an accelerometer simulator is needed to reproduce the linear accelerations. To provide the measuring environment for the seeker, a dark chamber containing the target simulator and turntables for simulating the missile/target relative information is needed. To ensure the time consistency of the running of real hardware and the numerical calculations, a real-time data acquisition and control system is needed to real-time collect data, process data, transmit signals and control devices (wind tunnel, model support rig, accelerometer simulator, target simulator, turntables). According to these requirements, a sample test system scheme for evaluating the missile's guidance control system is shown in Figure. 4.

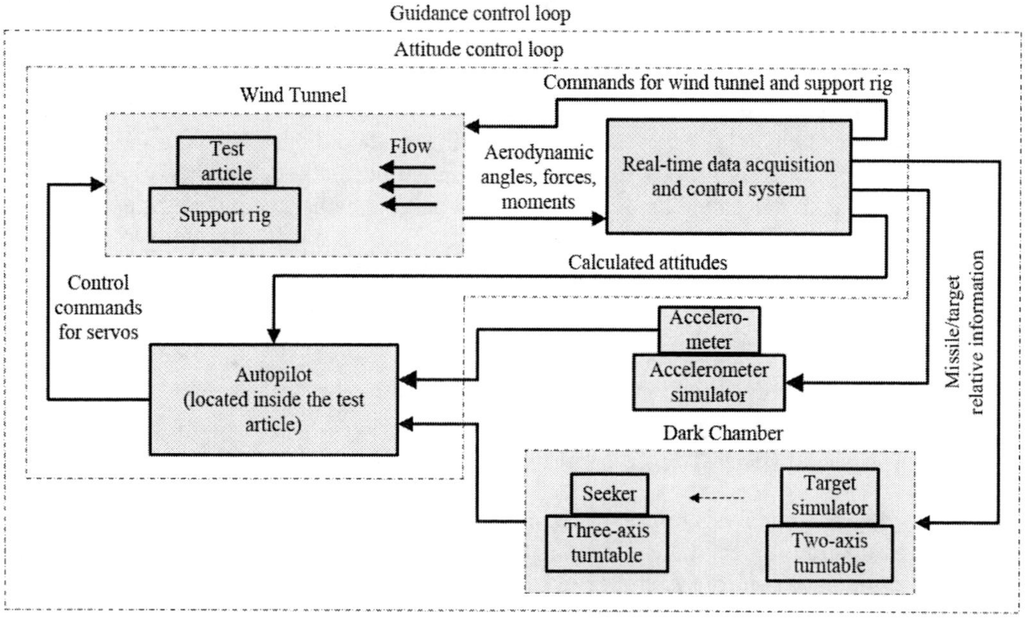

Figure 4. A sample WT-HILS system scheme for the missile's guidance control system.

(2) Requirements of the test performing. To different FCS performance indexes, tests of different types should be performed. To evaluate the attitude stability performance indexes, such as the attitude control limitation index, attitude stability tests should be obtained. The inputs for this kind of tests should be a gradually increased or gradually decreased attitude input command. To evaluate the attitude control performance indexes, such as the attitude tracking indexes in Table 1 and spin recovery indexes [5], attitude control tests should be performed. The inputs for this kind of tests should be the pre-input attitude commands first for driving the aircraft into a specific attitude condition, and then the desired attitude input commands for changing the specific attitude to the desired attitude. To evaluate the heading agility indexes, heading agility tests should be performed. The inputs for this kind of tests should be multistep attitude input commands so as to obtain the time for changing one attitude to another attitude. To evaluate the guidance performance indexes, which are the target tracking indexes for the FCS in Figure 3, guidance tests should be performed. The inputs for this kind of tests should be the target positions. Test procedures for guiding how to perform these tests should be specially designed towards different test types. For example, to evaluate the spin recovery indexes, what attitude input commands should be so as to reproduce spin motions, what desired

attitude input commands should be so as to recover normal motions from spin motions, how to apply these commands and how to record data should be given in the test procedure [5].

4.2.2 Requirements of the data processing method

After the above tests, various data can be obtained. Some data can be directly used as the indexes for performance determination, such as the flight performance parameters (aerodynamic angles, velocity directional angles, attitude angles, aircraft positions, overloading), while some have to be processed to the indexes, such as the indexes Table 1. To illustrate what a data processing method should be adopted, the indexes in Table 1 are taken as an example. After an attitude control test, an attitude tracking curve (a sample curve is shown in Figure 5) can be obtained. From the curve, the φ_m, φ_e, and t_s can be read. Then, the indexes σ and e_{tr} should be calculated with the following equations:

$$\sigma = |\varphi_m - \varphi_e| / \varphi_e \times 100\% \tag{1}$$

$$e_{tr} = |\varphi_e - \varphi_i| / \varphi_i \times 100\% \tag{2}$$

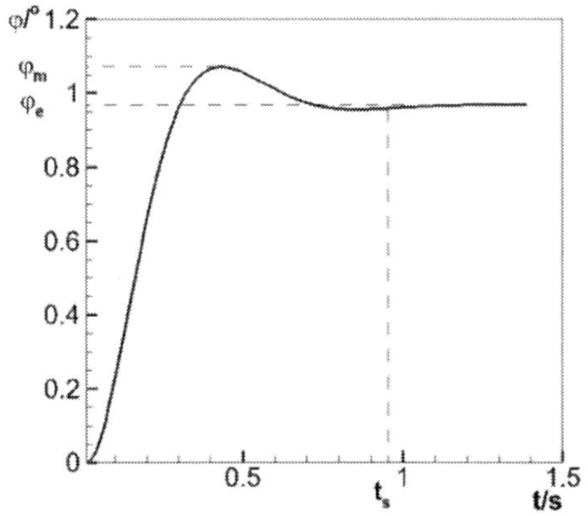

Figure 5. A sample attitude tracking curve.

4.2.3 Requirements of the performance determination method

After obtaining the indexes, performance determination methods are needed. These methods should give the index level criterions, which should specify different regions of performance levels allowed for the indexes. By comparing the indexes with their corresponding level regions, their performance levels can be determined, which will show how well the evaluated performances are.

Table 2. Sample level regions for the attitude tracking indexes in Table 1.

Performance indexes	**Level dividing**		
	Level 1 (Satisfying)	**Level 2 (Acceptable)**	**Level 3 (Controllable)**
e_{tr}	[0,1%]	(1%,2%]	(2%,3%]
t_s/s	(0,0.3]	(0.3,0.6]	(0.6,1]
σ	[0,45%]	(45%,100%]	(100%,145%]

The performance determination method for indexes in Table 1 is taken as an example here. The method sets the sample level regions of the attitude tracking indexes shown in Table 2. Supposing the value of e_{tr} is 0.7%. Comparing it with the region in Table 2, it is in the region of level 1. This shows the steady-state error of attitude tracking is satisfying.

5 Conclusions

The evaluation abilities and requirements of the WT-HILS for evaluation of FCS were analyzed and compared with the traditional HILS in this paper. The followings are summarizations.

(1) WT-HILS has three main features: limitation of aircraft translational motions in wind tunnels; limitation of wind tunnel airflow speed change; velocity directional angles (flight path angle, heading angle) cannot be reproduced in wind tunnels. Among the three features, velocity directional angles and translational motions can be numerically reconstructed in real-time with the data measured from wind tunnels, while the influence of speed change cannot be compensated numerically.

(2) Due to the limitation of airflow speed change, the WT-HILS evaluation abilities are limited: the guidance performances under steady flights, the stability and control performances under longitudinal short-term motions, lateral-directional motions and unsteady manoeuvres, and the heading agility indexes can be evaluated by WT-HILS, while the FCS performances relating to speed change, such as the guidance precision, longitudinal long-term control performances and flight path agility cannot be evaluated by WT-HILS. Then, the evaluation abilities among the WT-HILS, HILS and flight tests were compared.

(3) The technical content of a WT-HILS evaluation method includes the test method, the data processing method, and the performance determination method. According to the technical content, towards a sample missile's guidance control system, the requirements of the WT-HILS evaluation methods for this FCS were determined and examples of the test system schemes and the test performing schemes, the data processing methods, and the performance determination methods were given for illustrations.

References

1. Bacic M, Daniel R. Towards a low-cost hardware-in-the-loop simulator for free flight simulation of unmanned air vehicles. AIAA Modeling and Simulation Technologies Conference and Exhibit, AIAA 2005-6102, 2005.
2. Bacic M, Daniel R. Hardware-in-the-loop simulation of aerodynamic objects. AIAA Modeling and Simulation Technologies Conference and Exhibit, AIAA 2007-6465, 2007.
3. Papageorgiou G, Glover K. Two-degree-of-freedom control of an actively controlled wind-tunnel model. Journal of Guidance Control and Dynamics, 2002, 25(3): 510-516.
4. Richardson T S, Dubs A, Lowenberg M H, Jones C D. Wind-tunnel testing of a dynamic state-feedback gain scheduled control system. AIAA 2005-5976, 2005.
5. Sohi N P. Modeling of spin modes of supersonic aircraft in horizontal wind tunnel. 24th International Congress of the Aeronautical Sciences, 2004.
6. Gans N R, Dixon W E, Lind R, Kurdila A. A hardware in the loop simulation platform for vision-based control of unmanned air vehicles. Mechatronics, 2009, 19(2009): 1043-1056.
7. Hu J, Li Q. Primary investigation of the virtual flight testing techniques in wind tunnel. Journal of Experiments in Fluid Mechanics, 2010, 24(1): 95-99. (in Chinese)
8. Li H. Study on the similarity critiria and simulation method of the wind tunnel based virtual flight testing. PhD Thesis, China Aerodynamics Research and Development Center, 2012. (in Chinese)
9. Zhao Z L, Wu J Q, L H, Zhou W Q, Mao D Y, Yang H Y. Investigation of virtual flight testing technique based on 2.4m transonic wind tunnel. Acta Aeronautica et Astronautica Sinica, 2016, 37(2): 504-512. (in Chinese)

10. Nie B W, Zhu M H, Guo L L, Wen Y C, Jiang M. Key technique and design scheme of the wind tunnel virtual flight system. 2012 Chinese guidance, navigation and control conference (CGNCC2012), Beijing, 2012. (in Chinese)
11. Lawrence C, Mills B. Status update of the AEDC virtual flight testing development program. AIAA 2002-0168, 2002.
12. Magill J C, Cataldi P, Morency J R, Hammer D X, Burgess R, Jeter E. Demonstration of a wire suspension for wind-tunnel virtual flight testing. Journal of Spacecraft and Rockets, 2009, 46(3): 624-633.
13. Huang M, Wang Z W. A review of wind tunnel based virtual flight testing techniques for evaluation of flight control systems. International Journal of Aerospace Engineering, 2015(1), 1-22.
14. Huang M, Wang Z W, Guo Z Y, Niu Y B. Design of the wind tunnel based virtual flight testing evaluation method for flight control systems. Proc IMechE Part G: Journal of Aerospace Engineering, 2016, 0(0): 1-13.
15. Huang M, Wang Z W, Guo Z Y. Wind tunnel based virtual flight testing system and key technologies for the evaluation of flight control system. Journal of National University of Defense Technology, 2017, 39(2): 1-8. (in Chinese)
16. Huang M, Wang Z W, Guo Z Y, Zeng Q H, Niu Y B. Wind tunnel hardware-in-the-loop simulation techniques for flight control system evaluation. 2017 2nd International Conference on Control, Automation and Artificial Intelligence, 2017.
17. Zeng Q H, Guo Z Y. Unmanned flight control techniques and engineering. Beijing: National Defense Industry Press, 2011. (in Chinese)
18. Gao J Y, Li L Y, Feng Y C, etc. Aircraft flight qualities. Beijing: National Defense Industry Press, 2011. (in Chinese)

Research on LDPC Coding for Free Space Optical Communication with Modified Bessel-Gaussian Beams Based on K-distribution

Jin Wang[a], Fujiang Zeng, Qingqing Liang and Xiaofeng Zhang

China University of Geosciences, School of Mechanical Engineering and Electronic Information, Wuhan, China

Abstract. We investigate the performance of low-density parity-check (LDPC) coding for free space optical communication (FSO) with modified Bessel-Gaussian (MBG) beams based on K-distribution. Based on an extended Huygens–Fresnel principle, the receiver plane intensity is formulated and then we analyze the propagation characteristics of MBG beams traveling in a turbulent atmosphere. Numerical calculations show that the interaction of the transmission distance, refractive index structure constant, and wavelength will affect the imaginary term in the beam, allowing the beam to exhibit the properties of the MBG beams. To further improve the communication quality of the FSO system, we introduce a fast Quasi-Cyclic (QC) Low-Density Parity-Check (LDPC) code. Simulation results show that under weak and strong turbulence conditions, the performance of the QC-LDPC code becomes better and better as the code rate increases, and the bit error rate (BER) keeps decreasing as the code rate increases.

1 Introduction

Free Space Optical communication (FSO) is an optical communication technology that uses laser as an information carrier to propagate in free space to wirelessly transmit data for telecommunications or computer networking. FSO is characterized by ease of deployment, license-free, high bit rate, and immunity to electromagnetic interference, has attracted extensive research interests, and FSO is an important development direction of next-generation wireless communications [1,2]. Nevertheless, FSO are transmitted over atmospheric random channels, and atmospheric turbulence has a dispersion and absorption effect on the laser beam, this results in a weakened optical intensity and scintillation, eventually increased the bit error rate. In order to decrease the impact of atmospheric turbulence, we introduce the modified Bessel-Gaussian beams and the low-density parity-check (LDPC) code.

The modified Bessel-Gaussian beams has limited energy and is a physically achievable non-diffracting beams [3]. It can not only maintain its horizontal light intensity distribution within a limited propagation range, and it also has self-recovery performance when it is disturbed. A recent investigation indicates that Bessel beams possess a comparatively low scintillation index in all atmospheric regions when appropriate coherence parameters are selected.

In information theory, the low-density parity-check (LDPC) code is a linear error correcting code, a method of transmitting a message over a noisy transmission channel [4]. An LDPC is constructed using a sparse bipartite graph. LDPC codes are capacity-approaching codes, which means that

[a] Corresponding author : wangjin@cug.edu.cn

practical constructions exist that allow the noise threshold to be set very close (or even arbitrarily close on the binary erasure channel) to the theoretical maximum (the Shannon limit) for a symmetric memoryless channel.

In this paper, we discuss the effects of modified Bessel-Gaussian beams' parameters on FSO communication system. On the other hand, the performance of LDPC code for atmospheric turbulence channels are analyzed in this paper.

The remainder of this paper is organized as follows. In Section II, the modified Bessel-gaussian beams and the LDPC code is described in detail. Section III presents the numerical calculations and analysis. Finally, Section IV gives the conclusion of this paper.

2 Theoretical formulations

2.1 Modified Bessel-gaussian beams

The modified Bessel-Gaussian beams is an electromagnetic wave beams whose transverse electric field and irradiance distribution approximately satisfy the modified Bessel-Gaussian function. Modified Bessel-Gaussian beams is improved by Bessel beams and Gaussian beams. On a source plane whose radial and azimuthal coordinates are s and φ_s, the field of modified Bessel-Gaussian beams is given by [5]

$$u_s(s,\varphi_s)=I_n(a_B s)\exp\left(-k\alpha s^2\right)\exp\left(-jn\varphi_s\right) \tag{1}$$

Where the modified Bessel functions having order n, width parameter a_B, $k=2\pi/\lambda$ is the wave number with λ being the wavelength, $\alpha=1/\left(k\alpha^2\right)+j/2F_s$, where α_s and F_s, respectively, refer to radial Gaussian source size and focusing parameter, $j=\sqrt{-1}$.

Therefore, the formula for receiving light intensity at the receiving surface affected by the Bessel-Gaussian beam turbulence by the extended Huygens-Fresnel principle is [6]

$$\begin{aligned}&\left\langle I_r(r,\varphi_r)\right\rangle\\&=\left(\frac{k}{2\pi L}\right)^2\int_0^\infty\int_0^{2\pi}\int_0^\infty\int_0^{2\pi} ds_1 d\varphi_{1s} ds_2 d\varphi_{2s} u_s(s_1,\varphi_{1s})u_s^*(s_2,\varphi_{2s})s_1 s_2\\&\times\exp\left\{\frac{jk}{2L}\left[s_1^2-s_2^2-2rs_1\cos(\varphi_r-\varphi_{1s})+2rs_2\cos(\varphi_r-\varphi_{2s})\right]\right\}\\&\times\exp\left\{\left[-s_1^2-s_2^2+2s_1s_2\cos(\varphi_{1s}-\varphi_{2s})\right]/\rho_0^2\right\}\end{aligned} \tag{2}$$

where $*$ stands for the complex conjugate, ρ_0 measures the coherence length of a spherical wave propagating in the turbulent medium and will be given as $\rho_0=\left(0.545C_n^2k^2L\right)^{-3/5}$, C_n^2 is the refractive index structure constant indicating the turbulence strength.

When the beam diameter is larger than the turbulence, the intensity scintillation will be appeared. Under weak turbulent, the intensity scintillation is given by $\sigma_{\ln I}^2=0.496k^{7/6}L^{11/6}C_n^2$, and the strong turbulence is $\sigma_{\ln I}^2=1.23k^{7/6}L^{11/6}C_n^2$ [7].

2.2 Low-density parity-check (LDPC) code

In LDPC code, the minimum Hamming distance will increase linearly with the increase of the codeword length. Hence, the posteriori probability iterative decoding has the characteristic that the bit error rate decreases as the codeword length increases and Iterative decoding algorithm of LDPC code is close to the performance of Shannon limit [8,9]. LDPC codes can use random structure or algebraic method. The performance of random constructed LDPC code always is not ideal when using short code word length and the higher error flat layer would be easier to appear. The randomness of the structure would make it complex to decode. The algebraic structure of LDPC codes only exist in a specific code length and their bit rate at most of the time, have poor adaptability. In line with this, we applied a kind of LDPC code for fast encoding know as Quasi-cyclic LDPC Code (QC-LDPC). The QC-LDPC construction method is given as follows:

a) The rate is given as $R=(N-M)/N$, where N is the codeword length, M is information bit. It is important to note that the Matrix dimension is q=N/6, with prime number a and b is smaller than q.

b) The parity-check matrix H is constructed by sub-matrix: $H=[A \quad B]$

$$A=\begin{bmatrix} I_0 & I_a & I_{a^2} \\ 0 & I_0 & I_{a^2 b} \\ I_{b^2} & 0 & I_0 \end{bmatrix} \quad B=\begin{bmatrix} I_{a^3} & I_{a^4} & I_{a^5} \\ I_{a^3 b} & I_{a^4 b} & I_{a^5 b} \\ I_{a^3 b^2} & I_{a^4 b^2} & I_{a^5 b^2} \end{bmatrix} \tag{3}$$

Here, I_x consist of non-0 elements in A and B where $I_x = q \times q$ unit cyclic matrix which iterates each row of the identity matrix by x bits.

c) If A is singular, return to step 1 to re-select N. If A is non-singular, we can obtain the generator matrix G and a code named c_n that length is N by the characteristics of Parity Check Matrix.

$$G=\begin{bmatrix} A^{-1}\cdot \mathrm{B} & I_{N-M} \end{bmatrix}; c_n = u_n \cdot G \tag{4}$$

In the equation, $u_n = \begin{bmatrix} u_{n,0} & u_{n,1} & \cdots & U_{n,\mathrm{M}-1} \end{bmatrix}$ is a randomly selected information bit vector. The decoding algorithm according to the method in the literature [10].

3 Numerical calculations and analysis

In this section, we calculate the propagation properties of the modified Bessel-Gaussian beams in atmospheric turbulence and analyze the influence of main parameters on the received optical intensity. At the end, we analyzed the performance of LDPC code in FSO.

3.1 Optical intensity distribution

3.1.1 Source intensity

Initially we start with source intensity graphs. Fig 1 shows source plane intensity of modified Bessel-Gaussian beams at various orders. From fig 1, it is seen that higher order modified Bessel-Gaussian beams, that is $n>0$, will have well-like shape whose circumference is enlarged with increasing order. From fig 2, it shows that raising the Bessel width parameter a_B will have a similar role.

λ=1550$nm, \alpha_B = 200m^{-1}, \alpha_0 = 2.0cm, n = 0,1,2,5$

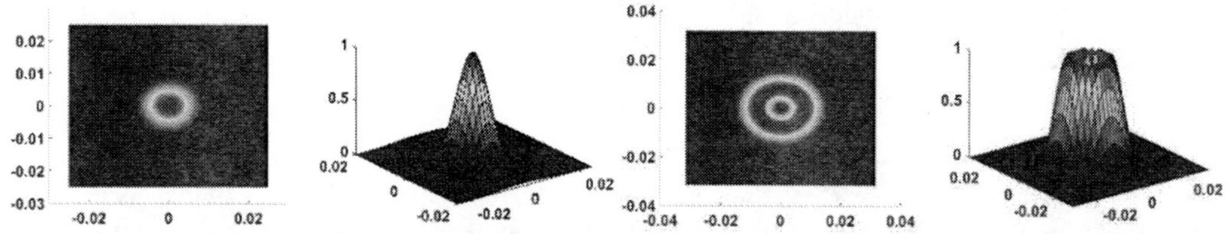

$n=0$ $n=1$

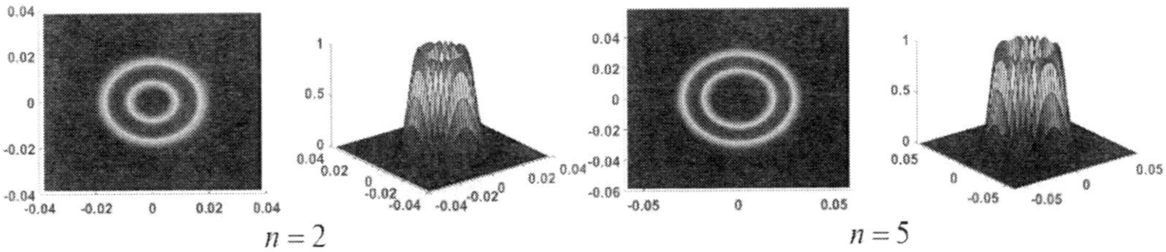

Figure 1. Source plane intensity pictures of modified Bessel-Gaussian beams at various orders.

$$\lambda=1550, n=1, \alpha_0=2.0cm, a_B=500,400,200,100m^{-1}$$

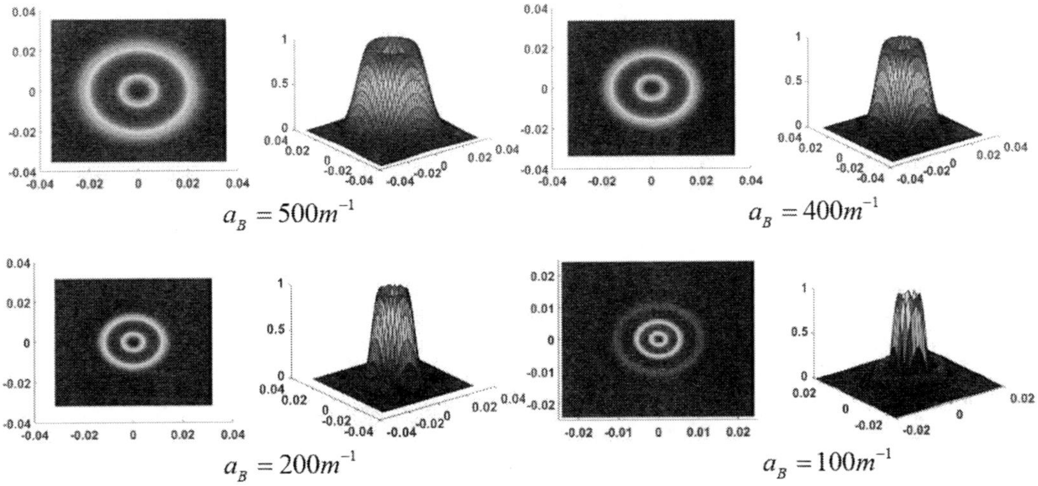

Figure 2. Source plane intensity pictures of modified Bessel-Gaussian beams at various width parameter.

3.1.2 Ideal receiver intensity

Now, receiver intensity graphs shall be offered. It is main describes the ideal receiver intensity in this section. Figure 3 displays the progress of modified Bessel-Gaussian beams along the propagation axis.

$$\lambda=1550, n=0, a_B=250m^{-1}, C_n^2=1\times10^{-15}m^{-2/3}$$

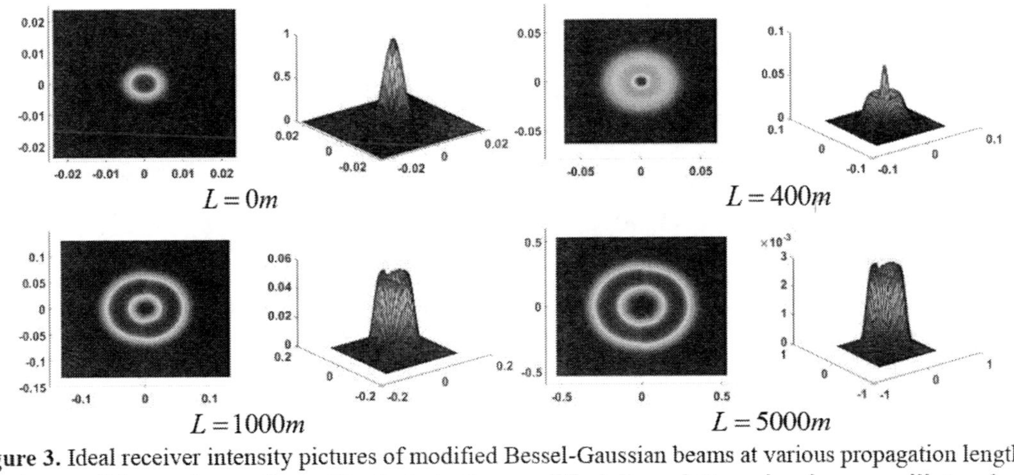

Figure 3. Ideal receiver intensity pictures of modified Bessel-Gaussian beams at various propagation lengths.

It is possible to detect from Fig 3 that, the modified Bessel-Gaussian beams will too lead to Gaussian beams after propagation in turbulent atmosphere.

Fig 4 exhibits ideal receiver intensity of modified Bessel-Gaussian beams at various C_n^2 values. Looking at Fig 4, it is possible to deduce that increasing the turbulence strength simply accelerates the process toward the eventual beam shape, thus a Gaussian beam is attained earlier.

$$\lambda = 1550, a_B = 200m^{-1}, L = 1km, n = 2$$

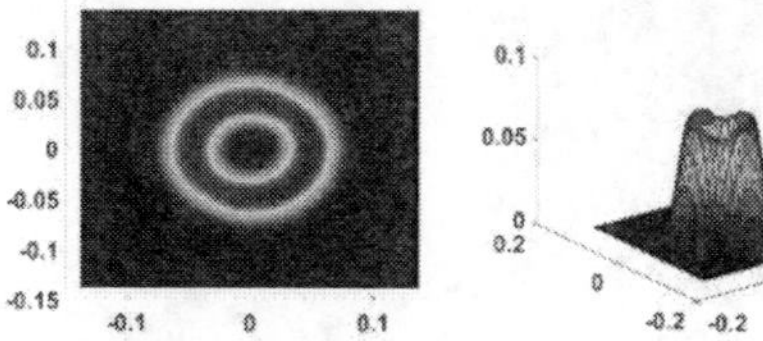

$C_n^2 = 1\times10^{-15}m^{-2/3}$

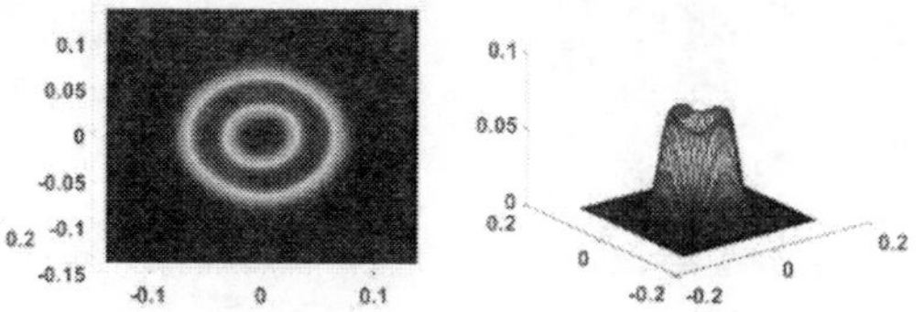

$C_n^2 = 1\times10^{-14}m^{-2/3}$

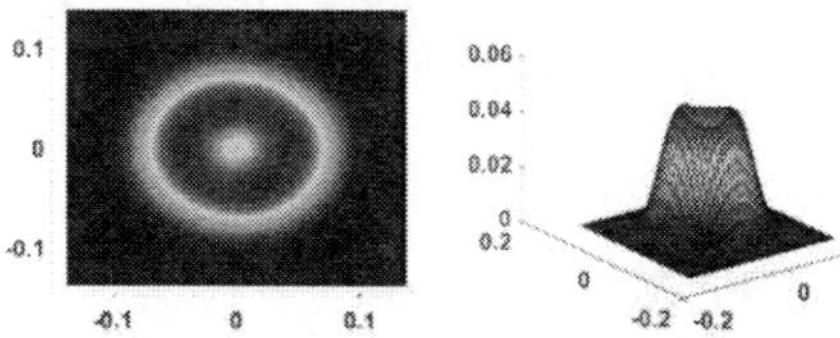

$C_n^2 = 1\times10^{-13}m^{-2/3}$

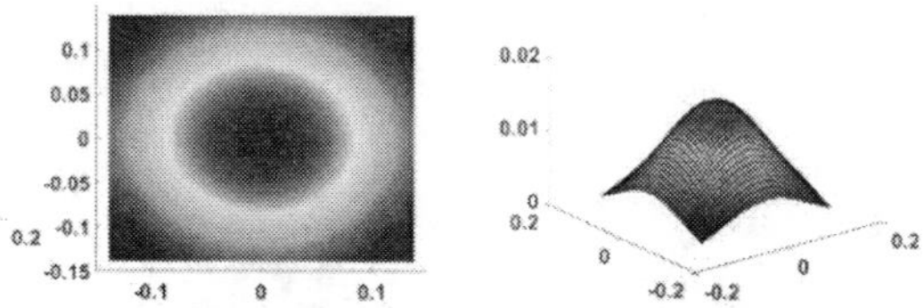

$C_n^2 = 1\times10^{-12}m^{-2/3}$

Figure 4. Ideal receiver intensity pictures of modified Bessel-Gaussian beams at various C_n^2 values.

$$a_B = 200m^{-1}, L = 1km, n = 2, C_n^2 = 1\times10^{-15}m^{-2/3}$$

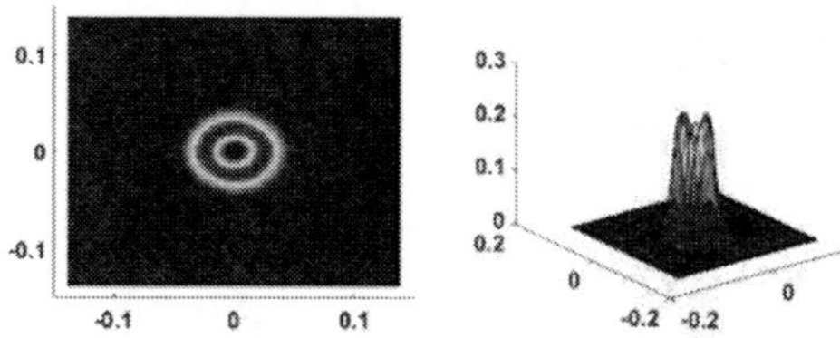

$\lambda = 800nm$

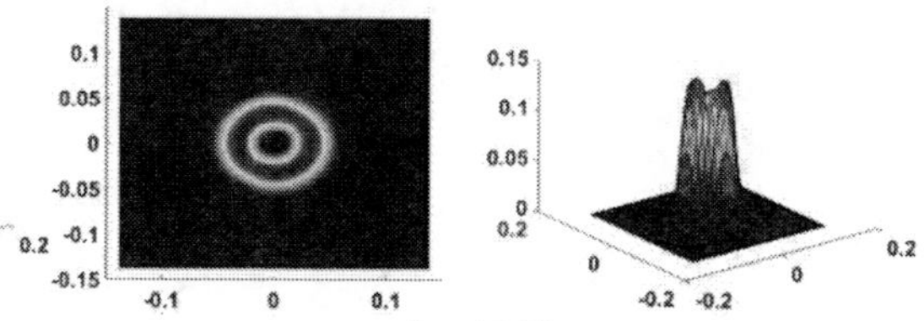

$\lambda = 1060nm$

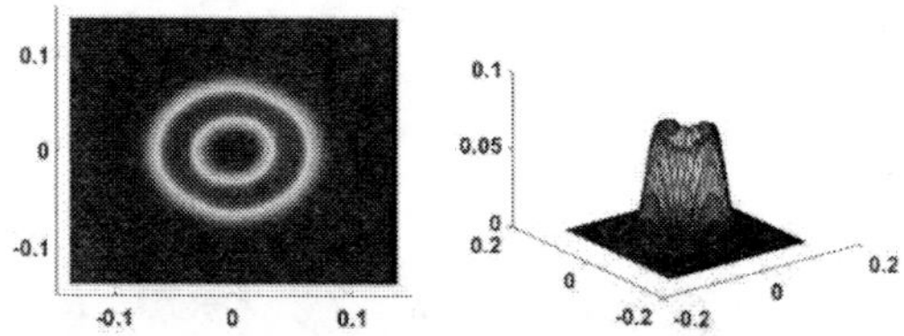

$\lambda = 1550nm$

Figure 5. Ideal receiver intensity pictures of modified Bessel-Gaussian beams at various wave lengths.

Next ideal receiver intensity of modified Bessel-Gaussian beams at various wavelengths are examined. Fig 5 shows that as wavelength increases, the optical intensity was improved.

3.1.3 Actual receiver intensity

Now, the actual receiver intensity graphs that consider the scintillation shall be offered. Fig 6 exhibits actual receiver intensity pictures of modified Bessel-Gaussian beams at various wave lengths. From Fig 6, know that increase in wavelength will develop the intensity.

$$a_B = 200m^{-1}, L = 1km, n = 2, C_n^2 = 1\times10^{-15}m^{-2/3}$$

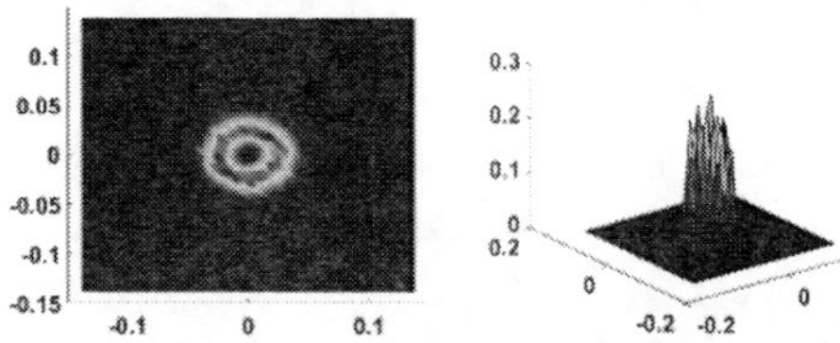

$\lambda = 800nm$

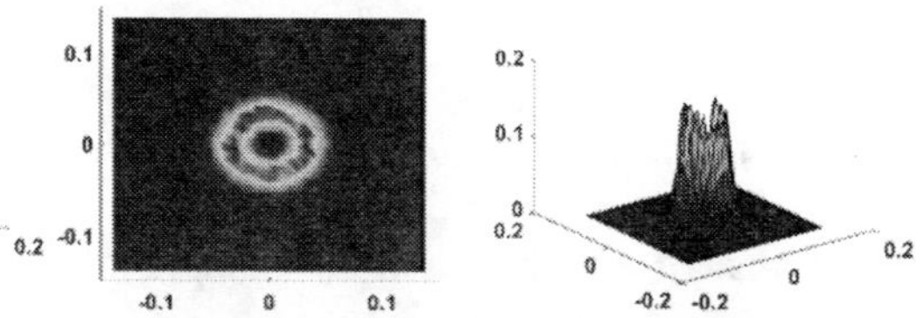

$\lambda = 1060nm$

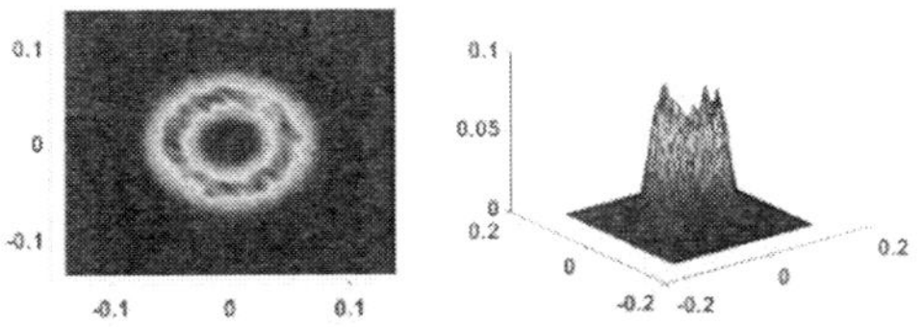

$\lambda = 1550nm$

Figure 6. Actual receiver intensity pictures of modified Bessel-Gaussian beams at various wave lengths.

$$\lambda\text{=}1550, a_B = 200m^{-1}, L = 1km, n = 2$$

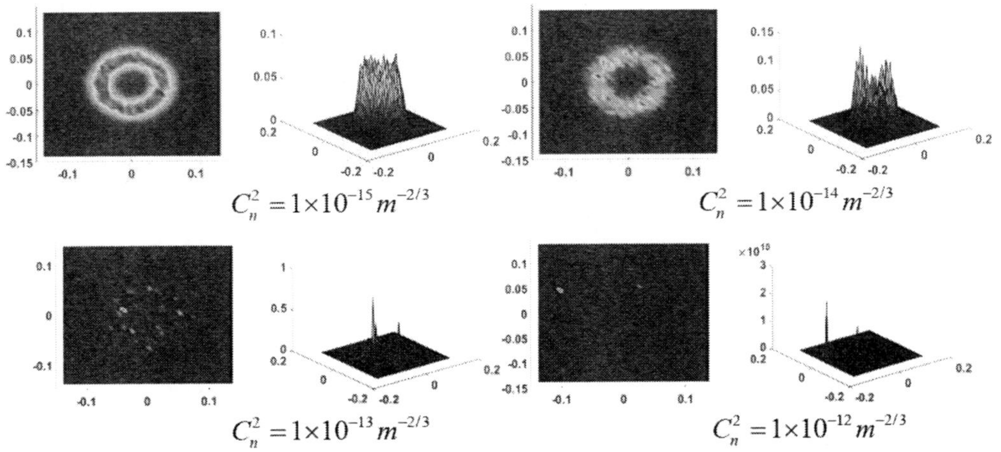

Figure 7. Actual receiver intensity pictures of modified Bessel-Gaussian beams at various C_n^2 values.

Fig 7 shows the actual receiver intensity of modified Bessel-Gaussian beams at various C_n^2 values. It is seen that, in the case of weak turbulence, the intensity distribution is very regular. Under the strong turbulence of $C_n^2 = 1\times10^{-12}$, the receiver intensity has lost its shape, and the optical beams become unrecognizable, which has caused FSO communication interruption.

$$\lambda\text{=}1550, n = 0, a_B = 250m^{-1}, C_n^2 = 1\times10^{-15}m^{-2/3}$$

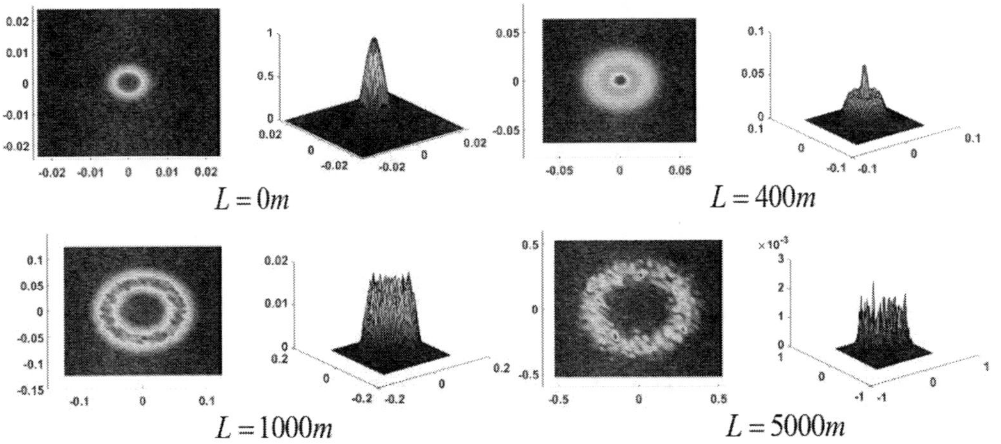

Figure 8. Actual receiver intensity of modified Bessel-Gaussian beams at various propagation lengths.

Fig 8 is the actual receiver intensity of modified Bessel-Gaussian beams at various propagation lengths. It shows that, the received beam is in a great shape and the intensity fluctuation is small when in a short propagation length. With the increase of transmission distance, the optical intensity becomes smaller and fluctuations increase.

3.2 LDPC code performance analysis

When the modified Bessel-Gaussian beam is transmitted into the atmosphere, the presence of atmospheric turbulence would have an impact on the FSO system. Hence, we introduce the bit error rate(BER) performance of LDPC codes under turbulence in this section.

Fig 9 shows the effect of bit rate on performance of the QC-LDPC code. From Fig 9, we can know that with the increase of the code rate, the error rate of the QC-LDPC code constantly decreases, indicating that the performance of the QC-LDPC code becomes better and better.

Then we give out the influence of iteration times on performance of QC-LDPC codes. Fig 10 shows that increasing the number of decoding iterations, the performance of bit error rate reduction QC-LDPC code will be improved. However, when the number of iterations is large enough, only the delay and complexity of the system is increased as it increased, while the performance of the QC-LDPC code will no longer improve.

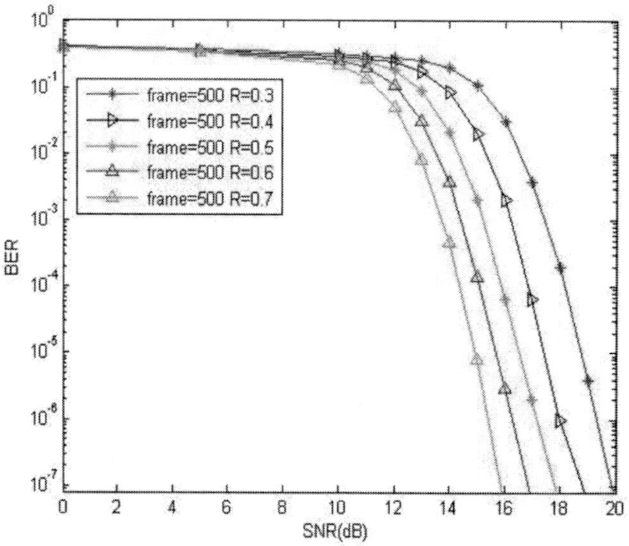

Figure 9. Effect of Bit Rate on Performance of QC-LDPC code

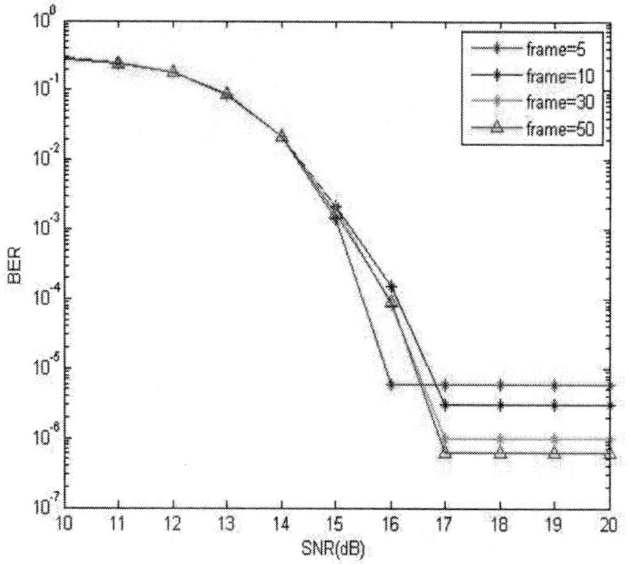

Figure 10. Influence of iteration times on performance of QC-LDPC.

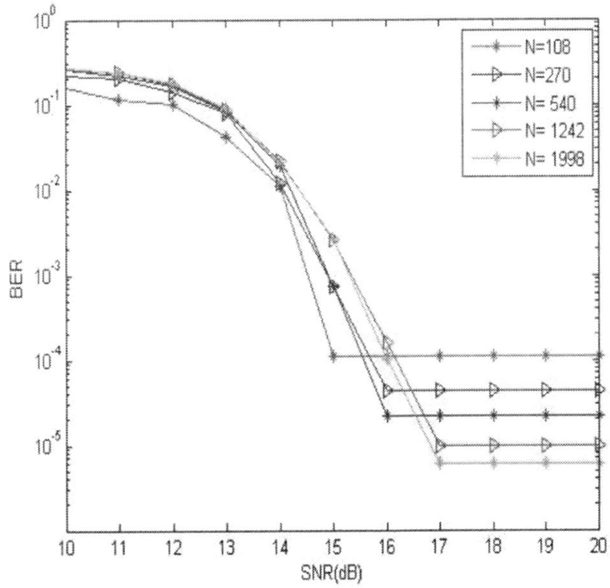

Figure 11. Influence of Code Length on Performance of QC-LDPC.

From the Fig 11, it is seen that the long code error performance of QC-LDPC code is better than that of short code. The code performance under a certain code length has a certain limit. As the code length increases, the complexity of encoding and decoding also increases, and the coding performance will be closer to the limit, and its performance will improve relatively as the code length increases.

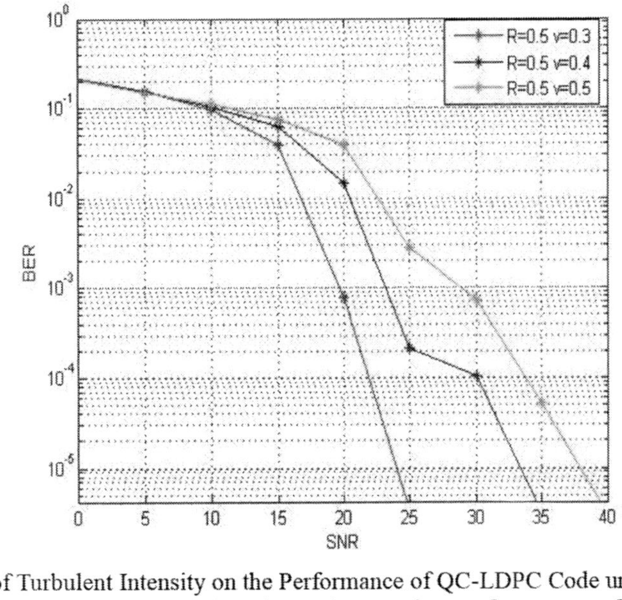

Figure 12. The Effect of Turbulent Intensity on the Performance of QC-LDPC Code under Weak Turbulence

Then we exhibit the effect of turbulent intensity on the performance of QC-LDPC Code under weak turbulence. Fig 12 shows that as the turbulence intensity increases, the bit error rate continues to increase and the performance of the QC-LDPC code becomes worse.

Now, the effect of turbulent intensity on the performance of QC-LDPC code under strong turbulence condition shall be offered. It is known that under certain conditions, the smaller the BER, the better the QC-LDPC code performance as the turbulence intensity decreases from the Fig 13.

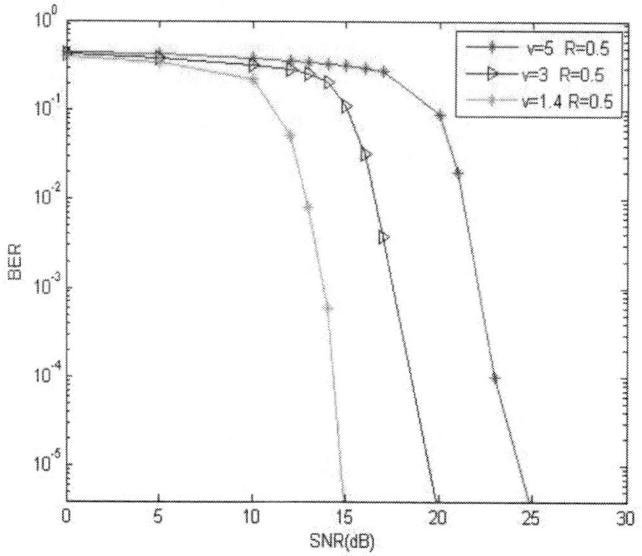

Figure 13. The Effect of Turbulent Intensity on the Performance of QC-LDPC Code under Strong Turbulence

4 Conclusions

For FSO communication system, we propose modified Bessel-Gaussian beams and LDPC codes to reduce the influence of atmospheric turbulence. We conclude that the order affects the Bessel part of the beam and changes the beam profile and outer ring, and that width parameter will affect the weights occupied by the Gaussian and Bessel parts of the beam. When the width increases, the beam will be biased towards the Bessel beam's shape and characteristics, whereas the width will be biased toward the Gaussian beam. In addition, increase in the transmission distance and refractive index structure constants will make the intensity of light scintillation more intense, both of which have an influence on the beam profile and shape. Under weak turbulence, as wavelength increases, it will improve the light intensity fluctuation. Finally, we find out that the interaction of the transmission distance, refractive index structure constant, and wavelength will affect the imaginary term in the beam, allowing the beam to exhibit the properties of the modified Bessel-Gauss beam. For the LDPC codes, we discover that under weak and strong turbulence conditions, the performance of the QC-LDPC code becomes better and better as the code rate increases, and the bit error rate keeps decreasing as the code rate increases.

Reference

1. H. Tian, L. Guo, P. Wang, et al. Optics Express, **25**, 19995-20011 (2017).
2. K. Davaslioglu, E. ÇAğlral, M. Koca, Optics Express, **18**, 16618-27(2010).
3. H.T. Eyyuboğlu, F. Hardalaç, Optics & Laser Technology, **40**, 343-351 (2008).
4. Myung S, Yang K, Kim J. IEEE Trans.inf.theory, **51**, 2894-2901 (2005).
5. Y. Zhu, Int. J. Mod Phys B, **30**, 1650193 (2016).
6. H.T. Eyyuboğlu, Appl. Phys. B, **88**, 259-265 (2007).
7. L.C. Andrews, R.L. Phillips. Bellingham: SPIE Press, (1998).
8. D.J.C. MacKay, IEEE Trans. on Information Theory, **45**, 399-431(1999).
9. Ivan B. Djordjevic, OPT. COMMUN. NETW, **2**, (2010).
10. J. Wang, C.P. Udeh, Free-Space Laser Communication and Atmospheric Propagation XXIX. International Society for Optics and Photonics,1009609. (2017).

Modeling and Simulation of Mission Planning Problem for Remote Sensing Satellite Imaging

Yao Pan[1,a], Zhong Ming Chi[1], Qi Long Rao[1], Kai Peng Sun[1] and Yi Nan Liu[1]

[1]*Shanghai Institute of Satellite Engineering, Shanghai 201109, China*

Abstract. Mission planning problem for remote sensing satellite imaging is studied. Firstly, the time constraint satisfaction problem model is presented after analyzing the characteristic of time constraint. Then, An optimal path searching algorithm based on the discrete time window is proposed according to the non-uniqueness for satellite to mission in the visible time window. Simulation results verify the efficiency of the model and algorithm.

1. Introduction

Remote sensing satellite can receive the information from the Earth sent by satellite borne senor in space, which have advantages of no constraint on any country, wide coverage, etc. They are widely applied in military reconnaissance, resources exploration, weather forecast and so forth. How to fulfill maximum imaging missions cooperatively through the effective strategy is the current problem of mission planning for remote sensing satellite[1].

The manoeuvrability of rolling and pitching can extend the visible time window for satellite to target compared with the satellite just rolling[2], and the imaging time window becomes a continuous selectivity variable. To tackle the problem of mission planning for remote sensing satellite , the characteristic of time constraint is analyzed. The time constraint satisfaction problem model is presented, and an optimal path searching algorithm is proposed.

2. Analysis of time constraint

Basic concepts in connection with time constraint are defined as follows:

- **Definition 1** Visible time window means the time window visible for satellite to target all the time. The Euler Angle for satellite to target is bijective with time in visible time window.
- **Definition 2** Imaging time window means the time window imaging for satellite to target. The Euler Angle for satellite to target is invariant in imaging time window.

Figure 1 shows the visible time window and imaging time window for satellite to target.

There are many similar constraints between satellite rolling and pitching and satellite just rolling, such as rolling angle constraint, pitching angle constraint, visible time window constraint and so forth[3-4]. No more details here. In this paper, time constraint is the main constraint to be discussed[5].

The characteristics of time constraint for remote sensing satellite mission planning problem are as follows:

---

[a] Corresponding author e-mail： panyaopaper@outlook.com

- The non-uniqueness of imaging time window for satellite to target. The imaging time window can be confirmed by imaging start time and imaging duration, and the imaging start time for satellite to target can be selected from the visible time window randomly.
- The non-uniqueness of imaging Euler Angle for satellite to target. The Euler Angle for satellite to target is changed with time in the visible time window, while it is invariant in the imaging time window. All illustrated in figure 1.
- The non-uniqueness of satellite attitude transition time between neighbouring targets. The attitude transition time between neighbouring targets can be confirmed by the imaging Euler Angles.

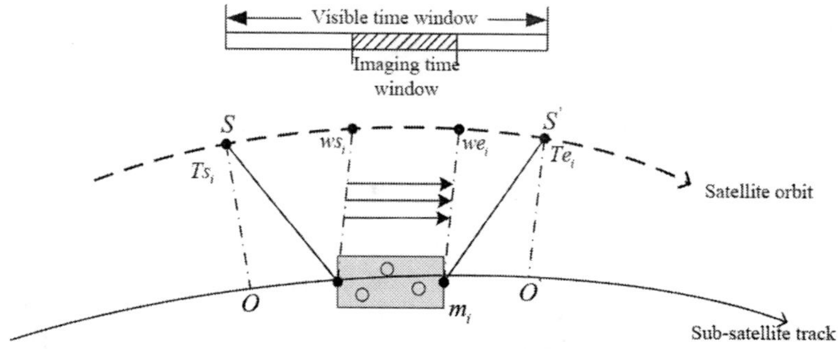

Figure 1. The visible time window and imaging time window for satellite to target

To confirm the imaging time window and imaging Euler Angle is the key problem in this article, and the imaging time window can be determined by imaging duration, while the imaging Euler Angle can be determined by imaging start time. So the problem is transformed into determining the imaging duration and imaging start time.

3. Time constraint satisfaction problem model

Time constraint satisfaction problem can be indicated by $\langle V, D, C \rangle$ triples[6], where V is the set of imaging time windows for targets, D is the set of visible time windows for targets, and C is the set of constraint conditions, including time window constraint, drift angle adjusting time constraint, maximum imaging time constraint in single orbit, and attitude transition time constraint.

In the remote sensing satellite mission planning problem, the variables are involved as follows:

- The visible time window is given by $TW_i = [Ts_i, Te_i]$ for mission i, where Ts_i is the visible start time and Te_i is the visible end time.
- The imaging time window is given by $tw_i = [ws_i, we_i]$ for mission i, where ws_i is the imaging start time and we_i is the imaging end time.
- The imaging duration is assumed by $Tcon_i$ for mission i.
- The drift angle adjusting time is treated as a constant T_Q, and the maximum imaging time in single orbit is assumed by $TIMG_{\max}$.
- The maneuver rate of rolling and pitching are denoted as ϑ and ρ separately.

Based on the above analysis of the problem, this paper establishes the time constraint satisfaction problem model (TCSPM) for remote sensing satellite mission planning, the mathematical description of the model and explanation are as follows:

- E is the object of the optimization, which means to minimize the total energy consumed by satellite during the mission planning, including the energy consumed by maneuver and imaging. That is,

$$E = \min(T_{moving} \times P_{moving} + T_{imaging} \times P_{imaging}) \tag{1}$$

where T_{moving} means the total time of attitude maneuver, $T_{imaging}$ means the total time of imaging, and P_{moving} and $P_{imaging}$ are average power of maneuver and imaging.

- The constraint conditions consisting of equation (2)-(5):

$$\forall i \in [1, n], Ts_i \le ws_i \le we_i \le Te_i \tag{2}$$

$$T_Q = constant \tag{3}$$

$$T_{imaging} = \sum_{i=1}^{n} Tcon_i \le TIMG_{\max} \tag{4}$$

$$we_i + \max\left\{\frac{|\beta_i - \beta_j|}{\vartheta} + \frac{|q_i - q_j|}{\rho}\right\} + t_s \le ws_j - T_Q \tag{5}$$

where β_i and β_j are the rolling angles of mission i and j, q_i and q_j are the pitching angles of mission i and j, t_s is the attitude stabilization time.

Equation (2) means the imaging time window for satellite to mission should be contained in the visible time window; Equation (3) shows the drift angle adjusting time is set to a constant[7]; Equation (4) is to ensure that the total imaging time of the satellite in single orbit could not exceed the upper limit; Equation (5) shows the required attitude transition time between neighbouring targets[8-9] (see figure 2).

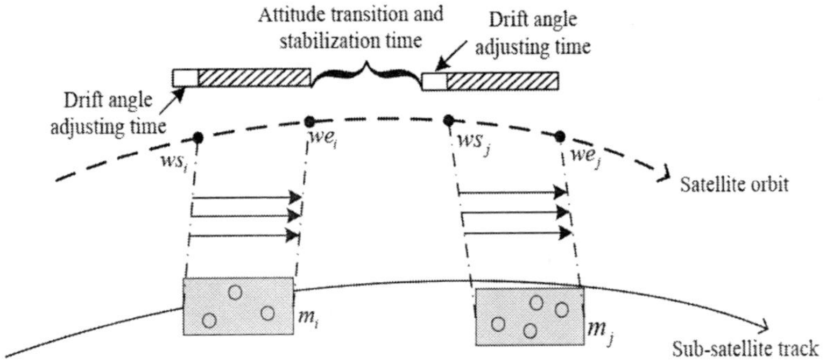

Figure 2. The constraint of attitude transition time for satellite

4. Optimal path searching algorithm

An optimal path searching algorithm based on the discrete time window is designed in this paper. Firstly, the visible time windows for planning missions need to be discrete according to certain fraction, generating a plurality of time points. Then, each time point is taken as the imaging start time. Lastly, the mature Hopfield neural network algorithm is used to search for the optimal path through continuous updating[10], and get the minimum energy consumption path.

Assumed that $\boldsymbol{M} = \{m_1, m_2, \cdots\cdots, m_n\}$ is the set of mission need to be planned, and $\boldsymbol{TW} = \{[Ts_1, Te_1], [Ts_2, Te_2], \cdots\cdots, [Ts_n, Te_n]\}$ is the set of visible time windows. The discrete fraction for visible time window is denoted as a. The steps of the algorithm are as follows:

Step 1: Sorting the planning missions according to the visible time order for satellite to missions.
Step 2: Parting the visible time windows of the planning missions by discrete fraction a, and the visible time window is divided into a parts, getting a time points, and taking each time point as imaging start time for satellite to the mission (see figure 3).

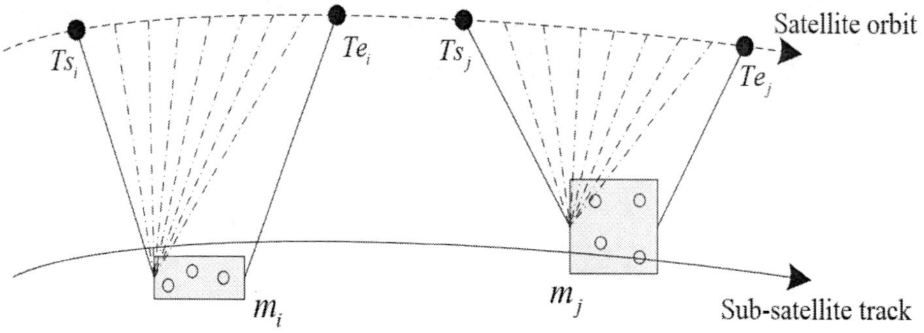

Figure 3. Discrete of visible time window

Step 3: Calculating the Euler Angle and imaging duration for each time point, and defining it as,

$$point(i,j) = (i, Ts_i + \frac{Te_i - Ts_i}{a} \times (j-1), \beta_{ij}, q_{ij}, Tcon_{ij}) \tag{6}$$

where i is the mission *ID*, *j* is the time point number, $Ts_i + \frac{Te_i - Ts_i}{a} \times (j-1)$ is the imaging start time, and β_{ij} , q_{ij} and $Tcon_{ij}$ are the rolling angle, pitching angle and imaging duration for satellite to mission.
Step 4: Starting from each time point of the first mission according to the time point number, it is extended to the feasible time point of the following mission satisfied the constraint conditions.
Step 5: Using Hopfield neural network algorithm update the time point path from first mission to the last one according to the minimum energy principle, and getting the optimal path minimized energy and consumed constraint conditions.

Figure 4 shows flow of the optimal path searching algorithm.

5. Simulation example

The parameters of satellite and other constraints in simulation are shown in table 1 and table 2.

Table 1. Parameters of remote sensing satellite

Major semi axis(km)	eccentricity	FOV(deg)	Pitching range(deg)	Rolling range(deg)	Manoeuvrability(deg/s)
7800	0.02	16.3	± 40	± 50	1.1

Table 2. Parameters of other constraints

Average power of attitude maneuver P_{moving} (w)	Average power of imaging P_{moving} (w)	Drift angle adjusting time T_Q (s)	Maximum imaging time in single orbit $TIMG_{max}$ (s)
640	984	30	600

A single orbit track was selected as the simulation scene, and the simulation start time was 03:41:17 March 23, 2016 (UTCG), the simulation end time was 05:28:32.6 March 23, 2016 (UTCG).

The set of planning missions was presented with trips segmented by circular Areatarget[11] and missions clustered by point targets from $t_1 \sim t_{18}$ [12]. All shown in table 3 and table 4.

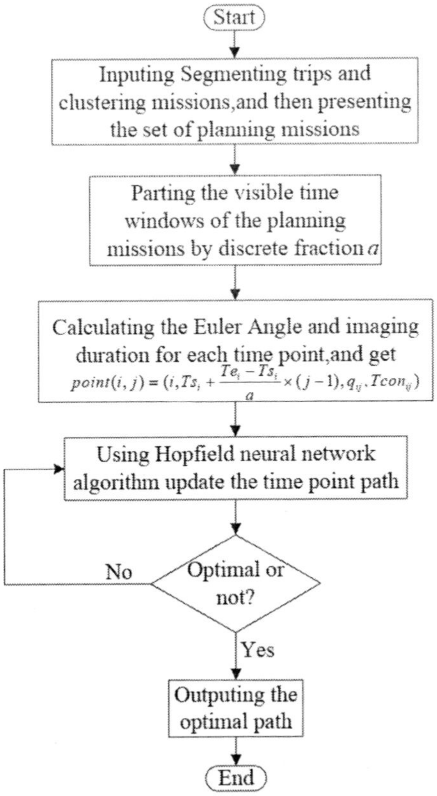

Figure 4. Flow of the optimal path searching algorithm

Table 3. Position of targets

Target ID	Longitude and latitude	Note	Target ID	Longitude and latitude	Note
t_1	$(21.06^\circ, 42.56^\circ)$		t_{10}	$(136.8^\circ, 34.46^\circ)$	
t_2	$(7^\circ, 49^\circ)$		t_{11}	$(129^\circ, 33^\circ)$	
t_3	$(10^\circ, 49^\circ)$		t_{12}	$(141.39^\circ, 15.04^\circ)$	
t_4	$(8.63^\circ, 50.21^\circ)$		t_{13}	$(145.21^\circ, 14.13^\circ)$	
t_5	$(82^\circ, 45^\circ)$		t_{14}	$(144.75^\circ, 13.4^\circ)$	
t_6	$(90^\circ, 47^\circ)$		t_{15}	$(148.14^\circ, -22.9^\circ)$	
t_7	$(128^\circ, 35^\circ)$		t_{16}	$(152.77^\circ, -25.4^\circ)$	
t_8	$(132^\circ, 34.62^\circ)$		t_{17}	$(152.2^\circ, -27.88^\circ)$	
t_9	$(121^\circ, 24^\circ)$		t_{18}	$(151.2^\circ, -33.13^\circ)$	
Areatarget	$(85.6^\circ, 54.7^\circ)$	Radius=550km			

Table 4. Set of planning missions

Mission type	Mission ID	Visible time window	Imaging time window	Rolling angle(deg)	Pitching angle(deg)
Segmenting trips	m_1	\	04:07:58~04:11:01	-15.983	31.23
	m_2	\	04:12:42~04:15:51	-32.233	-28.36
Clustering missions	m_3	03:55:33~04:00:16	\	-0.89	\
	m_4	04:08:35~04:17:08	\	-44.925	\
	m_5	04:20:13~04:24:09	\	10.555	\
	m_6	04:27:08~04:33:45	\	21.95	\

As in table 4, the imaging Euler Angles for segmenting trips are already confirmed, so its time window needn't to be discrete.

The searching capability of the proposed optimal path searching algorithm is related to the degree of discretization for visible time window. For the traditional mission planning methods, the greater the degree of discretization, the better the astringency of the optimal solution, and the computing resources on the ground can also meet the requirements. As for the autonomous mission planning, computing resources is limited for satellite, and if the degree of discretization is set too high, it will increase the amount of computation. On the contrary, if the degree of discretization is set too small, it will not guarantee the astringency of the optimal solution. This is also the difference between the autonomous mission planning on the satellite and the traditional mission planning on the ground.

Therefore, in this paper, the simulation was analyzed through setting different numbers to find a relatively suitable number of discrete fraction. And then, the suitable number was taken as the degree of discretization to get the relatively optimal solution, which can reduce the amount of computation, but also can guarantee the astringency of the solution.

Different numbers of discrete fraction were 5, 10, 20, 50, and the 5 fraction was taken as an example to introduce the solving process of the algorithm.

Firstly, the visible time windows of the planning missions were parted by 5 discrete fractions, and then the imaging pitch angle and imaging duration for each time point were calculated by using STK. Table 5 shows the information of each time point.

Table 5. Information of each time point

Mission ID	Visible time window	Time point	Imaging start time	Imaging duration(s)	Pitching angle(deg)	Rolling angle(deg)
m_3	03:55:33~04:00:16	$point(3,1)$	03:55:33	38.7	34.85	-0.89
		$point(3,2)$	03:56:29.6	36	20.87	
		$point(3,3)$	03:57:26.2	34.6	2.8	
		$point(3,4)$	03:58:22.8	35.7	-16.2	
		$point(3,5)$	03:59:19.4	39.1	-31.5	
m_4	04:08:35~04:17:08	$point(4,1)$	04:15:51	40.4	-27.4	-44.925
		$point(4,2)$	04:16:6.4	34.5	-28.6	
		$point(4,3)$	04:16:21.8	32.2	-30.1	
		$point(4,4)$	04:16:37.2	28.1	-31.4	

Mission ID	Visible time window	Time point	Imaging start time	Imaging duration(s)	Pitching angle(deg)	Rolling angle(deg)
		$point(4,5)$	04:16:52.6	24.4	-32.6	
m_5	04:20:13~04:24:09	$point(5,1)$	04:20:13	52.5	34.5	10.555
		$point(5,2)$	04:21:0.2	52.3	25.4	
		$point(5,3)$	04:21:47.4	48.7	15	
		$point(5,4)$	04:22:34.6	48	2.5	
		$point(5,5)$	04:23:21.8	41.3	-10.4	
m_6	04:27:08~04:33:45	$point(6,1)$	04:27:8	32.6	36.4	21.95
		$point(6,2)$	04:28:27.4	42	23	
		$point(6,3)$	04:29:46.8	50.2	4.55	
		$point(6,4)$	04:31:6.2	67	-15.2	
		$point(6,5)$	04:32:25.6	84.4	-31.4	

According to the information of each time point in table 5, the optimal path consumed minimum energy was calculated by using Hopfield neural network algorithm. That is,

$$O \rightarrow point(3,3) \rightarrow m_1 \rightarrow m_2 \rightarrow point(4,4) \rightarrow point(5,5) \rightarrow point(6,3) \quad (7)$$

Assumed that the initial Euler Angle for satellite starts from zero, and the Euler Angle for satellite along the direction of sub-satellite track is positive when rolling to left or pitching forward, and negative when rolling to right or pitching backward. Table 6 is the command sequence of attitude maneuver for satellite in the optimal path.

Table 6. Command sequence of attitude maneuver

Mission ID	Imaging time window	Rolling angle(deg)	Pitching angle(deg)	Maneuver time(s)
m_3	03:57:26.2~03:58:0.8	-0.89	2.8	3.321
m_1	04:07:58~04:11:1	-15.983	31.23	39.171
m_2	04:12:42~04:15:51	-32.233	-28.36	68.256
m_4	04:16:37.2~04:17:5.3	-44.925	-31.4	14.159
m_5	04:23:21.8~04:24:3.1	10.555	-10.4	68.832
m_6	04:29:46.8~04:30:37	21.95	4.55	23.711
Total imaging time(s)	544.2	Total attitude maneuver time(s)		217.45

As in table 6, when the discrete fraction is 5, the total imaging time for satellite is 544.2 seconds satisfied the maximum imaging time in single orbit that the total imaging time of the satellite in single orbit could not exceed 600 seconds. The total attitude maneuver time is 217.45 seconds, and the total energy consumed by optimal path is 674.66 kJ.

According to the above process, the total imaging time, total attitude maneuver time and total energy consumed in optimal path for 10, 20, and 50 fractions were calculated. All shown in table 7 and table 8.

It can be seen from table 8, with the increase of discrete fraction, the total imaging time for satellite to planning missions remained steady, the total attitude maneuver time dropped off, and the energy consumed reduced gradually too. Figure 5 shows the relationships between the total energy consumed by satellite in optimal path and the degree of discretization for time window.

Table 7. The optimal path for different fraction

Fraction	Optimal path
5	$O \to point(3,3) \to m_1 \to m_2 \to point(4,4) \to point(5,5) \to point(6,3)$
10	$O \to point(3,4) \to m_1 \to m_2 \to point(4,7) \to point(5,9) \to point(6,6)$
20	$O \to point(3,4) \to m_1 \to m_2 \to point(4,14) \to point(5,17) \to point(6,12)$
50	$O \to point(3,10) \to m_1 \to m_2 \to point(4,33) \to point(5,42) \to point(6,29)$

Table 8. The calculating results for different fraction

Fraction	Total imaging time (s)	Total attitude maneuver time (s)	Total energy consumed (kJ)
5	544.2	217.45	674.66
10	541.9	208.41	666.61
20	542.2	204.65	664.50
50	542.7	203.64	664.35

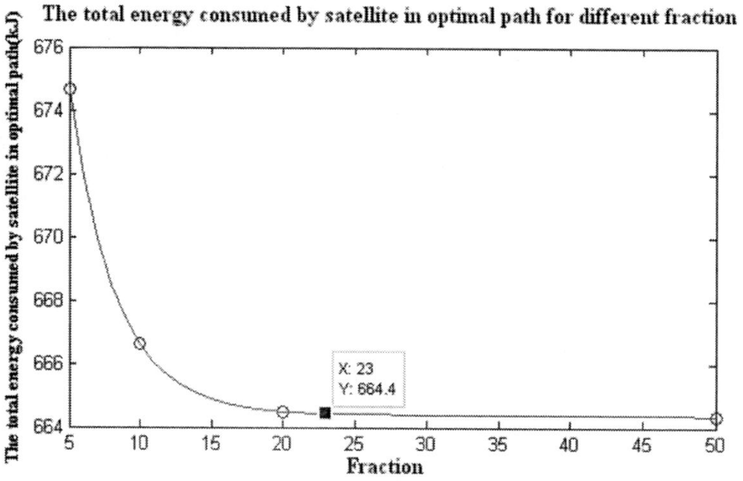

Figure 5. The total energy consumed by satellite in optimal path for different fraction

As in figure 5, with the increase of discrete fraction, the total energy consumed by satellite is convergent, and when the fraction comes to 23, the total energy consumed has remained unchanged basically, the total energy keeps 644.40 kJ. Therefore, the best discrete fraction can be set to 23, and the corresponding optimal path is,

$$O \to point(3,6) \to m_1 \to m_2 \to point(4,15) \to point(5,19) \to point(6,14) \quad (8)$$

Table 9 shows the command sequence of attitude maneuver for the best discrete fraction.

Table 9. Command sequence of attitude maneuver for the best discrete fraction

Mission ID	Imaging time window	Rolling angle(deg)	Pitching angle(deg)	Maneuver time(s)
m_3	03:56:34.5~03:57:2.8	-0.89	19.2	18.081
m_1	04:07:58~04:11:1	-15.983	31.23	24.411
m_2	04:12:42~04:15:51	-32.233	-28.36	68.256
m_4	04:16:37.9~04:17:6.3	-44.925	-31.3	14.069
m_5	04:23:18.4~04:23:58.7	10.555	-9.7	69.372
m_6	04:30:42.9~04:31:38	21.95	-9.6	10.346
Total imaging time(s)	542.1	Total attitude maneuver time(s)		204.45

According to the data in table 9, the total energy consumed for the best fraction 23 was calculated to be 664.27 kJ, and there is little difference compared with the result as 664.40 kJ obtained by figure 5. Therefore, the result calculated for fraction 23 can be taken as the optimal solution of the mission planning problem.

Figure 6 shows the attitude maneuver strategy of satellite. Assumed that the attitude maneuver is carried out immediately after completing the imaging missions for satellite.

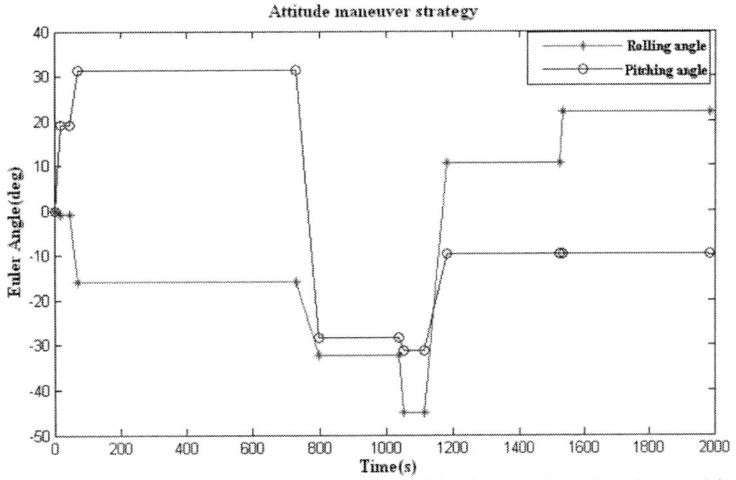

Figure 6. Attitude maneuver strategy of satellite for mission planning problem

6. Conclusion and future work

The mission planning problem for remote sensing satellite imaging is studied from the time constraint. The time constraint satisfaction problem model is presented, and an optimal path searching algorithm based on the discrete time window is proposed. Simulation results show that the proposed algorithm can meet the requirement of minimum consumed energy, but also can reduce the computation and improve the speed of mission planning.

Considering that the best discrete fraction is variable even for the same satellite, and the future work is analysing the influence factors of the best discrete fraction from orbit altitude, orbit inclination, satellite mass, satellite inertia, etc, and establishing the relationship between the best fraction and influence factors.

Acknowledgments

This work was supported by the National Key Research and Development Program of China under Grant 2016YFB0500203, and the National Natural Science Foundation of China under Grants 11602145 and 51601116.

References

1. Wang Chong, Li Jun, Jing Ning, Wang Jun and Chen Hao. A Distributed Cooperative Dynamic Task Planning Algorithm for Multiple Satellites Based on Multi-agent Hybrid Learning, *Chinese Journal of Aeronautics* **24** , 493-505 (2011)
2. Xiang Reng Xiang. Research on Selecting and Scheduling Observation of Agile Satellites , *National University of Defence Technology* (Changsha) **2** , 13-15 (2010)
3. Gerard Verfaillie and Thomas Schiex . Solution Reuse in Dynamic Constraint Satisfaction Problem , *Proceeding of the Twelfth Conference of the American Association of Artificial Intelligence* (Washington) **1** , 307-312 (1994)
4. D Song, A Frank and Van Der Stappen, etc . An Exact Algorithm Optimizing Coverage Resolution for Automated Satellite Frame Selection , *Proceeding of the IEEE International Conference on Robotics and Automation* , 63-70 (2004)
5. Lian Zhen Yu, Tan Yue Jin and Yan Zhen Zhen . Temporal Reasoning Technology for AEOS Scheduling , *Systems Engineering and Electronics* **35** , 1206-1211 (2013)
6. Ren Ming Kang . Adaptive Ant Colony Optimization for Constraint Satisfaction Problem , *Northeastern University* (Shenyang) **2** , 9-10 (2014)
7. Yu Tao, Xu Shu Yan and Han Cheng Shan . Batch-type Real-time Adjustment for Drift Angle of Space Camera , *Optics Precision Engineering* **17** , 1908-1914 (2009)
8. Michel Lemaitre, Gerard Veraillie, Frank Jouhaud and Jean Michel Lachiver . Selecting and Scheduling Observations of Agile Satellites , *Aerospace Science and Technology*l **6** , 367-381 (2002)
9. Guo Hao, Qiu Di Shan and Huang Wei . Simulation to Impact of Attitude Changing Duration on Tasks Scheduling of An Agile Imaging Satellite , *Journal of the Academy of Equipment Command & Technology* **23** , 70-74 (2012)
10. Yu Sheng Wei . *Case Analysis and Application of MATLAB Optimization Algorithm* ed Cui Jing (Beijing: Tsinghua University Press) **18** , 262-271 (2015)
11. Pan Yao, Chi Zhong Ming, Rao Qi Long and Luo Da . Dynamic Segmenting Method of Polygon Target Based on FOV for Remote Sensing Satellite Imaging , *Spacecraft Engineering* vol **26** , 38-42 (2017)
12. Guo Hao, Wu Guo Hua and Qiu Di Shan . Intensive Task Clustering Method for Agile Imaging Satellites , *Systems Engineering and Electronics* **34** , 931-935 (2012)

Architecture Design of Aviation Fault-tolerant Computer Based on ARINC659 Bus Technology

Gang An[1,a], Yu Li[2,b] and Xin Li[3,c]

[1] *School of management, Xi'an Jiaotong university, Xi'an, CN*
[2] *System Integration Center, FAI AVIC,Xi'an, CN*
[3] *Landing Gear System Research Institute, FAI AVIC, Xi'an, CN*

Abstract. The ARINC659 backplane bus is suitable for high safety and high reliability requirements of aircraft on-board computer communication systems. This paper analyzes the structure of ARINC 659 serial backplane bus and the bus fault tolerance mechanism. On the basis of backplane bus, a 4 degree of aviation fault-tolerant computer is designed. Moreover, the computer architecture and computer system of the instruction branch and monitoring branch are designed in the computer channel. The fault-tolerant management of the computer is realized by bus fault tolerance, redundancy voting between computers and the monitoring of the instruction and monitoring branches.

1. Introduction

With the rapid development of avionics, the aircraft relies more and more on Avionics systems. In order to realize the coordination of each air system, grasp and deal with the changing data information at any time, achieve the sharing of resources, so that the entire avionics system and integrated, and can efficiently and reliably work, then it must rely on high speed, reliable, aviation data bus standard[1].

Airborne data bus technology is one of the most important key technologies of Avionics Integration in modern aircraft. With the development of avionics system, new requirements have been put forward for airborne data bus, which has promoted the development of airborne data bus, and the airborne data bus determines the degree of integration of avionics systems. As the skeleton and nerve of avionics system, data bus plays an important role in avionics system, and it promotes each other synchronously with the development of avionics technology[2].

The ARINC 659 avionics data bus by the Aeronautical Radio Incorporation in 1993 to meet the bus standard proposed for efficient aircraft information transmission requirements, and has been used in commercial aviation, aircraft such as the Boeing 777 aircraft information management system. ARINC 659 is a kind of air bus with high data throughput, high fault tolerance and data certainty. The transmission rate is 60Mb/s and the coding efficiency is 98%. There is no need to transmit address information in transmission data, and 4 channel signals are used for error detection. The fault tolerance is better than the traditional double redundancy, and the complexity is lower than the

[a] An_gang@163.com;
[b] kingliyu@sina.com;
[c] 370713464@qq.com

traditional 4 redundancy. It can be used in the field of aviation, navigation, spaceflight and other bad working conditions.

2. ARINC659 Bus technology

2.1. Bus structure

The ARINC 659 bus specification is a standard for the development of digital data information transmission between modular Modular Avionics devices and field replaceable modules, and is a specification for the requirements of LRM. This specification describes the electrical characteristics and bus characteristics required for the integrated backplane avionics bus. Specifications specify the protocol, timing, speed, and physical characteristics of the bus. The ARINC 659 bus is a serial backplane bus, which reduces the hardware resources, simplifies the mode of operation and the bus monitoring mode.

The ARINC 659 bus is a half duplex, dual - dual serial data bus. Divided into A, B bus, X bus and Y containing each of the two sets of bus, there are AX, AY, BX, BY bus redundant information transmission. Each bus contains an independent clock line and two data lines, while transmitting two bits of data, so the full bus contains 12 signal lines. Bus monitoring is double redundancy, through the method of election input and cross comparison detection, the fault detection is carried out to ensure the high reliability data transmission[3].

The clock frequency of the ARINC 659 bus is 30MHz. At the same time, two bit data is transmitted, and the maximum throughput is close to 60 megabits per second (Mbps). The ARINC-659 bus uses the BTL bus driver, and the BTL driver is an open collector device with line or capability. The ARINC-659 bus structure is shown in figure 1.

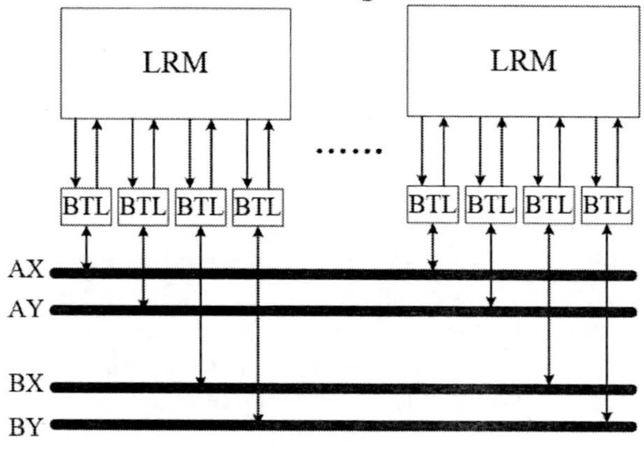

Figure 1. Bus structure

2.2. Bus fault-tolerance

The advantage of serial transmission lies in: Reduce hardware, simplify full concurrent monitoring, and therefore improve the reliability of the bus. To improve availability, each bus interface unit BIU (Bus Interface Unit) is connected to the two bus. Each line replaceable unit using dual LRM bus interface unit, cross check between the bus interface unit between four bus also cross check, the online replaceable unit LRM with dual self-test capability. Cross validation between four buses improves the availability of data.

By accessing the TDPA (Table Driven Proportional Access) protocol by table driven ratio, the ARINC 659 bus makes a deterministic division of time and space. It is controlled by commands stored in the nonvolatile command table memory in each bus interface unit BIU.

The ARINC 659 bus is a dual - dual configuration bus consisting of dual bus pairs (A and B). Each bus pair consists of a "X" and a "Y" bus. Each bus (Ax, Ay, Bx, and By) has a separate clock and two data lines, capable of transmitting two bit data at the same time. Thus, a complete bus group is composed of 12 bus transmission lines. The ARINC 659 bus is organized as shown in figure 2.

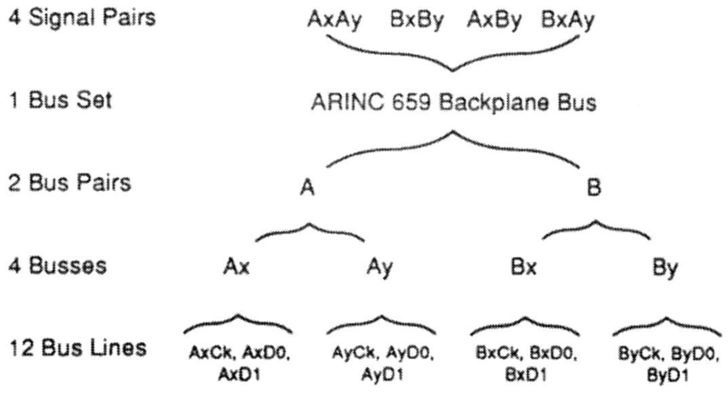

Figure 2. Bus constitution

Each online replaceable unit, LRM, has two bus interface units, BIU (BIUx and BIUy). BIUx transmits through the X bus transmission line, and the BIUy transmits through the Y bus transmission line. Each bus interface unit receives the data information of all four buses. Each bus is driven by an independent transceiver in the LRM to prevent single faults from adversely affecting more than one bus (for example, each online replaceable module has four transceivers).

By reasonably designing the physical layer, the complexity and overhead of message structure are reduced:

- Each individual clock eliminates the preamble that is used for clock phase synchronization.
- Double bus comparison eliminates the need for cyclic redundancy check (CRC) and other error control fields.
- The table driven protocol addressing save implicit contains the address field.
- The timing function of the table driven proportional access protocol eliminates the start and end characters.

The use of these technologies makes the ARINC 659 bus more efficient than other bus protocols. These technologies also make minimum transmission delay, which is very important for real-time systems.

When the receiver performs error detection and correction over the redundant bus, the receiver performs the following comparisons: Ax=Ay, Bx=By, Ax=By, and Bx=Ay. The bus pairs in these alignments are called signal pairs. Comparisons between Ax=Bx and Ay=By are invalid because these signals come from the same hardware source and may have related errors from the same source. Because of the error detection of four effective signal pairs, the fault tolerance of ARINC 659 bus is better than the traditional double redundancy, and it has less complexity than the traditional four redundancy.

The ARINC 659 bus has been designed as a backplane bus that can correct all single path errors and detect all double faults. Errors in all single bus interface units can be detected by the ARINC 659 bus. All parts of the bus system are fully monitored. All operations are double operations. Relying on multi-point double detection, the integrity of the operation path is guaranteed.

3. Computer architecture design

3.1. Overall architecture

In order to achieve the security and reliability level of the system, the computer architecture must consider the reliability, security and other aspects of the computer system. Redundant technology, as well as a framework for supporting resource sharing and refactoring, is the best way to ensure security and reliability metrics at the moment[4]. The open, modular architecture provides excellent support for system refactoring, system extensions, and system maintenance.

High security 4x2 computer fault tolerant redundancy configuration architecture, 4 redundancy computer system consists of 4 identical chassis components, each box contains a channel, the system is divided into 4 channels, numbered A, B, C, D. Each channel adopts two CPU boards with non similar design, and realizes the task function of the system arrangement by instruction and monitoring[5].Each channel of the computer is centered on the ARINC 659 serial backplane bus, and each function board is connected to the bus in the form of resources. Each function board is an intelligent node, which contains a processor or microcontroller to regulate and monitor the signal on the function board. All the functions are transmitted by serial bus in digital way. The CPU board is the system function computing resources, which consists of a time triggered bus to work, the calculation results on the data sending buffer, to other functions in the corresponding time slot transmission, bus time slot division as shown in figure 3.

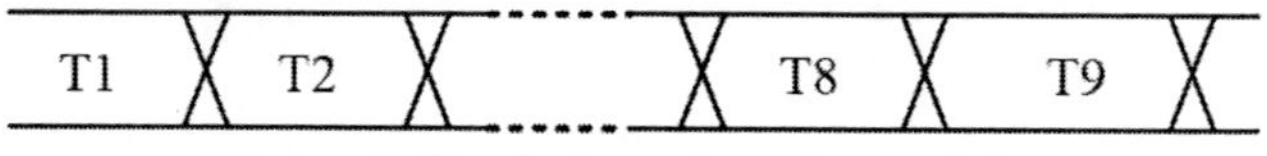

Figure 3. Backplane time slot division

Under the control of the board micro controller, the interface function board carries on the signal collection work at all times, and obtains the signal of the sensor in real time, and can carry on the data transmission at any time. At the same time, the micro controller responds to the commands transmitted by the bus in real time and performs the corresponding operation[6].

Data exchanges between redundant channels are performed by means of CCDL. Data transmission CCDL directly with the serial backplane bus connection, serial backplane bus data of four redundant channels of all connected, namely the backplane bus data is forwarded to the other case through the form of LVDS, while receiving other case data sent, the redundant system architecture in a unified "virtual". On. Each function board system between can exchange information, each function module is the system can be shared and resource reconfiguration of resources, this system can reach, to the function module for dynamic resource reconfiguration and system fault tolerance system to achieve the minimum unit. The block diagram of the computer system is shown in figure 4.

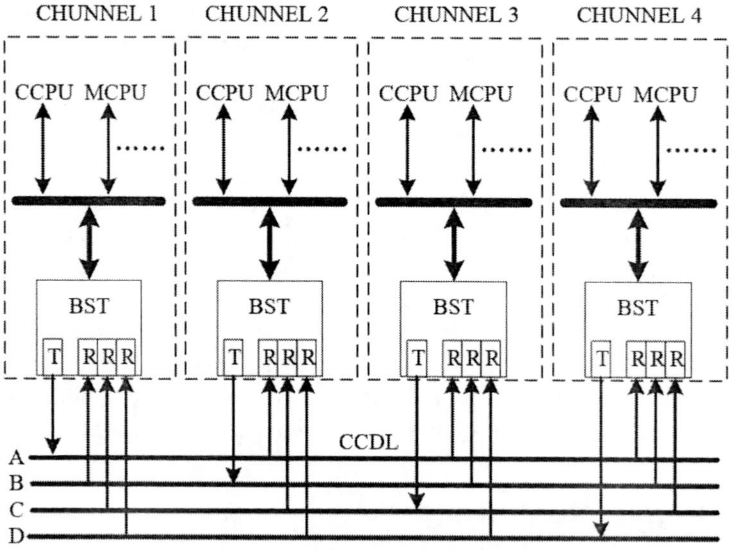

Figure 4. Computer system structure

3.2. Computer composition

High security fault-tolerant computer, each channel internal resource composition of the same, single channel computer module units, as shown in Figure 5, described in detail below:

- The instruction processor (CCPU) module: Calculation and management of the core, the main task is to control instruction solution and redundancy management, the basic work is the process of signal of various kinds of sensor redundancy management, and then use the voting results of control instruction calculation, then the output signal from the external command, I/O module to complete the corresponding control action.
- Monitoring processor (MCPU) module: synchronous acquisition and operation with CCPU board, but not external output instruction. Monitoring channel only transmits test data, and monitors all input and output data of CCPU. The output instruction signal of the monitoring channel is used to detect whether the instruction channel is out of order.
- Discrete quantity input output (DIO) module: discrete digital signal input and output processing.
- Analog input (AIN) module: analog signal input processing.
- Analog output (AOUT) module: analog signal output processing.
- Multi bus interface (MBI) module: according to the different uses and requirements of the computer, such as: MIL-1553B bus interface, ARINC429 bus interface, AFDX bus interface, and some necessary serial bus interface.
- Bus bridge transmission (BST) module: responsible for the exchange of the redundancy between computers and ARINC 659 bus data sharing, the backplane mode conversion for the cable model, in the form of LVDS is forwarded to the other LRM, is a continuation of the bus signal, the back signal in the conversion of computer level and transmission, to achieve more than cross communication data link the degree of inter channel (CCDL).
- Power supply (PS) module: 2-redundant power supply.
- Backplane (BP): ARINC 659 backplane bus as communication medium, module signal interconnection.

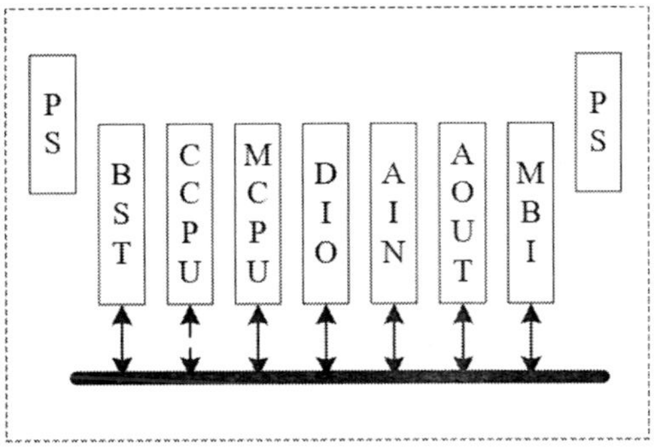

Figure 5. Computer internal composition

3.3. Redundancy management

In order to meet the requirements of high reliability, high security, aviation computer, improve the ability of fault, in the existing basic components on the basis of the reliability, redundancy design and redundancy management technology can greatly enhance the ability of fault tolerance. Redundancy management refers to the simultaneous execution of several units in a system that perform the same task, so that if a unit fails, it can be switched. This method is now widely used, and it can greatly improve the reliability of the system[7].

The working principle of monitoring put is: a link in the system using a plurality of identical components or tasks, and then put their output results of monitoring, out of which the most reasonable result to the next step, in order to achieve shielding failure of components or tasks, improve the reliability of the data output. Relatively common voting algorithms include arithmetic, average voting, majority voting, median voting, etc.[8].

Before the signal is transmitted into the computer channel, the first monitoring voting is carried out, i.e., the input signal is voted. The computer collects analog signals of each sensor or discrete signals of other controllers as input. Each of the redundant computers collects analog or discrete quantities for CCDL and votes through a voter. After comparing the signal with the maximum and minimum threshold to determine the validity of the signal, the result is taken as a median output. The discrete quantity is determined by the corresponding truth table in the logical circuit, and the result is 1 or 0.

After the input signal is voted, the result is computed by the computer, and then the second monitoring vote is carried out. The signals are computed simultaneously on the CCPU and MCPU in the same LRM, and then the results are compared. If the results are the same, the calculation results are correct. Outputs the result through the CCPU to the bus, carries on the third surveillance voting.

CCPU outputs the calculated results, and compares the results of the four channel CCPU with the output signals. The voting algorithm of the output signal is the same as that of the input signal voting algorithm. The voting process is shown in figure 6.

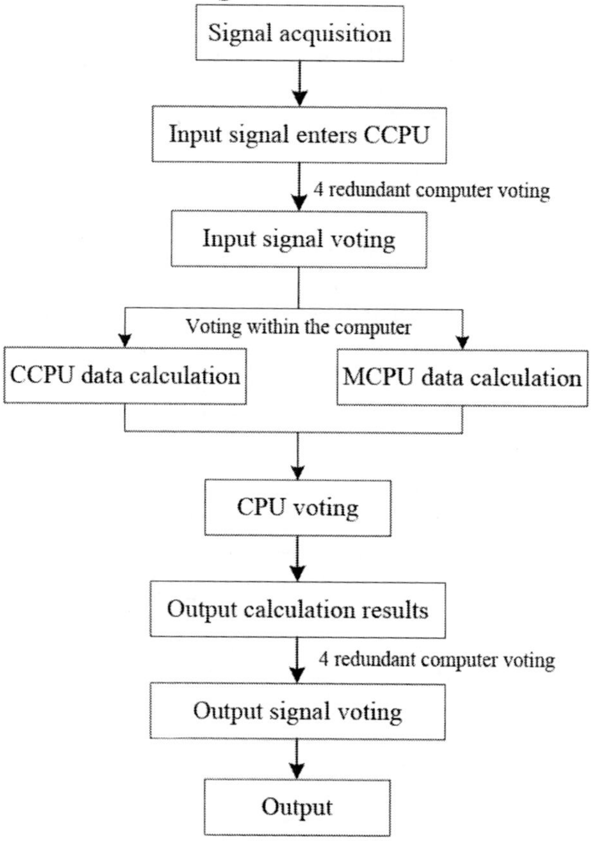

Figure 6. Voting process

- CCPU collecting input discrete signals and analog signals, and signal processing.
- Monitor and vote the collected input signals. The effective value of the selected value in the analog signal is calculated by the average value, and the final voting result is obtained. The discrete signal is processed according to the logic circuit corresponding to the truth table.
- Pass the result of voting to CCPU and MCPU for calculation. Since CCPU and MCPU are dual redundant, the results are compared and the results are valid or invalid.
- The effective data is transmitted directly to the next processing unit or device, and the output results are voted on the next computing node. The output vote of the computing node is the input vote of the next computing node.

Voting for 4 channel computers relies entirely on CCDL implementation. The 4 computers are represented by W, X, Y, and Z respectively.

During the operation of the 4-redundancy master computer, each computer itself is W, and the identity of the remaining two computers is X, Y, Z. Thus, the correspondence between A, B, C, D, four computers and W, X, Y, and Z is as follows:

- Computer A : W — A,X — B,Y — C,Z —D.
- Computer B: W — B,X — D,Y — C,Z — A.
- Computer C: W — C,X — A,Y — B,Z — D.
- Computer D: W — D,X — A,Y — B,Z — C.

In the above relation, the cross - flow process is designed as shown in figure 7.

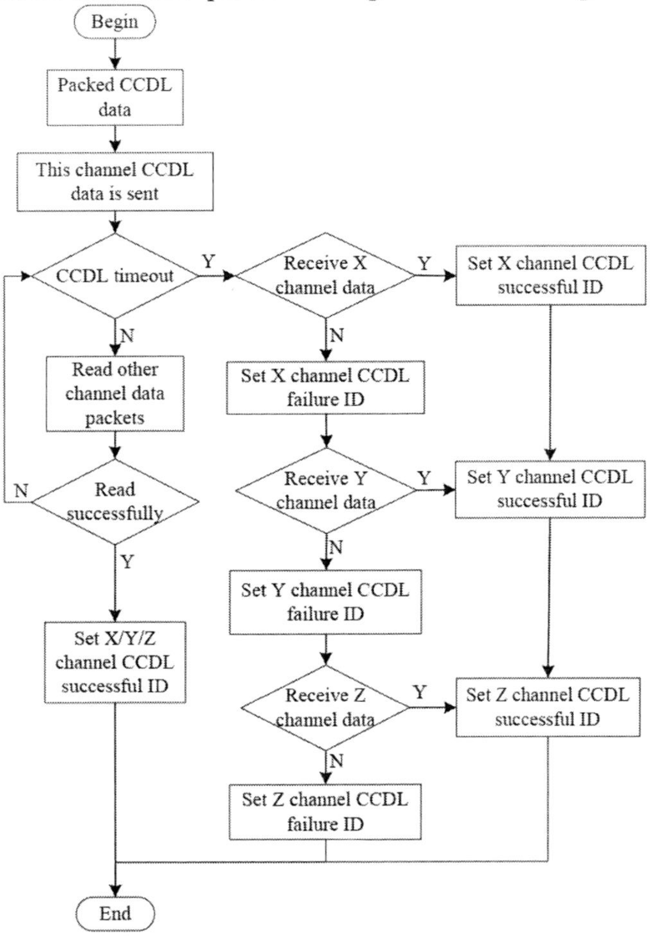

Figure 7. Cross channel data transmission process

The signal voting is reconstructed by the monitoring result of the monitor, determined and excluded the fault signal, and announced the system to input the channel at this time[9]. The basic voting principle of analog quantity in this system is:

- The 4 channel signal is effective, the voter takes the mean of this cycle, the second and the small.
- The 3 channel signal is effective, The voter takes the median of the 3 signals in this cycle.
- The 2 channel signal voting, the voter takes the mean of the 2 signals in this cycle.
- The 1 channel signal voting, when the corresponding sensor only has an effective channel, the input of the effective channel is the output of the voter.

Vote between CCPU and MCPU. Because the CCPU and MCPU double redundancy, the input signal through the input after the vote and input to the CCPU and MCPU are calculated, calculation results, through the MCPU bus to transmit the data to CCPU, CCPU will receive the data and the calculation results are compared, the same calculation shows that the results are valid, the calculation results by bus transfer to the BST, otherwise the calculated data for failure data. The voting process of CCPU and MCPU is shown in figure 8.

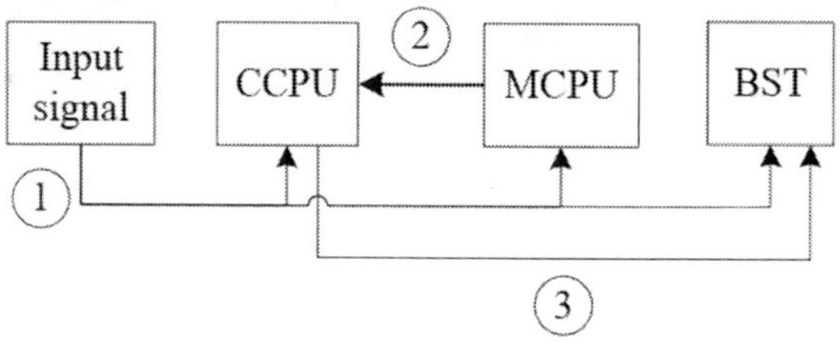

Figure 8. Voting process of CCPU and MCPU

- Input signals for input data voting, the voting results will be passed to CCPU and MCPU.
- MCPU and CCPU simultaneous calculation, MCPU will transfer the results of the calculation to CCPU, compare each other to vote.
- CCPU will vote data results transferred to BST for lower level voting.

4. Conclusions

The ARINC659 bus has high data throughput, strict fault isolation, data transmission, encoding efficiency, data transmission process does not need to transmit the address information, has the advantages of good fault tolerance, suitable for high security, high reliability and high integrated aircraft for computer communication system. The ARINC659 bus as the computer motherboard bus communication standard, 4 redundant computer design is composed of command channel and monitoring channel with self monitoring ability. The computer carries on the signal conversion through the BST, thus realizes the CCDL communication between computers. The signal input and instruction output between 4 redundant computers are designed, and the two level monitoring voting surface is used to decide the input and output information. And through the MCPU set in the computer, the CCPU instruction operation results are monitored in real time, thus ensuring the correctness of the single computer channel instruction. The fault tolerance performance of ARINC659 bus redundancy analysis, computer architecture design and matching computer redundancy management strategy, which can satisfy the requirement of high security, high reliability and fault tolerance requirements of airborne computer.

References

1. Wang Yong, Yu Hongkun. Airborne computer system. Beijing: National Defense Industry Press. 2008.
2. Yao Xudong. Design and implementation of avionics system simulation platform based on virtual instrument. Shanghai Jiao Tong University, 2012.
3. Airlines Electronic Engineering Committee, ARINC SPECIFICATION 659 BACKPLANE DATA BUS [C].DEC.27,1993
4. Huo Man. Prospects for the development of integrated avionics technology. Avionics technology,2003(3):12-17
5. Zhi Chaoyou, Airborne data bus technology and its application. National defense industry press. Jan.2009
6. Xu Zezhong, Lei Xun, Hu Rong. A study on the development of the new generation integrated avionics system. Avionics Technology. 2001,32(4):11-18.
7. Gao Jinyuan, Xia Jie. Principle of computer control. Beijing: Tsinghua University press, 2006.
8. Yang Wei. Fault tolerant flight control system. Xi'an: Northwestern Polytechnical University press. 2007.
9. Wang Yunming. Standards, processes, methods, and tools for embedded software development. Journal of computer technology and development. 2006.10.

Robust Optimization in Airlines' Hybrid Channels Coordination with Service-Level Constraints

ZHANG Feifei[a], HU Rong and ZHANG Junfeng

Mailbox 578, No.29 Jiangjun Avenue, Jiangning District, Nanjing, Jiangsu Province, China

Abstract. This essay considers the influence of the service level in the research of the robust optimization of airlines' hybrid channels coordination with uncertain demands. In the essay, the service level is taken as a constraint in the model that is established based on the characteristics of channels coordination and scenario analysis. In this way, it can make the airlines' hybrid channels coordination more effectively. Then three-dimensional numerical experiments of the robust optimization model are performed and the results show that the model proposed is robust to uncertain demands and fit for airlines' hybrid channels coordination; Airlines should focus on promoting the direct channel and passengers' stickiness; agents should focus on reducing the cost of the channels, and they should carry out improve-service-level cooperation to attract passengers; airlines and agents also should cooperate with each other to obtain win-win strategies.

1. Introduction

Recently, some China's major airlines such as Air China and China Southern Airlines have a statement successively with China's third major OTA Qunar.com to suspend cooperation, it shows that the channels competition between airlines and agents has been heating up and chaotic airlines' hybrid channels need effective coordination urgently.

At home and abroad, the research of airlines' hybrid channels management and robust optimization method have achieved fruitful results. Cheng, Lee, Shomali took American airlines as the object and studied the coordination problem of Internet direct channel and traditional marketing channel of airlines under the condition of electronic market [1]. Zhang, Ding, Sun carried out the research on strategy selection of airlines' hybrid channels by using evolutionary game theory [2]. Saha [3] and David, Adida [4] studied the coordination of dual-channel supply chain on the perspective of channel structure and channel competition. Xu and Huang [5-6] established the single-objective and multi-objective robust operation model for electronic supply chain with uncertainty demand. Then Xu and Huang [7] established the multi-objective fuzzy programming model for operation of supply chain by using the fuzzy set method and robust optimization method.

Service level is widely used in the field of supply chain. Liu, Dong and Zhang discussed the single buyer's optimal purchasing decision in two-period when facing opportunity constraints of service level [8]. Sun and Zhang mainly studied the strategy for the inventory cost of the traditional channels with service-level constraint. First in connection with the characteristics of the traditional charnels, they built the new inventory control model with service-level constraint considerations and then proposed the corresponding algorithm and steps for its solution [9]. Shen, Zhang and Qing considered

[a] Corresponding author: 1207923321@qq.com

a kind of supply chain system. The impacts of the service-level constraint on supply chain's profits and the effect of differentiated buy back contract were analyzed by numerical analysis [10].

These essays have conducted fruitful research on the question of airlines' hybrid marketing channel coordination and some qualitative analysis about service level and have achieved lots of useful results. However, at present, the study of management and optimization in airlines mainly focus on aviation operation problems like airlines flight schedules or airlines crew pairing, it still lack of robust optimization research on the coordination of the airlines' hybrid channels under the circumstance of fully Internet "+" and lack of the quantitative consideration of service level in the field of aviation. According to the present situation, the general scheme of this essay is as follows.

2. Modeling

2.1. Structure of channels

With the continuous development of Internet technology, the airline's hybrid channels are undergoing continuous changes. Airline's hybrid channels refer to the network marketing channels and traditional marketing channels co-existing and airlines' hybrid channels can be divided into two types of direct channels and distribution channels according to the subjects of the channel, the marketing channels that tickets are directly sold from airlines to passengers are known as direct channels; the channels that tickets are indirectly sold from ticket agents and other subjects to passengers are known as distribution channels. There are several sub-categories in direct channels and distribution channels.

The structure of channels coordination system is presented in Figure 1.

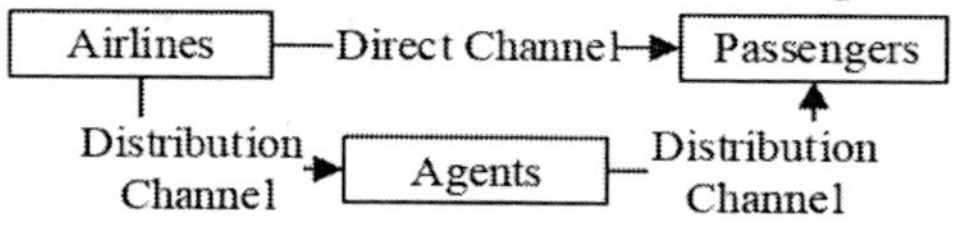

Figure 1. Airlines' hybrid channels coordination system.

2.2. Notations

The air tickets, airlines, passengers and agents in the coordination system are expressed separately with the subscripts i , j , k , l .

e_{ijk} , per unit main business income for airlines j sells air ticket i to passenger k (include the income of tickets, luggages, and other income got by airlines); $e_{l,ik}$, per unit main business income for agent l sells air ticket i to passenger k ; c_{ijk} , per unit variable cost for airlines j sells air ticket i to passenger k ; $c_{l,ij}$, per unit variable cost for airlines j copes with agent l ; $c_{l,ik}$, per unit variable cost for agent l sells air ticket i to passenger k ; r_{ij} , preparation cost for airlines j produces air ticket i ; f_{jk} , fixed cost for airlines j satisfies (some) demands of passenger k ; $f_{l,j}$, fixed cost for airlines j copes with agent l ; $f_{l,k}$, fixed cost for agent l copes with passenger k C_{ij} , maximum transportation capacity of air ticket i of airlines j ; T_j , total transportation capacity of airlines j ; d_{ik} , demands of air ticket i from passenger k ; q_{ik} , per unit penalty for the quantity of air ticket i does not satisfy the demands of passenger k ; x_{ijk} , quantity of air ticket i that passenger k purchases from airlines j ; $x_{l,ik}$, quantity of air ticket i that passenger k purchases from agent l ; $z_{jk} \in \{0,1\}$, if airlines j satisfies some demands of passenger k , then $z_{jk} = 1$, otherwise $z_{jk} = 0$; u_{ik} , quantity of air ticket i that does not satisfy the demands of passenger k ; $y_{ij} \in \{0,1\}$, if airlines j offers air ticket i , then $y_{ij} = 1$, otherwise $y_{ij} = 0$;

Assuming there are s kinds of possible scenarios of the passengers' uncertain demands in the coordination system which can use a set $G_s = \{1, 2, \cdots, s\}$ to represent. The probability of scenario s is p_s and $\sum_{s \in G_s} p_s = 1$. The demands of air ticket i from passenger k under scenario s is d_{ik}^s. Assuming x_{ijk}, $x_{l,ik}$, z_{jk} are control variables, the other decision variables are effective for all scenarios. Under these conditions, for each scenario s, the quantity of air tickets purchased by passengers from airlines and agents and whether the airlines adopts direct selling or distribution strategy will depend on different demands of the scenario, these parameters can represent with x_{ijk}^s, $x_{l,ik}^s$ and z_{jk}^s. In addition, u_{ik}^s represents the quantity of air ticket i that does not satisfy the demands of passenger k under scenario s.

In the model, there are two factors worthy of attraction: ①Air tickets have been realized in fully electronic and open inventory to all channels, so there is no products transportation and inventory problem when air tickets in the whole coordination system; ② In this essay, the service level indicates the degree of the satisfaction with the market demands. Considering the limit, the service level of the airlines can use α ($\alpha \in [0,1]$) to represent and the service level of the agents can use β ($\beta \in [0,1]$).

2.3. Robust optimization model

Objective function (maximize the total profits of airlines' hybrid channels coordination):

$$\max P = \sum_{s \in G_s} p_s \left[\sum_i \sum_j \sum_k e_{ijk} x_{ijk}^s + \sum_l \sum_i \sum_k e_{l,ik} x_{l,ik}^s - \sum_i \sum_j \sum_k c_{ijk} x_{ijk}^s - \sum_l \sum_i \sum_j c_{l,ij} x_{l,ik}^s - \sum_i \sum_j r_{ij} y_{ij} - \sum_j \sum_k f_{jk} z_{jk}^s - \sum_l \sum_j f_{l,j} - \sum_l \sum_k f_{l,k} - \sum_l \sum_i \sum_k c_{l,ik} x_{l,ik}^s - \sum_i \sum_k q_{ik} u_{ik}^s \right] \tag{1}$$

In the objective function, $\sum_i \sum_j \sum_k e_{ijk} x_{ijk}^s$ represents total main business income of production for all airlines under scenario s, $\sum_l \sum_i \sum_k e_{l,ik} x_{l,ik}^s$ represents total main business income of all agents under scenario s. Other constraints are as follows.

Total capacity of i:

$$\sum_k x_{ijk}^s + \sum_l x_{l,ik}^s - y_{ij} C_{ij} \le 0, \forall s, i, j \tag{2}$$

Total capacity of j:

$$\sum_i \sum_k x_{ijk}^s + \sum_l \sum_i \sum_k x_{l,ik}^s \le T_j, \forall s, j \tag{3}$$

The supply of airlines does not exceed the total demands of passengers:

$$x_{ijk}^s \le \sum_k z_{jk}^s d_{ik}^s, \forall s, i, j \tag{4}$$

The supply of agents does not exceed the total demands of passengers:

$$x_{l,ik}^s \le \sum_k d_{ik}^s, \forall s, i, l \tag{5}$$

The bottom limit the distribution channel is t:

$$\sum_s \sum_l \sum_i \sum_k x_{l,ik}^s \ge t \left(\sum_s \sum_i \sum_k x_{ijk}^s + \sum_s \sum_l \sum_i \sum_k x_{l,ik}^s \right), \forall j \tag{6}$$

Constraint of the service level of airlines:

$$\frac{\sum_{j} x_{ijk}^{s}}{d_{ik}^{s}} \geq \alpha, \forall s, i, k \tag{7}$$

Constraint of the service level of agents:

$$\frac{\sum_{l} x_{ikl}^{s}}{d_{ik}^{s}} \geq \beta, \forall s, i, k \tag{8}$$

Because the constraints of (7) and (8) are non-linear, supposing that $b = \frac{1}{d_{ik}^{s}}$, and then the constraints of (7) and (8) are rewritten as linear constraints:

$$b \sum_{l} x_{ikl}^{s} \geq \alpha, \forall s, i, k \tag{9}$$

$$b \sum_{l} x_{ikl}^{s} \geq \beta, \forall s, i, k \tag{10}$$

Constraint of demand and supply:

$$\sum_{j} x_{ijk}^{s} + \sum_{l} x_{l,ik}^{s} + u_{ik}^{s} \geq d_{ik}^{s}, \forall s, i, k \tag{11}$$

3. Numerical experiments and analysis

In order to carry out the numerical experiments of the robust optimization model mentioned above more objectively, two sets of data with certain and uncertain demands are set to carry out comparative numerical analysis, so as to test the robustness of the model and find out the effect of some values changes of some parameters in the model on the results of the model.

3.1. Robustness verification

Taking the questionnaire research results of passengers' trip purpose of Li, Zhu, and Chen as reference [11], assuming that there are 3 kinds of passengers in the system. Under the condition of uncertain demands, assuming that each kind of passengers have four kinds of demands scenarios with known probability, these scenarios and the probability of each scenario are shown in Table 1.According to the above principle and actual cost data obtained from the airlines' marketing department, assuming that the fixed cost and per unit variable cost for airlines copes with three agents are 25 and 4, 3, 2; preparation cost of airlines is 48; maximum transportation capacity is 1100; the bottom limit of proportion of the distribution channel is 60%. The data of income is shown in Table 2 to Table 5.

Table 1. Scenarios of demands

Scenarios	Prob.	Business	Leisure	Others
1	0.1	174	160	144
2	0.2	178	195	144
3	0.3	384	355	239
4	0.4	552	471	360
Certain	1	389	350	259

Table 2. Total main business income under scenario s

	Business	Leisure	Others
Airlines	64	53	46
Agents	57	60	40

Table 3. Per unit variable cost for airlines copes with passengers

	Business	Leisure	Others
Variable cost	18	10	20
Penalty	58	30	40

Table 4. Per unit variable cost for agents cope with passengers

Agents	Business	Leisure	Others
1	27	32	38
2	28	34	30
3	23	31	25

Table 5. Fixed cost to satisfy passengers' demands

	Business	Leisure	Others
Airlines	38	30	18
Agents 1	33	25	15
Agents 2	45	33	32
Agents 3	44	34	26

By LINGO software, keeping the other parameters unchanged and taking the service level α and β as 50%, when passengers' demands are certain and uncertain, maximum total profits of the coordination system is 20125 and 21765.It can be seen from the results the deviation rate of total costs of the coordination system is 8.15%, that is, for the given uncertain demands scenarios, the optimal solution of the model is close to the optimal solution when the demands is certain. And after verification, for any demands scenario $s \in G_s$, the optimal solution is always feasible.

3.2. Numerical experiments of the proportions of distribution channel

The proportions of tickets sold from direct and distribution channel are sum up to 1. Nowadays, the proportion of the distribution channel will have a falling trend. Setting the value of t from 100% to 0% to carry out numerical experiments, remaining other values in section 3.1 to observe the trend of maximum total profits of coordination system. The results are presented in Figure 2.

From Figure 2, we can see that the deviation rates of total profits show three-stage change trend with the value of t decrease. (1) When value of t is from 100% to 95% and from 5% to 0%, the deviation rate is near to 0, that is, decrease of the proportion of distribution channel basically does not affect the robustness of coordination system. But, due to some historical reasons, the two kinds of conditions of distribution channel mentioned above basically will not appear in real airlines' hybrid channels coordination system. (2) When value of t is from 95% to 50%, the deviation rates show a rising trend and peak in 50%, that is, the robustness of coordination system shows a falling trend and the falling trend will reach the valley when the proportion of direct channel is 50%. (3) When value of t is from 50% to 5%, the deviation rates show a falling trend.

These conclusions can be seen, for airlines and agents, based on the current distribution of the status quo and the airline's drastic reduction in the proportion of distribution channels, it is recommended that agents do not over reject the airline's reduced distribution channel ratio strategy, because the appropriate reduction in the proportion of distribution channels will also give agents robust advantages, it is a better strategy for agents. Therefore, agents can work closely with airlines, " retreat in order to advance", in the "sacrifice" of the appropriate reduction of distribution channels, get common pursuit of the best strategy of robustness with airlines, access to coordinated and win-win situation.

Next, setting the value of t from 100% to 0% to carry out numerical experiments, meanwhile remaining other values unchanged in section 3.1 to observe the trend of maximum total profits and airlines' and agents' profits of coordination system under certain and uncertain demands scenarios. The results of numerical experiments are shown in Figure 3 and Figure 4.

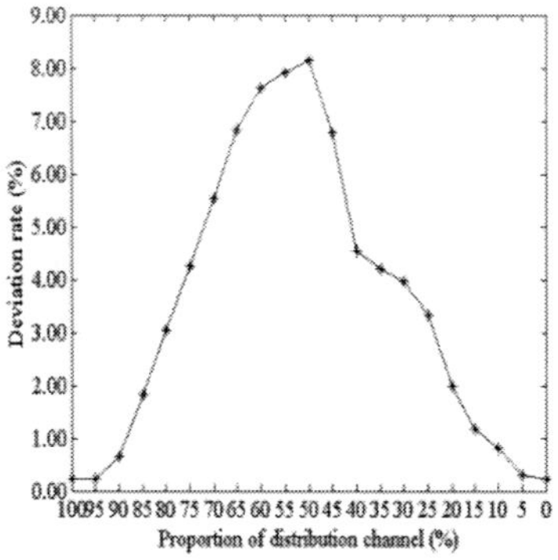

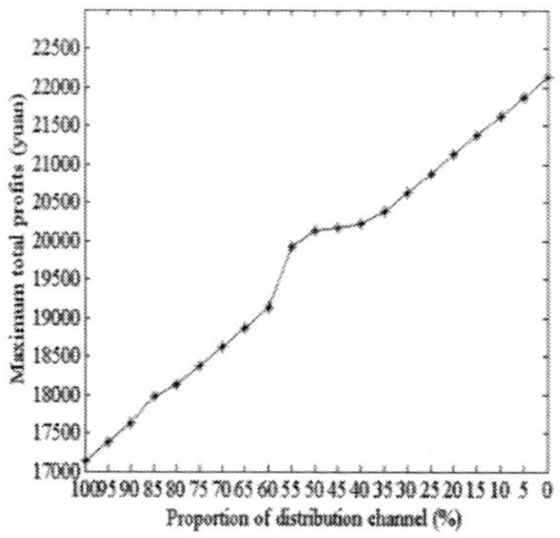

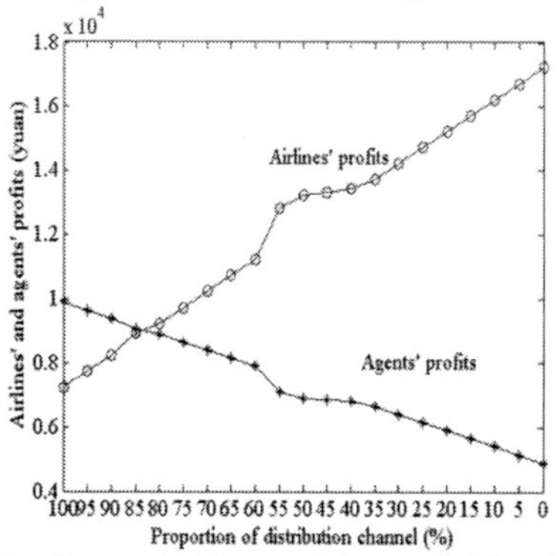

Figure 2. Change diagram of deviation rates **Figure 3.** Change diagram of total profits. **Figure 4.** Change diagram of airlines' and agents' profits.

It can be seen from Figure 3 that when the proportion of distribution channel decreases, the maximum total profits show a continued rising trend. It can be seen from Figure 4 that when the proportion of distribution channel decreases, the airlines' profits show the same trend with the total profits and the agents' profits show a continued falling trend.

For airlines, with the falling proportion of distribution channel, the higher proportion of direct channel, the higher total profits and airlines' profits will be. Therefore, airlines are supposed to spare no effort to improve the proportion of direct sales so as to seek maximize profits. For agents, with the falling proportion of distribution channel, the higher proportion of direct channel, the higher total profits will be but the lower agents' profits will be. Therefore agents should grasp the advantage of high proportion of distribution channel.

In summary, taking both the system robustness and total profits into account, when the proportion of distribution channel is from 100% to 50%, the deviation rates show a rising trend and peak in 50% and the maximum total profits increase all the time, that is, the robustness of coordination system shows a falling trend and the profits show a rising trend. Airlines and agents needs to make a choice between the system robustness and total profits. When the proportion of distribution channel is from 50% to 0%, the deviation rates show a falling trend and the maximum total profits still increase, that is, the robustness of coordination system and profits both show a rising trend. Airlines and agents should carry out deeper cooperation with each other and jointly promote the development of aviation market.

3.3. Numerical experiments of the level of service

With the development of the market, passengers' requirements on the service level are getting higher and higher, and the service level directly affects the revenue of airlines and agents. The following examines the impact of service level on system robustness.

Setting the service level of airlines and agents from 0% to 100% to carry out numerical experiments, meanwhile remaining other values unchanged in section 3.1. The results of numerical experiments are shown in Figure 5.

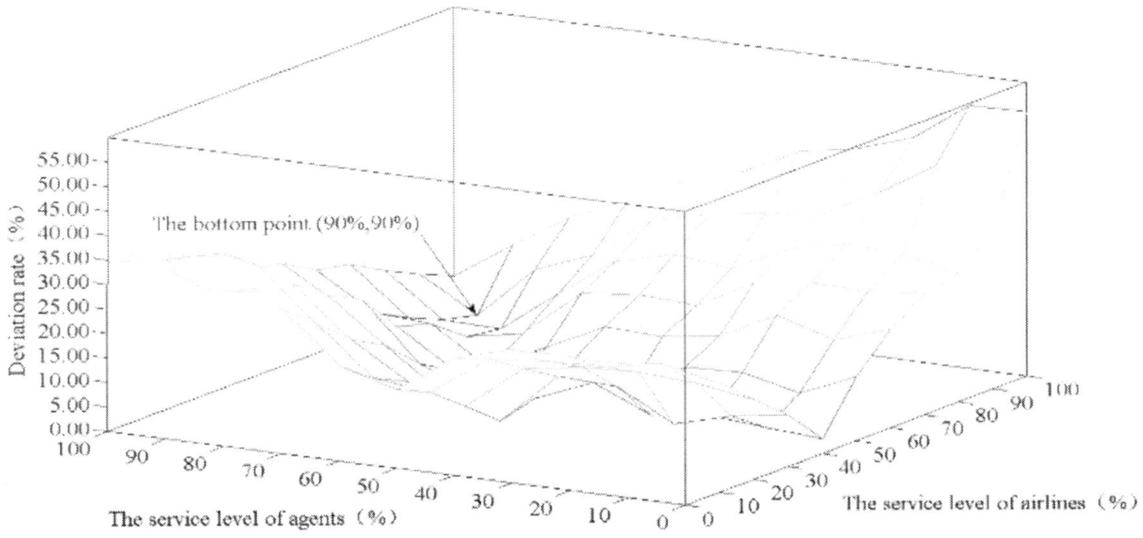

Figure 5. Change diagram of deviation rates of total profits with the change of the service level.

It can be seen from Figure 5 that when the service level of airlines remains the same, the deviation rates show the trend of first falling and then rising with the increase of service level of agents and when the service level of agents remains the same, the deviation rates show the same trend of first falling and then rising with the increase of service level of airlines. And the bottom point is when the service level of airlines and agents are both 90%, that is, the robustness of channels coordination is highest at that time.

Therefore, for airlines, when their service level is less than 90%, airlines should improve the service level to raise the proportion of direct channel through providing more seats by multivariate Internet direct channels, launching preferential package tickets and other measures, as well as increasing the cost of service. When their service level is greater than 90%, airlines can reduce the cost of service and coordination of channels, strategy of putting high proportion of preferential package tickets in the sales area where the service level is high can continue to use.

For agents, the increase of the service level is a double-edged sword. On the one hand, compared to the passengers of airlines, the passengers of agents have lower loyalty of channels, hence agents should improve the service level by providing more personalized user experience than airlines at the expense of more costs. On the other hand, agents can also obtain maximum benefits to offset the cost above by putting high proportion of preferential package tickets in the sales area where the service level is high.

4. Conclusion

By performing numerical experiments of the proportions of distribution channel and the level of service, with the proportion of distribution channel continues to decrease and the level of service continues to increase, ①airlines should focus on raising the proportion of direct channel and passengers' satisfaction through multivariate service and products; ②agents should improve the service level by providing more personalized user experience than airlines at the expense of more service costs and carry out deeper cooperation with airlines; ③airlines and agents should cooperate on service and experience and other aspects to attract passengers, and put high proportion of preferential package tickets in the sales area where the service level is high. It will help airlines' hybrid channels coordination system have higher robustness and more orderly, so as to reach win-win situation of airlines and agents.

But with the service requirements of passengers are higher and higher and the demands are increasingly diversified, how to enrich the quantification of service level are matters of concern.

Acknowledgments

This work was supported by the Graduate Open Funds of Nanjing University of Aeronautics and Astronautics in 2017 (kfjj20170708) and the Fundamental Research Funds for the Central Universities (NS2015068) and Social Science Fund of Jiangsu Province (15GLC001).

References

1. Cheng K, Lee Z H, Shomali H. Airline firm boundary and ticket distribution in electronic markets [J]. *International Journal of Production Economics*, 2012, **137**(1):137-144.
2. Zhang L, Ding X D, Sun Z F. An game model of airline tickets marketing channel [J]. *Mathematics in Practice and Theory*, 2014, **44**(19):155-161.
3. Saha S. Channel characteristics and coordination in three-echelon dual-channel supply chain[J]. *International Journal of Systems Science*, 2016, **47**(3):740-754.
4. David A, Adida E. Competition and Coordination in a Two-Channel Supply Chain[J]. *Production & Operations Management*, 2015, **24**(8):1358–1370.
5. Xu J W, Huang X W. A multi-objective robust operation model for supply chain with uncertain demands under electronic marketplace[J]. *Systems Engineering*, 2006, **24**(5):1-7.
6. Xu J W, Huang X W. Robust operation model for electronic supply chain with uncertain demands[J]. *Journal of Northeastern University (Natural Science)*, 2007, **28**(3):441-444.
7. Xu J W, Huang X W. A multi-objective fuzzy programming operation model for supply chain with uncertain price and demand[J]. *Statistics & Decision*, 2014, **32**(15):51-54.
8. Liu B, Dong M, Zhang D L. Analysis of Two-Period News-Vendor Problems with Service-Level Constraints [J]. *Journal of Shanghai Jiaotong University*, 2015, **49**(8):1244-1248.
9. Sun L, Zhang S. Study on Off-line Channel Inventory Model with Service Level Constraint Consideration [J]. *Logistics Technology*, 2014, **33**(10):254-257.
10. Shen C L, Zhang X X, Qing Z Q. Supply Chain Contracts Coordination with Consumer's Strategic Returns Policies under the Service Level Constraint [J]. *Chinese Journal of Management Science*, 2010, **18**(4):56-64.
11. Li H T, Zhu Y W, Chen W. The analysis of domestic air passenger transport market structure and consuming behavior [J]. *China Civil Aviation*, 2010, **120**(12):47-50.

MATEC Web of Conferences 179, 03027 (2018) https://doi.org/10.1051/matecconf/201817903027
2MAE 2018

Optimization Design of Thermal Test Waveguide Link for Communication Satellite

SHAO Yu-jie[1,a], Yu Zhong-xi[2], DU Chun-lin[1] and WANG Jing[1]

[1]*Thermal Test Technology and Product Department, Beijing Institute of spacecraft environment engineering , 104 Youyi Street Haidian,Beijing, PLC.*
[2]*Shandong Institute of Space Electronic Technology, No.513, Hangtian Road, Yantai, Shandong Province, PLC.*

Abstract. Problems in structure of water load power meter and waveguide assemblies of the communication satellite transponder thermal test. This paper analyzes water load layout and reasons of signal attenuation, and then it proposes an optimization that put water load detached from satellite's board. To ensure the reason of waveguide signal loss, this paper discusses the internal electromagnetic field distribution and currents on waveguide wall. the result of signal loss value showed that the new optimization can decrease signal attenuation value and shorten the process of crafts implementation.

1 Introduction

According to the statistics, as of March 2016, China put hundreds of satellites in orbit so that ranked the third in the world only preceded by America and Russia. At present, most of our orbiting satellites were related to information transmission[1], for example, the optical remote sensing satellite, military reconnaissance satellite, meteorological satellite, space exploration satellite and so on. However, the communication link usually use microwave to achieve signal transmission between satellites to satellites, and satellites to ground stations. In order to verify thermal design of the satellite correctly and the transponder subsystem work properly, it's Important to test transponder system on the ground[2].

After determining the radio frequency connection and power absorption scheme in the chamber, experimental designer set the cable and waveguide connection mode according to the sub-system of each frequency transponder. The frequency band absorption scheme mostly adopted water-loads absorption mode for thermal test or connected a high-power attenuation device outside the chamber. At present, the domestic general RF power absorption scheme needs to take into account the water-load. Therefore, the design of waveguide path and subsequent installation has very high requirements. Especially in the thermal test for large-scale communications and experimental satellite, such as high-throughput systems, have a certain impact on the longevity of satellites.

This article compared the measured values of electrical performance in different layouts, and thus improves the design of waveguides and water-loads. It aims to illustrate the implementation of the new solution and its significant advantages in reducing waveguide length, reducing signal attenuation, and reducing floor weight.

[a] Corresponding author: doriapompom123@163.com

2 Thermal test RF link scheme design

2.1 RF transmission in thermal test

Most of the satellite transponder thermal tests use metal rectangular waveguides to design transmission links. This paper takes the design of some satellite thermal test waveguide as an example. Firstly, the water cooling load layout method and the thermal test waveguide path were designed. Secondly, a brief demonstration and analysis for the design of the internal structure and the wall current of the guided mode were carried out.

2.2 Waveguide Path Design and Analysis in thermal test

There were 32 microwave loads would be installed in a satellite thermal test and must be fixed on the satellite boards. If all the models used as the previous models (directly fixed on the floor), it will be difficult to connect the water cooling load layout. We proposed two plans to solve this problem.

a. According to the design of the repeater, the forward outputs could be divided into two groups, and then the signal power synthesized by the external output multiplexer and then connected the water-load. However, the newly-manufactured output multiplexers needed to be redesigned and debugged with a period of at least 3 months.

b. The problem of fixing the water-load was solved by increasing the adapter plate. This program required 6~8 adapter plates, but the design time of the adapter plate and waveguide could be reduced within 1 month.

After considering the degree of difficulty and progress time, Option b was more feasible than Option a.

2.3 The Theory of Internal Field Calculation of Metal Rectangular Waveguide

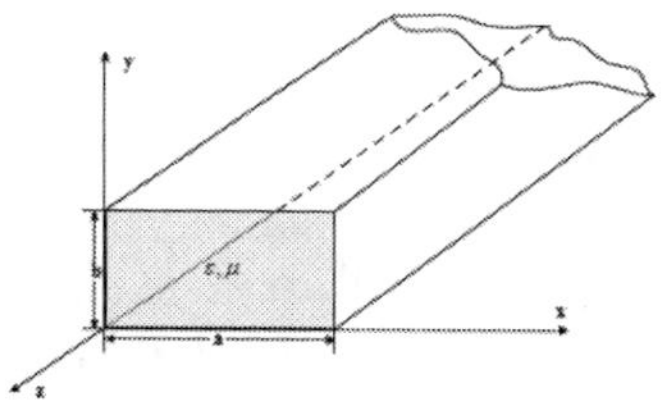

Figure 1. WR51 Waveguide（a=12.96mm,b=6.48mm）

As shown as figure 1, according to Maxwell Differential form of equation,

$$\left.\begin{aligned} \nabla \times \mathrm{E} &= -\frac{\partial B}{\partial t} \\ \nabla \times \mathrm{H} &= -\frac{\partial D}{\partial t} + J \\ \nabla \times \mathrm{B} &= 0 \\ \nabla \times \mathrm{D} &= \rho \end{aligned}\right\} \tag{1}$$

$$\left.\begin{aligned} \mathrm{D} = \varepsilon \mathrm{E} &= \varepsilon_0 \varepsilon_r E \\ \mathrm{B} = \mu \mathrm{H} &= \mu_0 \mu_r H \\ J &= \sigma E \end{aligned}\right\} \tag{2}$$

Boundary conditions,

$$E_x = 0, E_n = \frac{\sigma}{\varepsilon_0}, H_t = i, H_n = 0$$

According to (1), (2), the TE_{10} electric field component and magnetic component of no loss infinity WR51 waveguide could be solved,

$$\left.\begin{aligned} E_x = E_z = 0 \\ E_y = E_0 sin\left[\frac{\pi x}{a}\right] e^{j(\omega t-\beta z)} \\ H_x = -\frac{\beta}{\omega\mu} E_0 sin\left[\frac{\pi x}{a}\right] e^{j(\omega t-\beta z)} \\ H_y = 0 \\ H_z = -j\frac{\pi}{\omega a\mu} E_0 cos\left[\frac{\pi x}{a}\right] e^{j(\omega t-\beta z)} \end{aligned}\right\} \tag{3}$$

In formula(3), Phase constant $\beta = 2\pi/\lambda_g$;

Waveguide wavelength $\lambda_g = \lambda_0\sqrt{1 - (\lambda_0/\lambda_c)^2}$;

Critical wavelength $\lambda_c = 2a$;

Free space wavelength $\lambda_0 = c/f$

In formula(3),WR51 only exist E_y in mode TE_{10}. In flat xy,$E_y = E_0 \sin(\pi x/a)$. The electric field intensity was only related to x axis and it changed as sine law. When y=0,x=a, then $E_y = 0$. When x=a/2, E_{max}. When energy travels in the z direction,E_y would be in traveling wave state in z direction. And in profile of x=a/2, E_y sine distribution in z direction. The magnetic field in WR51 had H_x, H_z Component. Magnetic lines were distributed in the flat xz. Affinity relationship of E_y and $H_x$decided the energy travelled in z direction. When E_y and H_x in same direction, E_y and H_x reached maximum. When in x direction, E_y and H_x in same direction, sine distribution, the same as in cross section and longitudinal section. In z=0 section, H_z cosine changed in x direction. And in position of x=0,x=a,H_z was maximum, in position of x=a/2, $H_z = 0$.

2.4 Metal Rectangle Waveguide Wall Current Calculation Principle

Wall current size and direction were determined by the tangential magnetic field near the wall,

$$J_\delta = \hat{n} \times H_{tan} \tag{4}$$

Rectangular waveguides almost working in theTE_{10} mode. According to(3)(4)
(y=0) (y=b), $\hat{n} = \pm y$,

$$J_\delta = \hat{y} \times [\hat{x}H_x + \hat{z}H_z] = \hat{x}H_x - \hat{z}H_z = \left[H_{10}cos\left(\frac{\pi x}{a}\right)\hat{x} - j\frac{\beta a}{\pi}H_{10}sin\left(\frac{\pi x}{a}\right)\hat{z}\right]e^{j(\omega t-\beta z)}$$

$$J_\delta = \hat{y} \times [\hat{x}H_x + \hat{z}H_z] = -\hat{x}H_x + \hat{z}H_z = \left[-H_{10}cos\left(\frac{\pi x}{a}\right)\hat{x} + j\frac{\beta a}{\pi}H_{10}sin\left(\frac{\pi x}{a}\right)\hat{z}\right]e^{j(\omega t-\beta z)}$$

$\hat{n} = \hat{x}$,

$$J_\delta = \hat{x} \times \hat{z}H_Z = -\hat{y}H_Z = H_{10}e^{j(\omega t-\beta z)}\hat{y}$$

$\hat{n} = -\hat{x}$,

$$J_\delta = -\hat{x} \times \hat{z}H_Z = \hat{y}H_Z = -H_{10}e^{j(\omega t-\beta z)}\hat{y}$$

Therefore, when in mode TE_{10}, the wall current only had J_y, in same size and direction. In the up and down wall, the current consist of J_x and J_z. In the same position –x, the wall current in same size but in opposite direction[3].

3 Optimization and data comparison

3.1 Pre-optimization plan and test results

This paper chosed 12 Rf links among the 32 RF links in some satellite. According to plan b in 2.2, the number of waveguide as following table 1 shows,

Table 1. Quantity statistics for Pre-optimization plan

Port No.	Straight	E Bend	H Bend	Twist or Soft
101	5	1	2	0
102	6	2	2	1
103	4	2	1	0
104	5	1	1	0
105	6	2	2	0
106	4	1	1	1
107	4	2	0	1
108	3	2	1	0
109	2	1	1	0
110	4	2	1	1
111	5	1	1	1
112	3	2	1	0

To verify that the waveguide assembly met the electrical performance requirements and that the waveguide assembly needed to be tested prior to testing. The test included VSWR and Insertion Loss. The data were shown in the table 2.

Table 2. Component electrical performance test data

Port No.	Insertion Loss Index (dB/m)	VSWR Index	VSWR DATA	IL DATA(dB/m)
101	0.25	1.2:1	1.55	0.58
102	0.25	1.2:1	1.69	0.59
103	0.25	1.2:1	1.45	0.4
104	0.25	1.2:1	1.53	0.32
105	0.25	1.2:1	1.72	0.45
106	0.25	1.2:1	1.68	0.58
107	0.25	1.2:1	1.72	0.54
108	0.25	1.2:1	1.39	0.54
109	0.25	1.2:1	1.35	0.58
110	0.25	1.2:1	1.92	0.61
111	0.25	1.2:1	1.67	0.42
112	0.25	1.2:1	1.38	0.51

Although this program could meet the current thermal test requirements of transponders the design was not only too complicated, but also cost lots of time. The water-load would be fixed to the satellite's board, it can affect satellite's life. In order to simplify the tooling and shorten the test design cycle, a test scheme was put forward to mount the water-cooled load directly on the adjustable bracket. The solution could reduce the VSWR and insertion loss value of the waveguide assembly. And it also helped water-load completely depart from the satellites. The new method could effectively solve the interference problem between the transponder thermal test waveguide assembly and the on-board equipment.

3.2 Structural optimization and analysis

The new scheme minimized the waveguide design path. Water-cooled load which was fixed on the adjustable bracket could have at least 10mm adjustment margin in three dimensions. Reduced use of flexible or twisted waveguides,could greatly suppressed the attenuation of waveguide components [4]. After accounting and testing the new path, the number of waveguide components and electrical performance test results were shown in the table 3, and the VSWR data were shown in table 4.

Table 3. Optimized water-cooled load loop waveguide specification statistics

Port No.	Straight	E Bend	H Bend	Twist or Soft
101	1	1	1	0
102	2	1	0	0
103	2	0	1	0
104	2	1	1	0
105	2	0	1	0
106	1	0	0	0
107	1	1	0	0
108	1	1	0	0
109	2	1	0	0
110	2	1	0	0
111	1	1	0	0
112	2	1	0	0

Table 4. Optimized component electrical performance test data

Port No.	Frequency(GHz)	VSWR DATA	IL DATA(dB/m)
101	17-21	1.16	0.16
102	17-21	1.08	0.31
103	17-21	1.14	0.35
104	17-21	1.14	0.37
105	17-21	1.06	0.40
106	17-21	1.07	0.31
107	17-21	1.11	0.33
108	17-21	1.18	0.21
109	17-21	1.10	0.66
110	17-21	1.03	0.33
111	17-21	1.15	0.32
112	17-21	1.04	0.29

Compared with table 2, it was obviously to found, VSWR and insertion loss values were greatly reduced, and the program effectiveness could be verified.

4 Actual design study

Based on the research in the previous section, the water-cooled load distribution optimization design was performed for a certain type of satellite transponder thermal test waveguide assembly[5]. In order

to fixed 32 water-loads on board, Satellites needed to be opened on the east and west boards for the adapter plate during thermal reform if we chose original plan. It was estimated that approximately 500 test waveguides were needed. In addition, it was also necessary to design a microwave-loaded circulating water pipeline with a long implementation cycle and high operational risk.
The adventure of optimization design had 4 aspects,

- Reducing the output of experimental waveguide designs.
- Main line work auxiliary line.
- Reduce design interface
- Reduce test preparation difficulty and implementation risk.

5 Conclusions

By analysing the current design of the power absorbing waveguide in the transponder sub-system tank, the factors influencing signal attenuation were described in detail. Based on the original scheme, an adjustable bracket scheme that installs the water cooling load away from the satellite panel was proposed and compared. After the signal standing wave and insertion loss in the original scheme and optimization scheme, the optimized scheme has great advantages in reducing the signal attenuation value, reducing the number of used waveguide components, reducing the use of twisted waveguides and flexible waveguides, and even not using them. At the same time, the optimized solution can reduce the weight of satellite panels and increase the service life of satellites by solving the problems of complicated waveguide design and complicated installation.

References

1. Gan Zhngmin, Zhang Gengxin. Current development of satellite communications technology[A], Journal on Communications,2006,27(8)
2. Yang Yong, Liu Bo, Wang Jue, Tian Ye. The thermal vacuum test for telecommunication satellite[A], Spacecraft Environment Engineering,2013, 30(3),275-279
3. Li Yuankai. The Research of Water Load Power Meter with Waveguide Structure[J].
4. Xue Renwei, Xue Yaoming, Yu Boying. High power flexible rectangular waveguide[J]. Option Fiber & Electric Cable, 2005,(5):18-20
5. Li Te, Hao Erjuan, Li Zaijin,Wang Yong, Lu Peng, Qu Yi. Optimization of waveguide structure for high power 1060nm laser[J], Infrared Millim Waves, 2012,(3),226-230

Research on Applicability of Hot-Bulb Anemometer under Low Pressure

Li Xiyuan[1,2,a], Yin Xiaofang[1], Hou Yaqin[1], Gao Qinghua[1] and Wang jing[1]

[1] *Beijing Institute of Spacecraft Environment Engineering, Beijing 100094, China*
[2] *Beihang University, Beijing 100191, China*

Abstract. In Mars and other deep space exploration missions, the planetary atmosphere makes the difference in heat transfer characteristics on the planetary surface and on-orbit environment. In order to achieve the purposes of thermal model correction and spacecraft verification in extreme environment, Mars rover needs to be tested in a simulated Mars environment including low pressure, solar heat flux, wind speed and background temperature. Thus, wind speed should be measured at multiple points in the Mars rover thermal test. In a general Mars rover thermal balance test, the requirement for wind speed control and measurement is 0-15m/s under 700Pa pressure. The current anemometer for industrial use is mainly based on the dynamic pressure, heat or ultrasound. They have a small signal and need to be recalibrated at low pressures. In this paper, a constant heat flux hot-bulb anemometer model has been built using dimensionless number analysis method, with which the anemometer response under low pressure has been calculated. A series of calibration test has been employed to verify the model in space environment chamber. The two methods above reached a similar result, which demonstrates the effectiveness of the analysis.

1. Introduction

On the surface of Mars, the atmospheric pressure is about 700Pa; the main component is carbon dioxide, the temperature is about -120~20℃, and the wind speed is 0-15m/s. The low pressure convective makes the Mars rover face a different environment from the earth surface and the on-orbit environment. In order to achieve the purpose of thermal model correction, early failure screening, performance verification and other targets in extreme environments, low pressure, solar heat flux, wind speed and background temperature need to be simulated in a Mars rover thermal test. The thermal test of the Mars rover is usually carried out inside a large or medium space environment simulation chamber [1][2]. The cold and black background and ground contact heat transfer are achieved by the adjustable temperature shroud, the 700Pa pressure is simulated by a pressure control system, and a fan is employed to blow up the wind. The wind speed in the chamber should be measured in real time across the whole test. The current anemometer for industrial use is mainly based on the dynamic pressure, heat, and ultrasound.

Dynamic pressure anemometers include wind cup, pitot tube, five-hole probe and other forms, pitot tube and five-hole probe has been widely used in low-pressure wind speed measurement, but in the 700Pa environment, the dynamic pressure brought by the wind speed of 0-15m / s is only about 0.01Pa ~ 10Pa. Although the industrial micro pressure sensor using can achieve the resolution of 10⁻

[a] Corresponding author: lxy_422@msn.com

[3]Pa, they often have a larger size, heavier weight and can be hardly worked under low pressure. multi-point wind speed measurement and wind direction measurement will make the measurement system becomes hugely complicated [3].Thermal anemometers include hot wire, hot bulb, hot film and other forms, the wind speed is calculated from the sensor probe heat loss[4].Thermal anemometers are mounted on multiple Mars rovers such as the Netlander, Viking and Beagle 2. Some thermal anemometers can measure wind direction through thermal wake technique [5][6]. The thermal anemometer has the advantages of simple structure, lightweight, low power consumption, compactness and no moving parts. However, the disadvantages are also apparent: 1) calibration is crucial, and the correlation cannot be obtained by analytical model accurately; 2) As time goes on, calibration will drift slowly because of line resistance changed. 3) The change of gas temperature and pressure will also lead to the signal change. The measuring accuracy of the sensor depends on the measurement accuracy of the gas temperature and pressure. Because commercial product s can be hardly used in Mars environment, the present thermal anemometers used by the Mars rovers are often custom product. The ultrasonic anemometer utilizes the characteristic that the velocity of the sound wave is only related to the gas composition and temperature. Ultrasonic anemometers do not need to be calibrated under different pressure. The main drawback is that ultrasound attenuates significantly at low pressures, making it difficult to screen out signals from noise. In Mars explore, newly designed transducers are usually used instead of existing products [7][8]. Also, the ultrasonic anemometers often have a large volume, making it difficult to use in multi-point wind speed measurement.

To sum up, under the extremely low pressure, the primary method of wind speed measurement is a customized wind speed sensor at present. This paper aims at the applicability of existing industrial thermal anemometers under low pressure. The heat transfer model of the constant heat flux hot-bulb sensor under different pressure is built by the dimensionless number analysis method. The probe output under different pressures is analyzed. A test system has been developed to verify the analysis. Similar results have been obtained from the test and the dimensionless number analysis method. The correctness of the rue number analysis method has been verified, which provides a reference for the future wind speed measurement in the ground test of the Mars rovers.

2. Convection heat transfer model of hot-bulb anemometer

2.1. Simulation model

In this paper, a constant heat flux hot-bulb anemometer is selected as the research object, its diameter is 0.6mm, and the power is 0.08W. The internal structure of the probe is shown in Figure 1:

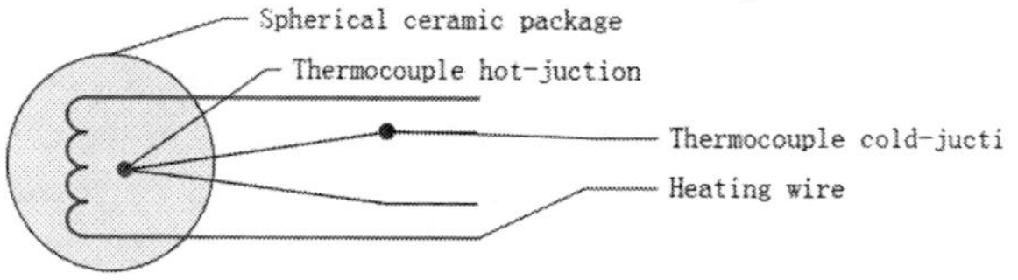

Figure 1. Probe model

When the sensor works, a constant current is applied to the heating wire to produce a constant power. In this case, the temperature of the cold junction of the thermocouple can be considered as the temperature of the fluid, and the temperature of the hot junction is hot bulb temperature. At this time, the convection heat transfer coefficient varies with the wind speed. It also leads to the difference of bulb surface temperature and difference of the thermoelectric potential. Therefore, the wind speed can be obtained by measuring the thermoelectric potential. In this paper, a customized millivolt transmitter was employed to measure the probe output.

2.2. Dimensionless number analysis

A convective heat transfer model for the probe has been built by dimensionless number analysis method. The external heat transfer of the probe can be divided into three forms: radiation heat transfer, convection heat transfer, and heat conduction through the probe lead and the support. In the heat balance state, The heating power is equal to the sum of the probe's heat loss:

$$Q_{Power} = Q_{con} + Q_{cond} + Q_{rad} = ks(T_s - T_e) + hs(T_s - T_e) + \varepsilon s\sigma(T_s^4 - T_e^4) \quad (1)$$

Where Q_{Power} is heating power on the probe(W), Q_{con} is the heat loss through conduction(W), Q_{cond} is the heat loss through convection (W), Q_{rad} is the heat loss through radiation (W), k is the equivalent heat transfer coefficient for dissipating heat flow through the form of heat conduction(W/(m^2•K)), s is the surface area of the hot bulb(m^2), h is the convection heat transfer coefficient(W/(m^2·K)), ε is the emissivity of the hot bulb surface, σ is Stefan-Boltzmann's constant, and the value is 5.67×10^{-8}W/(m^2·K^4), T_s is the temperature of the hot bulb, T_e is the temperature of the gas. When conduction and radiation heat transfer is negligible, the equation can be simplified as follow:

$$C_1 Q_{Sensor} = hs(T_s - T_e) \quad (2)$$

Where Q_{Sensor} is the total heat flow on the anemometer, C_1 is the ratio of heat flow on the hot bulb to the total heat flow. Reynolds number near the hot bulb probe can be calculated as:

$$Re = \frac{vl}{\upsilon} = \frac{vl\rho}{\mu} \quad (3)$$

Where v is gas flow velocity(m / s), l is the characteristic length of the hot bulb(m), its value is 0.0006m, υ is the kinematic viscosity of the gas (m^2/s), can be calculated by dynamic viscosity μ (Pa·S)·and the density ρ (kg/m^3). The dynamic viscosity does not vary with the density of the gas, and the density can be calculated through ideal gas formula by gas pressure and temperature.

According to the calculation, when the wind speed is among 0-15m/s under ambient pressure, the Reynolds number is about 20~500, while in the 700Pa environment, the Reynolds number is only 0.1 ~ 2 when the wind speed is among 0-15m/s. There are a variety of empirical formulas that give the relationship between Reynolds and Nusselt numbers. Kramers, Whitaker, Yuge, Vilet, and Raithby [9] have presented their expressions for different ranges, but most of them are suitable for medium and high Reynolds numbers. For Reynolds number among 0.1-100, most of the formulas have more than 50% error. The Whitaker correlation is closest to the experimental result at low Reynolds number[10]:

$$Nu = 2 + (0.4Re^{0.5} + 0.06Re^{2/3})Pr^{0.4} \quad (4)$$

Where Pr is Prandtl number. Convection heat transfer coefficient h can be expressed by Nusselt number:

$$h = \frac{Nu\lambda}{l} \quad (5)$$

Where λ is thermal conductivity (W/(m·K)). It changes little when the gas pressure changes. Therefore, the temperature difference between the sphere and the incoming gas can be expressed as:

$$\Delta T = \frac{C_1 Q_{Sensor} l}{Sf_\lambda(T_e + \Delta T)\left(2 + \left(0.4 \times \left(\frac{vl\rho}{f_\mu(T_e + \Delta T)}\right)^{0.5} + 0.06 \times \left(\frac{vl\rho}{f_\mu(T_e + \Delta T)}\right)^{0.5}\right) f_{Pr}(T_e + \Delta T)^{0.4}\right)} \quad (6)$$

Where T_e is incoming gas temperature (℃), f_λ , f_μ , f_{Pr} are functions of physical parameters change with temperature, respectively. When considering the thermocouple output signal and the temperature was linear, the above equation can be simplified as:

$$\Delta u = \frac{C\,Q_{Sensor}\,l}{Sf_{\lambda}(T_e + \Delta T)\,(2 + (0.4 \times (\frac{vl\rho}{f_{\mu}(T_e + \Delta T)})^{0.5} + 0.06 \times (\frac{vl\rho}{f_{\mu}(T_e + \Delta T)})^{0.5})f_{Pr}(T_e + \Delta T)^{0.4})} \tag{7}$$

Where constant C includes the thermocouple temperature coefficient and constant C_1, the surface area error, a small amount of radiation, thermal conductivity and other parameters that difficult to obtain by analysis are also embodied in the constant. The constant C can only be solved by the experimental method.

2.3. Natural Convection and Forced Convection Intensity Analysis

There are both natural convection and forced convection around the thermal anemometer probe. For a constant heat flux probe, natural convection intensity also changes with the probe surface temperature. In a wind measurement test, the force convection should be the primary heat transfer forms around the probe to minimize the errors caused by natural convection. The intensity of natural convection and forced convection can be evaluated by dimensionless number Gr / Re^2:

$$\frac{Gr}{Re^2} = \frac{\frac{ga_v \Delta t\, l^3}{v^2}}{(\frac{vl}{v^2})^2} = \frac{ga_v \Delta t\, l}{v^2} \tag{8}$$

Where Gr is Grashov number, g is the gravitational acceleration (m/s^2), Δt is the temperature difference between the hot bulb and the ambient gas, a_v is the body expansion coefficient (K^{-1}), can be expressed as follows:

$$a_v = -\frac{1}{\rho}\left(\frac{\partial \rho}{\partial T}\right)_p = -\frac{1}{\rho}\left(\frac{\partial\left(\frac{P}{RT}\right)}{\partial T}\right)_p = -\frac{1}{\rho} \times \frac{P}{R} \times -\frac{1}{T^2} = \frac{1}{T} \tag{9}$$

Where ρ is density(kg/m^3), P is pressure(Pa), T is temperature (K), R is the gas constant. As the formula shows, body expansion coefficient a_v is independent of pressure. Therefore, the Gr / Re^2 is only related to gravitational acceleration, characteristic temperature, temperature difference, characteristic length and flow velocity, of which gravitational acceleration, and characteristic length and flow velocity are set value, temperature difference can be calculated by equ.5. Characteristic temperature is the average temperature of the incoming gas and the temperature of the hot bulb. The flow is dominated by force convection when Gr / Re^2 is below 0.01, is dominated by natural convection when Gr / Re^2 is higher than 10. When Gr / Re^2 is between 0.01 and 10, both force convection and natural convection should be considered in the analysis.

3. Test

According to the literature [3], there are three basic methods for simulating the wind speed at a low pressure of about 1000 Pa, the rotation method, the closed wind tunnel method, and the open wind tunnel method. In the rotation method, the rotating platform drives the wind probe to rotate, the fluid medium is not moving, and the relative motion between the probe and the fluid medium is the wind speed. The fundamental principle is shown in Figure 2. In general, the cantilever and the rotating platform are set in a space environment simulation chamber, and the chamber is filled with the gas with desired pressure and temperature. The rotation method has the advantages of low cost, simple structure, pressure and gas composition stable. In order to verify the dimensionless number analysis model, a rotating calibration system has been built based on a median-size space environment simulation chamber, as shown in Figure 2:

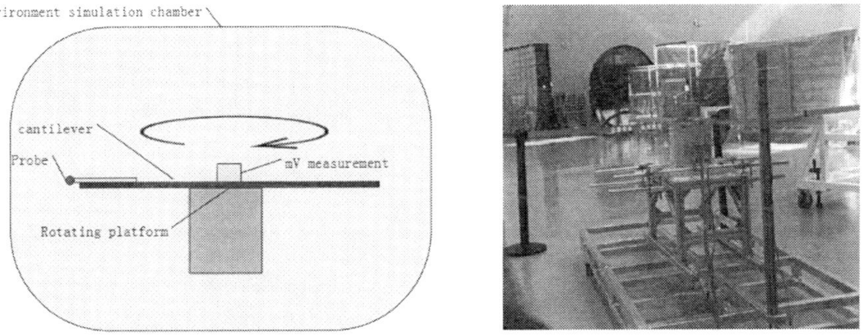

Figure 2. Rotating wind speed simulation system

As can be seen from the figure above, the hot bulb probe is fixed on the end of the cantilever. Wind speed around the probe can be calculated by the rotating platform angle speed and length of the cantilever. The 12-bit precision transmitter measures the millivolt signal and transfers the data to the computer outside the chamber by RS485. It is set on the rotating platform and can work in vacuum environment directly. The acquisition resolution is 0.01mv. The probe's constant heat flow was applied by a programmable power outside the chamber, the total power of the anemometer probe was 0.08W. In this paper. The probe was tested in space environment chamber under ambient pressure, 40000Pa, and 700Pa and the background and fluid temperature was ambient temperature. The output signal under ambient pressure is compared with the signal predicted by the dimensionless number analysis method, as shown in Figure 3($C \approx 36.10$), Gr / Re^2 under different wind speed was shown in Figure 4. Because of the aerodynamic drag under ambient pressure, the maximum simulated line speed is 11m/s.

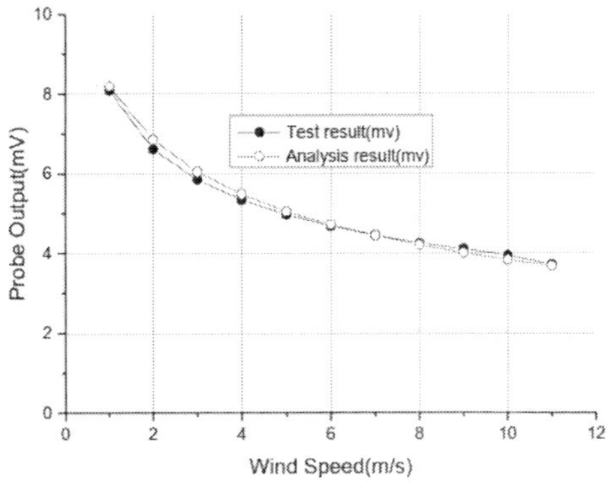

Figure 3. Probe output and dimensionless number analysis result under ambient pressure

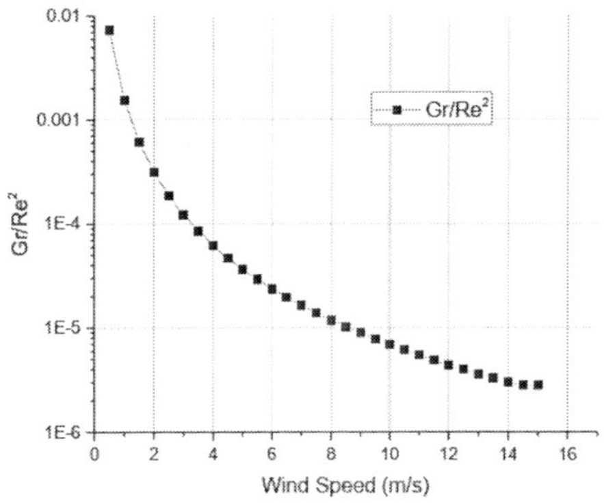

Figure 4. Gr / Re^2 at different winds speed under ambient pressure

As can be seen from the figure, under the ambient pressure, the dimensionless number analysis results were in good agreement with experimental data under ambient pressure. The Gr / Re^2 were below 0.01 in all test case, so that the force convection occupied a dominant position in the convection around the hot bulb probe. The probe output and analysis result under 40000Pa and 700Pa are shown in Figure 5. Gr / Re^2 under 700Pa are shown in Figure 6:

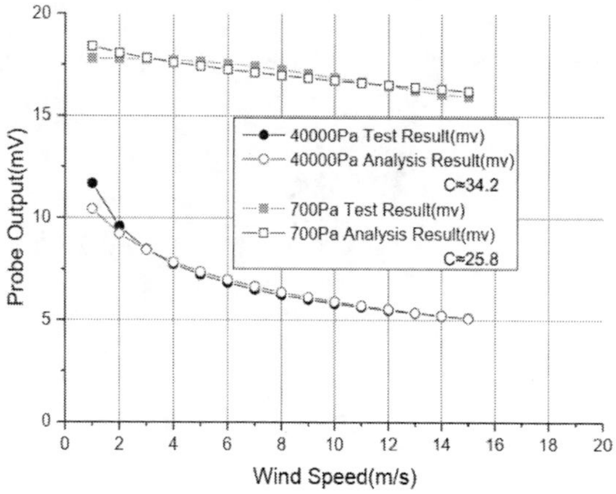

Figure 5. Probe output and analysis result under 40000Pa and 700Pa

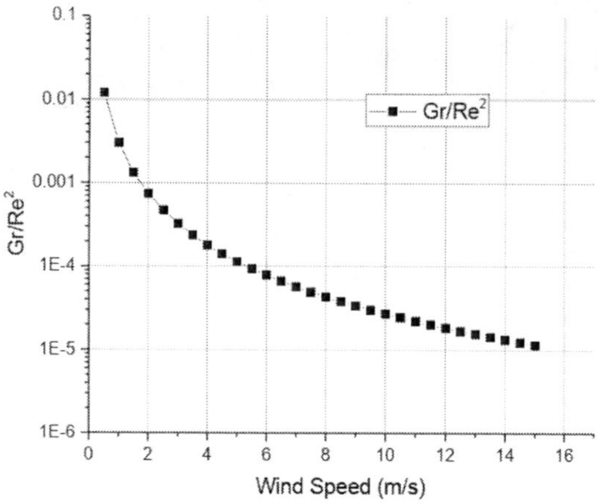

Figure 6. Gr / Re^2 at different winds speed under 700Pa pressure

As can be seen in Figure 6, Gr / Re^2 is still less than 0.01 for the wind speed above 1m/s. Therefore, the forced convection still occupies the dominant position in the total convective heat transfer, and natural convection will not bring additional measurement error. As the output signal and analysis result are seen in Figure 5, the dimensionless number analysis can be consistent with the test results by modifying the constant C. The constant C is about 0.041 at 40000Pa, and it deviates 7.1% from the ambient pressure result, but, the deviation is more than 44.9% under 700Pa pressure. The change is mainly due to the change of Nu-Re correlation with pressure. This result is similar to the experimental and analytical results in [11]. Thus, anemometers probe using under low pressure should be corrected with the experimental data. Meanwhile, because of the background temperature dependence of the radiation heat transfer, when using constant heat flow hot bulb anemometer under low pressure, the calibration environment should have similar radiation background, or radiation insulation material should be set around the probe to minimize the errors caused by radiation heat transfer. Also, as shown in Figure 6, the sensitivity of the sensor at different wind speeds is about 0.01-0.2mv/(m·s^{-1}) under 700 Pa. When the wind speed measurement is performed at a low pressure, the millivolt signal data transmitter should also meet the accuracy requirements.

4. Conclusion

Aiming at the problem of wind speed measurement under low pressure, this paper establishes a model of wind speed and output signal for industrial hot-bulb anemometer by using the dimensionless number analysis method. A rotating low-pressure wind speed calibration system has been built based on median size space environment simulation chamber. The system has been tested under ambient pressure, 40000Pa and 700Pa respectively. The sensor output under different wind speed conditions was obtained. The dimensionless number analysis model was corrected by the experimental data. The simulation model and the experiment achieved a similar result. In this paper, the dimensionless number analysis model for constant heat flow hot bulb probe can be directly applied to the output signal prediction at a pressure above 40000Pa and can be corrected by experimental data for 700Pa low-pressure environment. The analysis method can be used to evaluate the output of different anemometers in different environments, and provide a reference for the development of the corresponding test in future.

References

1. Ransome T, Pesket S, Toplis G. Thermal Balance Testing of the Beagle2 Mars Lander[C]. International Symposium Environmental Testing for Space Programmes, 2001.
2. Johnson K. Simulation of Mars Surface conditions for Characterization of the Mars Rover Thermal Response[J].
3. Wilson C F. Measurement of wind on the surface of Mars[D]. Oxford, 2003.
4. Bruun H H. Hot-Wire Anemometry[J]. Oxford Univ Pr, 1995.
5. Chamberlain T E, Cole H L, Dutton R G, et al. Atmospheric measurements on Mars - The Viking meteorology experiment[J]. Bulletin of the American Meteorological Society, 1976, 57(9).
6. Seiff A, Tillman J E, Murphy J R, et al. The atmosphere structure and meteorology instrument on the Mars Pathfinder lander[J]. Journal of Geophysical Research Planets, 1997, 102(E2): 4045–4056.
7. Kapartis S S R. Wind speed and direction measurement using Acoustic Resonance airflow sensing. FT Technologies, 1999.
8. Banfield D, Gierasch P J, Toigo A, et al. Mars Acoustic Anemometer[J], 2012.
9. Will J B, Kruyt N P, Venner C H. An experimental study of forced convective heat transfer from smooth, solid spheres[J]. International Journal of Heat & Mass Transfer, 2017, 109:1059-1067.
10. Dennis S C R, Walker J D A, Hudson J D. Heat transfer from a sphere at low Reynolds numbers[J]. Journal of Fluid Mechanics, 2006, 60(2):273-283.
11. Numata D, Anyoji M, Sugino Y, et al. Characteristics of Thermal Anemometers at Low-Pressure Condition in a Mars Wind Tunnel[C]// Aiaa Aerospace Sciences Meeting Including the New Horizons Forum and Aerospace Exposition. 2013.

The Experimental Platform of Frosting by Vapor on Cryogenic Surface in Vacuum Environment

Guo Qinliang[1,a], Qie Dianfu[1], Shang Yonghong[1], Wang Jing[1] and Bi Yanqiang[1]

[1]*Beijing Institute of Spacecraft Environment Engineering, China*

Abstract. This paper introduces an experimental platform for studying frosting on the cryogenic surface in vacuum environment. This platform contains vacuum system, cold surface temperature control system, image acquisition system, data acquisition system and vapor generation device. The cold surface temperature can be varied in the range of -185 to 0℃, the vacuum degree of the chamber can be controlled in the range of 1 pa to 1bar, and the flow rate of water vapour can be controlled accurately. The experimental platform can be used to study the micro-morphology and the thermal conductivity of frost layer.

1 Introduction

Frosting is a common phenomenon in nature, and it also exists in the engineering field of cryogenics, refrigeration and aerospace. When the cold surface temperature is lower than the freezing point temperature of the water vapor, the water vapor will frost on the cold surface. At present, in the field of aerospace, manned spacecraft has not realized the closed-loop water circulation yet. It will discharge a certain amount of water into the space in orbit. Water sublimator of space suit also releases some water vapor. The released vapor will be around the spacecraft, then it maybe frost on the cold surface of the spacecraft. It will have a bad impact on the spacecraft performance. For example, frosting on the radiating surface may affect the thermal control performance of spacecraft; frosting on the lens may significantly influence the camera optical properties; frosting on the key joint may affect the moving performance of the mechanism.

The phenomenon of frosting also exists in the spacecraft ground simulation test. There is inevitable a certain amount of water vapor inside the spacecraft simulation chamber, which will deposited a layer of frost on the cryogenic surface. If the frost is formed on the critical surface of the test piece, it will affect the original performance of the test piece, resulting in the inaccuracy of the test result and even the failure of the whole test.

This paper is aimed to build experimental platform of frosting in vacuum cryogenic environment for studying the micro-morphology and the thermal conductivity of frost layer. It has an important engineering value for the safe operation of the spacecraft. There are some studies on the topography of the frost in vacuum cryogenic environment. For ultra-low cold surface temperatures, the frost layer growth is controlled by the outer boundary layer mechanism. There is a tendency that the frost layer thickness decreases as the cold surface temperature decreases, which is totally different to the frost growth on the cold surfaces of ordinary-low temperature[1]. Xiaomin Wu[2,3] studied the frost

[a] Corresponding author: 1159497408@qq.com

formation on aluminum surface with grooved structures, and concluded that the frost prefers to grow at the top edges and corners of the structures instead of inside the grooves on the grooved surfaces. Boxe etc[4],measured the surface grain size and distribution, specific surface area, density, porosity of the frost under the pressure at 260Pa by using electronic microscope. The measured porosity of the frost layer is 0.31. Laufer[5] studied the frost film condensed by pure water vapor at 100K temperature, and found that the frost film had a needle structure. As the heating continued, it remained until the ice film completely disappeared. Mayer and Atteberry[6,7] found a similar needle-like structure on the surface of the frost film. Mayer[7] concluded that there were a large number of microscopic holes in the frost film, and the micropores provided a large amount of adsorption ratio (a few hundred m^2/g). As the ice film grew, the adsorption ratio decreased gradually. Schmitt[8] came to the similar conclusion that micro-pores are only a few microns in size. The effective thermal conductivity of porous media depends on the thermal conductivity of each phase and the porosity, and on the array formed by particles and pores[9,10]. Zou Tonghua [11] studied the frost deposited on the surface of the cold well coils of the freeze dryer with pure ice as the freeze-dried material. In his experiment, the cold trap temperature is -40℃, the vacuum degree is lower than 10Pa. Thermal conductivity of the frost layer was obtained by Fourier Law within the range of 0.15 to 0.25w/(m·K), that is considered to be 1/4 to 1/3 at the atmospheric pressure.

At present, there is less research on the frosting phenomenon in the vacuum environment, lack of detail image information of the microstructure of the frost layer. The variation range of the thermal conductivity of the frost layer is large, and the data comparison verification is lacking. Therefore, it is necessary to establish an experimental platform to study frost phenomenon in the vacuum cryogenic environment. Base on the platform, temperature data and morphology image of the frost will be obtained, and it will lay the foundation for the research in the future.

2 Experimental apparatus

The vacuum cryogenic experimental platform is shown in figure 1, the experimental system consists of five parts: cold surface temperature control system, image acquisition system, data acquisition system and vapor generation device.

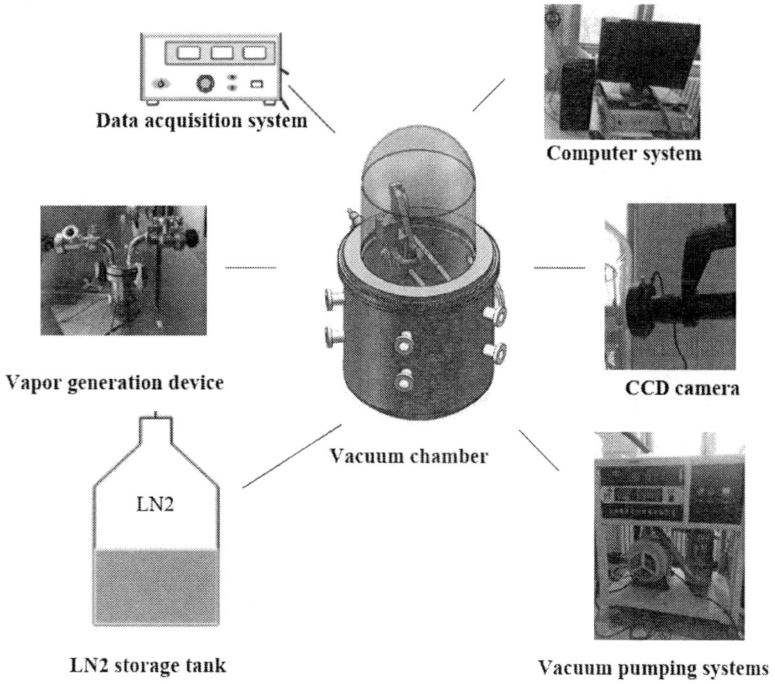

Figure 1. Schematic of the experimental platform

2.1 Vacuum system

The vacuum system mainly consists of vacuum chamber and vacuum pumping system. The vacuum chamber contains two parts: a stainless steel base and a quartz glass bell jar, which are connected by a flange. The effective volume of the chamber is$\Phi 380\times 800$mm.The quartz glass has such a good optical permeability that we can observe frost from outside the chamber by the microscopic camera. The vacuum pumping system is composed of dry pump and molecular pump, and the ultimate pressure is5×10^{-3}Pa.The vacuum degree can be controlled by the pumping rate and the outgassing rate. Figure 2 shows the vacuum degree changes during an experiment.

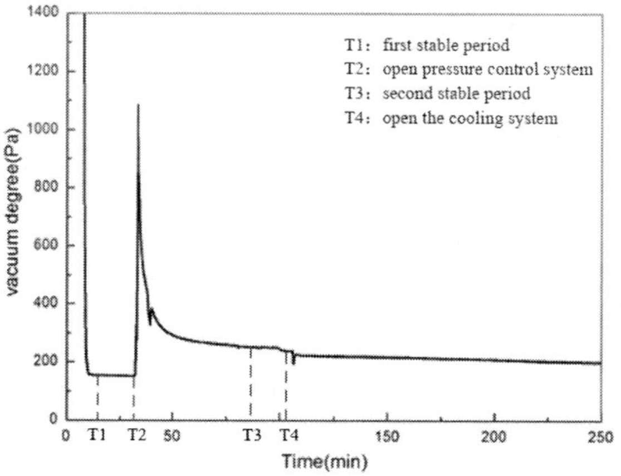

Figure 2. Vacuum degree as a function of time

2.2. Cold surface temperature control system

The surface temperature of the cold plate is controlled by the cooling system and the heating system. The cooling system has an open structure: liquid nitrogen flows out of the dewar, enters into cooling box to exchange heat with copper block, flows through cryogenic valve, is discharged into the atmosphere. The LN2 flux can be precisely controlled by the valve. The heating system consists of heater and program-controlled power. The cooling system and the heating system are regulated by a PID controller. By this way, the cold surface temperature can be controlled ranged from -185 to 0℃ with the fluctuating range of±0.3℃.The cold surface temperature in the experiment is shown in figure 3(a), 3(b) shows a locally amplified curve in steady state condition.

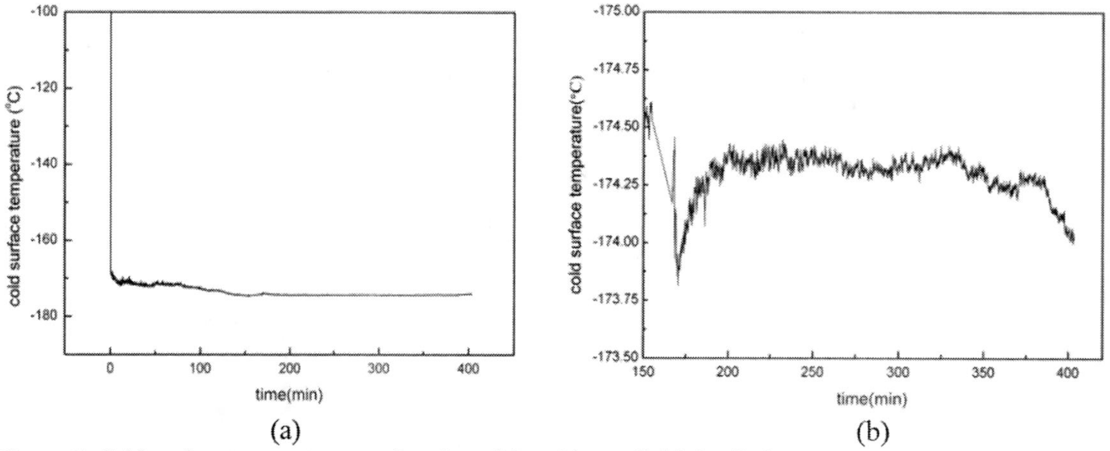

Figure 3. Cold surface temperature as a function of time.(a)overall (b) detail view

2.3 Vapor generation device

In this experiment, water vapor was generated by ice sublimation under low pressure. The vapor generation device is shown in figure4. Before the experiment, the pure distilled water is frozen into ice, then the prepared ice is added to the ice storage. The outer wall of the storage room is heated by the heater. After the experiment began, the ice began to sublimate as the pressure in the reservoir decreased, and the sublimation latent heat was provided by the heater, so that the water vapor could be produced continuously. The flow of vapor was controlled by the hand valve installed on the transmission line. The effect of wind speed on the experiment is not considered.

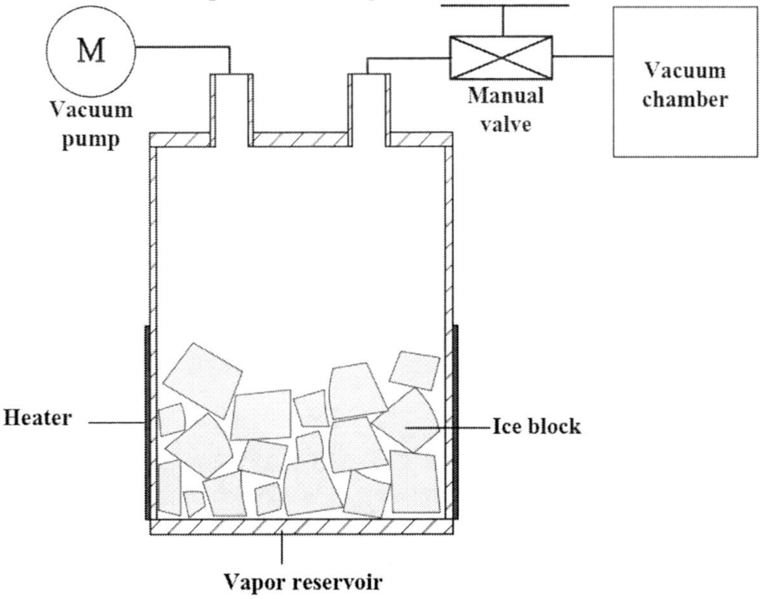

Figure 4. Schematic of the vapor generation device

2.4 Image acquisition system

The image acquisition system is used to measure the thickness of the frost layer and observe the morphology of the frost. The system consists of computer, image capture card, microscope and CCD camera (shown in figure5). The microscope is produced by Bo Hong optical instrument factory (the optical magnification can reach 4.5 times). The experiment uses the critical illumination for better observation. The light source has enough luminance to ensure the brightness of the image with no distortion on the frost surface temperature. The frosting image can be transported to the computer by the combination of photoelectric conversion and upper computer program. The change of frost can be recorded by the image capture card and be real-time monitored on the display.

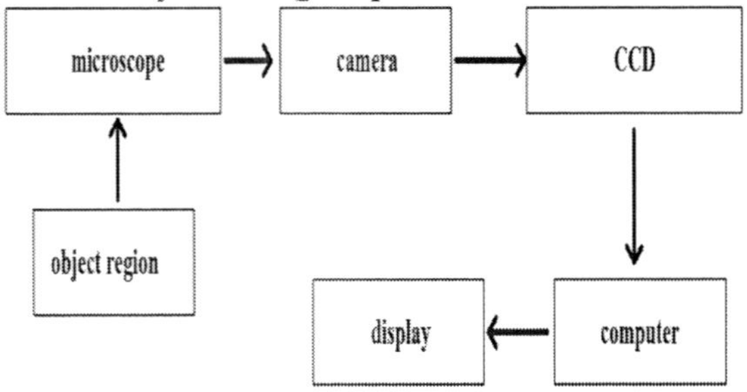

Figure 5. Schematic of the image acquisition system **Figure 6.** Image of frost structure by CCD camera

In the experiment, the camera has two directions: the horizontal one is used to observe the thickness and structure of the frost while the vertical one is used to observe the distribution of preliminary frost crystal formed on the cold surface (Shown in figure 6).

2.5 Data acquisition system

The cold plate, whose material is T2-copper, is shown in figure7. The plate is 70 millimeters in height, which are divided into two parts. The upper part of the cold plate, that is 50mm height, is the copper block. The top surface is used to frost, and the rest surfaces are arranged for the heating plates whose outside surfaces are covered with 10 mm thick PTFE insulation board to reduce heat leakage.

There are four thermocouples inserted into the holes to measure the top surface temperature, the other four thermocouples to measure the temperature of the block at the position of 20mm below the top surface. The lower part of the cold plate is the cold box, in which liquid nitrogen evaporate and cool the upper copper block.

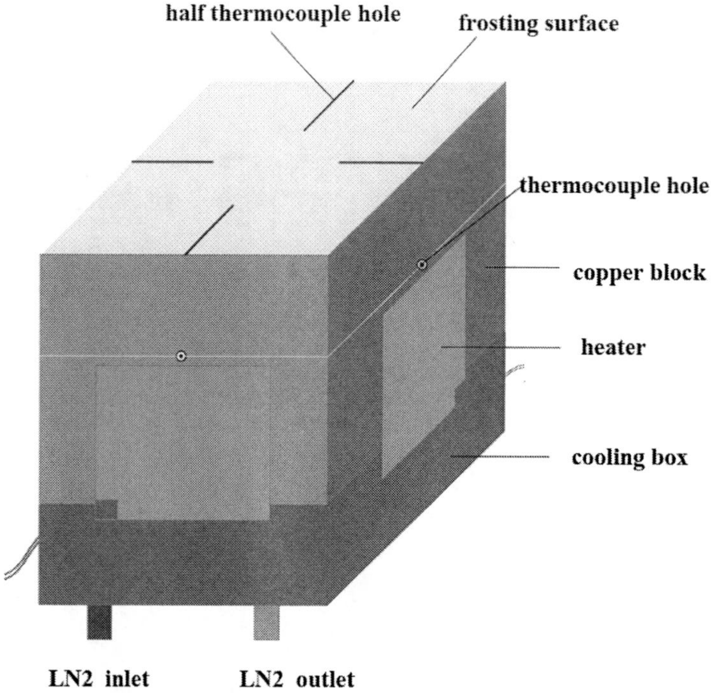

Figure 7. Structure diagram of the cold plate

From the above, we can obtain the temperature of the copper block, the top surface and the frost layer. Based on Fourier's law, the thermal conductivity of the frost can be calculated with the copper thermal conductivity known.

3 Conclusions

This paper introduced the principle and structure of the vacuum cryogenic experimental platform in detail. This experimental platform is aimed to study water vapor desublimation phenomenon. Through the test, the experiment conditions, including the cold surface temperature, the vacuum and the flow rate of water vapor, can be controlled accurately. The measurement system has the ability to observe the morphology of the frost and measure the thermal conductivity of the frost.

Acknowledgement

This work was supported by the National Natural Science Foundation of China (No.51506009).

References

1. L Li, Z Liu, Y Li, Y Dong. Frost deposition on a horizontal cryogenic surface in free convection[J]. International Journal of Heat and Mass Transfer,**113**(2017):166-175.
2. Xiaomin Wu, Fuqiang Chu, Qiang Ma, Bei Zhu. Frost formation and frost meltwater drinage characteristics on aluminum surface with grooved structures[J]. Applied Thermal Engineering, **118**(2017),448-454.

3. Xiaomin Wu, Fuqiang Chu, Qiang Ma. Frosting model based on phase change driving force[J]. International Journal of Heat and Mass Transfer,**110**(2017):760-767.
4. C.S. Boxe, B.R. Bodsgard, W. Smythe, M.T. Leu. Grain sizes, surface areas, and porosities of vapor-deposited H2O ices used to simulate planetary icy surfaces. Journal of Colloid and Interface Science **309** (2007) 412–418
5. Laufer, D., Kochavi, E., Bar-Nun, A., 1987. Phys. Rev. B **36**, 9219–9227.
6. Mayer, E., Pletzer, R., 1986. Nature **319**, 298–301.
7. Atteberry, C., 1999. MS. Thesis, University of Virginia.
8. Schmitt, B., Ocampo, J., Klinger, J., 1987. J. Phys. (Paris) **48**, C1,519–525.
9. Silvia Negrelli, Valter S. Nascimento Jr.A study of the effective thermal conductivity of frost formed on parallel plate channels[J]. Experimental Thermal and Fluid Science,**78**(2016), 301-308.
10. S. Negrell, C.J. Hermes, A semi-empirical correlation for the thermal conductivity of frost[J]. International Journal of Refrigeration,**58**(2015),243-252.
11. ZOU T H, XUE Y L, SUN Y. Experimental study on performance of frost on cold-trap used by vacuum freeze-drying machine[J]. Fluid Machinery,2009,37(4).

Method Based on SEFT-to-Petri for Safety Analysis of Software

Xu Sai-sai[1,a], Chen Jing[2], Sun Yu-ning[2], Gao Xin-rui[2], Wang Bo-han[2] and Wang Kun-long[2]

[1]*University of Science and Technology of China,230026, China*
[2]*Institute of Computer Technology and Application of Beijing, 100854, China*

Abstract. System safety is a vital non-functional requirement whose satisfaction is essential for system software. However, modern aerospace system software is more and more complicated, which results in a high complexity of analyzing system faults. With the increased acceptance of Model-based Systems Engineering as a new method for systems engineering, Model-based Safety Analysis is also proposed to formalize the task of safety analysis and automate the safety calculations. Our work is grounded on State/Event Fault Tree to analyze system faults and build functional model. Firstly, we can translate SEFT to state machine based on SysML with fault syntactic messages and match elements together with translating logic gates; after which, transforming state machine into Petri Net model by means of rigorous semantic relations to extract preliminary analytical model is deduced theoretically in this paper; finally, we can derive analyses of causes and results of faults from Petri Net model by adopting a set of mathematical and statistical analysis. Practically, we have also validated our work by a case study of an aeronautic control system to support this paper.

1 Introduction

Increasing system complexity results in an increase in complexity of the safety analyst's task to ensure that systems are safe. Thus, how to exactly build models to express a complicated control system and identify the safety-related fragments is an imminent issue in domain of avionics[1]. In theory, system safety uses systems theory and systems engineering approaches to prevent foreseeable accidents and to minimize the result of unforeseen ones. It is a planned, disciplined, and systematic approach to identifying, analyzing, and controlling hazards throughout the life cycle of a system in order to prevent or reduce accidents. Model-based Safety Analysis methods have been developed for formalising the work and subsequent automation of the safety calculatio [a] ns. However, these techniques use their own models that are not identical to the design models. Keeping consistency between these models often requires model-to-model transformations.

There exists varieties of difficulties in modelling and analysis,because the nature of a software with high complexity is still not identified clearly. FTA, i.e. Fault Tree Analysis[2], is a conventional modeling technique for describing safety and reliability analysis, which also plays a vital role in domains such as avionics, railways, maritime and automation. Nevertheless, FTA is not propitious to setting up software modelling for three factors as follows[3], (1) FTP consisted of a series of fault sub-

[a] Corresponding author : xusaisai@mail.ustc.edu.cn

trees with precise logics is not adequate to indicating a complex component system with its architectures; (2) FTP can reliably apply to domain safety analysis, but it is not a good idea to depict system behaviors; (3) FTP does not put an emphasis on conveying components with multi-state for its restrictive two-state. Although FTP can be not directly used in modelling, it is considerable and mentionable to combine this method with other mature modelling methodologies (e.g., Petri Net model) for quantized analysis.

State/Event Fault Tree (SEFT)[4] is assembled by Fault Tree (FT) and Component State (CS), which is not only able to conduct safety analysis but also build system model by explicating logic gates and elementary component. In this paper, we primarily obtain state machine based on SysML by extending SEFT, which requires logical corresponding relations; then, transforming state machine into Petri Net model is also a significant part of our work; at last, we can perform an automated safety analysis to get feedback and results on our analysis without the need for a safety specialist.

2 Theoretical Principle

This part briefly narrates ideological system of SEFT and discusses techniques of modeling grounded on SysML state machine. Moreover, software safety analysis framework based on SEFT is introduced in the rest of this chapter as well.

2.1 SEFT

Component Fault Tree (CFT) is an integration of both component and fault tree[5], which comprises a large number of independent structures together with system behaviors and fault messages[6]. Figure 1 shows internal elements and their mutual relations of a CFT whose interface is an interactive bridge with external environment. However, it is considerable that a state can not trigger off an event, but can be detonated by an event.

Then, we can transform CFT into SEFT which consists of a set of elements by mapping function, as shown in Figure 2.

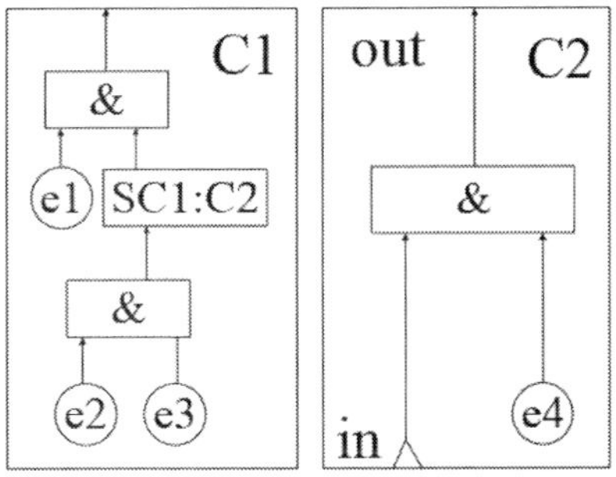

Figure 1. Component Fault Tree

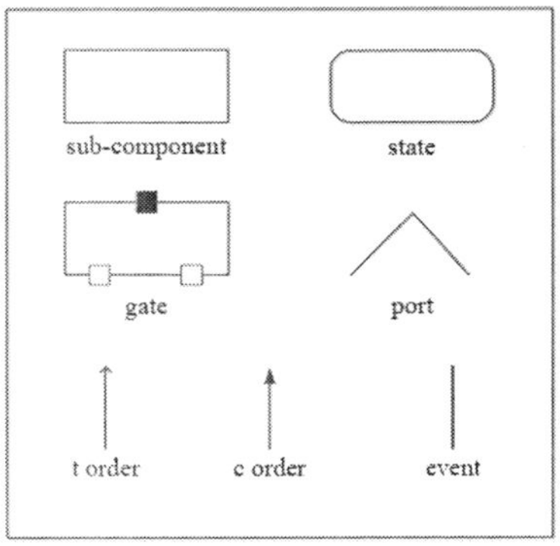

Figure 2. SEFT elements

2.2 SysML State Machine

State Machine[7] describes serial states involved by the instance of a class and a series of responses generated by events based on the object of a class through the whole life cycle of software. What is more, State Machine is a integration consisted of states, events and transitions.

2.3 Petri Net

The essence of petri net[8] indicates the flow of resources, which can build model grounded on real-time system and system function can be reflected by petri net. Software and Hardware Fault Petri Net (SHFPN) is one of Petri Net (PN) models which can be defined as six-tupes (P,T;F,K,W,M0), when expressing causality between faults and fault modes, especially to its formation mechanism. Moreover, petri net can be depicted graphically, as exhibited in Figure 3 and Figure 4.

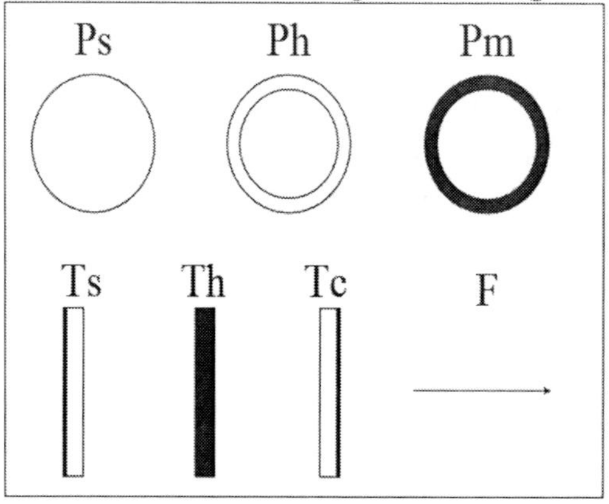

Figure 3. SHFPN elements

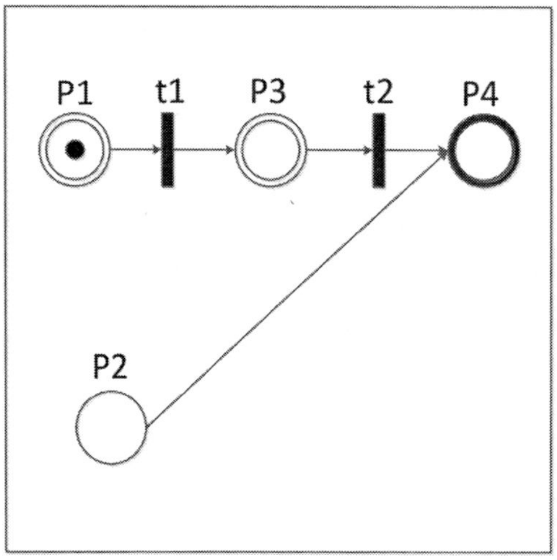

Figure 4. A laconic SEFT model

2.4 Analyical Framework

Existing safety engineering processes are normally based on standards that define a common framework for the derivation of safety requirements which combines hazard assessment and risk analysis techniques[9]. Our contextual framework is given by Figure 5 as below.

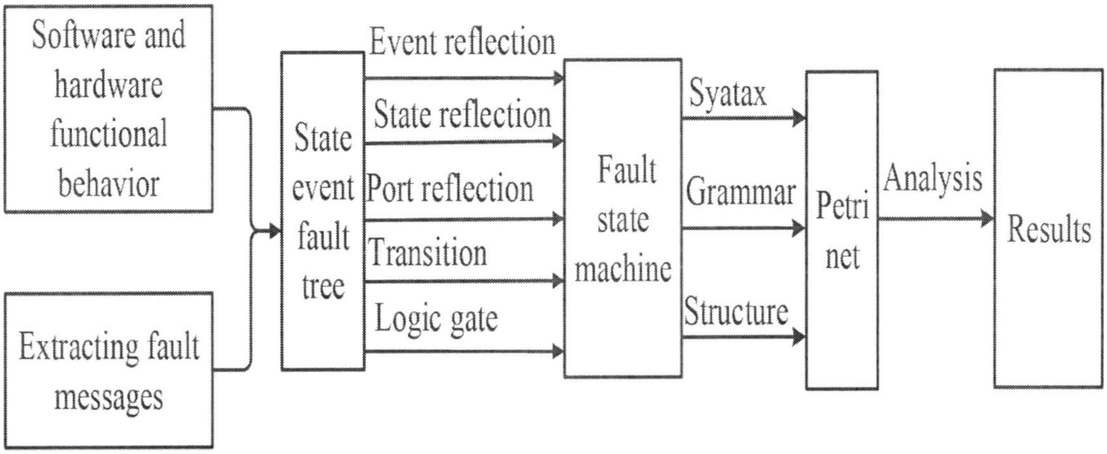

Figure 5. Software safety analysis framework based SEFT-to-Petri

3 TRANSFORMING SEFT INTO SYSML STATE MACHINE

3.1 transition-related algorithms

In order to get an accurate state machine, we have developed a model check algorithm to automatically satisfy a set of conditionss which is shown in Figure 6.

```
Algorithmic A: check the transformation conditions
Input: component[MaxC], NestComponent[MaxNC], port[MaXConnectedC], event[MaXConnectedC];
Function: check the transformation conditions
if (is SEFT) then begin
        Boolflag[3];
        Flag [0] = checkNull( component [i] ) ; / /Check component validity
        Flag [1] = checkcycle( NestComponent [j ] ) ; / /Check nest cycle
        Flag [2] = checkport( port [k] ) ; / /Check component and environment
        Flag [3] = checke2e( event [m] ) ; / /Check inexistence of transition
        Flag [4] = checkport2port( port [n] ) ; / /Check interaction
        end;
end if;
if ( Flag[0]& &Flag[1]& &Flag[2]& &Flag[3]& &Flag[4]) then begin
        Algorithmic B; Algorithmic C; Algorithmic D; Algorithmic E;
        end;
end if;
```

Figure 6. SEFT checking algorithm

The following excerpt is illustrative of reflection about state, event, port and transition to capture state machine from SEFT. There are four algorithms illustrated in Figure 7 delineated as follows via mapping function which is deduced by domain experts. If necessary, more detailed algorithms could be defined to deal with extremely complicated system.

```
Algorithmic C: mapping SEFT' s state diagram' s state
Input: state [MaxS] , event [MaxE] , port [MaxP] , transition [MaxT] ;
        ID: table Table ; flag;
Output:SMstate [MaxS] , SMevent [MaxE] , SMport [MaxP] , SMport [MaxT] ;
Function: mapping SEFT's state diagram's state
if (state) then begin
    for (i = 0; i < MaxS;i++){
        SMstate [i] = state [i] ;
        if (Init event fires the state[] ) then begin
            SMstate [i] .flag=1;
        Table. SMstate [i] . ID = Table . state [i] . ID;
    }
    end;
end if;
```

```
Algorithmic C: mapping SEFT' s events to state diagram' s events or states
Input: state [MaxS] , event [MaxE] , port [MaxP] , transition [MaxT] ;
        ID; table Table ; flag;
Output:SMstate [MaxS] , SMevent [MaxE] , SMport [MaxP] , SMport [MaxT] ;
Function: mapping SEFT's events to state diagram's events or states
if (event) then begin
    for (i = 0; i < MaxE;i++){
        SMevent [i] = event [i] ;
        if ( event [i] = = deterministic||exponential) then begin
            Add delay parameter;
            end;
        end if;
        Table. SMevent [i] . ID = Table . event [i] . ID;
    }
    end;
end if;
```

```
Algorithmic D: mapping SEFT's ports to state diagram's states or events
Input: state [MaxS] , event [MaxE] , port [MaxP] , transition [MaxT] ;
        ID; table Table; flag;
Output: SMstate [MaxS] ,SMevent [MaxE] ,SMport [MaxP] ,SMport [MaxT] ;
Function: mapping SEFT's ports to state diagram states or events
if (port) then begin
    for (i = 0;i<MaxP;i++){
        if (port [i] is event port)
            Add an event and transition to SM;
        end if;
        if (port [i] is state port)
            Add an state to SM;
        end if;
        Table.SMport [i] .ID = Table.eveportnt [i] .ID ;
        Port [i] . entry =export;
    }
    end;
end if;
```

```
Algorithmic E: the transformation elements to one SM
Input:state [MaxS] , event [MaxE] , port [MaxP] , transition [MaxT] ;
      ID; table Table; flag;
Output: SMstate [MaxS] , SMevent [MaxE] , SMport [MaxP] , SMport [MaxT]
Fucnction: mapping SEFT to SM
if ( transition) then begin
  for ( i = 0; i < Max T; i + + ){
    if ( transition [i] is state to state)
        SMtransition [i] = transition [i] ;
    end if;
    if ( transition [i] is state to event to state)
        Delete transition [i + 1] ;
        SMtransition [i] = transition [i] ;
    end if;
    if ( transition [i] is event [j] to event [k] )
        Delete transition [i] ;
    if ( event [j] = true)
        event [k] = true;
    end if;
end if;}
end;
end if;
```

Figure 7. The transformation elements to one SM

3.2 transition from logic gates to state machine

Generally, fault tree includes "AND" gate, "OR" gate and their combination. The rest of this section will describe the conversion between logic gates and state machine. "AND" gate demonstrated as "∧" represents interactive parts of several input terminals usually with two input terminals and one output terminal. "OR" gate illustrated as "∨" indicates that at least only one event occurs among multiple input events. As is shown in Figure 8, it clearly declares reflection from logic gates to state machine[10].

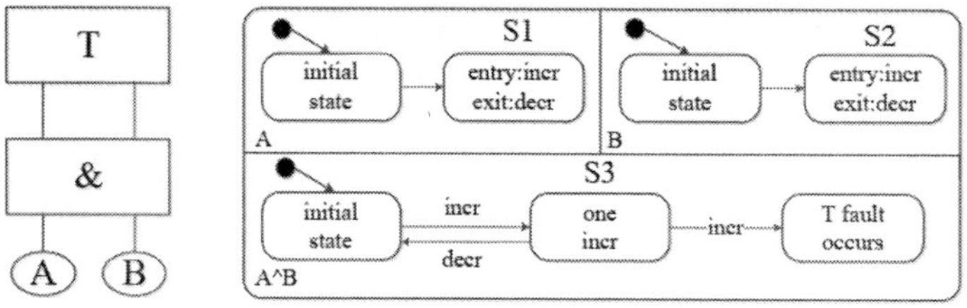

AND gate and transformed

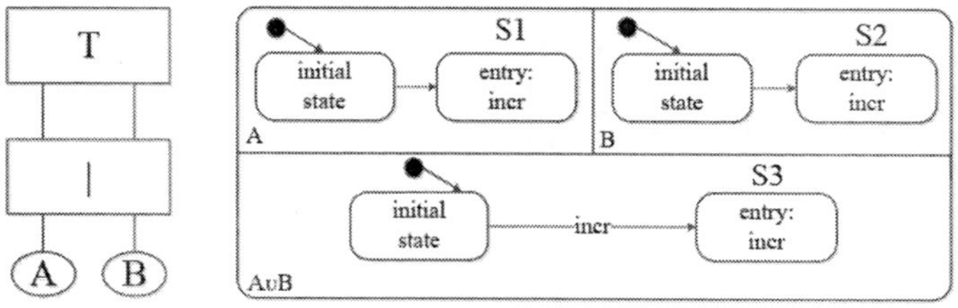

OR gate and transformed

Figure 8. Logic gates and transformed

In Figure 11, it represents the result of transition from AND gate and OR gate. Here more detailed messages of AND gate transition are interpreted , but the OR gate transition is not given in this section for same manner. For the derivation of the logic gates transition it is assumed that event A and event B are all fault element which must belong to the type of state-occurrence. When component A arrives a state s1, event incr will be detonated and accordingly component A∧B will arrive at state s3. However, once component A can't withdraw from state s1 and component B arrives at state s2, event incr will be touched off and component A∧B will arrive at state s4, which is consistent with Fault Tree in semantic categories.

4 Transition from SysML State Machine to Petri Net Model

Software and hardware fault petri net (SHFPN) [11] is another mode of state machine which is defined by six-tuples via mathematical constraints, i.e., SHFPN= $(P,T;F,K,W,M0)$. (1)P expresses fault state set, namely $P = P_S \cup P_H \cup P_M$, P_S is software fault tree set, namely $P_S = \{P_{S1}, P_{S2}, \cdots, P_{Sa}\}$, P_H represents hardware fault tree set, namely $P_H = \{P_{H1}, P_{H2}, \cdots, P_{Hb}\}$, P_M illustrates fault mode set. (2) T demonstrates event transition set, namely $T = \{t_1, t_2, \cdots, t_m\}$;for $t \in T, \exists\, p \in P_S$, get $p \in t^*$,which means software fault event, i.e., T_S; for $t \in T, \exists\, p \in P_H$, get $p \in t^*$,which shows hardware fault event, i.e., T_H; for $t \in T, \exists\, P_1 \in P_H \exists\, P_2 \in P_S$, get $P_1, P_2 \in t^*$,which stands for software and hardware fault event, i.e., T_C. (3)F defines as the flow of SHFPN, namely $F = P \times T \cup T \times P$. (4)K depicts the token, confirming whether an event occurs. (5)W shows the weight coefficient constrained by $W_{(p,t)} \equiv 1$. (6)M0 is defined as the initial fault state.

Additionally,for $p \in P_M$, if $t \in p^*$ *exists* and $t \in T_S \wedge t$ not $\in T_C$, we will name it software fault mode. Moreover, for $p \in P_M$, if $t \in p^*$ *exists* and $t \in T_H \wedge t$ not $\in T_C$, we will name it hardware fault mode. To transform fault state machine into petri net in Figure 12, we need further to perform the reflection of state-to-state and event-to-event[12].

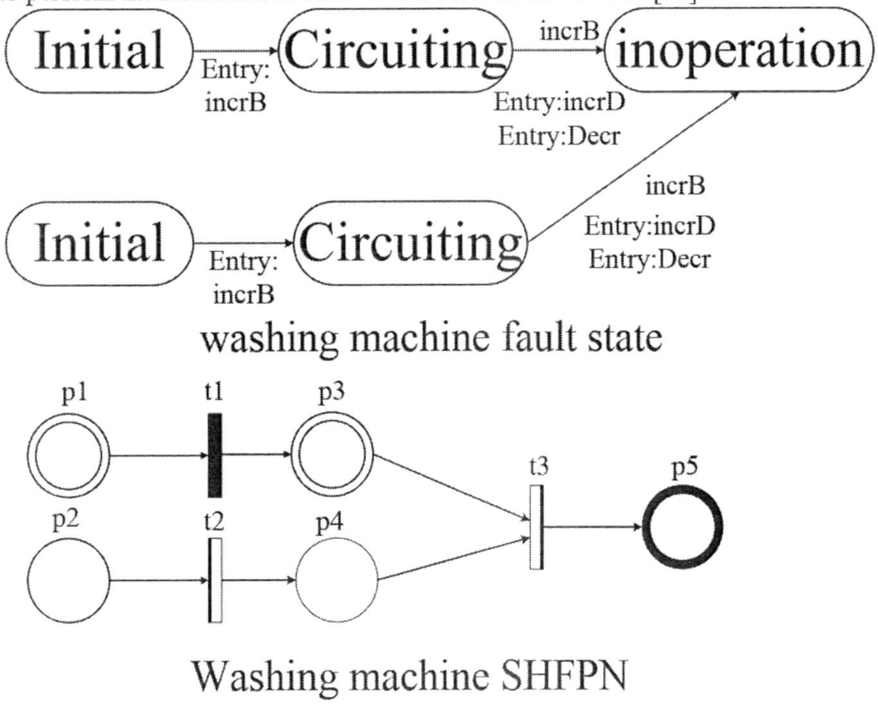

Figure 9. Transition from fault state machine to SHFPN based on washing machine

As is shown in Figure 9, an automatic washing machine often malfunctions or breaks down by hardware circuit fault and software program error which are described as corresponding states[13]. Besides, the event (such as incrB) is the trigger from one state to another state in the state machine above. According to consistency, it is considerable and reasonable for us to propose an algorithm depicted in Figure 10 fulfilling the transition from fault state machine to Software and Hardware Fault Petri Net (SHFPN).

```
Algorithmic F: mapping state diagram's state to SHFPN's state
Input: SM-state[Max S], SM-event[Max E]
Output: PNS-state[Max S], PNH-state[Max S], PNC-state[Max S],
        PNS-event[Max E], PNH-state[Max E],  PNC-state[Max E];
Function: transforming state diagram's state to SHFPN's state
If(state) then begin
    for(i=0, j=0, k=0, m=0; i<Max S; i++){
        if(state[i] instanceof software fault state) then begin
            PNS-state[j] = state[i];
            j++;
            end;
        end if;
        if(state[i] instanceof hardware fault state) then begin
            PNH-state[k]         = state[i];
            k++;
            end;
        end if;
        if(state[i] instanceof hardware fault state) then begin
            PNC-state[m] = state[i];
            m++;
            end;
        end if;
    }
    end;
    end if;
```

```
Algorithmic G: mapping state diagram's state to SHFPN's state
Input: SM-state[Max S], SM-event[Max E]
Output: PNS-state[Max S], PNH-state[Max S], PNC-state[Max S],
        PNS-event[Max E], PNH-state[Max E],  PNC-state[Max E];
Function: transforming state diagram's state to SHFPN's event
If(event) then begin
    for(i=0, j=0, k=0, m=0; i<Max E; i++){
        if(event[i] instanceof software fault state) then begin
            PNS-event[j]= event[i];
            j++;
            end;
        end if;
        if(event[i] instanceof hardware fault event) then begin
            PNH-event[k] = event[i];
            k++;
            end;
        end if;
        if(event[i] instanceof hardware fault event) then begin
            PNC-event[m] = event[i];
            m++;
            end;
        end if;
    }
    end;
    end if;
```

Figure 10. Mapping algorithm from SM to SHFPN

Figure 10 demonstrates that a mapping algorithm from State Machine (SM) to Software and Hardware Fault Petri Net (SHFPN) describes state reflection and event reflection[14]. Because SHFPN consists of three states, i.e., software fault state, hardware fault state and software-hardware fault state, more considerable focus is based on identifying these three states, as shown in the left part of Figure 10. Of course, SHFPN also contains three events, namely software event, hardware event and software-hardware event, which is extracted shown in the right part of Figure 10. So far, we have accomplished the transition from fault/event tree to petri net[15]. If you want to further analyze causes of faults,you can go far with existing theory of petri net.

5 Case Study

Automatic washing machine mainly contains software and hardware,the requirement of programmed control system[16] can be described as follows:

(1) In any case, washing machine's sheathing material should be insulated or the automatic washing machine should be isolated from other materials.

(2) When required,washing machine should work normally.

(3) When not required, washing machine must stop working.

According to this, fault analysis of washing machine is defined as a fault tree in Figure 11. As is illustrated in Figure 14, electrical faults in the electronics may lead to breakdown; Besides, errors of control block can generate system fault which will cause in-operation[17].

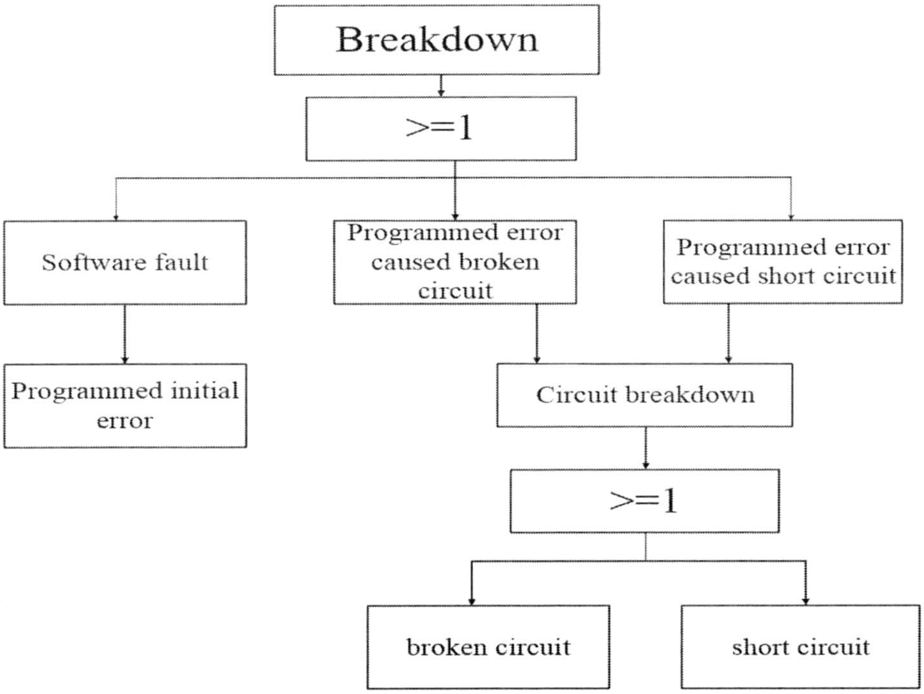

Figure 11. FT of washing machine

We can build the State/Event Fault Tree (SEFT) depicted in Figure 12, which graphically displays fault messages and related functional behavior[18].

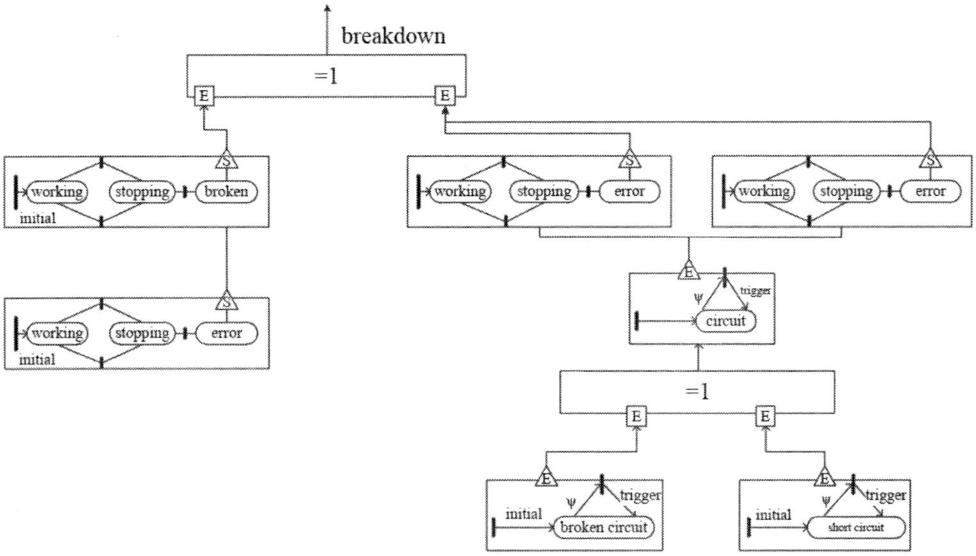

Figure 12. SEFT of washing machine

According to the mapping algorithm demonstrated in Figure 7 and the transition of logic gates, we can transform SEFT into Fault State Machine (FSM) illustrated in Figure 13.

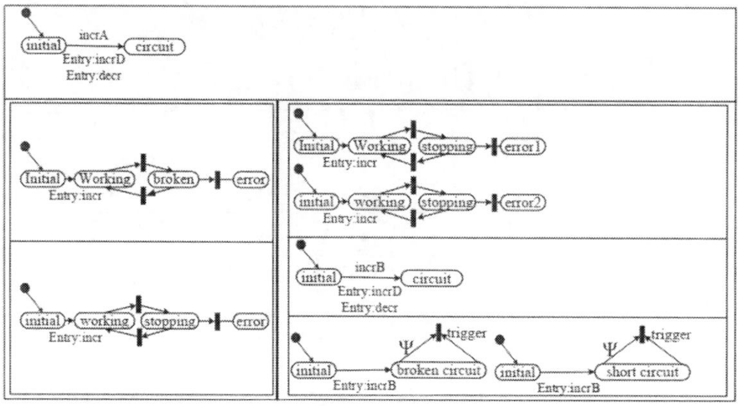

Figure 13. Fault SM of automatic washing machine

After that, we can get corresponding Software and Hardware Fault Petri Net (SHFPN) grounded on algorithm depicted in Figure 14.

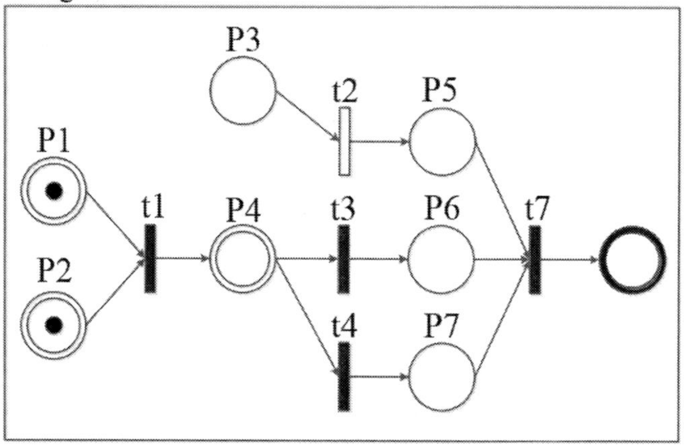

Figure 14. SHFPN of automatic washing machine

The following job is to obtain SHFPN's incidence matrix based on methodology of petri net as shown in Figure 15.

$$A1=\begin{pmatrix} -1 & 0 & 0 & 0 & 0 & 0 & 0 & 0 \\ -1 & 0 & 0 & 0 & 0 & 0 & 0 & 0 \\ 0 & -1 & 0 & 0 & -1 & 0 & 0 & 0 \\ 1 & 0 & -1 & -1 & 0 & 0 & 0 & 0 \\ 0 & 1 & 0 & 0 & 0 & -1 & -1 & 0 \\ 0 & 0 & 1 & 0 & 0 & 0 & -1 & 0 \\ 0 & 0 & 0 & 1 & 0 & 0 & -1 & -1 \\ 0 & 0 & 0 & 0 & 1 & 1 & 0 & 0 \end{pmatrix}$$

Figure 15. Incidence matrix of SHFPN

Result analysis:

We can get related analysis results, i.e., $P_{MC} = \{P_9\}$ which is a mode of software-hardware from corresponding incidence matrix above. In addition, we also have analyzed the fault mechanism whose results are described in Table 1.

Table 1. Fault mechanism analysis of P_9

Number	Fault causes	Means of transmission
1	P_1	t_1,t_3,t_7
2	P_2	t_1,t_4,t_7
3	P_3	t_2,t_7

6 Conclusion

Existing modeling language (such as UML and SysML) can clearly describe software system behavior and minutely express the change of system state, but their common defect is that they can't perform safety analysis. However, fault tree can express system fault behavior commendably, yet it is incapable of conveying functional behavior of system. Besides, methodology of petri net has been studied maturely, which has a set of holonomic theory of fault analysis. Therefore, it is complementary to combine fault tree with state machine based on SysML and petri net.

This paper firstly introduces the related theory providing basis of analysis. In addition, we talk about the fault tree and transition from fault tree to State/Event Fault Tree (SEFT), which is the start point of our subsequent work. Additionally, mapping algorithm from SEFT to FSM is significantly described as above. Last but not least, transforming FSM into SHFPN is the innovative point whose executable algorithm is displayed in Figure 13. Of course, if necessary, we can also take efforts to combine Markov Chans (MC) [19] with probability theory used in our paper which can enable us to analyse safety faultlessly.

References

1. C. Huo Wang, W. Ji, D. Wei, Acta Electronica Sinica, **31**, 12A (2003)
2. N. H. Roberts, W. E. Vesely, *Fault tree handbook*, (Government Printing Office, 1987)
3. B. Kaiser, C. Gramlich, Reliability Engineering & System Safety, **92**, 11 (2007)
4. B. Kaiser, *State event trees: a safety and reliability analysis technique for software controlled systems* (2007)
5. S. Mahmood, R. Lai, Y. Soo Kim, Information and Soft-w are Technology, **47**, 10 (2005)
6. B. Kaiser, P. Liggesmeyer, Proceedings of the 8th Australian Workshop on Safety Critical Systems and Software, **33**, Australian Computer Society, Inc. (2003)
7. M. Hause, Fifteenth European Systems Engineering Conference, **9** (2006)
8. M. Xiao Dong, Z. Peng, Chinese Journal of Computers, **6**, 1 (2015)
9. M. Jing, Z. Li, M. Shao Dong, International Information and Application Forum,**36**,4 (2009)
10. X. Jing Wen, Z. Ji En, S. Yue Yue, Computer Application and Software,**24**,22 (2009)
11. Z. Yan, M. Hong, Journal of Soft-w are, **20**, 6 (2009)
12. D. Wei, W. Ji, Q. Zhi Chang, Journal of Software, **14**, 4 (2003)
13. Z. Hong Lin, Z. Chun Yuan, L. Dong, Chinese Journal of Computers, **35**, 2 (2012)
14. P. Feiler, A. Rugina, Carnegie-mellon UnivPittsburgh Pa Software Engineering Inst (2007)
15. P. Fenelon, J. A. Mc Dermid, Journal of Systems and Software, 21, 3 (1993)
16. M. Hecht, A. Lam, and C. Vogl, In Engineering of Complex Computer Systems, 16th IEEE International Conference on (2011)
17. K. M. Hansen, A. P. Ravn, H. Rischel, ACM SIGSOFT Software engineering Notes, **16**, 5 (1991)
18. W. Si Qi, H. Zhi Qiu, *College of Computer Science and Technology*, (Nanjing University of Aeronautics and Astronautics, 2016)
19. Y. Jin, L. Xun, A. Jian xun, Y. Guang Yu, L. Ping, Journal of University of Chinese Academy of Sciences,**1**,4(2013)

AUTHOR INDEX

AUTHOR INDEX

EDP Sciences – Web of Conferences
17, Avenue du Hoggar
Parc d'Activité de Courtabouf
BP 112
F-91944 Les Ulis Cedex A
France

ISBN 978-1-5108-6802-1

CURRAN ASSOCIATES INC.
proceedings.com

Nanophotonics and Micro/Nano Optics International Conference (NANOP 2016)

EPJ Web of Conferences Volume 139 (2017)

Paris, France
7 - 9 December 2016

edp sciences